PRINCIPLES OF
ENGINEERING ECONOMIC ANALYSIS

PRINCIPLES OF
ENGINEERING ECONOMIC ANALYSIS

JOHN A. WHITE
School of Industrial and Systems Engineering
Georgia Institute of Technology
Atlanta, Georgia

MARVIN H. AGEE
Department of Industrial Engineering and Operations Research
Virginia Polytechnic Institute and State University
Blacksburg, Virginia

KENNETH E. CASE
School of Industrial Engineering and Management
Oklahoma State University
Stillwater, Oklahoma

JOHN WILEY & SONS
New York   Santa Barbara   London   Sydney   Toronto

Library of Congress Cataloging in Publication Data:

White, John A   1939–
   Principles of engineering economic analysis.

   Includes indexes.
   1. Engineering economy.   I.   Agee, Marvin H.,
1931–   joint author.   II.   Case, Kenneth E., joint
author.   III.   Title.
TA177.4.W48   658.1'55   77-4663
ISBN 0-471-01773-6

Printed in the United States of America

10 9 8 7 6 5 4 3 2 1

PREFACE

This textbook is for people who are interested in learning the principles involved in analyzing economic investment alternatives. Although the material is useful for accounting, economics, finance, and management students, its primary audience will be engineering and engineering technology students. Courses for which the book is intended are variously titled engineering economy, economic analysis, industrial economy, economic decision analysis, engineering decision analysis, and engineering economic analysis.

Over the past 16 years we have taught engineering economic analysis to more than 2500 persons, including undergraduate and graduate students from engineering and nonengineering academic programs. Additionally, we have taught refresher courses for professional engineering registration examinations and continuing education short courses, consulted, and worked in industrial positions that required the application of the material here. Based on that experience, we perceived a need for a textbook that presented a unified treatment of economic analysis principles. We also felt the need to address the subject from a *cash flow viewpoint*.

The following are some of the important features of this textbook.

1. A cash flow approach is taken throughout, including the treatment of replacement problems.
2. A systematic eight-step approach is given for performing a comparison of investment alternatives.

v

3. The study period must be specified explicitly instead of unknowingly employing a least common multiple of lives study period.

4. Comprehensive treatments are provided of:
   a. Cost concepts, including cost terminology, cost estimation, accounting principles, and nonmonetary considerations.
   b. Income taxes, including capital gains and losses, depreciation recapture, depletion, and investment tax credit considerations.
   c. Economic analyses in the public sector, emphasizing benefit-cost analysis procedures.
   d. Break-even, sensitivity, and risk analyses, with the latter including analytical and simulation approaches.
   e. Decision models, including decisions under risk and uncertainty and decision tree analyses.

5. Over 130 worked-out examples and approximately 300 problems are provided, with answers to even-numbered problems given at the end of the text.

6. Section numbers are provided for material covered by each exercise to facilitate a self-study program.

7. Modeling uniform, gradient, and geometric series, modeling changing interest rates, modeling equity amounts in loan payments, and modeling inflation effects are presented to emphasize the importance of *modeling* cash flow series instead of using more conventional cookbook approaches.

8. Extensive tables of compound interest factors are provided, including both discrete and continuous compounding, discrete and continuous flows, and gradient and geometric cash flow series.

9. Methods of comparing alternatives include present worth, annual worth, future worth, payback period, savings/investment ratio, internal rate of return, and external rate of return methods.

10. Depreciation is covered in conjunction with the discussion of income taxes, since depreciation is not a cash flow and it influences cash flows through taxation.

11. Replacement analysis is covered in the discussion of comparison of alternatives, since a cash flow analysis of replacement problems is essentially no different than any other type of investment alternative.

12. The data have been packaged in an eight-chapter format in the sequence most natural to the coverage of the material.

13. Only material that is required in the performance of engineering economic analyses is presented, since it is believed that including tangential material not normally covered by the instructor has a deleterious effect on the student.

14. A detailed Instructor's Manual has been prepared that includes detailed solutions to all exercises, sample examinations and visual aids for presenting the material in the text.

The manuscript has been tested in the classroom. The orientation of the material and its organization have been influenced by the recommendations of numerous colleagues and undergraduate and graduate students.

The first five chapters have been developed from our collective experience of teaching the subject to engineering sophomores from all disciplines; the material in the remaining three chapters has been covered in a senior-level elective course. Although elementary calculus is a sufficient prerequisite for the first six chapters, an introductory course in probability theory is suggested as a prerequisite for the final two chapters.

Many people influenced the development of this textbook. In particular, we have benefited from either taking coursework from or teaching with John R. Canada, Walter J. Fabrycky, Richard S. Leavenworth, William T. Morris, and Gerald J. Thuesen, who have authored or coauthored one or more texts on the subject. Our approach to the subject has also been influenced by our associations with Richard H. Bernhard, William E. Biles, Leland T. Blank, James R. Buck, Thomas P. Cullinane, Carl B. Estes, Gerald A. Fleischer, David R. Freeman, Lynwood A. Johnson, Raymond P. Lutz, Leon F. McGinnis, M. Wayne Parker, Gunter P. Sharp, Donald R. Smith, G. T. Stevens, and H. G. Thuesen. Therefore, the influence of others will be evident in various places in the book. Because it is impossible to credit the many individual contributions, we simply acknowledge gratefully their influence on our understanding of the subject.

Finally, we thank our wives Mary Lib, Barbara, and Lynn, for their patience and understanding during the project. They constantly demonstrated considerable insight on the subject by the gentle reminder, "If you know so much about investments, why aren't we rich?"

John A. White
Marvin H. Agee
Kenneth E. Case

CONTENTS

CHAPTER THREE
TIME VALUE OF MONEY OPERATIONS

CHAPTER FOUR
COMPARISON OF ALTERNATIVES

CHAPTER FIVE
DEPRECIATION AND INCOME TAX CONSIDERATIONS

CHAPTER SIX
ECONOMIC ANALYSIS OF PROJECTS IN THE PUBLIC SECTOR

CHAPTER SEVEN
BREAK-EVEN, SENSITIVITY, AND RISK ANALYSES

CHAPTER EIGHT
DECISION MODELS

PRINCIPLES OF
ENGINEERING ECONOMIC ANALYSIS

CHAPTER ONE
INTRODUCTION

1.1 BACKGROUND

The subject of this book is *money* and how to spend it wisely. For some, that is adequate motivation to give serious attention to its contents. However, others must be motivated in other ways (e.g., tests, homework, grades). What are your reasons for reading this textbook? If you are a student, then it is quite likely that you are enrolled in an economic analysis course and this is the assigned text. If you are not a student, then you probably are engaged in performing economic analyses and are interested in learning what we have to say about the subject. Admittedly, some will read this book who do not fall into the above categories, however, we expect the majority of the readers to be either economic analysis students or professionals.

The subject matter that we treat is variously referred to as *economic analysis, engineering economy,* and *economic decision analysis,* among other names. Traditionally, the application of economic analysis techniques in the comparison of engineering design alternatives has been referred to as *engineering economy.* However, the emergence of a widespread interest in economic analysis in public sector decision making has brought about greater use of the more general term, *economic analysis.*

Our approach to the subject will be a *cash flow approach.* A cash flow

1

occurs when money actually changes hands from one individual to another or, for that matter, from one organization to another. Thus, money received and money dispersed (spent) constitute cash flows. We also will emphasize the fundamentals of economic analysis without dwelling on some of the philosophical issues associated with the subject; for those interested in pursuing the latter, references are provided.

1.2 THE PROBLEM-SOLVING PROCESS

Economic analyses are typically performed as a part of the overall problem-solving process. In designating a new or improved product, a manufacturing process, or a system to provide a desired service, the "problem solver" is involved in performing the following steps.

1. Formulation of the problem.
2. Analysis of the problem.
3. Search for alternative solutions to the problem.
4. Selection of the preferred solution.
5. Specification of the preferred solution.

The *formulation of the problem* involves the establishment of its boundaries, and it is aided by taking a black box approach. As Krick [3] describes the black box, an originating state of affairs (state A) and a desired state of affairs (state B) exist. Also, a transformation must take place in going from state A to state B, as depicted in Figure 1.1. More than one method of performing the transformation from state A to state B exists and there is unequal preferability of these methods. The solution to the problem is visualized as a black box of unknown, unspecified contents having input A and output B.

The *analysis of the problem* consists of a relatively detailed phrasing of the characteristics of the problem, including restrictions and the criteria to be used in evaluating alternatives. Considerable fact gathering is involved, including the real restrictions on the problem, which must be satisfied. Consequently, any budget, quality, safety, personnel, environmental, and service level constraints that may exist are identified.

The *search for alternative solutions to the problem* involves the use of the

State A ⟶   ⟶ State B

FIGURE 1.1. Black box approach.

engineer's creativity in developing feasible solutions to the design problem. A number of aids to stimulate one's creativity are provided by Krick [3]. In particular, he suggests that you should:

1. Exert the necessary effort.
2. Not get bogged down in details too soon.
3. Make liberal use of the questioning attitude.
4. Seek many alternatives.
5. Avoid conservatism.
6. Avoid premature rejection.
7. Avoid premature acceptance.
8. Refer to analogous problems for ideas.
9. Consult others.
10. Attempt to divorce your thinking from the existing solution.
11. Try the group approach.
12. Remain conscious of the limitations of the mind in the process of idea generation.

Unfortunately, the natural tendency is to mix the evaluation of alternatives with the generation of alternative solutions. As a result, the search process usually terminates with the development of the first alternative, which can be justified economically. By separating the search process from the selection process, we enhance the chance of generating a number of alternatives that can be justified economically.

The *selection of the preferred solution* consists of the measurement of the alternatives, using the appropriate criteria. The alternatives are compared with the constraints and infeasible alternatives are eliminated. The benefits produced by the feasible alternatives are then compared. Among the criteria considered for choosing among alternatives is the economic performance of each alternative.

The *specification of the preferred solution* consists of a detailed description of the solution to be implemented. Predictions of the performance characteristics of the solution to the problem are included in the specification.

Example 1.1

As an illustration of the application of the problem-solving approach, consider the case of a locally owned electrical power generating company faced with an air pollution problem that had to be solved. The power plant was using low-grade coal with sulfur content of 1.9% by weight, and the air pollution control board for the state directed the power plant to comply with newly adopted standards for air quality. A consulting engineer was hired to study the problem and recommend the best course of action.

The engineer met with representatives of the air pollution control board, the power company, county and city officials, and suppliers of various fuel alternatives and associated power generating equipment. An analysis of the problem yielded an agreement to use a 20-year planning horizon, forecasts of power requirements over the planning horizon, and a specification of the criteria to be used in evaluating the alternatives. Specifically, a benefit-cost analysis was to be performed; a cost of $150/ton of sulfur pollutant released into the air was to be employed in the analysis.

A search for alternative solutions to the problem resulted in a consideration of bunker C fuel oil, low-sulfur furnace fuel oil, and natural gas. The methods described in Chapter Six were employed to compare the alternatives using a benefit-cost analysis; bunker C fuel oil was the recommended energy source.

Based on the report provided by the consulting engineer, the city contracted for the installation of storage tanks for the fuel oil, piping, pumps, and the other equipment necessary for converting the furnaces from coal to fuel oil.[1]

Example 1.2

As a second illustration of the problem-solving procedure, a leading manufacturer of automotive bearings was faced with the need to expand its distribution operations. A study team was formed to analyze the problem and develop a number of feasible alternative solutions. After analyzing the problem and projecting future distribution requirements, the following alternatives were arrived at:

1. Consolidate all distribution activities and expand the existing distribution center, located in Michigan.
2. Consolidate all distribution activities and construct a new distribution center, location to be determined.
3. Decentralize the distribution function and build several new distribution centers geographically dispersed in the United States.

After considering the pros and cons of each alternative, the officers of the company directed the study team to pursue the second alternative.

An extensive plant location study was performed. The location study resulted in five candidate locations being selected for final consideration. The criteria used to make the final selection included:

1. Land cost and availability.
2. Labor availability and cost.

[1] For additional discussion of the case study on which the above was based, see [1].

3. Proximity to supply and distribution points (present and future).

4. Taxes (property and income) and insurance rates.

5. Transportation (access to rail and interstate highways).

6. Community attitudes.

7. Building costs.

Based on site visits to each location, the officers of the company selected a site in Alabama and directed the study team to develop design alternatives for the material handling system to be used in the distribution center.

Applying the problem-solving procedure, four alternatives were obtained for evaluation. The first alternative involved the use of pallet racks, lift trucks, and flow racks; the second alternative included the use of an automated stacker crane system, lift trucks, conveyors, and flow racks; the third alternative suggested narrow-aisle, guided picking machines, high-rise shelving, conveyors, driverless tractor trains, and lift trucks; the fourth alternative consisted of an automated stacker crane system, a rail guided picking vehicle system, a sortation and accumulation conveyor system, and high-rise, narrow-aisle lift trucks.

A planning horizon of 10 years was used in performing an economic analysis of each design alternative. The fourth alternative was the most economical and was recommended to top management. Based on the detailed presentation and the economics involved, management approved a budget of $9 million to implement the recommendations of the study team.

As demonstrated by the preceding illustrations, engineers and systems analysts must be prepared to defend their solutions to problems. Economic performance is among the criteria used to evaluate each alternative. Monetary considerations seldom can be ignored. If economics are not considered in the criteria used in the final evaluation of the alternatives, they are usually involved in an initial screening of the alternatives. In fact, money can be a constraint on the alternatives that can be considered, as well as the basis for the final selection.

This textbook treats in detail the step in the problem-solving process that involves the selection of the preferred solution. The process of measuring cash flows and benefits and the consideration of multiple objectives in selecting the preferred design are also treated.

In comparing alternatives, the *differences* in the alternatives will be emphasized. Consequently, the aspects of the alternatives that are the same normally will not be included in the analysis.

1.3 A FUNDAMENTAL CONCEPT: THE TIME VALUE OF MONEY

A fundamental concept underlies much of the material covered in the text: *money has a time value*. The *value* of a given sum of money depends on *when* the money is received.

Example 1.3 ————————

To illustrate the concept of the time value of money, suppose a wealthy individual approaches you and says, "Because of your outstanding ability to manage money, I am prepared to present you with a tax-free gift of $100. However, if you prefer, I will postpone the presentation for one year, at which time I will guarantee that you will receive a tax-free gift of $X." Would you choose to receive the $100 now or the $X one year from now if $X equaled (1) $100, (2) $110, (3) $200, (4) $1000?

In presenting this situation to numerous students, no students preferred receiving $X in case 1), very few students preferred $X in case 2), most students preferred receiving $200 a year from now, and all students indicated a preference for $1000 a year from now. The point is that the value of $100 one year from now was perceived to be less than the value of $100 at present. For most students, the value of $110 one year from now was believed to be less than the value of $100 at present. Only a few students felt that $200 a year from now was less valuable than $100 at present. All students believed $1000 a year from now was more valuable than $100 at present. Thus, for each individual student, some value (or range of values) of $X exists for which one would be indifferent between receiving $100 now versus receiving $X a year from now. If, for example, one is indifferent for $X equal to $125, then we would conclude that $125 occurring one year from now has a *present value of* $100 *for that particular individual*.

Example 1.4 ————————

To continue our consideration of different cash flow situations, examine closely the two cash flow profiles given in Table 1.1. Both alternatives involve an investment of $10,000 in ventures that last for four years. Alternative A involves an investment in a minicomputer by a consulting engineer who is planning on providing computerized design capability for clients. Since the engineer anticipates that competition will develop very quickly if the plan proves to be successful, a declining revenue profile is anticipated.

Alternative B involves an investment in a land development venture by a group of individuals. Different parcels of land are to be sold over a four-year period. The land is anticipated to increase in value. There will be differences in the sizes of parcels to be sold. Consequently, an increasing revenue profile is anticipated.

The consulting engineer has available funds sufficient to undertake either

TABLE 1.1 Cash Flow Profiles for Two Investment Alternatives

	CF		
EOY	A	B	(A – B) Difference
0	–$10,000	–$10,000	$0
1	+ 7,000	+ 1,000	+6,000
2	+ 5,000	+ 3,000	+2,000
3	+ 3,000	+ 5,000	–2,000
4	+ 1,000	+ 7,000	–6,000

investment, but not both. The cash flows shown are cash flows after taxes and other expenses have been deducted. Both investments result in $16,000 being received over the four-year period; hence, a net cash flow of $6,000 occurs in both cases.

Which would you prefer? If you prefer Alternative B, then you are not acting in a manner consistent with the concept that money has a time value. The $6000 difference at the end of the first year is worth more than the $6000 difference at the end of the fourth year. Likewise, the $2000 difference at the end of the second year is worth more than the $2000 difference at the end of the third year.

Example 1.5

As another illustration of the impact of the time value of money on the preference between investment alternatives, consider investment alternatives C and D, having the cash flow profiles depicted in Figure 1.2. The cash flow

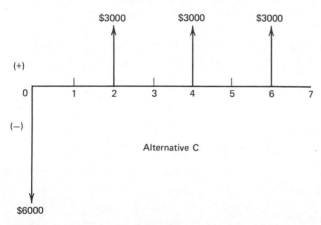

FIGURE 1.2. Cash flow diagrams for Alternatives C and D.

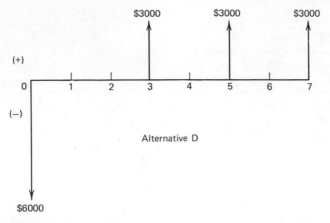

FIGURE 1.2 (*cont'd*)

diagrams indicate that the positive cash flows for Alternative C are identical to those for Alternative D, except that the former occur one year sooner; both alternatives require an investment of $6000. *If exactly one of the alternatives must be selected*, then Alternative C would be preferred to Alternative D, based on the time value of money.

Example 1.6

A third illustration of the effect of the time value of money on the selection of the preferred investment alternative is presented in Figure 1.3. Either Alternative E or Alternative F must be selected; the only differences in the performance characteristics of the two alternatives are economic differences. As shown in Figure 1.3, the economic differences reduce to a situation in which the receipt of $100 is delayed in order to receive $200 a year later. For

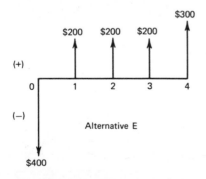

FIGURE 1.3. Cash flow diagrams for Alternatives E and F.

8 • INTRODUCTION

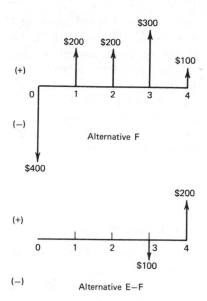

FIGURE 1.3 (*cont'd*)

this illustration, we would conclude that most of the students polled earlier would prefer Alternative E to Alternative F, since most of the students preferred $X when it equaled $200.

1.4 ECONOMIC JUSTIFICATION MAY NOT BE ENOUGH

Even though the text concentrates on the economic aspects of investment alternatives, note that in many instances economic justification may not be enough! Decisions can be quite different from recommendations. Managers typically have multiple criteria to be considered in reaching a final decision about the alternative to be adopted. Among the factors to be considered are: quality, safety, environmental impact, community attitudes, labor-management relationships, cash flow position, risks, system reliability, system availability, system maintainability, system operability, system flexibility, impact on personnel levels, training requirements, comparisons with competitors, impact on the different units within the organization, ego, customers' preferences, capital requirements, and economic justification.

The list of criteria can be imposing. However, perhaps the most important ingredient that separates the selected alternative from the *runners-up* is the salesmanship of the individual who presents the alternative to the manager. It is not unusual for a manager to adopt a weaker alternative because of the persuasive powers of the person who presented the alternative.

As Lutz [4] points out, *what is said* to support a project justification and *how it is said* influence the ultimate disposition of the funding request. As he puts it, "Communication content may outweigh the technical aspects of the proposal. The decision as to whether or not a project is funded often hinges on how successfully you communicate with the decision makers using their rules, words and decision criteria. Structuring your proposal for ease of acceptance is imperative and should be closely examined to try and increase the likelihood of obtaining capital budgeting dollars." Lutz argues that managers will often accept and implement a weak alternative because they:

1. Recognize the existence of a problem.
2. Understand the recommended solution.
3. Know how to implement the recommendation.
4. Anticipate immediate improvements with positive financial results.

A tendency exists for problem solvers to spend too much time trying to develop the recommended solution to the problem and too little time in determining the best way to sell the solution to management. Effective communication is essential; in particular, it is important to communicate with management in a way that can be understood and accepted [2].

It is important to write a proposal that includes an *executive summary*. The executive summary normally reduces the performance characteristics of the recommendation to a few pages. Technical aspects of the recommended solution and the technical details of the economic analysis are usually provided in the main report for the manager who wishes to obtain additional understanding of the information in the executive summary. A face-to-face presentation is also quite important in achieving effective communication.

Communication techniques such as the use of visual aids, voice control, dress, structure and clarity of expression are necessary to sell the proposed solution. However, it is probably more important to know your audience well and to prepare your proposal using their language. Since the language of managers is often the language of finance, Chapter Two introduces the financial terminology used by managers.

1.5 OVERVIEW OF THE TEXT

As an overview, we have organized the material in a manner consistent with the logical sequence of steps followed in performing an economic evaluation of investment alternatives. Chapter Two provides a discussion of cost concepts and includes discussions of elements of *costs*, measurement of cash flows, incremental costs, future costs, sunk costs, intangibles, irreducibles, nonquantifiables, standard costs, direct costs, indirect costs, fixed costs, variable costs, and opportunity costs. Given a feel for how the data are to be obtained for an economic analysis, Chapter Three presents some important fundamental concepts involving the *time value of money*. In fact, Chapter Three provides the foundation for the remainder of the book; therefore the reader must understand the time value of money operations presented in this chapter.

Chapter Four uses the time value of money operations in *comparing investment alternatives*. Measures of economic worth such as present worth, annual worth, future worth, payback period, rate of return, and savings/investment ratio are used in comparing mutually exclusive investment alternatives.

Depending on the interests of the reader, a choice might be made between Chapters Five and Six. Chapter Five addresses the issue of *income taxes* and their incorporation in economic analyses; Chapter Six treats *benefit-cost analysis*. Those involved primarily in the private sector will probably choose to study Chapter Five; those involved primarily in the public sector will probably choose to study Chapter Six. When one is involved in both the public and the private sectors, both chapters are appropriate!

Chapter Seven presents supplementary analysis techniques. In particular, the effects of risk and uncertainty on the analysis of economic investment alternatives are considered. In modeling the effects of risk and uncertainty mathematically, two approaches are taken: a prescriptive (normative) approach and a descriptive approach. A prescriptive model prescribes the action that, in some sense, is optimal; a descriptive model describes the behavior of the situation modeled. Chapter Seven provides a descriptive treatment of the effects of risk and uncertainty by presenting the subjects of *break-even*, *sensitivity*, and *risk analyses*. Prescriptive models are presented in Chapter Eight. Subjects considered include *decision making* under risk and uncertainty.

BIBLIOGRAPHY

1. Fleischer, G. A., *Benefit-Cost Analysis Applied to Air Pollution Control: A Case Study Based on the Dixon, Tiller County/U.S.A. Reference Community*, National Communicable Disease Center, Bureau of Disease Prevention and Environmental Control, Public Health Service, U.S. Dept. of H.E.W., Atlanta, Ga., 1968.

2. Klausner, R. F., "Communicating Investment Proposals to Corporate Decision-Makers," *The Engineering Economist*, 17 (1) 1971, pp. 45–55.

3. Krick, E. V., *An Introduction to Engineering and Engineering Design*, Wiley, 1965.

4. Lutz, R. P., "How to Justify the Purchase of Material Handling Equipment to Management," *Proceedings of the 1976 MHI Material Handling Seminar*, The Material Handling Institute, Pittsburgh, Pa., 1976.

CHAPTER TWO
COST CONCEPTS

2.1 INTRODUCTION

Engineering economic analysis is primarily concerned with comparing alternative projects on the basis of an economic measure of effectiveness. A nonrecurring initial investment, recurring operating expenses and possibly revenues, and a future resale value are usually associated with each project. This comparison process has a variety of cost terminologies and cost concepts, and it will be helpful to present them prior to the discussion in Chapter Four of the economic measures of effectiveness for comparing alternative projects. To implement the discussion of cost terminology, a typical production situation will now be described.

Let us assume that the business of a small manufacturing firm is job-shop machining. That is, the firm produces a variety of products and component parts according to customer order. Any given order may be for quantities of as few as five parts or as many as several hundred parts. The firm has periodically received orders to manufacture a part, which we will identify as Part No. 163H, for the Deetco Corporation. The part has been produced in a four-stage processing sequence consisting of (1) sawing bar stock to length, (2) machining on an engine lathe, (3) machining on an upright drill press, and (4) packaging. The unit cost to produce Part No. 163H by this sequence has been $15, where the unit cost consists of the major cost elements of direct labor, direct materials, and burden (prorated costs for insurance, taxes, electric power, marketing expenses, etc.). The firm is now in the process of negotiations with the Deetco Corporation to obtain a contract for producing 10,000 of these parts over a period of four years, or an average of 2500 units/year. A contract for this volume of parts is highly desirable but, in order to obtain the contract, the firm must lower the unit cost.

An engineer for the firm has been assigned to determine production methods to lower the unit cost. After study, the engineer recommends the purchase of a small turret lathe. With the turret lathe, the processing sequence of Part No. 163H would consist essentially of (1) machining bar stock on the turret lathe and (2) packaging. The estimated unit cost for Part No. 163H by this production method would be $10. Furthermore, the production rate by the new method would be increased over the old method because the turret lathe would replace the sawing, engine lathe, and drill press operations.

If the turret lathe is purchased, the saw, engine lathe, and drill press would not be sold but would be kept for other jobs for which the firm may receive orders. The turret lathe would be reserved for the production of Part No. 163H, but about 25% excess production capacity could be devoted to other jobs.

The incremental investment required to purchase the turret lathe and the new tooling required, as well as installing the machine, is $30,000. A freight charge of $400 would also be incurred. The functional life of the turret lathe is about 25 years, but the firm follows the policy of depreciating new machine tools over a 10-year period. The estimated salvage value of the turret lathe at the end of the 10-year planning period is $15,000. If the maximum unit price that the Deetco Corporation will pay for Part No. 163H is, say, $14, should the firm accept the contract for 10,000 parts and then purchase the turret lathe in order to execute the contract?

We will not answer the question in this chapter, but have cited the example to illustrate one type of decision with which this text is concerned. In fact, additional information would be required if the question were to be answered using the methodology presented in Chapter Four. However, you can readily appreciate that certain cost figures in the above example must be determined or estimated before any rational decision can be reached as to the purchase of the turret lathe. The cost elements contributing to the installed first cost of $30,000, the unit production cost of $10, and the salvage value of $15,000 would be determined or estimated from information obtained from a variety of sources. These sources might typically be production records, accountant's records, manufacturer's catalogs, publications from the U.S. Government Printing Office, etc. The engineer or system analyst should therefore be familiar with cost terminology, cost factors, and cost concepts as used by different specialists if effective economic comparisons and intelligent recommendations are to be made.

2.2 COST TERMINOLOGY

Because both cost definitions and cost concepts are included in this section, clarity will be achieved by the use of five categories: (1) life-cycle costs, (2) past and sunk costs, (3) future and opportunity costs, (4) direct, indirect and burden costs, and (5) fixed and variable costs.

2.2.1 Life-Cycle Costs

The life-cycle costs for any item, whether it is a product, project, or system, can be viewed as the summation of expenditures on the item from the birth to the death of the item. In less dramatic terms, life-cycle costs may include engineering design and development costs, fabrication and testing costs, operating and maintenance costs, and disposal costs. Life-cycle costs may also be expressed as the summation of acquisition, operation, maintenance, and disposal costs. Thus, life-cycle cost terminology may vary from author to author, but the basic meaning of the term is clear.

This textbook is predominantly concerned with the economic justification of engineering projects, the replacement of existing projects or capital assets, and the economic comparison of alternative projects. For the purpose of these types of analyses, we will define life-cycle costs to consist of (1) first cost (or initial investment), (2) operating and maintenance costs, and (3) disposal costs. There is obviously a time period involved for the life cycle and upon disposal, the item will have a salvage value.

The *first cost* of an item is considered the total initial investment required to get the item ready for service; such costs are usually nonrecurring during the life of the item. The term *item* should be interpreted in the general sense as a machine, a unit of equipment, a building, a project, a system, and so forth. For the purchase of a numerically controlled machine tool, Steffy et al. [12] state that the first cost of the machine tool may consist of the following major elements: (1) the basic machine cost, (2) costs for training personnel, (3) shipping and installation costs, (4) initial tooling costs, and (5) supporting equipment costs. For some other item, a different set of first-cost elements may be appropriate, but the first cost of an item normally involves more cost elements than just the basic purchase price. Whether the first cost elements are aggregated or maintained separately depends on income tax considerations and whether or not a before-tax or after-tax economic analysis is desired. Certain income tax laws and depreciation methods are presented in Chapter Five, and further discussion on this particular point is therefore deferred.

Operating and *maintenance* costs are recurring costs that are necessary to operate and maintain an item during its useful life. Operating costs usually consist of labor (direct and/or indirect), burden or overhead items (fuel or electric power, insurance premiums, inventory charges, etc., which will be discussed later), and possibly indirect materials (such as lubricants). It is usually assumed that these costs are annual, but maintenance costs may or may not be on a recurring, annual basis. That is, a regular, annual schedule of minor or preventive maintenance may be followed, or it could be policy that maintenance is performed only when necessary, such as when a major overhaul is required. In most cases, the maintenance policy would consist of both preventive maintenance and maintenance on an "as-needed" basis. In any case, repair and upkeep result in costs that must be recognized in the economic analysis of engineering projects.

When the life cycle of an item has ended, *disposal costs* usually result. Disposal costs may include labor costs for removal of the item, shipping costs, or special costs; an example of special costs is the disposal of hazardous materials. Although disposal costs may be incurred at the end of the life cycle, most items have some monetary value. This value is the *market* or *trade-in value* (i.e., the actual dollar worth for which the item may be sold at the time of disposal). Then, after deducting the cost of disposal, the net dollar worth at the time of disposal is termed the *salvage value.*

The market value, the disposal costs, and the salvage value are usually not known with certainty and therefore must be estimated. For an item that satisfies the United States Internal Revenue Service (IRS) definition of a capital asset and that decreases in value over time through physical deterioration, the IRS has approved various depreciation methods that can serve to estimate the rate of deterioration and consequent decrease in value of the asset. The value of the capital asset at the end of a given accounting period during the asset's life is termed the *book value.* The book value is thus an estimate of the market value; further discussion on depreciation accounting is deferred until Chapter Five. *Scrap value*, on the other hand, refers only to the value of the material of which the item is made. For example, a two-year-old automobile may have a scrap value of $200 but a market value of $3500. A distinction between these terms is not particularly important for evaluating potential investment projects—and salvage value will be used as the general term to denote the end-of-life value. For example, a trade-in value of $3000 minus disposal costs of $500 equals a net salvage value of $2500 for some item.

The life cycle obviously involves a time horizon, and the end of an item's life may be judged from a functional or "useful" point of view. The useful life of an item is based primarily on economic considerations and is generally shorter than the functional life. For example, an engine lathe may remain functionally useful for 40 years or more, but, because of periodic advancements in machine design technology, newer engine lathes have higher production rates; the economically useful life of an engine lathe is therefore judged to be 10 to 15 years. The economic life of an item is usually a matter of company policy that is greatly influenced by income tax considerations. In this textbook, useful life will mean economically, not functionally, useful.

2.2.2 Past and Sunk Costs

Past costs are historical costs that have occurred for the item under consideration. *Sunk costs* are past costs that are unrecoverable. The distinction is perhaps best made through examples. Assume that an investor purchases 100 shares of common stock in the ABC Corporation through a broker at $25 per share. In addition, the investor pays $85 in brokerage fees and other charges. Two months later and before receiving any dividend payments, the purchaser resells the 100 shares of common stock through the same broker at $35 per share minus $105 for selling expenses. The purchaser realizes a net profit of $810 ($3500 – $2500 – $85 – $105) on these transactions. At the time of sale, the $2500 and $85 are past costs, but because these are recovered after the sales transaction, sunk costs are not incurred. If, on the other hand, the investor were to sell the 100 shares two months after purchase and the market price were $20 per share, with a $70 charge for selling fees, the investor would incur a capital loss of $655 ($2000 – $2500 – $85 – $70). In this instance, some of the past costs would be recovered, but the $655 capital loss would be a sunk cost. If the investor reasons that the market price will decline further or if he or she simply needs the money, the $655 sunk cost should be ignored if the shares are to be sold for $20 each. However, sunk costs are not totally irrelevant to a present decision. They may qualify as capital losses and serve to offset capital gains or other taxable income and thus reduce income taxes paid in the year the loss was incurred. Examples in Chapter Five will illustrate this point. Also, past costs and sunk costs provide information that can improve the accuracy of estimating future costs for similar items.

Another example of sunk costs is a situation where an item of machinery is purchased for $10,000 and the salvage value at the end of five years is estimated as $5000. The annual decrease in value, or depreciation charge, is estimated as $1000. The $1000 annual cost of depreciation is a cost of production that, in theory, should be allocated to the output of the machinery. After allocating this and other manufacturing costs, general and administrative costs, and marketing costs to each unit of production, the total unit cost is determined. A profit percentage is then added to each unit of production in order to arrive at the unit selling price. Thus, when a unit is sold, a portion of each sales dollar returns a portion of the depreciation expense. In this illustration, it is assumed that sales will return, or recover, the total estimated depreciation expense of $5000 (first cost minus estimated salvage value) for the five-year period. However, if the machinery has a market value of only $2000 at the end of five years, there is a $3000 ($5000 – $2000) sunk cost if the machinery is sold for $2000. The $3000 capital loss represents an error in estimating the rate of depreciation, and the owner cannot insist that the machine is worth $5000 when the market value for the five-year-old machine is, in fact, only $2000. If the machine is kept, it is argued that the true value being kept is thus only $2000.

2.2.3 Future and Opportunity Costs

If the reference point at the present is $t = 0$, then all costs that may occur in the future are termed *future costs*. These future costs may be operating costs for labor and materials, maintenance costs, overhaul costs, and disposal costs. In any case, by virtue of occurring in the future, these costs are rarely known with certainty and must therefore be estimated. This is, of course, also true for future revenues or savings if these are involved in a given project. Estimates of future costs or revenues are uncertain and subject to error, and analysis is simplified if certainty of future costs, revenues, or savings is assumed. This assumption is made until Chapter Seven, where concepts of probability are introduced.

The cost of foregoing the opportunity to earn interest, or a return, on investment funds is termed an *opportunity cost*. This concept is best explained by means of illustrations. For example, if a person has $1000 and stores this cash in a home safe, the person is foregoing the opportunity to earn interest on the money by establishing a savings account in a local bank that pays, for example, 5% annual compound interest. (Of course, investments other than savings accounts are possible.) For a one-year period, the person is foregoing the opportunity to earn $(0.05) (\$1000) = \$50$. The $50 amount is thus termed the opportunity cost associated with storing the $1000.

A similar illustration of an opportunity cost is to assume that a person has $5000 cash on hand. This amount is considered *equity capital* if the $5000 was not borrowed (i.e., there is no debt obligation involved). The person has available secure investment opportunities such as establishing a personal savings account in a commercial bank or purchasing other financial instruments. From the available investment opportunities, suppose the optimum combination of risk (security level) and interest yield on the investment results in an 8% annual simple interest time-value of money (simple interest is explained more fully in Chapter Three). Thus, the investment of $5000 would yield $(0.08) (\$5000) = \$400$ each year. If the person, instead of investing the $5000, purchases an automobile for the same amount for personal use, the person will forego the opportunity to earn $400 interest per year. The $400 amount is again termed an annual opportunity cost associated with purchasing the automobile.

The same logic applies in defining an annual opportunity cost for investments in business and engineering projects. The purchase of an item of production machinery with $20,000 of equity funds prevents this money from being invested elsewhere with greater security or higher profit potential. This concept of *opportunity cost* is fundamental to the study of engineering economy and is a cost element that is included in virtually all methodologies for comparing alternative projects. In Chapter Four, the concept of opportunity costs will be discussed under the heading of *minimum attractive rate of return* (MARR).

Some individuals define the MARR as the *cost of capital*. As used here, the term *cost of capital* refers to the cost of obtaining funds for financing projects through debt obligations. These funds are usually obtained from external sources by (1) borrowing money from commercial banks or other organizations (e.g., insurance companies and pension funds) and (2) issuing bonds. These debt obligations are normally long term, as opposed to short-term obligations for the purchase of supplies and raw materials. The debt obligations result in interest payments on, say, a monthly, quarterly, semiannual, or annual basis. The interest payments are thus a cost of borrowed capital. Financing projects through issuing bonds is a method of obtaining capital funds that is probably less known to the reader than borrowing money from a commercial bank at a stated interest rate. Some elaboration of bonds is therefore appropriate.

Bonds are issued by various organizational units—partnerships, corporations (profit or nonprofit), governmental units (municipal, state, and federal), or other legal entities. The sale of bonds represents a legal debt of the issuing organization; bonds are generally secured by its assets. Examples are mortgage bonds or collateral bonds. Debenture bonds, on the other hand, are promissory notes or just a promise to pay. In any case, the purchaser of a bond has legal claim to the assets of the issuing unit but has no ownership privileges in the issuing unit. Purchasers of the common or preferred stock of an organizational unit do hold ownership status but may or may not have voting privileges, depending on the stipulations of the particular stock issue. In the sense that bonds are debt obligations and not ownership shares, bonds are considered a more secure investment than either common or preferred stock. This statement should not be taken as a universal truth, however, since the security level for a bond or a stock depends on many factors, economic and otherwise; the principal factor is the financial soundness of the issuing unit. Further details on interest payments on bank loans and interest payments on issued bonds are covered in Chapter Three.

Another method of financing engineering projects is through the use of *equity funds* (i.e., through ownership capital or cash that is debt free). The use of such funds incurs an opportunity cost, as mentioned previously, and will again be discussed in Chapter Four as the minimum attractive rate of return (*MARR*) for an investment project. However, the point is that the cost of capital is the cost of borrowed funds. Chapter Six further discusses the cost of capital for financing public utility projects.

2.2.4 Direct, Indirect, and Burden Costs

It will be helpful to provide definitions of direct, indirect, and burden costs in the context of a manufacturing environment. A typical cost structure for manufacturing, adapted from Ostwald,[1] is provided in Figure 2.1.

[1] Phillip F. Ostwald, *Cost Estimating for Engineering and Management*, Prentice-Hall, 1974, p. 55.

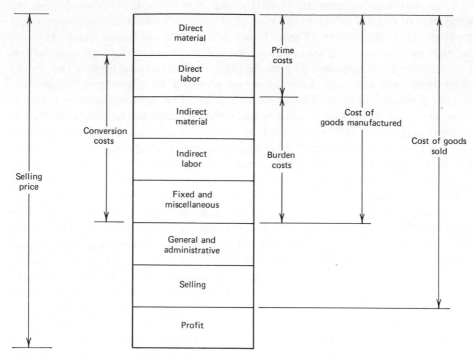

FIGURE 2.1. A cost structure for manufacturing.

The *cost of goods sold*, as shown in Figure 2.1, is the total cost of manufacturing a product or providing a service. An amount of profit is then added to this total cost to arrive at a selling price. Such a cost structure is helpful in arriving at a unit cost, which is a primary objective of cost accounting. The term *cost of goods sold*, as used here, has a different meaning from the term *cost of goods sold* used in general accounting practice, particularly for retail businesses. As will be mentioned later in the section on accounting principles, general accounting defines the cost of goods sold to be: beginning-of-the-year inventory plus purchases minus end-of-the-year inventory. Different meanings for the same terminology are unfortunate, but they do occur in the literature on accounting, and you are cautioned on this point. To simplify the treatment of the total cost of goods or services sold (as defined by Figure 2.1), the major cost elements can be defined as direct material, direct labor, and burden costs.

Direct material and labor costs are the costs of material and labor that are easily measured and conveniently allocated to a specific operation, product, or project.

Indirect costs for both labor and material, on the other hand, are either very difficult or impossible to assign directly to a specific operation, product, or project. Alternately, the expense of directly assigning such costs is prohibitive, and actual direct costs are therefore considered to be indirect.

As an example of these different cost elements, suppose the raw material for a given part is a rectangular gray iron casting. The casting is milled on five sides, the unmachined surface is painted and air-dried, then four through holes are drilled and tapped. These finished parts are stacked in open wooden boxes, 30 per box, and are delivered to a local customer.

In this example, the direct labor required per part to machine, paint, and package is probably readily determined. The labor required to receive the raw materials, handle parts between work stations, load boxes onto a truck, and deliver material to the customer is less easily identified and assigned to each part. This labor would be classified as indirect labor, especially if the labor in receiving, handling, shipping, and delivery is responsible for dealing with many different parts during the normal workday. The unit purchase price of the gray iron blocks is an identifiable direct material cost. The paint used per part may or may not be easily determined; if it is not, it is an example of indirect material cost. Also, any lubricating oils used during the machining processes would be an indirect material cost, not readily assigned on a per part basis.

Burden (or overhead) costs consist of all costs of manufacturing other than direct material and direct labor. A given firm may identify different burden categories such as factory burden, general and administrative burden, and marketing expenses. Furthermore, burden amounts may be allocated to a total plant, departments within a plant, or even to a given item of equipment. Typical specific items of cost included in the general category of burden are: indirect materials, indirect labor, taxes, insurance premiums, rent, maintenance and repairs, depreciation, supervisory and administrative personnel, and utilities (water, electric power, etc.). It is the task of cost accounting to assign a proportionate amount of these costs to various products manufactured or to services provided by a business organization.

2.2.5 Fixed and Variable Costs

Fixed costs do not vary in proportion to the quantity of output. General administrative salaries, taxes and insurance, rent, building depreciation, and utilities are examples of cost items that are usually invariant with production volume and hence are termed *fixed costs*. Such costs may be fixed only over a given range of production; they may then change and be fixed for another range of production. *Variable* costs vary in proportion to quantity of output. These costs are usually for direct material and direct labor.

Many cost items have both fixed and variable components. For example, a plant maintenance department may have a constant number of maintenance personnel at fixed salaries over a wide range of production output. However, the amount of maintenance work done and replacement parts required on equipment may vary in proportion to production output. Thus, total annual maintenance costs for a plant over several years would consist of both fixed and variable components. Indirect labor, equipment depreciation, and electrical power are other cost items that may consist of fixed and variable components. Determining the fixed and variable portion of such a cost item may not be possible; if it is possible, the expense of establishing detailed measurement techniques and accounting records may be prohibitive. A comprehensive discussion on this issue is outside our scope, and you are referred to books on general cost accounting for further reading.

Certain total costs (TC), then, can be expressed as the sum of fixed (FC) and variable costs (VC). As an example, the total annual cost for operating a personal automobile for a given year might be expressed as

$$TC(x) = FC + VC(x)$$

where $x =$ miles per year. Costs for insurance, license tags, depreciation, certain maintenance, and interest on borrowed money if the automobile were financed are essentially fixed costs, regardless of the miles traveled per year. Expenses for gasoline, oil, tire replacements, and certain maintenance are proportional to, or functional with, the mileage per year. One could argue, however, that depreciation expenses are comprised of both fixed and variable components. Arbitrarily assigning numerical values to the total cost function, assume that

$$TC(x) = \$650 + \$0.08x$$

is a valid relationship for a given year in question (the expression is restricted to a given year, since actual depreciation expenses, and hence the fixed expenses, vary from year to year). This relationship assumes linearity in the additivity of the cost components and in the constant rate of \$0.08/mile for the variable cost term. The variable cost component is often a nonlinear relationship but, for simplicity, linearity is assumed here. Figure 2.2 graphically illustrates the total cost function.

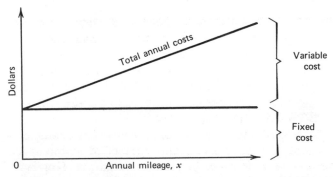

FIGURE 2.2. Total annual costs as a function of annual mileage.

2.3 BREAK-EVEN ANALYSIS

Now let us consider Figure 2.2 as a total cost function for a production line in a manufacturing firm where the output from the line is a single product. Furthermore, let it be assumed that each unit of production can be sold for $R and that the total revenue (TR) is a linear function of the production quantity:

$$TR(x) = \$Rx$$

Adding this functional relationship to Figure 2.2 and modifying the terminology for this example yields Figure 2.3.

It is noted from Figure 2.3 that the total annual revenue equals the total annual cost at point A or at an annual production volume of x^* units. Thus, at x^*,

$$TR(x^*) = TC(x^*)$$
$$= FC + VC(x^*)$$

and x^* is termed the annual production volume required in order to *break even*. Certain important observations can now be made. If the production volume is less than x^*, an annual net loss will occur, the amount of which is equal to $TC(x) - TR(x)$, evaluated for a particular value of x. By the same token, if the production volume is greater than x^*, then an annual net revenue or profit will result (the shaded region of Figure 2.3). The amount of annual profit is equal to $TR(x) - TC(x)$, evaluated for a particular value of x.

It is generally desirable to have a "low" break-even value. For the general example of Figure 2.3, this can be accomplished in three independent ways: (1) increase the slope of the total revenue line, (2) decrease the slope of the variable cost line, and (3) decrease the magnitude of the fixed cost line. Increasing the slope of the total revenue line means increasing the selling price of the product,

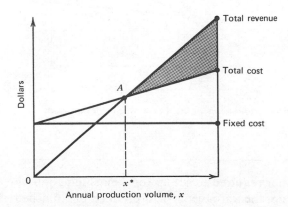

FIGURE 2.3. Relationship between total revenue and total cost as a function of annual production volume.

which may be a poor marketing strategy in a competitive market environment where sales would be lost. Fixed costs, although not literally "fixed," are difficult to reduce. Thus, reducing variable costs for direct material and labor usually offers the greatest opportunity to the engineer or analyst for profit improvement.

The concept of break-even analysis is, of course, general. Assuming that a break-even point exists, then, for two relationships $y = g(\cdot)$ and $y = h(\cdot)$, which are functions of a single variable x, the value of x for break-even, say x^*, may be determined from equating $g(x) = h(x)$ and solving for x^*. The concept can be extended to more than two functions of a single variable, say $y = h(x)$, $y = g(x)$, and $y = t(x)$. If these are all linear functions, then Figure 2.4 depicts two of the possible results.

In Figure 2.4b, there is no unique break-even value of x involving all three functional relationships. The linear equations $y = h(x)$ and $y = t(x)$ intersect at point B or $x = x_1^*$, which is then the break-even value for these two relationships. Point C, or $x = x_2^*$, is the break-even value for $y = h(x)$ and $y = g(x)$. Point D, or $x = x_3^*$, is the break-even value for $y = g(x)$ and $y = t(x)$.

The concept of break-even analysis also extends to nonlinear functions, with one or more break-even values, and functions of more than a single variable, which may be of linear or nonlinear form. However, examples and problems dealing only with functions of a single variable will be presented in this chapter.

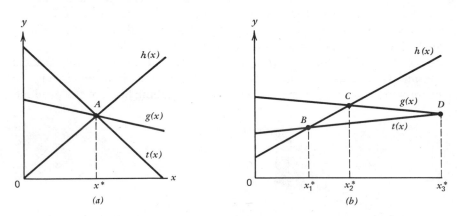

FIGURE 2.4. Graphical representation of break-even point. (a) A single break-even value. (b) Two break-even values.

Example 2.1

The cost of tooling and direct labor required to set up for a machining job on a turret lathe is $300. Once set up, the variable cost to produce one finished unit consists of $2.50 for material and $1 for labor to operate the lathe. For simplicity, it is assumed these are the only relevant fixed and variable costs. If each finished unit can be sold for $5, determine the production quantity required to break even and the net profit (or loss) if the lot size is 1000 units. Letting $x =$ the production volume in units, then

$$TR(x) = TC(x) = FC + VC(x)$$

and

$$\$5x = \$300 + (\$2.50 + \$1.00)x$$

Solving yields $x = x^*$ (the break-even value), or

$$x^* = \$300/\$1.50 = 200 \text{ units}$$

For a production output of 1000 units, the net profit, P, is calculated to be

$$P = \$5(1000 \text{ units}) - (\$2.50 + \$1.00)(1000 \text{ units}) - \$300$$
$$= \$1200$$

Example 2.2[2]

This example concerns the decision of selecting between two alternative methods of processing crude oil in a producing oil field, where the basis for the decision is the number of barrels of crude oil processed per year. The two methods of processing the crude oil are (1) a manually operated tank battery or (2) an automated tank battery. The tank batteries consist of heaters, treaters, storage tanks, etc., that remove salt water and sediment from crude oil prior to its entrance to pipelines for transport to an oil refinery.

For each alternative, fixed costs and variable costs are involved. Fixed costs include items such as pumper labor, maintenance (fixed over the production quantity of interest), taxes, certain energy costs (power to operate control panels and motors in continuous operation), and, for manual tank batteries, a cost for oil "shrinkage." Variable costs for chemical additives, heating, and noncontinuous operating motors are proportionate with the volume of oil being processed. This relationship is assumed to be linear over the production quantity of interest, since the data given in Tables 2.1 and 2.2 are based on a production quantity of 500 barrels/day.

In addition to the fixed and variable costs given in Table 2.1 for the automatic tank battery operation, other annual fixed costs are:

$$D_1 = \text{annual cost of depreciation and interest}$$
$$= \$3082$$

---

[2] The example is taken, with slight modification, from Ferguson and Shamblin [3] by permission of the publisher.

TABLE 2.1 Cost Data for Automatic Tank Battery Operations

Fixed cost
Control panel power	$0.15/day
Circulating pump power (3 horsepower)	0.82/day
Maintenance	1.00/day
Meter calibration	0.40/day
Chemical pump power (1/4 horsepower)	0.32/day
Total	$2.69/day
or $982/year ($2.69/day × 365 days)	

Variable cost
Pipeline pump (5 horsepower @ 50% utilization)	$0.63/day
Chemical additives (7.5 quarts/day)	3.75/day
Inhibitor (2 quarts/day)	1.00/day
Gas (10.8 MCF/day × $0.0275/MCF)	0.30/day
Total	$5.68/day
or $0.01136/barrel based on 500 barrels/day	

TABLE 2.2 Cost Data for Manual Tank Battery Operation

Fixed cost
Chemical pump power	$0.16/day
Circulating pump power	0.82/day
Total	$0.98/day
or $358/year ($0.98/day × 365 days)	

Variable cost
Chemical additives (7.5 quarts/day)	$3.75/day
Gas	0.30/day
Total	$4.05/day
or $0.00810/barrel based on 500 barrels/day	

M_1 = annual cost of maintenance, taxes, and labor

$$= \$5485$$

Letting x = the number of barrels of oil processed per year, the total annual cost, $TC_1(x)$, for the automatic tank battery operations is given by

$$TC_1(x) = FC_1 + VC_1(x)$$
$$= (\$982 + \$3082 + \$5485)$$
$$+ \$0.01136x$$
$$= \$9549 + \$0.01136x$$

In addition to the fixed and variable costs given in Table 2.2 for the manual tank battery operations, other annual fixed costs are:

$$D_2 = \text{annual cost of depreciation and interest}$$
$$= \$2017$$

$$M_2 = \text{annual cost of maintenance, taxes, and labor}$$
$$= \$7921$$

Then, the total annual cost, $TC_2(x)$, for the manual tank battery operation is given by

$$TC_2(x) = FC_2 + VC_2(x)$$
$$= (\$358 + \$2017 + \$7921)$$
$$+ \$0.00810x$$
$$= \$10,296 + \$0.00810x$$

By equating the two total cost functions, the break-even production volume can be determined as

$$TC_1(x) = TC_2(x)$$
$$\$9549 + \$0.01136x = \$10,296 + \$0.00810x$$
$$x^* = 229,141 \text{ barrels/year}$$

The interpretation of the break-even point in this example is that $x^* = 229,141$ barrels/year is the *point of indifference* between the choice of the two alternatives. If production volume is less than x^*, then the first alternative, or the automatic tank battery operation, would be preferred. For instance, if $x = 200,000$ barrels/year, then $TC_1(x) = \$11,821$ and $TC_2(x) = \$11,916$. Similarly, if production volume is greater than x^*, the manual tank battery operation would be preferred.

Example 2.3

A firm manufactures small machines for compression molding of plastics. These machines are used primarily in metallurgical laboratories to produce a plastic mounting base for metal specimens that are subsequently polished, etched, and microscopically analyzed.

The firm is presently selling 20 machines/month, which is about 60% of plant capacity with annual fixed costs of $75,000. These fixed costs are estimated to remain at this figure if production were increased to 90% of capacity. The molding machines are now selling for $2500 each, with unit variable costs of $1800.

By investing about $100,000 in an advertising campaign directed toward foreign markets, the firm estimates that 10 additional machines/month could be sold, but at a selling price of only $2000. By utilizing some present direct labor idle time, it is determined that variable costs could be reduced to

$1600/unit for the additional 10 machines/month. How many years would be required before the firm could recover the original $100,000 marketing expenditure?

Since 20 units/month or 240 units/year is 60% of capacity, then 240/0.60 = 400 units/year is 100% of capacity. A 90% capacity is therefore 360 units/year. An additional 10 machines/month added to the present 20 machines/month results in 360 units/year, and the annual fixed costs would remain at $75,000 with the increased production. Fixed costs are therefore not affected.

Then, for the machines sold on the foreign market, a unit profit of $2000 – $1600 = $400 or an annual profit of $400/unit × 120 units = $48,000 would result. The $100,000 investment would thus be recovered in $100,000 ÷ $48,000 = 2.0834 years or 25 months.

In the example, it was necessary to determine only the revenues and costs associated with the additional 10 machines/month (i.e., the *incremental* factors). As practiced in Chapters Four, Five, Six and Seven, an incremental analysis will be a problem-solution approach and, as such, will be concerned with isolating only the costs and revenues that are relevant to a given decision. Incremental costs and revenues may also be termed *marginal costs* and *revenues*. However, the general term *marginalism* is a broader economic concept that is treated extensively in standard texts on economic theory. This theory will not be presented in this textbook, but additional examples of incremental analysis are given in the chapters mentioned previously.

2.4 ESTIMATION

The estimation of future events or outcomes of present actions taken is obviously a fact of life for every individual, group or organization. Family budgeting, weather forecasts, market forecasts of demand for consumer products, and predicting the annual national revenues from income taxation are only a few examples of the almost limitless number and variety of estimates that are made in personal and business lives. In this textbook, we are concerned with estimation in the specific context of factors relevant to comparing alternative engineering/investment projects and making a selection from these projects. The annual revenues or savings, the initial and annual recurring costs, the useful and/or actual life of a project, and the future salvage value of capital assets such as buildings and equipment that may be associated with a given project are rarely, if ever, known with certainty.

Many different terms pertain to the general subject of estimation. We will not attempt in this text to enumerate and explain all the terms exhaustively; selected terminology will be given throughout the book as needed to explain the topics of a given chapter or section. Furthermore, in-depth study of estimation procedures and the accuracy of such estimated values is the study of mathematical statistics and probability theory, about which a vast literature exists.

Chapters Three to Six are concerned with comparing alternative projects when the estimated values for relevant factors are single-valued or point estimates. In such cases, the single-valued estimates are considered certain to occur. Moving a step toward realism, an interval estimate could be made for the value of a given factor such as annual costs. That is, a high and low value or range might be estimated for the factor. Chapter Seven discusses factors that contribute to uncertainty in estimated values and considers deviations from single-valued or point estimates. Chapter Eight continues the discussion on decision making with the topics of risk and uncertainty, defining these terms and posing methodology for selecting an alternative from a set of feasible alternatives, each of which has uncertain outcomes.

Although it is difficult to state precisely in quantitative terms, there is a relationship between the accuracy of an estimate and the cost of making the estimate. Intuitively, as more detailed information is obtained as the basis for an estimate and as more mathematical preciseness is exercised in calculating the estimate, the more accurate the estimate should be. However, as the level of detail increases, the more cost is involved in making the estimate. Ostwald[3] has conceptualized this notion by the function

$$C_T = C(M) + C(E)$$

where

C_T = the total cost of making the estimate, dollars

$C(M)$ = the functional cost of making the estimate, dollars

$C(E)$ = the functional cost of errors in the estimate, dollars

As depicted in Figure 2.5, the total cost of making an estimate reaches a minimum value when the amount of detail reaches a value D_1. Quantitatively defining the amount of detail is at best difficult and may be a practical impossibility. However, this concept of the total cost of an estimate varying with the amount of detail involved in making the estimate is realistic and is important to the general subject of estimation. In the abbreviated discussion that follows on cost estimation techniques, it will be noted that the individual techniques are based on varying amounts of detail with implied differences in the cost of making the estimate.

The text by Ostwald [8] is a primary reference on the subject of cost estimating. He defines four categories of estimated items: operations, products, projects, and systems. The categories are based on the scope of activity involved, but the distinction among them is not sharp and unequivocal. An

[3] Op. cit., p. 455.

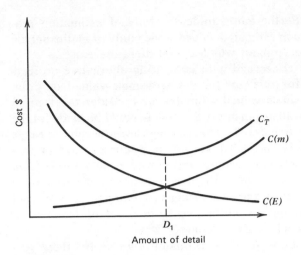

FIGURE 2.5. Cost of increasing detail.

operation is considered a basic or lowest-level activity. Machining, assembly, and painting are examples of "operations" for cost estimation purposes. Therefore, operations are subelements of manufacturing a product or providing a service. Producing a given product or providing a given service may be a subelement of a project, and a system may consist of several projects. A system may, for example, be defined as a total manufacturing plant (or several plants), a regional health care program of facilities and personnel, or a total program of national defense.

Generally, the larger the scope of the categorical item being estimated, the more difficult it is to make accurate estimates of the costs and revenues involved. Ostwald presents a comprehensive discussion of estimating techniques for all categories, and Vinton [14] provides procedural detail for estimating the cost of certain manufacturing operations and manufactured products.

2.4.1 Project Estimation

For all categories of estimated items previously mentioned, three principal classes of estimates, based on accuracy and degree of detail, may be defined: (1) order-of-magnitude estimates, (2) preliminary estimates, and (3) detailed estimates. *Order-of-magnitude estimates* are usually gross estimates based on experience and judgment and made without formal examination of the details involved. Accuracies of ±50%, relative to actual cost, can be expected.

Preliminary estimates are also gross estimates, but more consideration is given to detail in making the estimate than for order-of-magnitude estimates. Certain subelements of the overall task are individually estimated, engineering specifications are considered, etc. Estimating the cost for manufacturing new components or products before designs and production plans are complete is an example of preliminary cost estimation. This type of estimate enables process engineering and product engineering groups to compare alternative designs or manufacturing methods and assess the economic impact of them. An accuracy level of about ± 20% of actual cost can be expected with preliminary estimates.

Detailed estimates are expected to result in an accuracy level of ± 5% of actual cost. In preparing the estimate, each subelement of the overall task is considered, and an attempt is made to assign a realistic cost to the subelement. Pricing a product or contract bidding usually involves detailed estimates of the costs involved.

Estimating the cost of a project may involve all of the above types of estimates. The time available to make the estimate, the information available, the experience and knowledge of the persons making the estimate, and the type and scope of the project are all factors that affect the desired accuracy of the project's cost estimate and the cost of making the estimate. When a comparison of alternative projects is based on future cash flows, the cost and revenue figures are necessarily estimates that will vary from actual cost and revenue figures.

In subsequent chapters, examples illustrating the principles of engineering economic analysis will primarily involve the following estimated items; first cost (may include design, purchase, and installation costs), recurring operating and maintenance costs, recurring revenues or savings, disposal costs, useful life, salvage value, income taxes, cost of capital, and opportunity cost. These cost items have been discussed previously in this chapter under the section on cost terminology.

Estimates of the useful life may be based on IRS guidelines or general company policy. The IRS has an interest in the useful life estimate and publishes guidelines on lives of various depreciable items. It is conjectured that these IRS guideline estimates are based on functional life. Companies may adopt shorter lives than the IRS guideline estimates and, if reasonable, the company estimates are accepted by IRS.

If a company repeatedly buys a particular item of equipment, records may reveal that 10% of the items survive for four years, 20% survive for five years, 50% survive for six years, 15% survive for seven years, and the remaining 5% survive for eight years. The mean, or expected, life for any given item of equipment is thus equal to $[(0.10)(4) + (0.20)(5) + (0.50)(6) + (0.15)(7) + (0.05)(8)]$ or 6.15 years. Manufacturers of machinery and equipment are also excellent sources of information on the estimated life of their products. Such estimates would basically be functional life estimates, not economic life estimates.

In making an economic comparison of alternative projects, the analyst may only be interested in a particular period of time or planning horizon. This planning horizon may be (and often is) different from the useful life of one or more of the projects under consideration. For example, a job contract may cover a five-year period when the lives of the projects under consideration are ten years. The planning horizon adopted might then be five years, and only the relevant cost and revenue figures for the five-year period are used in the comparison. Further discussion on this point is deferred until Chapter Four.

2.4.2 General Sources of Data

There are many sources for providing data to make the various estimates required in comparing alternative investment projects. Morris [6] devotes a chapter to the general subject of gathering data for economic analysis purposes and cites both internal and external sources of information. Examples of sources within a firm are sales records, production control records, inventory records, quality control records, purchasing department records, work measurement and other industrial engineering studies, maintenance records, and personnel records. Some, all, or additional records are input to the accounting function of the firm, which then compiles various financial reports for management. A principal objective of the accounting system is to determine per-unit costs for direct materials, direct labor, and overhead involved in manufacturing a product or providing a service. Since overhead costs, by definition, cannot be allocated as direct charges to a given product or service, overhead costs are prorated among the various products or services by somewhat arbitrary methods, some of which are presented later in the chapter. Thus, a caution is raised concerning the use of cost accounting data for estimating the cost of overhead items associated with a project. Nevertheless, the accounting system can and usually does serve as an important, if not primary, internal source of detailed estimates on operating costs, maintenance costs, and material costs, among others.

Sources of data external to the firm may be grouped into two general classes: published information that is generally available, and information (published or otherwise) available on request. Available published information includes the vast literature of trade journals, professional society journals, U.S. government publications, reference handbooks, other books, and technical directories. Information not generally available except by request includes many sources listed in the previous category. For instance, many professional societies and trade associations publish handbooks, other books, special reports, and research bulletins that are available on request. Manufacturers of equipment and distributors of equipment are excellent sources of technical data, and most will readily supply this information without charge. Additionally, various government agencies, commercial banks (particularly holding companies involved in leasing buildings and equipment), and research organizations (commercial, governmental, industrial, and educational) may be sources of data to aid the estimating process.

2.5 ACCOUNTING PRINCIPLES

As already mentioned, the engineering economist should have some understanding of basic accounting practice and cost accounting techniques in order to obtain data from the firm's accounting system. If accounting is classified into general accounting and cost accounting, then cost accounting is judged the most important to the engineering economist as a source of data for making cost estimates pertinent to engineering projects. Cost accounting will therefore receive the greater emphasis in this text; in either case, the treatment of accounting is cursory and is directed toward fundamental accounting concepts instead of comprehensive accounting detail.

According to Niven and Ohman[7], an accounting system is concerned with the four primary functions of *recording, classifying, summarizing,* and *interpreting* the financial data of an organization, whether or not the organization is a business. The discussion to follow assumes a business organization, and, in virtually all such organizations, general accounting information is summarized in at least two basic financial reports:

1. A *balance sheet*, or statement of financial conditions, provides a summary listing of the assets, liabilities, and net worth accounts of the firm as of a particular date.
2. A *profit and loss statement* (or *income statement*) shows the revenues and expenses incurred by the firm during a stated period of time—a month, quarter, or year, for example.

2.5.1 Balance Sheet

Records of financial transactions and a variety of internal reports put into the accounting system information on sales and other revenues and the expenses incurred in obtaining the revenues. Revenues and expenses for a specified period are then summarized on the profit and loss statement. The net profit or loss resulting for this period is then "posted" to the balance sheet for an update of the financial report. Thus, although the two basic financial reports provide different financial pictures of the firm, they are directly related. Before illustrating this fact, discussion of certain terminology is necessary.

The items listed on a balance sheet are usually classified into three main groups: assets, liabilities, and net worth items. Subgroups may also be identified, such as current and fixed assets or liabilities. *Assets* are properties owned by the firm, and *liabilities* are debts owed by the firm against these assets. The dollar difference between assets and liabilities is the *net worth* of the business, which measures the investment made by the owner or owners of the business plus any accumulated profits left in the business by the owners or stockholders. A fundamental accounting equation is thus defined:

$$\text{assets} - \text{liabilities} = \text{net worth}$$

Rewriting, we have

$$\text{assets} = \text{liabilities} + \text{net worth}$$

and the usual format of a balance sheet follows the equation in this form.

Current assets are the properties that can be readily converted into cash; an arbitrary period of one year is usually assumed as a criterion for conversion. Similarly, *current liabilities* are the debts that are due and payable within one year from the data of the balance sheet in question. *Fixed assets*, then, are the properties owned by the firm that are not readily converted into cash within a one-year period, and *fixed liabilities* are long-term debts due and payable after one year from the date of the balance sheet. Typical current-asset items are cash, accounts receivable, notes receivable, raw material inventory, work in process, finished goods inventory, and prepaid expenses. Fixed-asset items are land, buildings, equipment, furniture, and fixtures. Items that are typically listed under current liabilities are accounts payable, notes payable, interest payable, taxes payable, prepaid income, and dividends payable. Fixed liabilities may also be notes payable (or bonds payable), mortgages payable, and so forth. Net-worth items appearing on a balance sheet are less standard and, to a degree, depend on whether the business is a sole proprietorship, a partnership, or a corporation. The size of the corporation is also an influencing factor on item designation. However, items such as capital stock, retained earnings, or earned surplus appear under the net worth group. An example of a balance sheet for a hypothesized firm is exhibited in Table 2.3.

The sample balance sheet in Table 2.3 is balanced (i.e., assets = liabilities + net worth) and it gives a statement of the financial condition of the organization as of a specific date—the close of an accounting period. Although depreciation will be covered in Chapter Five, note the fixed asset portion of the balance sheet. For example, the building originally cost $200,000, and depreciation expenses have been charged annually, so that the total depreciation charges, as of the date of the balance sheet, have been $50,000 or the amount entered as the depreciation reserve for the building. The first cost of the depreciable asset (building, in this case) minus the amount in the depreciation reserve equals the *book value.* The book value is an estimate of the salvage value of the asset if it were sold at the present time. A similar explanation applies for the fixed asset of equipment; in this balance sheet, the account is an aggregate for all equipment owned by the company instead of an individual listing of the equipment, which could be the case, depending on accounting practice.

TABLE 2.3 Sample Balance Sheet

Jax Tool and Engineering Company, Inc.
Balance Sheet
December 31, 19__

Assets

Current assets		
Cash	$ 25,000	
Accounts receivable	115,000	
Raw materials	8,500	
Work in process	7,000	
Finished goods inventory	3,000	
Small tool inventory	12,500	
Total current assets		$171,000
Fixed assets		
Land	$ 30,000	
Building	$200,000	
Less: Depreciation Reserve	50,000	150,000
Equipment	$750,000	
Less: Depreciation Reserve	150,000	600,000
Office equipment	10,000	
Total fixed assets		$790,000
Total assets		$961,000

Liabilities and Net Worth

Current liabilities		
Accounts payable	$ 32,000	
Taxes payable	15,000	
Total current liabilities		$ 47,000
Fixed liabilities		
Mortgage loan payable	$130,000	
Equipment loan payable	350,000	
Total fixed liabilities		$480,000
Total liabilities		$527,000
Common stock	$325,000	
Retained earnings	80,000	
Earned surplus (for current year)	29,000	
Total equity		$434,000
Total liabilities and equity		$961,000

2.5.2 Income Statement

The second basic financial report compiled by the accounting system is the *income statement* or *profit and loss* statement. For the current accounting period, the income statement provides management with (1) a summary of the revenues received through sales and investments, (2) a summary of the expenses incurred to obtain the revenues, and (3) the profit or loss resulting from business operations during the current accounting period. The format of the income statement varies, and the revenue and expense items depend on the type of business involved. Thus, the income statement given in Table 2.4 is illustrative only and still concerns our hypothesized Jax Tool and Engineering Company. Let us further assume that the period of time covered is one year, which has ended as of the date of the balance sheet given in Table 2.3.

The format for this income statement is basically that of Schedule C of IRS form 1040, and is an oversimplification of such a statement for even a small corporation. For example, the *cost of goods sold* item may be considerably more detailed to reflect multiple product lines, the depreciation items may be detailed to a greater extent to reflect multiple classes of assets, and several

---

TABLE 2.4 Sample Income Statement

Jax Tool and Engineering Co., Inc.
Income Statement
Year Ended December 31, 19__

Sales		$1,200,000
Less cost of goods sold:		
Inventory, January 1, 19__	$ 26,000	
Plus purchases	432,000	
	$458,000	
Less inventory, December 31, 19__	44,000	414,000
Gross profit		$ 786,000
Less expenses:		
Direct labor	$420,000	
Depreciation—building	10,000	
Depreciation—equipment	30,000	
Repairs and maintenance	41,500	
Indirect labor	218,000	
Utilities	9,800	
Supplies, tooling	1,700	$ 731,000
Net profit before income taxes		$ 55,000
Less income taxes		26,000
Net profit (posted to earned surplus)		$ 29,000

common expense items—such as employee benefits contributions, insurance premiums, and advertising—do not appear in the sample income statement.

2.5.3 Accounting Ratios

Recall that one purpose of the accounting system is to interpret the financial data of an organization. One technique for interpretation is the calculation of various accounting ratios. Calculating and using these ratios are not common practices of the engineering economist. However, they are presented here primarily for general interest and, because the engineering economist must converse with accountants, to provide further accounting terminology.

A ratio to assess the firm's working-capital condition is the *current ratio*, defined as

$$\text{current ratio} = \frac{\text{current assets}}{\text{current liabilities}}$$

Using Table 2.3, the current ratio for the Jax Company is $171,000/$47,000 = 3.6383, which implies that current assets would cover short-term debts approximately 3.64 times. Current ratios of 2 to 3 are common.

The current ratio assumes that the current assets of inventories are convertible to cash within one year; therefore, a more conservative ratio of the liquidity of the company is the *acid-test ratio*, which is defined as

$$\text{acid-test ratio} = \frac{\text{current assets} - \text{inventories}}{\text{current liabilities}}$$

For Table 2.3, the acid-test ratio is ($171,000 - $31,000)/$47,000 = 2.9787. In this calculation, all current assets except cash and accounts receivable are considered inventories.

A ratio that measures the financial strength of the firm is the *equity ratio*:

$$\text{equity ratio} = \frac{\text{stockholder's equity}}{\text{total assets}}$$

For Table 2.3, the equity ratio is $434,000/$961,000 = 0.4516; thus, 45.16% of the Jax Tool and Engineering Company is owned by the stockholders. (The $80,000 retained earnings and $29,000 earned surplus items are also owned by the stockholders.) This particular company has considerable debt, and the financial strength of the company is questionable.

The above three ratios can be computed using balance sheet data. A common accounting ratio calculated from income statement data is the *operating ratio*:

$$\text{operating ratio} = \frac{\text{total revenues}}{\text{total expenses}}$$

Using Table 2.4, the operating ratio is $1,200,000/$1,145,000 = 1.048, where the total expense figure includes the cost of goods sold but excludes the income taxes paid. (Income taxes may or may not be included in calculating the ratio.) An operating ratio greater than 1.0 indicates that a gross profit (before income taxes in the example used) is being made and is, therefore, desirable. Such operating ratios are perhaps more meaningful when they are computed for different product lines or different plants within a multiplant corporation. They can then be used by management to assess the effectiveness of individual product lines or plants.

The *income ratio* is defined as

$$\text{income ratio} = \frac{\text{net profit}}{\text{total revenue}} \times 100\%$$

and, using the after-tax data from Table 2.4, is calculated to be $29,000/$1,200,000 × 100% = 2.417%. This figure for the Jax Company is somewhat low for a manufacturing firm and is more characteristic of large retail organizations, particularly grocery corporations. For these, the 2.417% may, in fact, be high. In any case, this ratio indicates the net after-tax profit margin on gross revenues; management would be particularly interested in trends (rising, decreasing, or static) of the ratio.

The net after-tax profit for a given year can also be used to calculate a "gross" rate of return on investment. That is,

$$\text{gross rate of return} = \frac{\text{net profit}}{\text{total investment}} \times 100\%$$

There are many versions of this ratio, and they can all be criticized from an engineering economist's point of view, since the time value of money is ignored. The time value of money is the basis for most of the discussion in this textbook.

The above ratio is commonly used in the business world. Different versions of the ratio arise primarily because of different denominator terms used to calculate the ratio. What is the total investment, especially if the ratio is computed on an annual basis? Does the total investment include inventories, as well as buildings and equipment and land? For our purposes, the total investment of the Jax Company will be defined as all the fixed assets at their original cost. Thus, the gross rate of return for the accounting period covered by Table 2.4 is

$$R/R = \frac{\$\ 29,000}{\$790,000} \times 100\% = 3.67\%$$

which is a very low rate of return on investment and not typical of successful manufacturing firms.

2.5.4 Cost Accounting

The balance sheet and the income statement in Section 2.5.3 are considerably removed both in time and in detail from decisions at the usual engineering project level. More important to the engineering economist as a source of cost information is the cost accounting system within a particular firm. The firm may be involved in manufacturing or providing services, and if it is involved in manufacturing, production may be on a job-shop or process basis. There are some fundamental differences in cost accounting procedures for determining manufacturing costs versus determining the cost of providing a service; also, there are differences in accounting procedures if manufacturing is on a job-shop or process basis. In order to concentrate on basic principles instead of details, the cost accounting system assumed will be that of a job-shop manufacturing firm. Thus, the emphasis will be on determining the per-order costs for a job order.

The total cost of producing any job order consists of direct material, direct labor, and burden costs associated with the particular job order. An additional item of cost could be special tooling or equipment purchases strictly for the job order in question. This definition of total cost does not detail the burden cost into factory burden, general burden, and marketing expenses in order to simplify the presentation. Materials for a given job order may include purchased parts and in-house fabricated parts, and the cost for direct materials is determined primarily from purchase invoices. Questions of scrap allowances and averaging material costs, which fluctuate over time, present problems in obtaining accurate direct material costs, but determining such costs is reasonably straightforward. Direct labor time spent on a job order is normally recorded by operators on labor time cards, and the direct labor cost is determined by applying the appropriate labor cost rates. The labor rates, as determined by the accounting system, will normally include the cost of employee fringe benefits in addition to the basic hourly rate. Although accurately determining the direct labor cost for a given job is a major accounting problem, it is more readily determined than the burden cost.

Burden costs cannot be allocated as direct charges to any single job order and must, therefore, be prorated among all the job orders on some arbitrary basis. Common methods for distribution are:

1. The rate per direct labor hour.
2. A percentage of direct labor cost.
3. A percentage of prime cost (direct material plus direct labor cost).

If a single burden rate for the entire manufacturing firm is to be used, this would, of course, be an average burden rate for the entire factory, assuming the expense of providing and using an hour of factory facilities is about the same throughout the factory. To illustrate the three methods listed, it is assumed that a company experienced the following costs in the previous year 19__.

Total direct labor hours	120,000
Total direct labor cost	$480,000
Total direct material cost	$600,000
Total burden cost	$360,000

Then, the *burden rate per direct labor hour* would be

$$\text{rate} = \frac{\text{burden cost}}{\text{direct labor hours}}$$

$$= \frac{\$360,000}{120,000}$$

$$= \$3/\text{direct labor hour}$$

If a particular job order requires 100 hours of direct labor with an average rate of $4/hour and $850 of direct materials, then the total cost of the job would be computed as

Direct material cost			$ 850
Direct labor cost	= 100 hours × $4/hour =		400
Burden cost	= 100 hours × $3/hour =		300
Total cost			$1550

Determining the burden rate as a *percentage of direct labor cost* for this company will yield

$$\% = \frac{\text{burden cost}}{\text{direct labor cost}}(100\%)$$

$$= \frac{\$360,000}{\$480,000}(100\%)$$

$$= 75\%$$

For the same job above, the total cost would be computed as

Direct material cost	=	$ 850
Direct labor cost	=	400
Burden cost = (0.75)($400) =		300
Total cost		$1550

Determining the burden rate as a *percentage of prime cost* for this company will yield

$$\% = \frac{\text{burden cost}}{\text{direct labor cost} + \text{direct material cost}}(100\%)$$

$$= \frac{\$360,000}{\$1,080,000}(100\%)$$

$$= 33\tfrac{1}{3}\%$$

For the same job above, the total cost would be computed as

$$
\begin{aligned}
\text{Direct material cost} &= \$\ 850.00 \\
\text{Direct labor cost} &= 400.00 \\
\text{Burden cost} = (0.33)(\$1250) &= \underline{416.67} \\
\text{Total cost} &\ \$1666.67
\end{aligned}
$$

Determining the burden cost for a job order by the "rate per direct labor hour" method will yield the same result as the "percentage of direct labor cost" method. This is true in general and can readily be shown algebraically. The "percentage of prime cost" method will necessarily yield a different assignment of burden to a job order than the other two methods. The choice between these three methods is arbitrary; indeed, other methods are used by cost accountants in distributing burden costs to a given job order. The rate per direct labor hour method is perhaps most commonly used.

Whatever method is chosen from the above for distributing burden costs to job orders in a current year, the rates or percentages are based on the previous year's cost figures. Thus, burden rates may change from year to year within a particular firm.

Since an average burden rate for the entire factory may very well be too gross an estimate when actual burden costs differ among departments within the factory, cost accounting may determine individual burden rates for departments or cost centers. Furthermore, the hourly rates for direct labor may vary among these cost centers. A further refinement is to determine burden rates for individual machines within cost centers. Then, as particular job orders progress through cost centers (departments and/or machines), the direct labor time (or machine time) spent on the job order in the various cost centers is recorded, the appropriate labor or machine rates and burden rates are applied, and the total cost for the job is calculated.

The following example illustrates the variety of methods used to distribute burden to a given cost center; the total burden for the cost center will then be distributed by yet another method.

Example 2.4

The following information has been accumulated for the Deetco Company's two departments during the past year. (See table on following page.)

The Deetco Company distributes depreciation burden based on (1) the first cost of equipment in each department, (2) a zero salvage value of the equipment in 10 years, and (3) a constant annual (or straight-line) rate of decrease. All burden other than depreciation is first distributed to each department according to the number of employees in each department, and a burden rate per direct labor hour is computed.

What selling price should the company quote on Job Order D if raw material costs are estimated as $900, estimated direct labor hours required in

	Department A	Department B	Total
Direct materials cost	$72,000	$24,000	$96,000
Direct labor cost	$26,800	$14,600	$41,400
Direct labor hours	7,800	4,800	12,600
Number of employees	4	2	6
First cost of equipment	$25,000	$20,000	$45,000
Annual depreciation	$ 2,500	$ 2,000	$ 4,500
Other factory burden			$15,000
General burden			$35,000

Departments A and B are 30 hours and 100 hours, respectively, and profit is to be calculated as 25% of selling price?

For Department A, the total burden allocated is determined as

$$
\begin{array}{lll}
\text{Annual depreciation} & & = \$\ 2{,}500 \\
\text{Other factory burden} & = (4/6)(\$15{,}000) = & 10{,}000 \\
\text{General burden} & = (4/6)(\$35{,}000) = & \underline{23{,}334} \\
\quad \text{Total burden costs} & & \$35{,}834
\end{array}
$$

Thus, the burden rate for Department A per direct labor hour is

$$
\text{rate} = \frac{\$35{,}834}{7800 \text{ hours}}
$$

$$
= \$4.594/\text{direct labor hour}
$$

For Department B, the total burden allocated is determined as

$$
\begin{array}{lll}
\text{Annual depreciation} & & = \$\ 2{,}000 \\
\text{Other factory burden} & = (2/6)(\$15{,}000) = & 5{,}000 \\
\text{General burden} & = (2/6)(\$35{,}000) & \underline{11{,}666} \\
\quad \text{Total burden costs} & & \$18{,}666
\end{array}
$$

Thus, the burden rate for Department B per direct labor hour is

$$
\text{rate} = \frac{\$18{,}666}{4800 \text{ hours}}
$$

$$
= \$3.889/\text{direct labor hour}
$$

Then the estimated total cost for Job Order D is computed as

Direct material cost	= $ 900.00
Direct labor cost for Department A = ($26,800/7800)(30 hours) =	103.08
Burden cost for Department A = ($4.594)(30 hours) =	137.82
Direct labor cost for Department B = ($14,600/4800)(100 hours) =	304.17
Burden cost for Department B = ($3.889)(100 hours) =	388.90
Total costs	$1833.97

If x = the selling price of Job Order D, then

$$x = \text{total cost} + \text{profit}$$
$$= \$1833.97 + 0.25x$$

and

$$x = \frac{\$1833.97}{0.75} = \$2445.29$$

2.5.5 Standard Costs

Although the first task of cost accounting is to determine per-item or per-order costs, another major purpose of cost accounting is to interpret financial data so that management can (1) measure changes in production efficiency and (2) judge the adequacy of production performance. Establishing cost standards can be of great assistance in achieving these objectives. A standard-cost system involves in advance of manufacture (1) the preparation of standard rates for material, labor, and burden, and (2) the application of these rates to the standard quantities of material and labor required for a job order or for each production operation required to complete the job order.[4]

Since a process-type manufacturing firm, such as an oil refinery, outputs the same product (or a few products) over a long time period, cost standards are more readily determined for process firms than for job-shop firms, where the variety of output is large and varies with customer order. However, the number and type of production operations required to complete various job orders are finite for a given manufacturing firm. Each job order is, of course, made up of single units. Thus, a standard amount of material can be determined for each unit, and standard labor times and machine times can be determined for each unit. It is usually the responsibility of the work measurement function within the firm to determine these standard quantities. Then, by applying standard unit material costs and standard labor rates, standard unit costs for material, labor, and burden can be determined. The standard costs then serve as a basis for measuring production efficiency and performance over time. Deviations from standard costs may be caused by several factors, especially (1) raw material price variations and (2) actual quantities of material and labor used versus the standard amounts of these items. This latter factor is the one of primary concern in determining production efficiency and performance, measures of which provide information to management to aid in cost control.

[4] Frank W. Wilson and Philip D. Harvey (Editors), *Tool Engineers Handbook*, Second Edition, McGraw-Hill, 1959, pp. 2–23.

2.6 NONMONETARY CONSIDERATIONS

The principal emphasis of this textbook is on the use of logical methodology to choose an investment project from among several investment projects available to the decision maker, where the criterion for choice will be some single economic measure of effectiveness. However, the decision maker has several economic measures from which to select. For example, because of an increase in market demand, the management of a firm may face the choice between (1) going to a second shift of production using the same machinery, and (2) installing more highly mechanized machinery in order to meet the new production schedules. Clearly, a comparison of the costs for each of these alternatives over some planning horizon is in order and, one objective of management would no doubt be to meet production schedules at the lowest possible annual cost of production.

Although considerable investigation and study are required, many of the factors involved in a decision that may at first seem noneconomic can be expressed in (or reduced to) monetary values. For instance, in the above example, maintenance personnel may have to be trained in the installation and repair of the new machinery but, most likely, these training costs can be determined and charged to the "mechanization" alternative. However, other factors involved are not so easily reduced to monetary values and, indeed, some factors can not be so reduced. For example, the installation of the mechanized machinery would probably reduce the number of personnel required with the present production system, whereas the second shift alternative would result in additional employees. If persons are laid off or transferred to other jobs within the firm, their salaries could be considered annual savings accruing to the mechanized machinery alternative and, conversely, the salaries of additional employees integral with the second shift alternative would be annual costs. However, layoffs or transfers could have a deleterious effect on the morale of the remaining employees; on the other hand, adding new personnel could have a beneficial effect on morale. Assessing the cost or gain of these expected changes in morale is virtually impossible. Nevertheless, the potential effect on production that morale changes may have must be considered by management.

Factors that affect a decision but cannot be expressed in monetary terms are often called *intangibles* or *irreducibles*. Practically all real-world business decisions involve both monetary and intangible factors. In the above production example, assume that the alternative of a second shift results in a total annual cost of $100,000 for both shifts over a five-year planning horizon with no persons laid off or transferred or new employees hired. For the mechanized machinery alternative, there would be an estimated annual cost of $85,000 over the five-year planning horizon with six persons laid off. Thus, this oversimplified decision is reduced to two measures: a monetary annual cost and the single intangible factor of employee morale. If management chooses the second shift alternative, this would imply that management places a higher subjective utility value on "good employee morale" than on the $15,000 difference in annual costs between the two alternatives. That is, the objective of good employee morale is more important than the objective of lowest annual cost (at least for this difference of $15,000). Instances of decision situations where intangible considerations outweigh monetary ones are frequent, and the engineering economist should not become distraught at this fact. Within the hierarchy of management responsibilities for a firm, the higher the level of management, the more likely it is that intangible considerations will be given greater subjective weight. Engineering project proposals should therefore reflect a knowledge of intangibles that management will wish to consider.

Miller and Starr have made interesting observations in regard to the decision-making process.[5]

1. Being unable to satisfactorily describe goals in terms of one objective, *people customarily maintain various objectives.*

2. Multiple objectives are frequently in conflict with each other, and when they are, a *suboptimization* problem exists.

3. At best, we can only optimize as of *that time* when the decision is made. This will frequently produce a supoptimization when viewed in subsequent times.

4. Typically, decision problems are so complex that any attempt to discover *the* set of optimal actions is useless. Instead, people set their goals in terms of outcomes that are *good enough.*

5. Granted all the difficulties, human beings make every effort to be *rational* in resolving their decision problems.

Accepting the premise that multiple objectives are involved in a decision, it follows that different measures for these objectives may arise. For instance, two or more engineering project alternatives could each involve annual costs in dollars, weight in pounds, repairability values on an arbitrary index from 1 to 10, and so forth. The decision problem would be greatly simplified if the various measures could be transformed to a single measure, say, values on a "utility" scale. Then, each measure (dollars, pounds, etc.) for a given alternative could be converted to a utility value, the new utility values could be weighted by their relative importance, and the weighted utility values could be aggregated by an appropriate functional form. A single utility value would thus result for each alternative, and a selection would be made by choosing the alternative having the maximum utility value. This area of study is the study of utility theory and value measurement—very interesting, but not

[5] David W. Miller and M. K. Starr, *Executive Decisions and Operations Research*, Second Edition, Prentice-Hall, 1969, pp. 52–53.

without its controversy. It is outside the scope of this book, but selected references are given in the bibliography of this chapter. Some of these stress the mathematical theory [2, 4, 9, 10] and others offer an applied, but approximate, approach [1, 5, 11].

There will be some additional discussion on multiple objectives in Chapter Eight, but otherwise it will be assumed throughout this book that only a single economic measure of effectiveness is relevant in comparing alternative projects.

2.7 SUMMARY

In this chapter we provided an introduction to the language of accountants, financial analysts, and managers. To be successful in selling engineering designs, one must learn to communicate; to communicate effectively, all parties must speak the same language. Additionally, the data sources for economic analyses are often to be found in the accounting systems; hence, it is essential that you be familiar with accounting principles. To complete the coverage of cost concepts, a brief introduction to cost estimation was presented. Finally, the need for consideration of nonmonetary aspects in a decision problem was treated.

BIBLIOGRAPHY

1. Apple, James M., *Material Handling Systems Design*, Ronald Press, 1972, Chapters 13 to 15.

2. Chernoff, Herman, and Moses, Lincoln E., *Elementary Decision Theory*, Wiley, 1959.

3. Ferguson, Earl J., and Shamblin, James E., "Break-Even Analysis," *The Journal of Industrial Engineering*, *XVIII* (8), August 1967.

4. Fishburn, Peter C., *Decisions and Value Theory*, Wiley, 1964.

5. Kepner, Charles H., and Tregoe, Benjamin B., *The Rational Manager*, McGraw-Hill, 1965.

6. Morris, William T., *The Analysis of Management Decisions*, Revised Edition, Richard D. Irwin Co., 1964, Chapter 5.

7. Niven, William, and Ohman, Anka, *Basic Accounting Procedures*, Prentice-Hall, 1964, p. 2.

8. Ostwald, Phillip F., *Cost Estimating for Engineering and Management*, Prentice-Hall, 1974.

9. Raiffa, Howard, *Decision Analysis—Introductory Lectures on Choice Under Uncertainty*, Addison-Wesley, 1968.

10. Schlaifer, Robert, *Probability and Statistics for Business Decisions*, McGraw-Hill, 1959.

11. Starr, Martin Kenneth, *Product Design and Decision Theory*, Prentice-Hall, 1963, Chapters 1 and 6.

12. Steffy, Wilbert, Smith, Donald H., and Souter, Donald, *Economic Guidelines for Justifying Capital Purchases*, Industrial Development Division, University of Michigan, Ann Arbor, Michigan, 1973, p. 89.

13. Tarquin, Anthony J., and Blank, Leland T., *Engineering Economy*, McGraw-Hill, 1976.

14. Vinton, Ivan R. (Editor), *Realistic Cost Estimating For Manufacturing*, Society of Manufacturing Engineers, Dearborn, Michigan, 1968.

PROBLEMS

1. A gasoline-powered, special-purpose machine can be replaced by a similar new machine that is electrically powered. The incremental investment required would be $3000. The salvage value of the new machine is assumed zero at any time after purchase. It is estimated that a net annual savings of $500 in operating costs will result if the new machine is purchased. If the time value of money is assumed to be zero, how many years of savings are required to recover the incremental investment? (2.3)

2. An engine lathe or a turret lathe can be used to produce a job order of Part 173. If the job order is produced on the engine lathe, Machinist B performs all required operations (i.e., the tooling setup and the operation of the lathe during the machining cycle). On the engine lathe, a tooling setup time of 15 minutes and machining time of 30 minutes is required to produce each part. Machinist B earns $5/hour and the burden rate for each machining hour of engine lathe operation is $2.50. The tooling costs per lot for production on the engine lathe are estimated to be $200.
   If the job order is produced on the turret lathe, Machinist A must do the initial tooling setup for the job, and four hours are required for this. Once the setup is done, no further tooling setup per part is required, and Machinist B could operate the turret lathe for the machining cycles. Machinist A earns $8/hour. The machining time for Part No. 173 is 10 minutes on the turret lathe, and the burden rate for each machining hour is $3.20. The tooling costs per lot for production on the turret lathe is $350. For what job-order size would the turret lathe be economically preferred to the engine lathe? (2.3)

3. In a stable economic environment, the A. B. Jax Specialty Foundry Co. can produce a maximum of 1500 gray iron railroad car wheels per month and sell these for $125 each. Now, after prolonged labor problems in the coal industry, the foundry can only sell an average of 500 wheels/month at $80 each. The coal industry strike is temporary, but the future over the next six months to a year appears very uncertain.

The foundry has a depreciable investment in buildings and equipment of $300,000, and the value decreases at a rate of 4% each year.

At a production rate of 500 wheels/month, direct labor costs will increase from $30/wheel at maximum production rate to $36/wheel. Direct material costs per wheel will remain at $20. Other annual burden costs theoretically vary in linear fashion from $60,000 at zero output to $105,000 at maximum production capacity. Would you recommend that the A. B. Jax Specialty Foundry Co. shut down temporarily or continue to operate at this reduced capacity? (2.3)

4. A commercial machine shop regularly produces a stainless steel component for a major electronics manufacturer. The machine shop purchases the component from a nearby specialty steel company in semifinished condition, performs drilling and milling operations on the part, and ships it to the electronics firm.

The machining operations are those that can readily be performed on a tape-controlled drill press with a turret head. Thus, the management of the machine shop feels the purchase of such a machine to produce only this part is economically justified. An engineer is then assigned the task of determining the production quantity for break-even, assuming a time value of money (minimum attractive rate of return) equal to zero.

The engineer compiles the following information and cost estimates. The tape-controlled machine will have an installed first cost of $20,000, which includes the necessary electronic software, cutting tools and holders, and work-holding devices. Training of the operator for the machine is included in the purchase price. The economic life of the machine is assumed to be 10 years with a salvage value of $8000 at that time. The decrease in asset value is estimated at $1200/year and judged to be an annual fixed cost. Other fixed costs are $900/year. The steel parts are sold to the electronics firm for $7.20/unit. The variable unit costs are estimated as $0.85 for direct labor, $3.50 for direct material, and $1.70 for burden (excluding depreciation of the machine—the $1200/year fixed cost mentioned previously). What annual sales volume (number of parts) is required in order to break even on the machine purchase if linearity is assumed? (2.3)

5. A subsidiary plant of a major furniture company manufactures wooden pallets primarily for the parent company but also sells pallets to other industrial customers. The pallet produced is essentially of standard size and design. Minor modifications in pallet design occur but, relative to the standard design, the quantity sold is negligible and the subsidiary plant can be considered a single-product plant. The subsidiary plant has the capacity to produce 450,000 pallets/year. Presently, the plant is operating at 60% of capacity. The average selling price of a pallet is $6.25 with a variable cost per pallet of $5 (unit revenue and cost rates are linearly related to production quantity). At zero output, the subsidiary plant's annual fixed costs are about $254,000 and are approximately constant up to the maximum production quantity per year. (2.3)

(a) With the present 60% of capacity production, what is the expected annual profit or loss for the subsidiary plant?

(b) What annual volume of sales is required in order for the plant to break even?

(c) What would be the annual profit or loss if the plant were operating at 80% of capacity?

6. The Smoky Tobacco Company is currently experiencing decreased demand for one of their brands of nonfiltered cigarettes, Brand Hack. Sales of other nonfiltered brands remain stable, and filtered brands and pipe tobacco sales are increasing. The company estimates that Brand Hack sales will recover in the intermediate future of 5 to 10 years, and they have some desire, although not strong, to remain fully competitive in the nonfiltered cigarette market. However, business decisions are largely economic and the company wishes to address the question of whether to shut down the Brand Hack production line or continue to operate at reduced capacity for several years.

In the past, sales of Brand Hack have averaged 1 million cartons/year at $2.50/carton. Annual sales are presently 250,000 cartons.

The original investment in the production line is $500,000, and this value decreases at a rate of $25,000 each year. (For simplicity, a salvage value of zero is thus assumed.) Annual taxes (other than income taxes) and insurance costs are 5% of the original investment.

Direct material costs are $0.90/carton and direct labor costs are $0.30/carton. Burden costs, excluding maintenance, are $1.05/carton. Annual maintenance costs, which vary linearly with production volume, are $10,000 at zero output and $35,000 at 1 million cartons output.

Neglecting the consideration of converting the production line to a new brand and other intangible factors, should the Brand Hack production line be continued or shut down? (2.3)

7. A first production run of Component G is made with the following data compiled:

Direct material cost = $1.00/unit
Direct labor cost = 0.70/unit
Burden cost = $1.50/unit
Selling price = $4.00/unit

The above costs are based on all good units produced and no rejects or scrap. Parts are inspected only once, after all the manufacturing operations are performed but prior to shipping. During the production run, a scrap rate of 30% has occurred. What is the maximum scrap rate permissible in order to break even (i.e., total costs equal to total revenues)? (2.3)

8. A typical gasoline-powered farm tractor with PTO rating of 35 horsepower has an operating cost of about $1.42/hour of use (fuel, lubricants, oil filters, repairs, and maintenance), excluding labor cost, and an annual fixed cost of about $700 (for depreciation, insurances and taxes, housing, and opportunity costs).

Similar data for a diesel-powered tractor of the same size are $1.22/hour of use for operating cost and $786 annual fixed cost.[6]

(a) What is the number of operating hours per year for break-even between the two tractors? (2.3)

---

[6] Data for this problem taken from Publication 510 (revised February 1974) entitled "Farm Machinery Performance and Costs," Extension Division, Virginia Polytechnic Institute and State University, Blacksburg, Virginia.

(b) If the estimated number of operating hours per year is 1000, what annual savings are estimated if the diesel-powered tractor is purchased instead of the gasoline-powered tractor? (2.3)

9. K. Z. Moley purchased a small retail restaurant business and opened on July 1, 19__. At the date of opening, he had invested $8000 of equity funds with the following breakdown: $4000 in equipment, $3000 in inventory items, and $1000 in operating cash.

(a) Prepare a balance sheet for the business as of July 1, 19__. (2.5.1)

The data below summarize the gross sales and expenses for the restaurant during the first three-month period.

Gross sales	$13,500
Purchases	7,000
Salaries	3,000
Advertising expense	250
Rent expense	600
Expense for utilities	400

For the purchases, $6000 was paid with cash and $1000 is still owed. At the close of the three-month period, the end-of-period inventory is worth $2600.

(b) Prepare an income statement for the three-month period covered. (2.5.2)

(c) If, at the end of the three-month period on October 1, 19__, the net worth (or ownership) account is $9850, determine the amount of the cash account in order to balance the accounting equation as of October 1, 19__. (2.5.1)

10. A successful building contractor purchased a 150-acre farm to operate on a part-time basis and raise beef cattle. The purchase price of the farm was $75,000. The contractor paid $25,000 cash and financed the remainder over ten years at a 9% annual simple interest rate with a mortgage, payable to a local bank. Soon after the purchase of the farm, the contractor purchased cattle for $8000, paid $3000 in cash, and gave a promissory note to the seller for the balance. The note carried an 8% annual simple interest rate and was to be paid off within five years. During the first full year, the farm operation resulted in the following revenues and expenses.

Calves sold	$6500
Hay sold	200
Labor expenses	500
Expenses for machinery hired	2000
Veterinarian fees	175
Fertilizer purchased	1200
Property taxes	450
Expenses for repairs	375
Interest expenses	4900
Expenses for miscellaneous supplies	125

Prepare an income statement for the farming operation and determine the net profit (loss) before income taxes. (2.5.2)

11. An income statement and balance sheet for the WAC Company covering a calendar year ending December 31 are as follows.

Income Statement

Gross income from sales		$247,000
Less: cost of goods sold		138,800
Net income from sales		$108,200
Operating expenses		
Rent	$ 9,700	
Salaries	30,200	
Depreciation	5,800	
Advertising	4,300	
Insurance	1,500	$ 51,500
Net profit before income taxes		56,700
Less: income taxes		23,973
Net profit after income taxes		$ 32,727

Balance Sheet

Assets			Liabilities	
Cash		$94,227	Notes payable	$ 25,000
Accounts receivable		8,000	Accounts payable	6,000
Raw material inventory		10,000	Declared dividends	20,000
Work-in-process inventory		15,000	Total liabilities	$ 51,000
Finished goods inventory		18,500		
Land		30,000	*Net worth*	
Building	$80,000		Capital stock	$200,000
Less: reserve	$ 8,000	72,000	Earned surplus	32,727
Equipment	$40,000		Total net worth	$232,727
Less: reserve	$ 4,000	36,000		
			Total liabilities and	
Total assets		$283,727	net worth	$283,727

If land, building, and equipment are fixed assets, and notes payable is a fixed liability, compute (a) the current ratio, (b) the acid-test ratio, (c) the equity ratio, (d) the operating ratio, and (e) the income ratio for net profit after taxes for the WAC Company. (2.5.3)

12. An order for 5000 units of Part D-142 is received by the J. T. Kling Engineering Company, a small machine shop. The finished dimensions of the rectangular part are $1\frac{7}{8}$ inch $\times$ $1\frac{13}{16}$ inch $\times$ 4 inch (neglecting tolerances). The raw material for this part is purchased 2 inch $\times$ 2 inch $\times$ 8 foot SAE 1020 steel rectangular bar stock, with each unit costing $12 and yielding 20 part blanks. The basic manufacturing sequence, with standard machining times per part, and machine burden rates (per machining hour) is given below.

Operating	Standard Time per Part	Machine Burden Rate
Cutoff on power hacksaw	1 minute	$0.30/hour
Mill two sides; deburr	4 minutes	0.50/hour
Drill three 3/8-inch diameter holes	2 minutes	0.40/hour
Surface grind one side; deburr	2 minutes	0.35/hour
Package	0.25 minutes	—

The direct labor time per part is the same as the machine (or operation) time per part. The tooling cost for this job order is estimated to be $500. Excluding tooling costs and machine burden, other factory burden costs (for indirect labor, utilities, indirect materials, etc.) are $9/direct labor hour. If the average direct labor hour rate (including fringe benefits) is $5.75/hour, (a) determine the total estimated costs for the job order of 5000 units, and (b) the unit selling price if profit is to be 30% of the total cost. (2.5.4)

13. The welding department of a mining equipment manufacturing plant consists of four cost centers: manual arc welding (A), semiautomatic welding (B), furnace brazing and heat treating (C), and finishing (D). Some oxyacetylene cutting is also done in Center B. Assume that it is possible to allocate departmental burden expenses directly to each cost center and that the following data for the welding department were compiled last year by the accounting system.

Cost Center	Departmental Expenses	Direct Labor Hours	Direct Labor Cost	Direct Material
A	$10,500	10,000	$52,000	$8,000
B	6,800	4,000	16,000	8,000
C	4,600	1,500	4,500	3,000
D	2,400	2,800	6,000	2,000

Compute the burden rate (or rates) applicable by the following methods. (2.6)
(a) Blanket (departmental) percentage of direct labor cost.
(b) Blanket percentage of prime cost.
(c) Blanket hourly rate per direct labor hour.
(d) Percentage of direct labor cost for each cost center.
(e) Percentage of prime cost for each cost center.
(f) Rate per direct labor hour for each cost center.

14. Assume the welding department and four cost centers given in problem 13. The direct labor hours and costs for each cost center remain the same as problem 13, but new and additional data are given below (assume cost data are for the previous year).

Cost Center	Square Feet Occupied	Cost of Machinery	Number of Direct Labor Employees
A	900	$ 5,000	5
B	400	6,000	2
C	600	9,000	1
D	500	3,000	1
Total	2400	$23,000	9

Expenses other than for direct labor and materials chargeable to the welding department last year were

Maintenance	$ 4,000
Gas and electricity	10,000
Supervision and other indirect labor	24,000
Miscellaneous supplies	5,000
Equipment depreciation	2,300
Building depreciation	4,000

Determine a burden rate per direct labor hour for each cost center if the welding department expenses above are first allocated to each cost center as follows. (2.6)

(a) Maintenance expenses and equipment depreciation expenses are allocated according to the value of equipment (percent of total) in each cost center.

(b) Building depreciation expenses are allocated according to the floor space occupied by each cost center.

(c) Supervision and other indirect labor expenses are allocated according to the number of direct labor employees of each cost center.

(d) Supplies and gas and electricity expenses are allocated according to the number of direct labor hours for each cost center.

CHAPTER THREE
TIME VALUE OF MONEY OPERATIONS

3.1 INTRODUCTION

Design alternatives are normally compared by using a host of different criteria, including system performance and economic performance. Among the system performance characteristics that are of concern, quality, safety, and customer service considerations are of primary importance. Among the economic performance characteristics normally considered are initial investment requirements, return on investment, and the cash flow profile. Since the cash flow profiles are usually quite different among the several design alternatives, in order to compare the economic performances of the alternatives, one must compensate for the differences in the timing of cash flows.

As discussed in Chapters One and Two, the concept of the *time value of*

design alternatives. In this chapter we examine a number of mathematical operations that are based on the time value of money, with an emphasis on modeling cash flow profiles.

In order to provide motivation and to provide a familiar scenario, the treatment of time value of money operations will be presented primarily in the context of personal finance. However, each of the concepts examined in this chapter can and does occur in the business world.

3.2 INTEREST CALCULATIONS

In considering the time value of money, it is convenient to represent mathematically the relationship between the current or *present* value of a single sum of money and its *future* value. Letting time be measured in years, if a single sum of money has a current or *present* value of P, its value in n years would be equal to

$$F_n = P + I_n$$

where F_n is the accumulated value of P over n years, or the *future* value of P, and I_n is the increase in the value of P over n years. I_n is referred to as the accumulated *interest* in borrowing and lending transactions and is a function of P, n, and the *annual interest rate, i*. The annual interest rate is defined as the change in value for $1 over a one-year period.

Over the years, two approaches have emerged for computing the value of I_n. The first approach considers I_n to be a linear function of time. Since i is the rate of change over a one-year period, it is argued that P changes in value by an amount Pi each year. Hence, it is concluded that I_n is the product of P, i, and n, or,

$$I_n = Pin$$

and

$$F_n = P(1 + in)$$

This is called the *simple interest* approach.

The second approach used to compute the value of I_n is to interpret i as *the rate of change in the accumulated value of money*. Hence, it is argued that the following relation holds,

$$I_n = iF_{n-1}$$

and

$$F_n = F_{n-1}(1 + i)$$

This approach is referred to as the *compound interest* approach.

The approach to be used in any particular situation depends on how the interest rate is defined. Since practically all monetary transactions are based currently on compound interest rates instead of on simple interest rates, we will assume compounding occurs unless otherwise stated.

A convenient method of representing the time value of money is to visualize positive and negative cash flows as though they were generated by a borrower and a lender. In particular, suppose you loaned $1000 to an individual who agreed to pay you interest at a rate of 7%/year. The $1000 is referred to as the *principal* amount. At the end of one year, you would receive $1070 from the borrower. Thus, we might say that $1070 one year from now is worth $1000 today based on a 7% interest rate or, conversely, we might say that $1000 today has a value of $1070 one year from now based on a 7% interest rate.

If the individual borrowed the $1070 for an additional year, then you would be owed $1144.90, since the interest on $1070 for one year equals $(0.07) \times$ ($1070), or $74.90. Equivalently, borrowing $1000 for two years at 7% interest yields $1144.90 owed if interest is *compounded* annually. Compound interest involves the computation of interest charges during a time period based on the unpaid principal amount plus any accumulated interest charges up to the beginning of the time period. The interest amount due for a given interest period converts to principal for the purpose of calculating the interest amount due in the subsequent interest period. The relationship between the principal amount and the compounding of interest is given in Table 3.1.

Recall that whenever the interest charge for any time period is based only on the unpaid principal amount and not on any accumulated interest charges, simple interest calculations apply. The interest due for a given interest period does not convert to principal for the purpose of calculating the interest amount due in the subsequent interest period. In the preceding illustration, the amount owed at the end of two years would be the $1000 principal plus the interest charges of $(0.07)(\$1000) = \$70$ each year or $1140 total. Thus, in this example, the $1144.90 for the compounding case compares with the $1140 for the simple interest case. (In some cases, the difference is much more dramatic!)

TABLE 3.1 Illustrating the Effect of Compound Interest

End of Period	(A) Amount Owed	(B) Interest for Next Period	(C) = (A) + (B) Amount Owed for Next Period[a]	
0	P	Pi	$P + Pi$	$= P(1+i)$
1	$P(1+i)$	$P(1+i)i$	$P(1+i) + P(1+i)i$	$= P(1+i)^2$
2	$P(1+i)^2$	$P(1+i)^2 i$	$P(1+i)^2 + P(1+i)^2 i$	$= P(1+i)^3$
3	$P(1+i)^3$	$P(1+i)^3 i$	$P(1+i)^3 + P(1+i)^3 i$	$= P(1+i)^4$
$\vdots$	$\vdots$			
$n-1$	$P(1+i)^{n-1}$	$P(1+i)^{n-1} i$	$P(1+i)^{n-1} + P(1+i)^{n-1} i$	$= P(1+i)^n$
n	$P(1+i)^n$			

[a] Notice, the value in column (C) for the end of period $(n-1)$ provides the value in column (A) for the end of period n.

Example 3.1

Person A borrows $4000 from Person B and agrees to pay $1000 plus accrued interest at the end of the first year and $3000 plus the accrued interest at the end of the fourth year. What are the amounts for the two payments if 8% annual simple interest applies? For the first year, the payment is $1000 +

$(0.08)(\$4000) = \$1320$. For the fourth year, the payment is $\$3000 + (0.08)(\$4000 - \$1000)(3) = \$3720$.

3.3 SINGLE SUMS OF MONEY

To illustrate the mathematical operations involved in modeling cash flow profiles using compound interest, first consider the investment of a single sum of money, P, in a savings account for n interest periods. Let the interest rate per interest period be denoted by i and let the accumulated total in the fund n periods in the future be denoted by F. As shown in Table 3.1, assuming no monies are withdrawn during the interim, the amount in the fund after n periods equals $P(1+i)^n$. As a convenience in computing values of F (the future worth) when given values of P (the present worth), the quantity $(1+i)^n$ is tabulated in Appendix A for various values of i and n. The quantity $(1+i)^n$ is referred to as the *single sum, future worth factor* and is denoted $(F|P\ i,n)$. The expression $(F|P\ i,n)$ is read as the F, given P factor at $i\%$ for n periods. The above discussion is summarized as follows.

Let P = the equivalent value of an amount of money at time zero, or present worth

F = the equivalent value of an amount of money at time n, or future worth

i = the interest rate per interest period

n = the number of interest periods

Thus, the future worth is related to the present worth as follows:

$$F = P(1+i)^n \tag{3.1}$$

or equivalently,

$$F = P(F|P\ i,n) \tag{3.2}$$

A cash flow diagram depicting the relationship between F and P is given in Figure 3.1. Remember that F occurs n periods after P.

Example 3.2 ―――――――――――――――――――――――――

An individual borrows $1000 at 6% compounded annually. The loan is paid back after five years. How much should be repaid?
Using the compound interest tables in Appendix A for 6% and five periods,

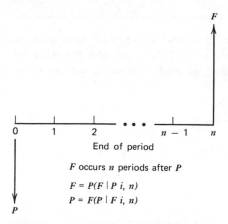

End of period

F occurs n periods after P

$F = P(F \mid P\ i,\ n)$

$P = F(P \mid F\ i,\ n)$

FIGURE 3.1. Cash flow diagram of the time relationship between P and F.

the value of the $(F|P\ 6,5)$ factor is found to be 1.3382. Thus,

$$F = P(F|P\ 6,5)$$
$$= \$1000(1 + 0.06)^5$$
$$= \$1000(1.3382)$$
$$= \$1338.20$$

The amount to be repaid equals $1338.20.

Since we are able to determine conveniently values of F when given values of P, i, and n, it is a simple matter to determine values of P when given values of F, i, and n. In particular, since

$$\boxed{F = P(1+i)^n} \tag{3.3}$$

on dividing both sides by $(1+i)^n$, we find that the present worth and future worth have the relation

$$\boxed{P = F(1+i)^{-n}} \tag{3.4}$$

or

$$P = F(P|Fi,n)$$

where $(1+i)^{-n}$ and $(P|Fi,n)$ are referred to as the *single sum, present worth factor.*

Example 3.3

To illustrate the computation of P given F, i, and n, suppose you wish to accumulate \$2000 in a savings account two years from now and the account pays interest at a rate of 6% compounded annually. How much must be deposited today?

$$P = F(P|F\ 6,2)$$
$$= \$2000(0.8900)$$
$$= \$1780.00$$

3.4 SERIES OF CASH FLOWS

Having considered the transformation of a single sum of money to a future worth equivalent when given a present worth amount and vice versa, we generalize that discussion to consider the conversion of a series of cash flows to present worth and future worth equivalents. In particular, let A_k denote the magnitude of a cash flow (receipt or disbursement) at the end of time period k. Using discrete compounding, the present worth equivalent for the cash flow series is equal to the sum of the present worth equivalents for the individual cash flows. Consequently,

$$P = A_1(1+i)^{-1} + A_2(1+i)^{-2} + \cdots + A_{n-1}(1+i)^{-(n-1)} + A_n(1+i)^{-n} \quad (3.5)$$

or, using the summation notation,

$$P = \sum_{k=1}^{n} A_k(1+i)^{-k} \quad (3.6)$$

or, equivalently,

$$P = \sum_{k=1}^{n} A_k(P|F\ i,k) \quad (3.7)$$

Example 3.4

Consider the series of cash flows depicted by the cash flow diagram given in Figure 3.2. Using an interest rate of 6%/interest period, the present worth equivalent is given by

$$P = \$300(P|F\ 6,1) - \$300(P|F\ 6,3) + \$200(P|F\ 6,4)$$
$$+ \$400(P|F\ 6,6) + \$200(P|F\ 6,8)$$
$$= \$300(0.9434) - \$300(0.8396) + \$200(0.7921)$$
$$+ \$400(0.7050) + \$200(0.6274)$$
$$= \$597.04$$

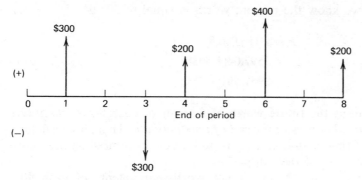

FIGURE 3.2. Series of cash flows.

The future worth equivalent is equal to the sum of the future worth equivalents for the individual cash flows. Thus,

$$F = A_1(1+i)^{n-1} + A_2(1+i)^{n-2} + \cdots + A_{n-1}(1+i) + A_n \qquad (3.8)$$

or, using the summation notation,

$$F = \sum_{k=1}^{n} A_k(1+i)^{n-k} \qquad (3.9)$$

$$= \sum_{k=1}^{n} A_k(F|P\ i,n-k) \qquad (3.10)$$

Alternately, since we know the value of future worth is given by

$$F = P(1+i)^n, \qquad (3.11)$$

substituting Equation 3.6 into Equation 3.11 yields

$$F = (1+i)^n \sum_{k=1}^{n} A_k(1+i)^{-k}$$

Hence,

$$F = \sum_{k=1}^{n} A_k(1+i)^{n-k} \qquad (3.12)$$

For the example under consideration, the worth of the series at time $t = 8$ is given by

$$
\begin{aligned}
F &= \$300(F|P\ 6,7) - \$300(F|P\ 6,5) + \$200(F|P\ 6,4) \\
&\quad + \$400(F|P\ 6,2) + \$200 \\
&= \$300(1.5036) - \$300(1.3382) + \$200(1.2625) \\
&\quad + \$400(1.1236) + \$200 \\
&= \$951.56
\end{aligned}
$$

Alternately, since we know the present worth is equal to 597.04

$$F = P(F|P\ 6,8)$$
$$= \$597.04(1.5938)$$
$$= \$951.56$$

Note that in computing the future worth of a series of cash flows, *the future worth amount obtained occurs at the end of time period n.* Thus, if a cash flow occurs at the end of time period n, it earns no interest in computing the future worth amount at the end of the nth period.

Obtaining the present worth and future worth equivalents of cash flow series by summing the individual present worths and future worths, respectively, can be quite time consuming if many cash flows are included in the series. However, with the development and widespread use of pocket calculators (some of which have the capability of performing compound interest calculations automatically), minicomputers, and time-sharing systems, it is not uncommon to treat all series in the manner described above.

When such hardware are not available, it is possible to use more efficient solution procedures if the cash flow series have one of the following forms.

Uniform Series of Cash Flows

$$A_k = A \qquad\qquad k = 1, \ldots, n$$

Gradient Series of Cash Flows

$$A_k = \begin{cases} 0 & k = 1 \\ A_{k-1} + G & k = 2, \ldots, n \end{cases}$$

Geometric Series of Cash Flows

$$A_k = \begin{cases} A & k = 1 \\ A_{k-1}(1+j) & k = 2, \ldots, n \end{cases}$$

3.4.1 Uniform Series of Cash Flows

A uniform series of cash flows exists when all of the cash flows in a series are equal. In the case of a uniform series the present worth equivalent is given by

$$P = \sum_{k=1}^{n} A(1+i)^{-k} \tag{3.13}$$

where A is the magnitude of an individual cash flow in the series.

Letting $X = (1 + i)^{-1}$ and bringing A outside the summation yields

$$P = A \sum_{k=1}^{n} X^k$$

$$= AX \sum_{k=1}^{n} X^{k-1}$$

Letting $h = k - 1$ gives the geometric series

$$P = AX \sum_{h=0}^{n-1} X^h \tag{3.14}$$

Since the summation in Equation 3.14 represents the first n terms of a geometric series, the closed form value for the summation is given by

$$\sum_{h=0}^{n-1} X^h = \frac{1 - X^n}{1 - X} \tag{3.15}$$

Hence, on substituting Equation 3.15 into Equation 3.14, we obtain

$$P = AX \left(\frac{1 - X^n}{1 - X} \right)$$

Replacing X with $(1 + i)^{-1}$ yields the following relationship between P and A.

$$P = A \left[\frac{(1 + i)^n - 1}{i(1 + i)^n} \right] \tag{3.16}$$

more commonly expressed as

$$P = A(P|A\ i,n) \tag{3.17}$$

where $(P|A\ i,n)$ is referred to as the *uniform series, present worth factor* and is tabulated in Appendix A for various values of i and n.

Example 3.5 ────────────────────────────────

An individual wishes to deposit a single sum of money in a savings account so that five equal annual withdrawals of $2000 can be made before depleting the fund. If the first withdrawal is to occur one year after the deposit and the fund pays interest at a rate of 7% compounded annually, how much should be deposited?

Because of the relationship of P and A, as depicted in Figure 3.3, in which

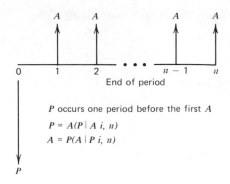

P occurs one period before the first A

$P = A(P|A\ i,\ n)$

$A = P(A|P\ i,\ n)$

FIGURE 3.3. Cash flow diagram of the relationship between P and A.

P occurs one period before the first A, we see that

$$P = A(P|A\ 7,5)$$
$$= \$2000(4.1002)$$
$$= \$8200.40$$

Thus, if \$8200.40 is deposited in a fund paying 7% compounded annually, then five equal annual withdrawals of \$2000 can be made.

Example 3.6 ─────────────────────────

In Example 3.5, suppose that the first withdrawal will not occur until three years after the deposit.

As depicted in Figure 3.4, the value of P to be determined occurs at $t = 0$, whereas a straightforward application of the $(P|A\ 7,5)$ factor will yield a single sum equivalent at $t = 2$. Consequently, the value obtained at $t = 2$ must be moved backward in time to $t = 0$. The latter operation is easily performed using the $(P|F\ 7,2)$ factor. Therefore,

$$P = A(P|A\ 7,5)(P|F\ 7,2)$$
$$= \$2000(4.1002)(0.8734)$$
$$= \$7162.23$$

Deferring the first withdrawal for two years reduces the amount of the deposit by $\$8200.40 - \$7162.23 = \$1038.17$.

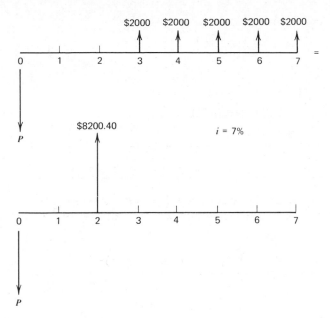

FIGURE 3.4. Equivalent cash flow diagrams.

The reciprocal relationship between P and A can be expressed as

$$A = P\left[\frac{i(1+i)^n}{(1+i)^{n-1}}\right] \qquad (3.18)$$

or as

$$A = P(A|P\ i,n) \qquad (3.19)$$

The expression $(A|P\ i,n)$ is called the *capital recovery factor* for reasons that will become clear in Chapter Four. The $(A|P\ i,n)$ factor is used frequently in both personal financing and in comparing economic investment alternatives.

Example 3.7

Suppose $10,000 is deposited into an account that pays interest at a rate of 7% compounded annually. If 10 equal, annual withdrawals are made from the account, with the first withdrawal occurring one year after the deposit, how much can be withdrawn each year in order to deplete the fund with the last withdrawal?

Since we know that A and P are related by

$$A = P(A|P\ i,n)$$

then

$$A = P(A|P\ 7,10)$$
$$= \$10,000(0.1424)$$
$$= \$1424$$

Example 3.8

Suppose that in Example 3.7 the first withdrawal is delayed for two years, as depicted in Figure 3.5. How much can be withdrawn each of the 10 years?

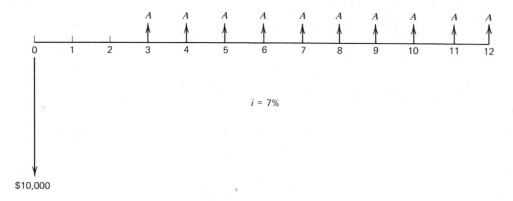

FIGURE 3.5. Cash flow diagram of deferred payment example.

The amount in the fund at $t = 2$ equals

$$V_2 = P(F|P\ 7,2)$$
$$= \$10,000(1.1449)$$
$$= \$11,449$$

Therefore, the size of the equal annual withdrawals will be

$$A = V_2(A|P\ 7,10)$$
$$= \$11,449(0.1424)$$
$$= \$1630.34$$

Thus, delaying the first withdrawal for two years increases the size of each withdrawal by $206.34.

The future worth of a uniform series is obtained by recalling that

$$F = P(1 + i)^n \qquad (3.20)$$

Substituting Equation 3.16 into Equation 3.20 for P and reducing yields

$$F = A\left[\frac{(1 + i)^n - 1}{i}\right] \qquad (3.21)$$

or, equivalently,

$$F = A(F|A\ i,n) \qquad (3.22)$$

where $(F|A\ i,n)$ is referred to as the *uniform series, future worth factor.*

Example 3.9

If annual deposits of $1000 are made into a savings account for 30 years, how much will be in the fund immediately after the last deposit if the fund pays interest at a rate of 8% compounded annually?

$$F = A(F|A\ 8,30)$$
$$= \$1000(113.283)$$
$$= \$113,283$$

The reciprocal relationship between A and F is easily obtained from Equation 3.21. Specifically, we find that

$$A = F\left[\frac{i}{(1 + i)^n - 1}\right] \qquad (3.23)$$

or, equivalently,

$$A = F(A|F\ i,n) \qquad (3.24)$$

The expression $(A|F\ i,n)$ is referred to as the *sinking fund factor,* since the factor is used to determine the size of a deposit one should place (sink) in a fund in order to accumulate a desired future amount. As depicted in Figure 3.6, F occurs at the same time as the last A. Thus, the last A or deposit earns no interest.

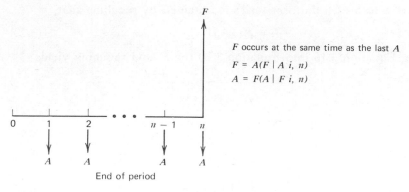

F occurs at the same time as the last A

$F = A(F \mid A\ i,\ n)$

$A = F(A \mid F\ i,\ n)$

End of period

FIGURE 3.6. Cash flow diagram of the relationship between A and F.

Example 3.10

If \$150,000 is to be accumulated in 35 years, how much must be deposited annually in a fund paying 8% compounded annually in order to accumulate the desired amount immediately after the last deposit?

$$A = F(A|F\ 8{,}35)$$
$$= \$150{,}000(0.0058)$$
$$= \$870$$

3.4.2 Gradient Series of Cash Flows

The gradient series of cash flows is depicted in Figure 3.7. The first positive cash flow occurs at the end of the second time period; each successive cash flow increases in magnitude by an amount G. An alternate and more convenient representation of the size of the cash flow at the end of period k is given by

$$A_k = (k-1)G \qquad k = 1, \ldots, n \tag{3.25}$$

The gradient series arises when the value of an individual cash flow differs by a constant, G, from the preceding cash flow. As an illustration, if an individual receives an annual bonus and the size of the bonus increases by \$100 each year, then the series is a gradient series. Also, operating and maintenance costs tend to increase over time because of both inflation effects and a gradual deterioration of equipment; such costs are often approximated by a gradient series.

The present worth equivalent of a gradient series is obtained by recalling

$$P = \sum_{k=1}^{n} A_k (1+i)^{-k} \tag{3.26}$$

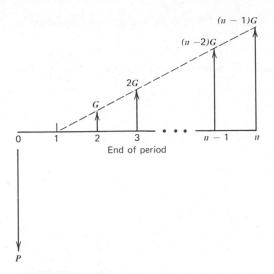

FIGURE 3.7. Cash flow diagram of the gradient series.

Substituting Equation 3.25 into Equation 3.26 gives

$$P = \sum_{k=1}^{n} (k-1)G(1+i)^{-k} \qquad (3.27)$$

or, equivalently,

$$P = G \sum_{k=1}^{n} (k-1)(1+i)^{-k} \qquad (3.28)$$

As an exercise you may wish to show that the summation reduces to

$$P = G\left[\frac{1-(1+ni)(1+i)^{-n}}{i^2}\right] \qquad (3.29)$$

Expressing the term in brackets in terms of interest factors already treated yields

$$P = G\left[\frac{(P|A\,i,n) - n(P|F\,i,n)}{i}\right] \qquad (3.30)$$

or, equivalently,

$$P = G(P|G\,i,n) \qquad (3.31)$$

where $(P|G\,i,n)$ is the *gradient series, present worth factor* and is tabulated in Appendix A.

A uniform series equivalent to the gradient series is obtained by multiplying the value of the gradient series present worth factor by the value of the $(A|P\,i,n)$ factor to obtain

$$A = G\left[\frac{1}{i} - \frac{n}{i}(A|F\,i,n)\right]$$

or, equivalently,

$$A = G(A|G\,i,n) \qquad\qquad (3.32)$$

where the factor $(A|G\,i,n)$ is referred to as the *gradient-to-uniform series conversion factor* and is tabulated in Appendix A. To obtain the future worth equivalent of a gradient series at time n, multiply the value of the $(A|G\,i,n)$ factor by the value of the $(F|A\,i,n)$ factor.

It is not uncommon to encounter a cash flow series that is the sum or difference of a uniform series and a gradient series. To determine present worth and future worth equivalents of such a composite, one can deal with each special type of series separately.

Example 3.11

An individual deposits an annual bonus into a savings account that pays 6% compounded annually. The size of the bonus increases by $100/year; the initial bonus was $300. Determine how much will be in the fund immediately after the fifth deposit.

A cash flow diagram for this example is given in Figure 3.8. Note that the cash flow series consists of the sum of a uniform series of $300 and a gradient series with G equal to $100. Converting the gradient series to a uniform series gives

$$A = G(A|G\,6,5)$$
$$= \$100(1.8836)$$
$$= \$188.36$$

(Notice that n equals 5 even though only four positive cash flows are present in the gradient series.)

The cash flow series given in Figure 3.8 is equivalent to a uniform series having cash flows equal to $300 + $188.36 or $488.36. Converting the uniform

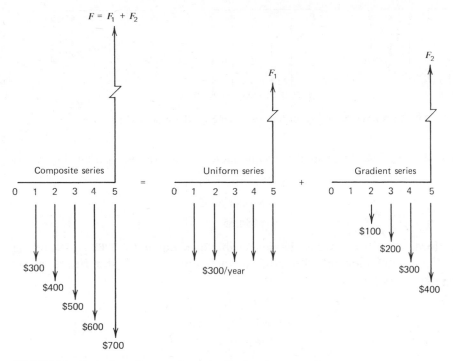

$$F = F_1 + F_2$$

Composite series = Uniform series + Gradient series

$$F_1$$

$$F_2$$

Composite series

0 1 2 3 4 5

$300
$400
$500
$600
$700

Uniform series

0 1 2 3 4 5

$300/year

Gradient series

0 1 2 3 4 5

$100
$200
$300
$400

FIGURE 3.8. Cash flow diagram for a gradient series example.

series to a future worth equivalent:

$$F = A(F|A\ 6,5)$$
$$= \$488.36(5.6371)$$
$$= \$2752.93$$

Thus, $2752.93 will be in the fund immediately after the fifth deposit.

Example 3.12 ————————————————————————

Five annual deposits are made into a fund that pays interest at a rate of 8% compounded annually. The first deposit equals $800; the second deposit equals $700; the third deposit equals $600; the fourth deposit equals $500; and the fifth deposit equals $400. Determine the amount in the fund immediately after the fifth deposit.

As depicted by the cash flow diagrams in Figure 3.9, the cash flow series can be represented by the difference in a uniform series of $800 and a gradient

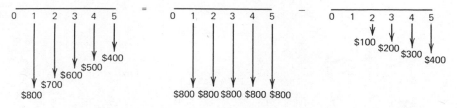

FIGURE 3.9. Cash flow diagrams for the decreasing gradient series example.

series of $100. The uniform series equivalent of the gradient series is given by

$$A = G(A|G\ 8,5)$$
$$= \$100(1.8465)$$
$$= \$184.65$$

Therefore, a uniform series having cash flows equal to $800 − $184.65, or $615.35, is equivalent to the original cash flow series. The future worth equivalent is found to be

$$F = A(F|A\ 8,5)$$
$$= \$615.35(5.8666)$$
$$= \$3610.01$$

3.4.3 Geometric Series of Cash Flows

The geometric cash flow series, as depicted in Figure 3.10, occurs when the size of a cash flow increases (decreases) by a fixed percent from one time period to the next. If j denotes the percent change in the size of a cash flow from one period to the next, the size of the kth cash flow can be given by

$$A_k = A_{k-1}(1+j) \qquad k = 2, \ldots, n$$

or, more conveniently,

$$A_k = A_1(1+j)^{k-1} \qquad k = 1, \ldots, n \tag{3.33}$$

The geometric series is used most often to represent the growth (positive j) or decay (negative j) of costs and revenues due to inflation or recession. As an illustration, if labor costs increase by 10% a year, then the resulting series representation of labor costs will be a geometric series.

The present worth equivalent of the cash flow series is obtained by substituting Equation 3.33 into Equation 3.6 to obtain

$$P = \sum_{k=1}^{n} A_1(1+j)^{k-1}(1+i)^{-k} \tag{3.34}$$

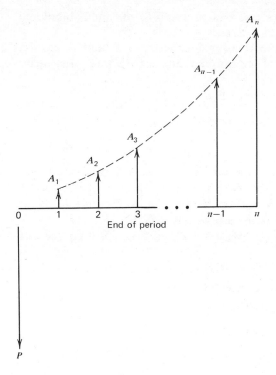

FIGURE 3.10. Cash flow diagram of the geometric series.

or

$$P = A_1(1+j)^{-1} \sum_{k=1}^{n} \left(\frac{1+j}{1+i}\right)^k \qquad (3.35)$$

As an exercise, you may wish to show that the following relationship results:

$$P = \begin{cases} A_1 \left[\dfrac{1-(1+j)^n(1+i)^{-n}}{i-j} \right] & i \neq j \\[4mm] \dfrac{nA_1}{1+i} & i = j \end{cases} \qquad (3.36)$$

or

$$P = A_1(P|A_1\ i,j,n) \qquad (3.37)$$

where $(P|A_1\ i,j,n)$ is the *geometric series, present worth factor* and is tabulated in Appendix A for various values of i, j, and n.

For the case of $j \geq 0$ and $i \neq j$, the relationship between P and A can be conveniently expressed in terms of compound interest factors previously considered.

$$P = A_1 \left[\frac{1 - (F|P\ j,n)(P|F\ i,n)}{i - j} \right] \qquad i \neq j, \qquad j \geq 0 \qquad (3.38)$$

Example 3.13

Labor costs have been increasing at an annual rate of 8%. A firm wishes to set aside funds to cover labor costs for the next five years. Determine how much must be set aside today if the money will be invested and will earn interest at a rate of 10%. The labor cost next year will be $50,000.

For this example, $A_1 = \$50,000$, $j = 8\%$, $i = 10\%$, and $n = 5$. The present worth equivalent will be

$$P = A_1(P|A_1\ 10,8,5)$$
$$= \$50,000(4.3856)$$
$$= \$219,280$$

If labor costs increase at an annual rate of 10%,

$$P = \frac{nA_1}{1 + i}$$
$$= \frac{5(\$50,000)}{1.1}$$
$$= \$227,272.73$$

The future worth equivalent of the geometric series is obtained by multiplying the value of the geometric series present worth factor and the $(F|P\ i,n)$ factor to obtain

$$F = \begin{cases} A_1 \left[\dfrac{(1 + i)^n - (1 + j)^n}{i - j} \right] & i \neq j \\ nA_1(1 + i)^{n-1} & i = j \end{cases} \qquad (3.39)$$

or

$$F = A_1(F|A_1\ i,j,n) \qquad (3.40)$$

where $(F|A_1\ i,n)$ is the *geometric series, future worth factor* and is tabulated in Appendix A. From Equation 3.3.9, notice that $(F|A_1\ i,j,n) = (F|A_1\ j,i,n)$.

Example 3.14 ───

An individual receives an annual bonus and deposits it in a savings account that pays 6% compounded annually. The size of the bonus increases by 5% each year; the initial deposit was $500. Determine how much will be in the fund immediately after the tenth deposit.

In this case, $A_1 = \$500$, $j = 5\%$, $i = 6\%$, and $n = 10$. Thus, the value of F is given by

$$F = A_1(F|A_1\,6,5,10) = A_1(F|A_1\,5,6,10)$$
$$= \$500(16.190)$$
$$= \$8095.00$$

The interest factors developed to this point are summarized in Table 3.2. Values of the factors given in Table 3.2 are provided in Appendix A.

TABLE 3.2 Summary of Discrete Compounding Interest Factors

To Find	Given	Factor	Symbol
P	F	$(1+i)^{-n}$	$(P\|F\,i,n)$
F	P	$(1+i)^n$	$(F\|P\,i,n)$
P	A	$\dfrac{(1+i)^n-1}{i(1+i)^n}$	$(P\|A\,i,n)$
A	P	$\dfrac{i(1+i)^n}{(1+i)^n-1}$	$(A\|P\,i,n)$
F	A	$\dfrac{(1+i)^n-1}{i}$	$(F\|A\,i,n)$
A	F	$\dfrac{i}{(1+i)^n-1}$	$(A\|F\,i,n)$
P	G	$\dfrac{1-(1+ni)(1+i)^{-n}}{i^2}$	$(P\|G\,i,n)$
A	G	$\dfrac{(1+i)^n-(1+ni)}{i[(1+i)^n-1]}$	$(A\|G\,i,n)$
P	$A_{1,j}$	$\dfrac{1-(1+j)^n(1+i)^{-n}}{i-j}$	$(P\|A_1\,i,j,n)^a$
F	$A_{1,j}$	$\dfrac{(1+i)^n-(1+j)^n}{i-j}$	$(F\|A_1\,i,j,n)^a$

$^a\, i \neq j$.

3.5 MULTIPLE COMPOUNDING PERIODS IN A YEAR

Not all interest rates are stipulated as annual compounding rates. For example, if the interest rate is stipulated as 6% compounded quarterly, then the interest period is a three-month period with 1.5% interest rate/interest period. Thus, if $1000 is borrowed at an interest rate of 6% compounded quarterly, then the amount owed at the end of five years or 20 interest periods is obtained as follows.

$$F = P(F|P\ 1.5,20)$$
$$= \$1000(1 + 0.015)^{20}$$
$$= \$1000(1.3469)$$
$$= \$1346.90$$

With quarterly compounding, $1346.90 is to be repaid.

Example 3.15 ────────────────────────────────

As another illustration of multiple compounding periods within a year, suppose interest is stated as 8% compounded quarterly. If you borrow $1000 for one year, how much must be repaid?

Using an interest rate of 2%/three-month period, after four interest periods the amount owed is given by:

$$F = P(F|P\ 2,4)$$
$$= \$1000(1.0824)$$
$$= \$1082.40$$

Thus, $1082.40 would be owed. Note that if interest were stated as 8.24% compounded annually,

$$F = P(F|P\ 8.24,1)$$
$$= \$1000(1.0824)$$
$$= \$1082.40$$

and the same amount would be owed.

────────────────────────────────

It can be concluded that 8% compounded quarterly is equivalent to 8.24% compounded annually. The rate of 8% is referred to as the *nominal interest rate*; the rate of 8.24% is referred to as the *effective annual interest rate*. The effective interest rate is defined as the annual compounding rate that is equivalent to the stated interest rate.

Letting r denote the nominal annual interest rate, m denote the number of compounding periods per year, i denote the interest rate per interest period, and i_{eff} denote the effective interest rate per year,

$$i_{\text{eff}} = \left(1 + \frac{r}{m}\right)^m - 1$$
$$= (1 + i)^m - 1$$

$$\boxed{i_{\text{eff}} = \left(F|P\frac{r}{m}, m\right) - 1} \qquad (3.41)$$

To illustrate with 8% compounded semiannually, r equals 0.08, m equals 2, and

$$i_{\text{eff}} = (1 + 0.04)^2 - 1$$
$$= (F|P\ 4,2) - 1$$
$$= 0.0816$$

Thus, 8% compounded semiannually is equivalent to 8.16% compounded annually (i.e., 8.16% is the effective annual interest rate when interest is 8% compounded semiannually).

Suppose interest is stated as 18% compounded monthly. The effective annual interest rate is given by

$$i_{\text{eff}} = (1 + 0.015)^{12} - 1$$
$$= (F|P\ 1.5,12) - 1$$
$$= 0.1956, \qquad \text{or} \qquad 19.56\%$$

Example 3.16

An individual borrowed $1000 and paid off the loan with interest after 4.5 years. The amount paid was $1250. What was the effective annual interest rate for this transaction? Letting the interest period be a six-month period, it is seen that the payment of $1250 and the debt of $1000 are related by the expression

$$F = P(F|P\ i,n)$$

Thus,

$$\$1250 = \$1000(F|P\ i,9)$$

or

$$\$1250 = \$1000(1 + i)^9$$

Dividing both sides by $1000 gives

$$1.25 = (1 + i)^9$$

Taking the logarithm of both sides yields

$$\log 1.25 = 9 \log (1 + i)$$

Dividing both sides by nine and taking the antilog of the result provides the relation

$$(1 + i) = 1.0251$$
$$i = 0.0251$$

Thus, the six-month interest rate is 2.51%, and computing the effective annual interest rate yields

$$i_{eff} = (1 + i)^2 - 1$$
$$= (1 + 0.0251)^2 - 1$$
$$= 0.0508$$

The effective annual interest rate for the loan transaction was approximately 5.08%.

Instead of using logarithms, we could have searched the interest tables for a value of i that yielded a value of 1.25 for the $(F|P\ i,9)$ factor. Since i would be found to equal approximately 2.5%, the computation of the effective annual interest rate would have followed the same procedure used to determine the effective annual interest rate for 5% compounded semiannually.

With the passage of the Truth-in-Lending Bill, anyone borrowing money must be informed of the total amount of interest paid and the *annual percentage rate* associated with the transaction. Depending on how the lender computes the annual percentage rate, the percentage quoted can be either the nominal annual interest rate, the effective annual interest rate, or something else entirely. To illustrate, many revolving charge accounts charge an interest rate of 1.5%/month, which is equivalent to 18% compounded monthly or 19.56% compounded annually. In many cases, the annual percentage rate is stated to be 18%.

Since, in personal financing, the annual percentage rate might be computed differently among the alternative sources of funds, the effective annual interest rate is a very good basis for comparing alternative financing plans. Subsequently, we will find that the effective annual interest rate is also useful when the timing of cash flows does not coincide with the end of interest periods.

3.6 CONTINUOUS COMPOUNDING

In the discussion of the effective interest rate in the previous section it was noted that as the frequency of compounding in a year increases the effective interest rate increases. Since monetary transactions occur daily or hourly in most businesses and money is normally "put to work" for the business as soon as it is received, compounding is occurring quite frequently. If one wishes to account explicitly for such rapid compounding, then *continuous compounding relations should be used.* Continuous compounding means that each year is divided into an infinite number of interest periods. Mathematically, the single payment compound amount factor under continuous compounding is given by

$$\lim_{m \to \infty} \left(1 + \frac{r}{m}\right)^{mn} = e^{rn}$$

where n is the number of years, m is the number of interest periods per year, and r is the nominal annual interest rate. Given P, r, and n, the value of F can be computed using continuous compounding as follows.

$$F = P\,e^{rn} \tag{3.42}$$

or

$$F = P(F|P\ r,n)_\infty \tag{3.43}$$

where $(F|P\ r,n)_\infty$ denotes the *continuous compounding, single sum, future worth factor.* The subscript ∞ is provided to denote that continuous compounding is being used. The interest tables for continuous compounding are given in Appendix B.

Example 3.17 ————————————————————————————

If \$2000 is invested in a fund that pays interest at a rate of 6% compounded continuously, after five years the cumulative amount in the fund will total

$$F = P(F|P\ 6,5)_\infty$$
$$= \$2000(1.3499)$$
$$= \$2699.80$$

Thus, a withdrawal of \$2699.80 will deplete the fund after five years.

The effective interest rate under continuous compounding is easily obtained using the relation

$$i_{\text{eff}} = e^r - 1 \tag{3.44}$$

or

$$i_{\text{eff}} = (F|P\,r,1)_\infty - 1 \qquad (3.45)$$

To illustrate, if interest is 8% compounded continuously, then the effective interest rate is given by

$$i_{\text{eff}} = (F|P\ 8,1)_\infty - 1$$
$$= 0.0833$$

Thus, 8.33% compounded annually is equivalent to 8% compounded continuously.

The inverse relationship between F and P indicates that

$$P = Fe^{-m} \qquad (3.46)$$

or

$$P = F(P|F\,r,n)_\infty \qquad (3.47)$$

where $(P|F\,r,n)_\infty$ is called the *continuous compounding, single sum, present worth factor.*

3.6.1 Discrete Flows

If it is assumed that cash flows are discretely spaced over time, then the continuous compound relations for the uniform, gradient, and geometric series can be obtained. Substituting e^{-m} for $(1+i)^{-n}$, $e^r - 1$ for i, and e^m for $(1+i)^n$ in the remaining discrete compounding formulas yields the continuous compound interest factors summarized in Table 3.3. Values for these factors are provided in Appendix B.

Example 3.18 ————————————————————————

To illustrate the use of the continuous compound interest factors, suppose $1000 is deposited each year into an account that pays interest at a rate of 6% compounded continuously. Determine both the amount in the account immediately after the tenth deposit and the present worth equivalent for ten deposits.

The amount in the fund immediately after the tenth deposit is given by the relation

$$F = \$1000(F|A\ 6,10)$$
$$= \$1000(13.2951)$$
$$= \$13,295.10$$

The present worth equivalent for ten deposits is obtained using the relation

$$P = \$1000(P|A\ 6,10)_\infty$$
$$= \$1000(7.2965)$$
$$= \$7296.50$$

TABLE 3.3 Summary of Continuous Compounding Interest Factors

To Find	Given	Factor	Symbol
P	F	e^{-rn}	$(P|F\,r,n)_\infty$
F	P	e^{rn}	$(F|P\,r,n)_\infty$
F	A	$\dfrac{e^{rn}-1}{e^{r}-1}$	$(F|A\,r,n)_\infty$
A	F	$\dfrac{e^{r}-1}{e^{rn}-1}$	$(A|F\,r,n)_\infty$
P	A	$\dfrac{e^{rn}-1}{e^{rn}(e^{r}-1)}$	$(P|A\,r,n)_\infty$
A	P	$\dfrac{e^{rn}(e^{r}-1)}{e^{rn}-1}$	$(A|P\,r,n)_\infty$
P	G	$\dfrac{e^{rn}-1-n(e^{r}-1)}{e^{rn}(e^{r}-1)^2}$	$(P|G\,r,n)_\infty$
A	G	$\dfrac{1}{e^{r}-1}-\dfrac{n}{e^{rn}-1}$	$(A|G\,r,n)_\infty$
P	A_1,c	$\dfrac{1-e^{(c-r)n}}{e^{r}-e^{c}}$	$(P|A_1\,r,c,n)_\infty^{a}$
F	A_1,c	$\dfrac{e^{rn}-e^{cn}}{e^{r}-e^{c}}$	$(F|A_1\,r,c,n)_\infty^{a}$
P	$\bar{A}$	$\dfrac{e^{rn}-1}{re^{rn}}$	$(P|\bar{A}\,r,n)$
$\bar{A}$	P	$\dfrac{re^{rn}}{e^{rn}-1}$	$(\bar{A}|P\,r,n)$
F	$\bar{A}$	$\dfrac{e^{rn}-1}{r}$	$(F|\bar{A}\,r,n)$
$\bar{A}$	F	$\dfrac{r}{e^{rn}-1}$	$(\bar{A}|F\,r,n)$

[a]$r \neq c.$

In the case of the geometric series the size of the kth cash flow will be assumed to be given by

$$A_k = A_{k-1}e^c \qquad k = 2, \ldots, n \qquad (3.48)$$

or, equivalently,

$$A_k = A_1 e^{(k-1)c} \qquad k = 1, \ldots, n \qquad (3.49)$$

where c is the nominal compound rate of increase in the size of the cash flow. The resulting expressions for the *continuous compounding, geometric series present worth factor* and the *continuous compounding, geometric series future worth factor* are given, respectively, by $(P|A_1\ r,c,n)_\infty$ and $(F|A_1\ r,c,n)_\infty$, as given in Table 3.3 for the case of $r \neq c$. As an exercise you may wish to derive the appropriate expressions when $r = c$.

Example 3.19

An individual receives an annual bonus and deposits it in a savings account that pays 5% compounded continuously. The size of the bonus increases each year at a rate of 6% compounded continuously; the initial deposit was $500. Determine how much will be in the fund immediately after the tenth deposit.

In this case, $A_1 = \$500$, $r = 5\%$, $c = 6\%$, and $n = 10$. Thus, the value of F is given by

$$F = A_1(F|A_1\ 5,6,10)_\infty$$
$$= \$500(16.4118)$$
$$= \$8205.90$$

The effect of continuous compounding increases the amount in the fund by $110.90.

If discrete flows occur during a year, it is necessary to define r consistent with the spacing of cash flows. To illustrate, suppose semiannual deposits are made into an account paying 6% compounded continuously. In this case the nominal semiannual rate would be $6\% \div 2$ time periods/year, or $r = 3\%$, and n would equal the number of semiannual periods involved.

Since the differences in discrete and continuous compounding are not great in most cases, it is not uncommon to see discrete compounding used when continuous compounding is more appropriate. The arguments given for this are that errors in estimating the cash flows will probably offset any attempts to be very precise by using continuous compounding and that the interest rate used in discrete compounding is actually the effective interest rate resulting from continuous compounding.

3.6.2 Continuous Flow

Thus far only discrete cash flows have been assumed. It was assumed that cash flows occurred at, say, the end of the year. In some cases money is expended throughout the year on a somewhat uniform basis. (Costs for labor, carrying inventory, and operating and maintaining equipment are typical examples.) Consequently, as a mathematical convenience, instead of assuming that money flows in discrete increments at the end of monthly, weekly, daily, or hourly time periods, it is assumed that money flows continuously during the time period at a uniform rate. Instead of having a uniform series of discrete cash flows of magnitude A, it assumed that a *total* of $\bar{A}$ dollars flows uniformly and continuously throughout a given time period. Such an approach to modeling cash flows is referred to as the *continuous flow* approach.

To illustrate the continuous flow concept, suppose you are to divide $1000 into k equal amounts to be deposited at equally spaced points in time during a year. The interest rate per period is defined to be r/k, where r is the nominal rate. Thus, the present worth of the series of k equal amounts is

$$P = \frac{\$1000}{k} \quad \left(P|A\frac{r}{k}, k \right)$$

or

$$P = \frac{\$1000}{k} \left[\frac{(1+(r/k))-1}{r/k(1+(r/k))^k} \right]$$

which reduces to

$$P = \$1000 \left[\frac{1}{r} - \frac{1}{r(1+(r/k))^k} \right]$$

Taking the limit of P as k approaches infinity gives

$$\lim_{k \to \infty} P = \$1000 \left(\frac{1}{r} - \frac{1}{re^r} \right)$$

or

$$P = \$1000 \left(\frac{e^r - 1}{re^r} \right)$$

In general, for n years,

$$P = \bar{A} \left(\frac{e^{rn} - 1}{re^{rn}} \right) \tag{3.50}$$

or

$$P = \bar{A}(P|\bar{A}\ r,n) \tag{3.51}$$

where $(P|\bar{A}\,r,n)$ is referred to as the *continuous flow, continuous compounding uniform series present worth factor* and is tabulated in Appendix B.

The remaining continuous flow, continuous compound interest factors are summarized in Table 3.3. Values of the factors are given in Appendix B for various values of r and n. With continuous flow and continuous compounding the continuous annual cash flow, $\bar{A}$, is equivalent to $Ar/(e^r - 1)$, when discrete flow and continuous compounding is used. Hence, the discrete flow equivalent of $\bar{A}$ is given by

$$A = \frac{\bar{A}(e^r - 1)}{r}$$

or

$$A = \bar{A}(F|\bar{A}\,r,1)$$

Example 3.20 ───

What are the present worth and future worth equivalents of a uniform series of continuous cash flows totalling \$1000/year for 10 years when interest is compounded continuously at a rate of 10%/year?
The present worth equivalent is given by

$$P = \$1000(P|\bar{A}\,10,10)$$
$$= \$1000(6.3212)$$
$$= \$6321.20$$

The future worth equivalent is given by

$$F = \$1000(F|\bar{A}\,10,10)$$
$$= \$1000(17.1828)$$
$$= \$17,182.80$$

Continuous flow and continuous compounding are concepts that are used to represent more closely the realities of business transactions. In fact, several of the computer programs developed by industries and governmental agencies and used to perform economic analyses incorporate both concepts. Despite such arguments for their use, the concepts have not been widely accepted among economic analysts. We have presented both concepts in anticipation that both will become more popular in the future through evolution.

3.7 EQUIVALENCE

Throughout the preceding discussion we have used the term *equivalence* without defining what was meant by the term. This was done intentionally in order to introduce subtly the notion that two cash flow series or profiles are *equivalent* at some specified interest rate, *k%*, if their present worths are *equal* using an interest rate of *k%*.

Example 3.21

The two cash flow profiles shown in Figure 3.11 are equivalent at 6%, since each has a present worth of $565.66. You may wish to verify this claim.

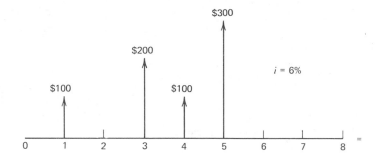

FIGURE 3.11. Cash flow diagrams for the equivalence example.

Of course, if the present worths of two cash flow profiles are equal at *k%*, their "worths" at any given point in time will be equal when using a discount (interest) rate of *k%*. Likewise, if two cash flow profiles are equivalent at *k%*, their respective uniform series equivalents will be equal when expressed over the same time period. Hence we could state that the two cash flow profiles shown in Figure 3.11 are equivalent at 6%, since each has a worth of $901.56 at the end of period eight; alternately, we could assert their equivalence at 6% on the basis that each can be represented by a uniform series of $91.09/period over the interval [1, 8].

Example 3.22

What single sum of money at $t = 6$ is equivalent to the cash flow profile shown in Figure 3.12 if $i = 5\%$?

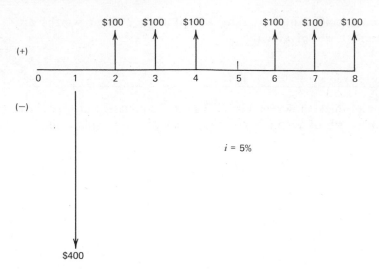

FIGURE 3.12. Cash flow diagram for the equivalence example.

The present worth of the cash flow profile is given by

$$P = -\$400(P|F\ 5,1) + \$100(P|A\ 5,3)(P|F\ 5,1)$$
$$+ \$100(P|A\ 5,3)(P|F\ 5,5)$$
$$= -\$400(0.9524) + \$100(2.7232)(0.9524) + \$100(2.7232)(0.7835)$$
$$= \$91.76$$

Moving \$91.76 forward in time to $t = 6$ gives

$$F = \$91.76(F|P\ 5,6)$$
$$= 91.76(1.3401)$$
$$= \$122.97$$

Thus, at 5% a positive cash flow of \$122.97 at $t = 6$ is equivalent to the cash flow profile shown in Figure 3.12.

Example 3.23 ————————————————————————

Using an 8% discount rate, what uniform series over five periods, [1, 5], is equivalent to the cash flow profile given in Figure 3.13?

The cash flow profile in Figure 3.13a consists of the difference in a uniform series of \$500 and a gradient series, with $G = \$100$. A uniform series

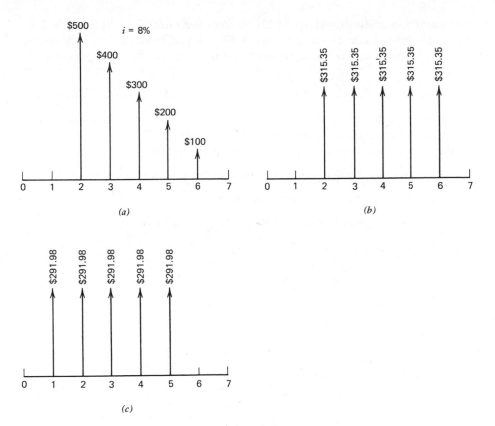

FIGURE 3.13. Cash flow diagrams for the equivalence example.

equivalent of the cash flow profile can be obtained for the interval [2, 6] as follows:

$$A = \$500 - \$100(A|G\ 8,5)$$
$$= \$500 - \$100(1.8465)$$
$$= \$315.35$$

The uniform series of $315.35 over the interval [2, 6] must be converted to a uniform series over the interval [1, 5]. Thus, *each* of the five cash flows must be moved back in time one period. The discounted value of $315.35 over one time period using an 8% discount rate is

$$P = \$315.35(P|F\ 8,1)$$
$$= \$315.35(0.9259)$$
$$= \$291.98$$

Consequently, a uniform series of $291.98 over the interval [1, 5] is equivalent to the cash flow profile given in Figure 3.13a. If you have doubts concerning the equivalence, compare their present worths using an 8% interest rate.

Example 3.24

Determine the value of X that makes the two cash flows, as given in Figure 3.14, equivalent when a discount rate of 6% is used.

Equating the future worths of the two cash flow profiles at $t = 4$ gives

$$\$200(F|A\ 6,4) + \$100(F|A\ 6,3) + \$100 = [\$200 + X(A|G\ 6,4)](F|A\ 6,4)$$

Cancelling $200(F|A 6,4) on both sides yields

$$\$100(3.1836) + \$100 = X(1.4272)(4.3746)$$

Solving for X gives a value of $67.

$i = 6\%$

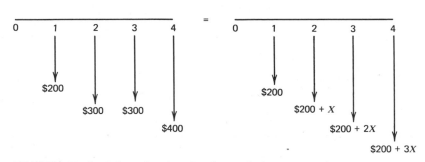

FIGURE 3.14. Cash flow diagrams for the equivalence example.

Example 3.25

For what interest (discount) rate are the two cash flow profiles shown in Figure 3.15 equivalent?

Converting each cash flow profile to a uniform series over the interval [1, 5], gives

$$-\$4000(A|P\ i,5) + \$1500 = -\$7000(A|P\ i,5) + \$1500 + \$500(A|G\ i,5)$$

or

$$\$3000(A|P\ i,5) = \$500(A|G\ i,5)$$

which reduces to

$$(A|G\ i,5) = 6(A|P\ i,5)$$

88 • TIME VALUE OF MONEY OPERATIONS

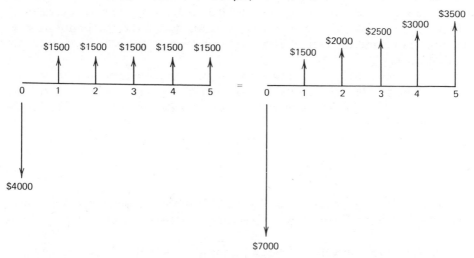

FIGURE 3.15. Cash flow diagrams for the equivalence example.

On searching through the interest tables at $n = 5$, it is found that the $(A|G\ i,5)$ factor is six times the value of the $(A|P\ i,5)$ factor for an interest rate between 12 and 15%. Specifically, with a 12% interest rate,

$$(A|G\ 12,5) - 6(A|P\ 12,5) = 1.7746 - 6(0.2774) = 0.1102$$

and, using a 15% interest rate,

$$(A|G\ 15,5) - 6(A|P\ 15,5) = 1.7228 - 6(0.2983) = -0.0670$$

Interpolating for i gives

$$i = 0.12 + \frac{(0.15 - 0.12)(0.1102)}{(0.1102 + 0.0670)}$$

or

$$i = 0.1386$$

Therefore, using a discount rate of approximately 13.86% will establish an equivalence relationship between the cash flow profiles given in Figure 3.15.

Example 3.26 _____

A firm purchases a machine for $30,000, keeps it for five years, and sells it for $6000. During the time the machine was owned by the company, operating and maintenance costs totaled $8000 the first year, $9000 the second year, $10,000 the third year, $11,000 the fourth year, and $12,000 the fifth year. The firm uses a 15% interest rate in performing economic analyses. Determine the single sum of money occurring at (1) $t = 0$ and (2) $t = 5$, which is equivalent to the cash flow history for the machine. Also, determine the uniform series occurring over the interval $[1, 5]$ that is equivalent to the cash flow profile for the machine.

In the economic analysis literature, the single sum equivalent at time zero for a cash flow profile is called the *present worth* or *present value* for the cash flow profile. We will use the term present worth and denote it as PW. For the example,

$$PW = -\$30{,}000 - \$8000(P|A\ 15{,}5) - \$1000(P|G\ 15{,}5)$$
$$+ \$6000(P|F\ 15{,}5)$$
$$= -\$59{,}609.57$$

Hence, a single expenditure of $59,609.57 at time zero would have been equivalent to the cash flows experienced during the ownership of the machine.

The single sum equivalent at the end of the life of a project is termed the *future worth* or *future value* for the project. We denote the future worth by FW and compute its value as follows:

$$FW = -\$30{,}000(F|P\ 15{,}5) - [\$8000 + 1000(A|G\ 15{,}5)](F|A\ 15{,}5)$$
$$+ \$6000$$
$$= -\$119{,}897$$

Alternately, the future worth can be obtained from the present worth:

$$FW = PW(F|P\ i{,}n)$$
$$= -\$59{,}609.57(F|P\ 15{,}5)$$
$$= -\$119{,}898.69$$

(The difference of $1.69 is due to round-off error in the tables.) Hence, a single expenditure of $119,897 at time five would have been equivalent to the cash flows associated with the machine.

A uniform series equivalent for a series of yearly cash flows is referred to as the *annual worth* or *equivalent uniform annual cost* for the project. The latter expression is most appropriate for the type example under consideration, since the resulting uniform series is a cost, not an income, series. The annual worth designation is AW; $EUAC$ denotes the equivalent uniform annual cost. We will use both designations throughout the text.

For the example problem, the $EUAC$ determination is performed as follows:

$$EUAC = \$30{,}000(A|P\ 15{,}5) + [\$8000 + \$1000(A|G\ 15{,}5)]$$
$$- \$6000(A|F\ 15{,}5)$$
$$= \$1778.20$$

Hence an annual expenditure of $1778.20/year for five years is equivalent to the cash flow profile associated with the machine investment. Alternately, the equivalent uniform annual cost can be obtained from either the present worth or the future worth.

$$EUAC = -PW(A|P\ i,n)$$
$$EUAC = -FW(A|F\ i,n)$$

Present worth, future worth, and annual worth computations are used often in comparing economic investment alternatives having different cash flow profiles. Consequently, we will have need for PW, FW, and AW calculations for the discussion of alternative comparisons in Chapter Four.

3.8 SPECIAL TOPICS

The previous discussion of compound interest methods and the concept of equivalence provides the foundation for the subsequent treatment of economic analysis. At this point sufficient background material has been covered for an understanding of the discussion in Chapter Four. However, before leaving the subject of time value of money, a number of related topics will be covered. A method for determining the amount of interest included in an individual loan payment is presented; methods for dealing with changing interest rates are treated; the situation where the spacing of cash flows and the compounding frequency differ is dealt with; methods are presented for analyzing situations involving an infinite number of cash flows; the use of compound interest methods in analyzing investments in bonds is discussed; methods are provided for coping with inflationary effects in compound interest calculations; and, finally, the concept of capital recovery cost is presented. Since each topic is treated independently, you may wish to study only those topics that are of particular interest.

3.8.1 Principal Amount and Interest Amount in Loan Payments

With both personal and corporate investments that are financed from borrowed funds, income taxes are affected by the amount of interest paid. Hence, it is quite important to know how much of each payment is interest and how much is reducing the principal amount borrowed initially. To illustrate this situation, suppose you borrowed $10,000 and paid it back using four equal annual installments, with interest computed at 10% compounded annually. The payment size is computed to be

$$A = P(A|P\ 10,4)$$
$$= \$10,000(0.3155)$$
$$= \$3155$$

The interest accumulation the first year is 0.10(10,000), or $1000. Therefore, the first payment consists of a $1000 interest payment and a $2155 principal payment. The unpaid balance at the beginning of the second year is $10,000 − $2155, or $7845; consequently, the interest charge the second year is 0.10($7845), or $784.50. Thus, the second payment consists of a $784.50 interest payment and a principal payment of $3155 − $784.50, or $2370.50. The unpaid balance at the beginning of the third year is $7845 − $2370.50, or $5474.50, so the interest charge the third year is 0.10($5474.50), or $547.45. Thus, the third payment consists of a $547.45 interest payment and a principal payment of $3155 − $547.45, or $2607.55. The unpaid balance at the beginning of the fourth year is $5474.50 − $2607.55, or $2866.95; the interest charge the fourth year is 0.10($2866.95), or $286.70. Thus, the fourth payment consists of a $286.70 interest payment and a principal payment of $3155 − $286.70, or $2868.30 (the principal payment in the fourth payment should equal the unpaid balance of $2866.95 at the beginning of the fourth year. The difference of $1.35 is due to round-off errors in computing the payment size.)

The amount of principal remaining to be repaid immediately after making payment $(k − 1)$ can be found by stripping off the interest on the remaining $n − k + 1$ payments; letting U_{k-1} denote the unpaid principal after making payment $k − 1$, we know that

$$U_{k-1} = A(P|A\ i, n − k + 1) \tag{3.52}$$

or

$$U_{k-1} = A \sum_{j=1}^{n-k+1} (1 + i)^{-j}$$

where A denotes the size of the individual payments and i represents the interest rate used to compute A.

The amount by which payment k reduces the unpaid principal will be designated E_k and is given by the relation, $E_k = U_{k-1} − U_k$. Hence,

$$E_k = A \sum_{j=1}^{n-k+1} (1 + i)^{-j} − A \sum_{j=1}^{n-k} (1 + i)^{-j} \tag{3.53}$$

or

$$E_k = A\left[\sum_{j=1}^{n-k} (1 + i)^{-j} + (1 + i)^{-(n-k+1)} \right] − A \sum_{j=1}^{n-k} (1 + i)^{-j}$$

Thus,

$$E_k = A(1 + i)^{-(n-k+1)} \tag{3.54}$$

or

$$\boxed{E_k = A(P|F\ i, n − k + 1)} \tag{3.55}$$

Recalling how the payment size is determined, we express E_k as

$$E_k = P(A|P\,i,n)(P|F\,i,n-k+1) \qquad (3.56)$$

$$\underset{1000\;.1175\qquad\quad .5645}{}$$

where P is the original principal amount borrowed, and E_k is the amount of payment k, which is an *equity* payment (i.e., payment against principal). Letting I_k be the amount of payment k, which is an interest payment, it is seen that

$$I_k = A - E_k$$

$$I_k = A[1 - (P|F\,i,n-k+1)] \qquad (3.57)$$

Example 3.27

To illustrate the use of the formulas for computing the values of E_k and I_k, suppose \$10,000 is borrowed at 6% annual interest and repaid with five equal annual payments. The payment size is found to be

$$A = \$10,000(A|P\ 6,5)$$
$$A = \$10,000(0.2374)$$
$$= \$2374$$

Thus

$$E_1 = \$2374(P|F\ 6,5) = \$1774, \qquad I_1 = \$600$$
$$E_2 = \$2374(P|F\ 6,4) = \$1880, \qquad I_2 = \$494$$
$$E_3 = \$2374(P|F\ 6,3) = \$1994, \qquad I_3 = 380$$
$$E_4 = \$2374(P|F\ 6,2) = \$2112, \qquad I_4 = \$262$$
$$E_5 = \$2374(P|F\ 6,1) = \$2240, \qquad I_5 = \$134$$

Example 3.28

An individual purchases a \$50,000 house and makes a down payment of \$10,000. The remaining \$40,000 is financed over a 25-year period at 7% compounded monthly. The monthly house payment is computed to be

$$A = \$40,000(A|P\ 7/12,300)$$
$$= \$282$$

The individual keeps the house for five years and decides to sell it. How much equity is there in the house?

The amount of principal remaining to be paid can be determined by computing the present worth of the remaining 240 monthly payments using a 7/12% interest rate, or

$$P = \$282(P|A\ 7/12,240)$$
$$= \$36,378$$

Thus, the individual's equity equals ($40,000 - $36,378) or $3622, plus the value of the down payment, or $13,622. It is interesting, (or perhaps depressing if you are a homeowner) to determine that over the five-year period the individual made payments totaling $16,920, of which $13,298 was for interest.

3.8.2 Changing Interest Rates

The preceding discussion assumed that the interest rate did not change during the time period of concern. Recent experience indicates that such a situation is not likely if the time period of interest extends over several years (i.e., more than one interest rate may be applicable). Considering a single sum of money and discrete compounding, if i_k denotes the interest rate appropriate during time period k, the future worth equivalent for a single sum of money can be expressed as

$$F = P(1+i_1)(1+i_2)\ldots(1+i_{n-1})(1+i_n) \tag{3.58}$$

and the inverse relation

$$P = F(1+i_n)^{-1}(1+i_{n-1})^{-1}\ldots(1+i_2)^{-1}(1+i_1)^{-1} \tag{3.59}$$

Example 3.29

Consider the situation depicted in Figure 3.16 in which an individual deposited $1000 in a savings account that paid interest at an annual compounding rate of 5% for the first three years, 6% for the next four years, and 7% for the next two years. How much was in the fund at the end of the ninth year?

Letting V_t denote the value of the account at the end of time period t, we see that

$$V_3 = \$1000(F|P\ 5,3)$$
$$= \$1000(1.1576)$$
$$= \$1157.60$$

Likewise,

$$V_7 = \$1157.6(F|P\ 6,4)$$
$$= \$1157.6(1.2625)$$
$$= \$1461.47$$

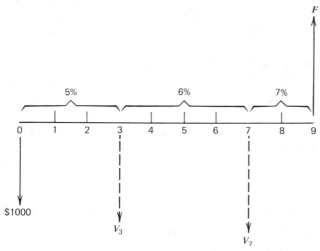

FIGURE 3.16. Cash flow diagram for a changing interest rate example.

Similarly,

$$F = V_9 = \$1461.47(F|P\ 7,2)$$
$$= \$1461.47(1.449)$$
$$= \$1673.24$$

Alternately, the amount in the account at the end of nine years is given by

$$F = \$1000(1.05)(1.05)(1.05)(1.06)(1.06)(1.06)(1.07)(1.07)$$
$$= \$1673.24$$

Extending the consideration of changing interest rates to series of cash flows, the present worth of a series of cash flows can be represented as

$$P = A_1(1 + i_1)^{-1} + A_2(1 + i_1)^{-1}(1 + i_2)^{-1} + \cdots + A_n(1 + i_1)^{-1}(1 + i_2)^{-1} \cdots (1 + i_n)^{-1}$$
$$(3.60)$$

The future worth of a series of cash flows can be given by

$$F = A_n + A_{n-1}(1 + i_n) + A_{n-2}(1 + i_{n-1})(1 + i_n) + \cdots$$
$$+ A_1(1 + i_2)(1 + i_3) \cdots (1 + i_{n-1})(1 + i_n)$$
$$(3.61)$$

Example 3.30 ─────────────────

Consider the cash flow diagram given in Figure 3.17 with the appropriate interest rates indicated. Determine the present worth, future worth, and uniform series equivalents for the cash flow series.

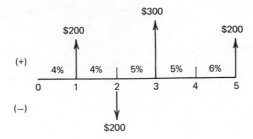

FIGURE 3.17. Cash flow diagram for a changing interest rate example.

Computing the present worth gives

$$P = \$200(P|F\ 4,1) - \$200(P|F\ 4,1)(P|F\ 4,1)$$
$$+ \$300(P|F\ 4,1)(P|F\ 4,1)(P|F\ 5,1)$$
$$+ \$200(P|F\ 4,1)(P|F\ 4,1)(P|F\ 5,1)(P|F\ 5,1)(P|F\ 6,1)$$
$$= \$200(P|F\ 4,1) - \$200(P|F\ 4,2)$$
$$+ \$300(P|F\ 4,2)(P|F\ 5,1) + \$200(P|F\ 4,2)(P|F\ 5,2)(P|F\ 6,1)$$
$$= \$200(0.9615) - \$200(0.9246) + \$300(0.9246)(0.9524)$$
$$+ \$200(0.9246)(0.9070)(0.9434)$$
$$= \$429.79$$

The future worth is given by

$$F = \$200 + \$300(F|P\ 5,1)(F|P\ 6,1) - \$200(F|P\ 5,2)(F|P\ 6,1)$$
$$+ \$200(F|P\ 4,1)(F|P\ 5,2)(F|P\ 6,1)$$
$$= \$200 + \$300(1.05)(1.06) - \$200(1.1025)(1.06)$$
$$+ \$200(1.04)(1.1025)(1.06)$$
$$= \$543.25$$

The uniform series equivalent is obtained as follows.

$$P = A(P|F\ 4,1) + A(P|F\ 4,2) + A(P|F\ 4,2)(P|F\ 5,1)$$
$$+ A(P|F\ 4,2)(P|F\ 5,2) + A(P|F\ 4,2)(P|F\ 5,2)(P|F\ 6,1)$$

$$\$429.79 = A[(0.9615) + (0.9246) + (0.9246)(0.9524)$$
$$+ (0.9246)(0.9070)(0.9434)]$$

$$4.396 A = 429.79$$
$$A = \$97.76$$

Thus, \$97.76/time period for five time periods is equivalent to the original cash flow series. As an exercise one may wish to solve the example problem assuming continuous compounding.

3.8.3 End-of-Period Cash Flows and End-of-Period Compounding

Thus far we have emphasized end-of-period cash flows and end-of-period compounding.[1] Depending on the financial institution involved, in personal finance transactions, savings accounts might not pay interest on deposits made in "the middle of a compounding period." Consequently, you should not expect answers obtained using the methods we describe to be exactly the same as those provided by the financial institution.

The beginning-of-period cash flows can be handled very easily by noting that the end of period k is the beginning of period $k + 1$. To illustrate, rental payments might be made at the beginning of each month. However, one can think of the payment made at the beginning of, say, March as having been made at the end of February.

In Chapter Four, end-of-year cash flows are assumed unless otherwise noted. It is realized that monetary transactions take place during a calendar year, but it is convenient to ignore any compounding effects within a year and deal directly with end-of-year cash flows.

We previously mentioned the possibility of cash flows occurring at intervals that were not the same as compounding intervals. Two possibilities might arise: money is compounded *more* frequently than the occurrence of cash flows and money is compounded *less* frequently than the occurrence of cash flows.

Example 3.31 ————————————————————————————

Suppose an individual makes monthly deposits into a fund that pays interest at a rate of 8% compounded quarterly. Depending on how the financial institution interprets the rate of "8% compounded quarterly," money deposited during a quarter might not earn any interest. We interpret 8% compounded quarterly to mean that deposits earn interest equivalent to 8% compounded quarterly. Hence, the monthly interest rate to be applied to the monthly deposits should be equivalent to 8% compounded quarterly. Letting i be the monthly rate, we note that the effective interest rate for $i\%$/month should be the same as for 8% compounded quarterly. Therefore

$$i_{\text{eff}} = (1 + i)^{12} - 1$$
$$= \left(1 + \frac{0.08}{4}\right)^4 - 1$$

[1] The exception to this was the treatment of continuous cash flows.

or

$$(1+i)^{12} = (1.02)^4$$

or

$$(1+i)^3 = 1.02$$

Thus,

$$1 + i = (1.02)^{1/3}$$
$$i = (1.02)^{1/3} - 1$$
$$i = 0.0066227$$

Consequently, a monthly interest rate of 0.66227% can be used to determine the balance in the fund at the end of any month.

Example 3.32

Suppose quarterly deposits are made into a fund that pays interest at a rate of 6% compounded monthly. By the same arguments used in the previous example, we know that

$$i_{eff} = (1+i)^4 - 1$$
$$= \left(1 + \frac{0.06}{12}\right)^{12} - 1$$

or

$$(1+i)^4 = (1.005)^{12}$$
$$i = (1.005)^3 - 1$$
$$= 0.015075$$

Thus, a quarterly interest rate of 1.5075% can be used to determine the balance in the fund at the end of any quarter.

3.8.4 Perpetuities and Capitalized Value

A specialized type of cash flow series is a perpetuity, a uniform series where the payments continue indefinitely. By virtue of being a special case, an infinite series of cash flows would be encountered much less frequently in the business world than a finite series of cash flows. However, for such very long-term investment projects as bridges, highways, forest harvesting, or the establishment of endowment funds where the estimated life is 50 years or more, an infinite cash flow series may be appropriate.

If a present value P is deposited into a fund at interest rate i/period so that a payment of size A may be withdrawn each and every period forever, then the following relation holds between P, A, and i.

$$Pi = A$$

Thus, as depicted in Figure 3.18, P is a present value that will pay out equal payments of size A indefinitely if the interest rate per period is i. The present value P is termed the *capitalized value* of A, the size of each of the perpetual payments.

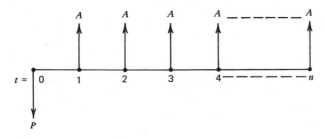

FIGURE 3.18. A finite cash flow series.

Example 3.33 ────────────────────────

What deposit at $t = 0$ into a fund paying 5 1/2% annually is required in order to pay out $600 each year forever? The solution is straightforward from

$$P = \frac{A}{i} = \$600/0.055 = \$10,909.10$$

By means of the subsequent example, let us now broaden the meaning of capitalized value to be that present value that would pay for the first cost of some project and provide for its perpetual maintenance at an interest rate i.

Example 3.34 ────────────────────────

Project ABC consists of the following requirements:

1. A $20,000 first cost at $t = 0$.
2. A $2000 expense every year.
3. A $10,000 expense every third year forever, with the first expense occurring at $t = 3$.

What is the capitalized cost of Project ABC if $i = 10\%$ annually?

It will be instructive to determine the capitalized cost of each requirement separately and then sum the results. First, the capitalized cost of "$10,000 every third year forever" may be determined from any of three points of view. One view is that a value P is required at the beginning of a three-year period such that, with interest compounded at 10% annually, a sum of $P + \$10,000$ will accrue at the end of the three-year period. Thus, $10,000 would be withdrawn; thereby leaving the value P to repeat the cycle indefinitely each three-year period. This logic is illustrated in Figure 3.19.

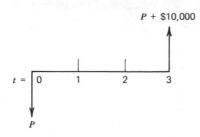

FIGURE 3.19. Capitalized cost of $10,000 every third year—first view.

Thus,

$$P(F|P \ 10,3) = P + \$10,000$$

or

$$1.3310P = P + \$10,000$$

and

$$P = \$30,211.48$$

which is the capitalized cost of the requirement. A second view is to reason that, for each three-year period, three equal deposits of size A are required that will amount to $10,000 at the end of the third year. Then, if these payments of size A occurred every year, $10,000 would be available every third year forever. The present value P that yields the required payment A is the solution, as shown in Figure 3.20.

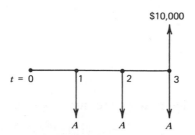

FIGURE 3.20. Capitalized cost of $10,000 every third year—second view.

Furthermore,
$$A = F(A|F\ 10,3) = \$10,000(0.3021)$$
$$= \$3021$$

Then,
$$P = \frac{A}{i} = \frac{\$3021}{0.10} = \$30,210.00$$

which is approximately the same answer as from the first approach. The difference of $1.48 is due to rounding in the interest factor calculations.

A third point of view for the $10,000 requirement is to consider the infinite series depicted in Figure 3.21, such that $P = A/i$ would yield the desired result. However, the value of i required is the effective interest rate for a three-year period. Thus,

$$i = [(1 + 0.10)^3 - 1] = 0.3310$$

and
$$P = \frac{\$10,000}{0.3310} = \$30,211.48$$

The capitalized cost of Project ABC is computed as:

1. Capitalized cost of $20,000 first cost   = $20,000
2. Capitalized cost of $2000 every year = $2000/0.10 = $20,000
3. Capitalized cost of $10,000 every third year   = $30,210
                    Total capitalized cost $70,210

The capitalized value of $70,210 would provide $A = Pi = \$70,210(0.10) = \$7021$ every year forever.

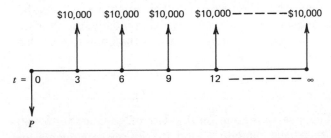

FIGURE 3.21. Capitalized cost of $10,000 every third year—third view.

3.8.5 Bond Problems

Although bonds are important financial instruments in the business world as investment opportunities, there are additional reasons for considering bond problems in an economic analysis textbook.

1. The issuance and sale of bonds is a mechanism by which capital may be raised to finance engineering projects.
2. Bond problems illustrate the notion of equivalence. That is, the purchase price of a bond is equivalent (has the same present value) to the returns from the bond at an appropriate compound interest rate. The returns from the bond consist of periodic interest payments to the bondholder and the redemption value, or sales price, of the bond.
3. A bond problem to calculate the *yield* on the bond's purchase price is analogous to the calculation of the *internal rate of return* on an investment. The internal rate of return method for comparing alternative investment projects will be presented in Chapter Four. Thus, problems involving the calculation of bond yield will introduce the internal rate of return method.

The latter two reasons are the primary ones for treating bond problems in this text. Subsequent discussion presents the appropriate terminology and illustrates the three types of problems that are possible.

An organizational unit desiring to raise capital may issue bonds totaling, say, $1 million, $5 million, $25 million, or more. A financial brokerage firm usually handles the issue on a commission basis and sells smaller amounts to other organizational units or individual investors. Individual bonds are normally issued in even denominations such as $500, $1000, or $5000. The stated value on the individual bond is termed the *face* or *par value*. The par value is to be repaid by the issuing organization at the end of a specified period of time, say 5, 10, 15, 20, or even 50 years. Thus, the issuing unit is obligated to *redeem* the bond at par value at *maturity*. Furthermore, the issuing unit is obligated to pay a stipulated *bond rate* on the face value during the interim between date of issuance and date of redemption. This might be 8% payable quarterly, 7 1/2% payable semiannually, 6 3/4% payable annually, etc. For the purpose of the problems to follow, it is emphasized that the bond rate applies to the par value of the bond.

Example 3.35

A person purchases a $5000, five-year bond on the date of issuance for $5000. The stated bond rate is "8% semiannually," and the interest payments are received on schedule until the bond is redeemed at maturity for $5000. The bond rate per interest period is $8\%/2 = 4\%$. Thus, the bond holder receives

$(0.08/2)(\$5000) = \$200$ payments every six months. A cash flow diagram for the duration of the investment is given in Figure 3.22, where time periods are six-month intervals.

It is noted from Figure 3.22 that the $5000 expenditure at $t = 0$ yields $200 each interest period for 10 periods and a $5000 redemption value at $t = 0$. Thus, the $5000 investment at $t = 0$ yielded the revenues from $t = 1$ through $t = 10$. What annual rate of return or interest rate did the $5000 investment yield?

One might intuitively answer the above question by stating that the $5000 investment was exactly returned (no loss or gain in capital) at redemption and, since during the interim an interest rate of 8% semiannual was received, then the yield *must* be 8% semiannual. Accepting this argument for the moment, one might further pose certain hypotheses concerning the transaction. For instance, it is hypothesized that the $5000 at $t = 0$ is equivalent (has the same present worth) to the revenue cash flows if the time value of money is 8% compounded semiannually. That this hypothesis is true is shown by the relation

$$P = A(P|A\ 4,10) + F(P|F\ 4,10)$$

or

$$\$5000 = (0.04)(\$5000)(8.1109) + \$5000(0.6756),$$

and

$$\$5000 = \$5000$$

A second hypothesis concerning the transaction is that if a premium above par value is paid for the bond at $t = 0$ and all revenue figures remain the same, then an annual yield (rate of return) less than the bond rate of 8% semiannual

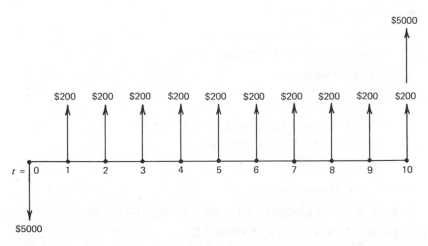

FIGURE 3.22. Cash flow for a $5000 bond.

will be received on the investment. For example, if $5500 were paid for the bond at $t = 0$, the present worth of the $5500 outgo is not equal (equivalent) to the present worth of the revenues at an 8% semiannual yield. This is shown by

$$\$5500 \neq (0.04)(\$5000)(P\,|\,A\ 4,10) + \$5000(P\,|\,F\ 4,10)$$

$$\$5500 \neq \$200(8.1109) + \$5000(0.6756)$$

$$\$5500 \neq \$5000$$

This result now raises the pertinent question, what *is* the annual yield on the investment if $5500 is paid for the bond and the revenues remain the same? Intuition may suggest that the yield will be less than 8% semiannual because the purchase value of $5500 decreases (a loss of investment capital) to $5000 on redemption. Furthermore, the semiannual payments of $200 are not equal to 4% of the $5500 purchase price. In order to answer the yield question more precisely, let us answer the alternate question of "what interest rate per period (or yield per period) will make the purchase price equivalent to the revenue cash flows?" That is,

$$\text{Present worth of outgo} = \text{Present worth of revenues}$$

at what interest rate per period? Alternately, what interest rate satisfies the following equation?

$$\$5500 = (0.04)(\$5000)(P\,|\,A\ ?,10) + \$5000(P\,|\,F\ ?,10)$$

The solution to the above equation gives the answer to the original question, and the task is to solve for the positive roots of the polynomial. However, an approximate solution to avoid such tedium is by trial and error. An iterative procedure follows.

For $i = 4\%$ (8% semiannual),

$$\$5500 \neq (0.04)(\$5000)(P\,|\,A\ 4,10) + \$5000(P\,|\,F\ 4,10)$$

$$\$5500 \neq \$200(8.1109) + \$5000(0.6756)$$

$$\$5500 \neq \$5000$$

For $i = 3\%$ (6% semiannual),

$$\$5500 \neq (0.04)(\$5000)(P\,|\,A\ 3,10) + \$5000(P\,|\,F\ 3,10)$$

$$\$5500 \neq \$200(8.5302) + \$5000(0.7441)$$

$$\$5500 \neq \$5426.50$$

For $i = 2\ 1/2\%$ (5% semiannual),

$$\$5500 \neq (0.04)(\$5000)(P\,|\,A\ 2.5,10) + \$5000(P\,|\,F\ 2.5,10)$$

$$\$5500 \neq \$200(8.7521) + \$5000(0.7812)$$

$$\$5500 \neq \$5656.42$$

TABLE 3.4 Bond Yield Interpolation

For $i =$	0.03	X	0.025
PW of revenues =	$5426.50	$5500	$5652.42

From the last two trials (for $i = 3\%$ and $i = 2\ 1/2\%$), the equivalent present worth of $5500 desired for revenues is bracketed, as shown in Table 3.4. Using the data of Table 3.4, we can solve for X by linear interpolation, or

$$\frac{0.03 - 0.025}{\$5426.50 - \$5652.42} = \frac{0.03 - X}{\$5426.50 - \$5500}$$

and

$$X = 0.02837 \quad \text{or} \quad 2.827\%$$

Thus, the equivalent yield on the $5500 investment is approximately 2.837%/period, or (2)(2.837%) = 5.674% semiannually, or an effective annual yield of $[(1 + 0.02837)^2 - 1](100\%) = 5.7545\%$. These figures support the *a priori* intuition of an annual yield less than 8% semiannual, and the second hypothesis is accepted.

Bond problems arise in economic analysis because many bonds trade daily through financial markets such as the New York Stock Exchange. Thus, bonds may be purchased for less than, greater than, or equal to par value, depending on the economic environment. They may also be sold for less than, greater than, or equal to par value. Furthermore, once purchased, bonds may be kept for a variable number of interest periods before being sold. A variety of situations can occur, but only three basic types of bond problems can occur. These will be presented after formalizing the discussion thus far. We now employ the following notation.

P = the purchase price of a bond

F = the sales price (or redemption value) of a bond

V = the par or face value of a bond

r = the bond rate per interest period

i = the yield rate per interest period

n = the number of interest payments received by the bondholder

$A = Vr$ = the interest payment received

The general expression relating these is

$$P = Vr(P|A\ i,n) + F(P|F i,n) \tag{3.62}$$

Now, the three types of bond problems follow:

1. Given P,r,n,V, and a desired i, find the sales price F.
2. Given F,r,n,V, and a desired i, find the purchase price P.
3. Given P,F,r,n, and V, find the yield i that has been earned on the investment.

Each of these cases is illustrated in the following examples.

Example 3.36

A $1000, 8% semiannual bond is purchased for $1050 at an arbitrary $t = 0$. If the bond is sold at the end of three years and six interest payments, what is the selling price to earn 6% nominal? The cash flow for this example is given by Figure 3.23.

FIGURE 3.23. Cash flow for a $1000 bond—determine sales price.

A mathematical statement of the cash flow depicted in Figure 3.23 follows.

$$P = Vr(P|A\ 3,6) + F(P|F\ 3,6)$$

or

$$\$1050 = (\$1000)(0.04)(5.4172) + F(0.8375).$$

Solving the above equation for F yields a value of $995.

Example 3.37

If a $1000, 8% semiannual bond is purchased at an arbitrary $t = 0$, held for three years and six interest payments, and redeemed at par value, what must the purchase price have been in order to earn a nominal yield of 10%? From the cash flow given in Figure 3.24,

$$P = Vr(P|A\ 5,6) + F(P|F\ 5,6)$$

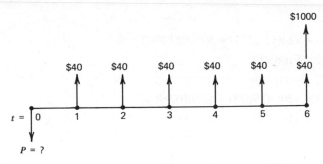

FIGURE 3.24. Cash flow for a $1000 bond—determine purchase price.

or

$$P = (\$1000)(0.04)(5.0757) + \$1000(0.7462)$$

and

$$P = \$949.23$$

Example 3.38 ―――――――――――――――――――――――――

If a $1000, 8% quarterly bond is purchased at $t = 0$ for $1020 and sold three years later for $950, (1) what was the quarterly yield on the investment, and (2) what was the effective annual yield? From the cash flow diagram in Figure 3.25?

$$\$1020 = (\$1000)(0.02)(P|A\,i,12) + \$950(P|F\,i,12)$$

It is then necessary to solve for the unknown i by trial and error as follows:

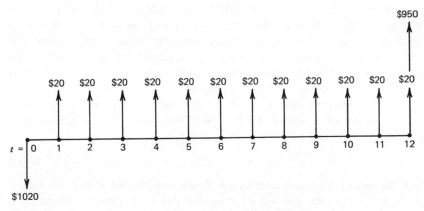

FIGURE 3.25. Cash flow for a $1000 bond—determine yield.

For $i = 1\%$,

$$\$1020 \neq \$20(11.2551) + \$950(0.8874)$$
$$\$1020 \neq \$1068.13$$

For $i = 1\ 1/2\%$,

$$\$1020 \neq \$20(10.9075) + \$950(0.8364)$$
$$\$1020 \neq \$1012.73$$

Then, interpolating gives

$$\frac{0.01 - 0.015}{\$1068 - \$1012.73} = \frac{0.01 - X}{\$1068.13 - \$1020}$$

$$X = 0.01434 \qquad \text{or} \qquad 1.434\% \text{ per quarter}$$

The effective annual yield is

$$[(1 + 0.01434)^4 - 1]100\% = 5.861\%$$

3.8.6 Inflationary Effects

The dynamic nature of the economy in the past decade has focused considerable attention on inflation and its effect on economic decision making. Alternative approaches that are typically used to account for inflationary effects include:

1. Express all cash flows in terms of "then-current" dollar amounts and combine the inflation rate with the interest rate.
2. Express all cash flows in terms of "constant worth" dollar amounts and use an interest rate without an inflation rate component.

The latter approach appears to be the method most preferred by practitioners. However, it is not uncommon to encounter claims that the interest or discount rate used in time value of money computations includes a component for inflation. If care is not taken to insure a proper accounting for inflation, one might partially account for inflation, rather than account for it in both the discount rate used and the cash flow estimates.

Letting j be the inflation rate, C_k be the "constant worth" value of a cash flow at the end of period k, and T_k be the "then-current" value of a cash flow at the end of period k, the following relation holds.

$$T_k = C_k(1 + j)^k \tag{3.63}$$

Thus, when the set of constant worth cash flows constitutes a *uniform series*, (i.e., $C_k = A$, $k = 1, \ldots, n$), the set of then-current cash flows constitutes a *geometric series* (i.e., $T_k = A_1(1 + j)^{k-1}$), where $A_1 = A(1 + j)$.

Using the first approach to account for inflationary effects, the present worth equivalent of a series of T_k cash flows is computed as follows.

$$P = \sum_{k=0}^{n} T_k (1+j)^{-k}(1+i)^{-k}$$

$$= \sum_{k=0}^{n} T_k (1+j+i+ij)^{-k}$$

$$= \sum_{k=0}^{n} T_k (1+d)^{-k} \tag{3.64}$$

where d is a discount rate equal to $i+j+ij$. The second approach computes the present worth equivalent using constant worth dollar amounts by noting that

$$P = \sum_{k=0}^{n} T_k (1+j)^{-k}(1+i)^{-k}$$

can be given as

$$P = \sum_{k=0}^{n} C_k (1+i)^{-k} \tag{3.65}$$

In the preceding sections of the chapter we implicitly assumed that either the cash flows were in the form of constant worth dollars or the discount rate used with then-current dollars included a component for inflation.

Example 3.39

To illustrate the two approaches in dealing with inflation, suppose $j = 5\%$, $i = 10\%$, $T_0 = -\$10{,}000$, $T_1 = \$1000$, $T_2 = \$3000$, $T_3 = \$5000$, and $T_4 = \$7000$. As shown in Table 3.5, the present worth equivalent for the situation is $-\$356$. In this instance, $d = 0.05 + 0.10 + 0.005 = 0.155$, or 15.5%. If the the present worth were computed using $d = 10\%$ and then-current dollar estimates, the present worth would have been \$1175 instead of the $-\$356$

TABLE 3.5 Present Worth Calculations Under Inflation

k	T_k	$(1+d)^{-k}$	$T_k(1+d)^{-k}$	C_k^a	$(1+i)^{-k}$	$C_k(1+i)^{-k}$
0	-$10,000	1.0000	-$10,000	-$10,000	1.0000	-$10,000
1	1,000	0.8658	866	952	0.9091	866
2	3,000	0.7496	2,249	2,721	0.8264	2,249
3	4,000	0.6490	2,596	3,455	0.7513	2,596
4	7,000	0.5619	3,933	5,759	0.6830	3,933
			-356			-356

$^a C_k = T_k(1.05)^{-k}$.

obtained considering inflation. Consequently, what appears to have been a profitable investment without considering inflation has a negative present worth when the effects of inflation are included in the analysis.

Although the second approach, involving constant worth dollar amounts, appears to be the simplest method of incorporating inflation in the analysis, estimating cash flows in terms of constant worth dollar amounts is not a trivial exercise. We mentioned previously that a uniform series of constant worth cash flows converts to a geometric series of then-current cash flows. You expect cash flows for, say, labor costs to increase over time; some portion, but not all, of the increase is due to inflation. Thus, it is difficult to factor out the portion of the increase that is due to inflation in order to provide inflation-free estimates of future cash flows.

Another aspect of inflation that tends to complicate the analysis is the differences that may exist in inflation rates for not only various types of cash flows, but also for different regions of the country and world. The firm that has numerous plants scattered throughout not only the same country, but also the world, must cope with the economic differences that exist among the locations, as well as the differences in inflation rates for labor, equipment, materials, utilities, and supplies.

As a further complication in dealing with inflation, the inflation rate tends to change from one time period to the next. If j_t denotes the inflation rate for period t and i_t denotes the interest rate for period t, the present worth can be expressed in then-current dollars as

$$P = \sum_{k=0}^{n} T_k \prod_{t=0}^{k} (1 + j_t)^{-1}(1 + i_t)^{-1} \tag{3.66}$$

in the case of discrete compounding and as

$$P = \sum_{k=0}^{n} T_k e^{-\sum_{t=0}^{k} (c_t + r_t)} \tag{3.67}$$

in the case of continuous compounding, where c_t is the nominal inflation rate for period t and r_t is the nominal interest rate for period t; in the above c_0, r_0, j_0, and i_0 are defined to be equal to zero.

Inflation is a much discussed, but little understood, subject in the area of economic investment alternatives. Some argue that inflation effects can be ignored, since inflation will affect all investment alternatives in roughly the same way. Thus, it is argued, the relative differences in the alternatives will be approximately the same with or without inflation considered. Others argue that the inflation rate during the past decade has been so dynamic that an accurate prediction of the true inflation rate and its impact on future cash flows is not possible. Another argument for ignoring explicitly the effects of inflation is that it is accounted for implicitly, since cash flow estimates for the future are made by individuals conditioned by an inflationary economy. Thus, it is argued, any estimates of future cash flows probably incorporate implicitly inflationary effects. A final argument for ignoring inflation in comparing investment alternatives involving only negative cash flows is that an alternative that is preferred by ignoring inflation effects will be even more attractive when effects of inflation are incorporated in the analysis.

The above arguments are certainly valid in some instances. However, counterarguments can be given for each. Consequently, before dismissing inflation effects as unnecessary in performing economic analysis, the individual situation should be considered closely.

3.8.7 Capital Recovery Cost

In the engineering economy literature there is frequent reference to the term *capital recovery cost*. If an investment of $\$P$ is made in an asset, the asset is used for n years, and disposed of for a salvage value of $\$F$, then the capital recovery cost, CR, is defined as

$$CR = P(A|P\,i,n) - F(A|F\,i,n) \qquad (3.68)$$

However, since

$$(A|P\,i,n) = (A|F\,i,n) + i \qquad (3.69)$$

then on substituting Equation 3.69 into Equation 3.68 we obtain

$$CR = (P - F)(A|F\,i,n) + Pi \qquad (3.70)$$

As will be noted in Section 5.4.4, the first term on the right hand side of Equation 3.70 is the annual sinking fund deposit for sinking fund depreciation; the second term is the opportunity cost due to $\$P$ being tied up in the asset. On the basis of Equation 3.70 capital recovery cost is often defined as the cost of depreciation plus a minimum return on the investment.

Alternately, Equation 3.68 can be given as

$$CR = (P - F)(A|P\,i,n) + Fi \qquad (3.71)$$

due to the relationship between the factors $(A|P\,i,n)$ and $(A|F\,i,n)$. Among the three methods of computing the capital recovery cost, Equation 3.71 appears to be the most popular. We tend to use Equation 3.68, since it is a direct application of the cash flow approach.

Example 3.40 ───

Consider an investment of $10,000 in a unit of equipment which lasts for 10 years and is sold for $1000. An interest rate of 10% is used. Determine the capital recovery cost using the three methods presented.

$$CR = \$10,000(A|P\ 10\%,10) - \$1000(A|F\ 10\%,10)$$
$$= \$10,000(0.1627) - \$1000(0.0627)$$
$$= \$1564.30$$

$$CR = (\$10,000 - \$1000)(A|F\ 10\%,10) + \$10,000(0.10)$$
$$= \$9000(0.0627) + \$1000$$
$$= \$1564.30$$

$$CR = (\$10,000 - \$1000)(A|P\ 10\%,10) + \$1000(0.10)$$
$$= \$9000(0.1627) + \$100$$
$$= \$1564.30$$

3.9 SUMMARY

In this chapter we developed the time value of money concept and defined a number of mathematical operations consistent with that concept. To motivate the discussion, we emphasized the subject from the viewpoint of personal financing. In subsequent chapters we apply the concepts developed in this chapter to the study of investment alternatives from the viewpoint of an ongoing enterprise.

PROBLEMS

1. Person A sells Person B a used automobile for $2000. Person B pays $500 cash down and gives Person A two personal notes for the remainder due. The principal of each note is therefore $750. One note is due at the end of the first year; the other note due at the end of the second year. The annual simple interest rate agreed on is 8 1/2%. How much total interest will Person B pay Person A? (3.2)

2. A debt of $1000 is incurred at $t = 0$. An annual simple interest rate of 8% on the unpaid balance is agreed on. Three equal payments of $388 each at $t = 1, 2, 3$ will pay off this debt and the relevant interest due. For *each* payment, what is (a) the payment on the principal, and (b) the interest amount paid? (3.2)

3. If $5000 is deposited at $t = 0$ into a fund paying 6% compounded per period, what sum will be accumulated at the end of eight periods, or $t = 8$? What would be the sum accumulated if the fund paid 5% compounded per period? (3.3)

4. If a deposit of $1000 at $t = 0$ amounts to $4300 at the end of the eighth compounding period, what value of i is involved? (3.3)

5. How long does it take a deposit in a 6% fund to triple in value? (3.3)

6. If a fund pays 6% compounded annually, what single deposit is required at $t = 0$ in order to accumulate $8000 in the fund at the end of the tenth year ($t = 10$)? (3.3)

7. How much money today is equivalent to $1000 in five years, with interest at 7% compounded annually? (3.3)

8. A person deposits $500, $1200, and $2000 at $t = 0, 1, 2$, respectively. If the fund pays 6% compounded per period, what sum will be accumulated in the fund at (a) $t = 2$ and (b) $t = 3$? (3.4)

9. How much should be deposited at $t = 0$, into a fund paying 10% compounded per period, in order to withdraw $700 at $t = 1$, $1500 at $t = 3$, and $2000 at $t = 7$ and the fund be depleted?

10. A person deposits $4000 in a savings account that pays 6% compounded semiannually. Three years later he deposits $4500. Two years after the $4500 deposit $2500 is deposited. Four years after the $2500 deposit, half of the accumulated funds is transferred to a fund that pays 7% compounded annually. How much money will be in each fund, six years after the transfer? (3.4)

11. A woman annually deposits $\$A_k$ in an account at time k, $k = 1, \dots, 20$, where

$$A_k = 10k(1.05)^{k-1}$$

If the fund pays 5%, how much is in the fund immediately after the seventeenth deposit?

Note.

$$\sum_{k=1}^{n} k = n(n+1)/2$$

$$\sum_{k=1}^{n} kx^k = \frac{(x-1)(n+1)x^{n+1} - x^{n+2} + x}{(x-1)^2}$$

(3.4)

12. What equal, annual deposits must be made at $t = 1, 2, 3, 4, 5, 6$ in order to accumulate $10,000 at $t = 6$ if "money is worth" 10% compounded annually? (3.4.1)

13. A debt of $1000 is incurred at $t = 0$. What is the amount of three equal payments at $t = 1, 2, 3$ that will repay the debt if "money is worth" 8% compounded per period? (3.4.1)

14. Five deposits of $300 each are made at $t = 1, 2, 3, 4, 5$ into a fund paying 8% compounded per period. How much will be accumulated in the fund at (a) $t = 5$, and (b) $t = 9$? (3.4.1)

15. If you know the values of the $(F|P,6,n)$ factor for $n = 1, 2, \ldots, 10$ show how you would determine the value for the $(A|P,6,10)$ factor (3.4.1)

16. A deposit of $\$X$ is placed into a fund paying 10% compounded annually at $t = 0$. If withdrawals of $\$3154.67$ can be made at $t = 1, 2, 3$, and 4 such that the fund is depleted with the last withdrawal, show that the value of X is $\$10,000$ by (a) use of the interest factor $(P|A\ 10,4)$, and (b) use of the interest factors: $(P|F\ 10,1)$, $(P|F\ 10,2)$, $(P|F\ 10,3)$, and $(P|F\ 10,4)$. (3.4.1)

17. A person deposits $\$1000$ in an account each year for five years; at the end of five years, half of the account balance is withdrawn; $\$2000$ is deposited annually for five more years, with the total balance withdrawn at the end of the fifteenth year. If the account earns interest at a rate of 5%, how much is withdrawn (a) at the end of five years, and (b) at the end of 15 years? (3.4.1)

18. With 5% interest compounded annually, how much money will be accumulated in a fund at the time of the fifth deposit, if equal deposits of $\$521.08$ are made annually? (3.4.1)

19. A person borrows $\$10,000$ at 8% compounded annually and wishes to pay the loan back over a five-year period with annual payments. However, the second payment is to be $\$500$ greater than the first payment; the third payment is to be $\$1000$ greater than the second payment; the fourth payment is to be $\$1500$ greater than the third payment; and the fifth payment is to be $\$2000$ greater than the fourth payment. Determine the size of the first payment. (3.4.2)

20. A debt of $\$X$ is incurred at $t = 0$ (purchase of land). It is agreed that payments of $\$5000, \$4000, \$3000$, and $\$2000$ at $t = 4, 5, 6$, and 7, respectively, will satisfy the debt if 10% compounded per period is the appropriate interest rate. By use of the *uniform gradient series factor* and any other relevant interest factor(s), determine the amount of the debt, $\$X$. (3.4.2)

21. By the use of a uniform gradient series formula and any other appropriate formulas, determine the future worth, F, at $t = 15$ of the following deposits: $\$1000$ at $t = 8$, $\$900$ at $t = 9$, $\$800$ at $t = 10$, and $\$700$ at $t = 11$. Assume the fund pays 20% compounded per period. (3.4.2)

22. Mr. Jones receives an annual bonus from his employer. He wishes to deposit the bonus in a fund that pays interest at a rate of 6% compounded continuously. His first bonus is $\$1000$. The size of his bonus is expected to increase at a rate of 4%/year. How much money will be in the fund immediately *before* the tenth deposit? (3.4.3)

23. An individual works for a company that pays an annual bonus, the size of which is based on experience with the company. After one year with the company the bonus equals $\$500$. The size of the bonus thereafter compounds at an annual rate of 4%. The individual decides to place half the bonus in a fund that pays 5% compounded annually.
    (a) How much money will be in the fund immediately prior to the sixth deposit?
    (b) What is the answer to (a) if the fund compounds at 4% annually? (3.4.3)

24. A man invests $10,000 in a venture that returns him $500(0.80)^{k-1}$ at the end of year k for $k = 1, \ldots, 20$. With an interest rate of 10%, what is the equivalent uniform annual profit (cost) for the venture? (3.4.3)

25. A man places $1000 in a fund at the end of 1964. He places end-of-year deposits in the fund until the end of 1983, when his last deposit is made. The fund pays 5% compounded annually. If the size of a deposit at the end of year k, A_k, equals $0.90A_{k-1}$, how much will be in the fund immediately after the last deposit? (3.4.3)

26. Ms. Smith deposits $1000 in a savings account in her local bank. The bank pays interest at a rate of 6% compounded semiannually. Three years after making the single deposit she withdraws half the accumulated money in her account. Five years after the initial deposit, she withdraws all of the accumulated money remaining in the account. How much does she withdraw five years after her initial deposit? (3.5)

27. A person wishes to make a single deposit P at $t = 0$ into a fund paying 8% compounded quarterly such that $1000 payments are received at $t = 1, 2, 3$, and 4 (periods are three-month intervals) and a single payment of $5000 is received at $t = 12$. What single deposit is required? (3.5)

28. A man borrows $20,000 at 8% compounded quarterly. He wishes to repay the money with 10 equal semiannual installments. What must be the size of the payment if the first payment is made one year after obtaining the $20,000? (3.5)

29. A woman borrows $2000 at 8%/year compounded monthly. She wishes to repay the loan with 12 end-of-month payments. She wishes to make her first payment three months after receiving the $2000. She also wishes that, after the first payment, the size of her payment be 10% greater than the previous payment. What is the size of her sixth payment? (3.5)

30. Monthly deposits of $100 are made into an account paying 4% compounded quarterly. Ten monthly deposits are made. Determine how much will be accumulated in the account two months after the last deposit. (3.5)

31. A person makes four consecutive semiannual deposits of $1000 in a savings account that pays interest at a rate of 6% compounded semiannually. How much money will be in the account two years after the last deposit? (3.5)

32. An individual borrows $5000 at an interest rate of 8% per year compounded semiannually and desires to repay the money with five equal end-of-year payments, with the first payment made two years after receiving the $5000. What should be the size of the annual payment? (3.5)

33. What is the effective interest rate for 6% compounded monthly? (3.5)

34. What monthly deposits must be made in order to accumulate $4000 in five years if a fund pays 9% compounded monthly? (Assume the first deposit is made at the end of the first month and the final deposit is made at the end of the fifth year.) (3.5)

35. What monthly payments are required in order to pay off a $20,000 debt in five years if the *nominal* rate of interest charged on the unpaid balance is 9% compounded annually.

36. A man borrows $20,000 at 8% compounded semiannually. He pays back the loan with four equal semiannual payments, with the first payments made one year after receiving the $20,000. What should be the size of each of the four payments? (3.5)

37. A man borrows $1000 and pays the loan off, with the interest, after two years. He pays back $1150. What is the effective interest rate for this transaction? (3.5)

38. A woman borrows $5000 at 1%/month. She desires to repay the money using equal monthly payments for 10 months. The woman makes four such payments and decides to pay off the remaining debt with one lump sum payment at the time for the fifth payment. What should the size of this payment be if interest is truly compounded at a rate of 1%/month?

39. Mr. Wright borrows $8000 from a bank that charges interest at 6% compounded semiannually. Mr. Wright is to pay the money back with six equal payments. However, the first payment is to be made immediately on receipt of the $8000. Successive payments are spaced one year apart.
    (a) Determine the size of the equal annual payment.
    (b) At the time of the fourth payment, suppose Mr. Wright decides to pay off the loan with one lump sum payment. How much should be paid? Include the fourth payment. (3.5)

40. Approximately how long will it take a deposit to triple in value if money is worth 8% compounded semiannually? (3.5)

41. Assume a person deposits $2000 now, $1000 two years from now, and $5000 five years from now into a fund paying 8% compounded semiannually.
    (a) What sum of money will have accumulated in the fund at the end of the sixth year?
    (b) What equal deposits of size A, made every six months (with the first deposit at $t = 0$ and the last deposit at the end of the sixth year), are equivalent to the three deposits stated above? (3.5)

42. A man borrows $1000 from the Shady Deal Finance Company. He is told the interest rate is merely 1.7%/month, and his payment is computed as follows.

    Payback period = 30 months
    Interest = 30(0.017)($1000) = $510
    Credit investigation and insurance = $20
    Total amount owed = $1530
    Payment size = $1530/30 = $51 per month

    What is the approximate interest rate for this transaction? (3.5)

43. Operating and maintenance costs for a production machine occur continuously during the year. If the total annual operating and maintenance cost is $8000 and money is worth 15% compounded continuously, what single sum of money at the present is equivalent to five years of operating and maintenance costs? (3.6)

44. Labor costs occur continuously during the year. At the end of each year a new labor contract becomes effective for the following year. Let $\bar{A}_j$ denote the cumulative labor cost occurring uniformly during year j, where $\bar{A}_j = 1.05\bar{A}_{j-1}$. If money is worth 10% compounded continuously and $\bar{A}_1 = \$25,000$, determine the present worth equivalent for five years of labor costs. (3.6)

45. A person borrows $10,000 and wishes to pay it back with 10 equal annual payments. What will the size of the payments be if the interest charge is 10% compounded (a) annually, (b) semiannually, and (c) continuously? (3.6.1)

46. Operating and maintenance costs for year k are given as $C_k = (e^{0.05})C_{k-1}$, $k = 2, 3, \ldots, 15$, with $C_1 = \$1000$. Determine the equivalent uniform annual operating and maintenance cost based on continuous compounding with a nominal interest rate of 10%. (3.6.1)

47. Four equal, quarterly deposits of $1000 each are made at $t = 0, 1, 2,$ and 3 (time periods are three-month intervals) into a fund that pays 8% compounded *continuously*. Then, at $t = 7$ and $t = 10$, withdrawals of size $\$A$ are made so that the fund is depleted at $t = 10$. What is the size of the withdrawals? (3.6.1)

48. Semiannual deposits of $500 are made into a fund paying 8% compounded continuously. What is the accumulated value in the fund after 10 such deposits? (3.6.1)

49. A firm buys a new computer that costs $100,000. It may either pay cash now or pay $25,000 down and $10,000/year for 10 years. If the firm can earn 5% on investments, which would you suggest? (3.7)

50. A firm brings out a small computer that it hopes to sell to small businesses. The computer sells for $10,000, but potential customers will need financing assistance in order to buy it. Three financing plans are being considered.
    (a) Pay $10,000 in cash one year after receiving the computer.
    (b) Pay $2275 a year for five years with the first payment made one year after receiving the computer.
    (c) Pay $1340 a year for 10 years with the first payment made two years after receiving the computer.
    What plan should a small business choose if it requires a 6% return on its investments? Why? (3.7)

51. It is desired to determine the size of the uniform series over the time period $[2, 6]$ that is equivalent to the cash flow profile shown below using an interest rate of 10%. (3.7)

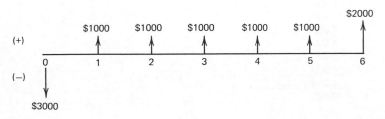

52. Given the following cash flows, what single sum at $t = 6$ is equivalent to the given data. Assume $i = 8\%$. (3.7)

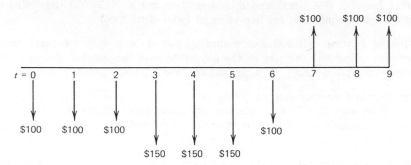

53. A machine is purchased at $t = 0$ for $20,000 (including installation costs). Net annual revenues resulting from operating the machine are $6000. The machine is sold at the end of 10 years, $t = 10$, for $2000. The cash flow series for the machine is equivalent to what single sum, X, at $t = 6$ if money is worth 10% compounded continuously? (3.7)

54. Assume the following two investments plans.
   (a) Purchase for $4386.68 (a negative cash flow) and receive (1) $400 at the end of each six months for four years ($t = 1, 2, \ldots, 8$), and (2) a single payment of $2000 at the end of the fourth year.
   (b) Purchase for $4091.14 and receive (1) $900 at the *beginning* of each year for the four-year period, and (2) a single payment of X at the end of the fourth year.
   If money is worth 6% compounded semiannually, what is the value of X so that the two investments are equivalent? (3.7)

55. A man invests $5000 and receives $500 each year for 10 years, at which time he sells out for $1000. With a 10% interest rate, what equal annual cost (profit) is equivalent to the venture? (3.7)

56. Consider the following cash flow series.

EOY	CF
0	−$10,000
1	3,000
2	3,500
3	4,000
4	4,500
5	5,000
6	5,500
7	6,000
8	6,500

At 10% annual compound interest, what uniform annual cash flow is equivalent to the above cash flow series? (3.7)

57. A person borrows $1000 from a bank at $t = 0$ at 8% simple interest for two years. He pays the total interest due for the two-year period at $t = 0$ and thus receives $840 at $t = 0$. If he pays back $1000 at $t = 2$, the person is, in effect, paying an interest rate of $X\%$ compounded annually. Solve for X. (3.7)

58. Assume payments of $2000, $5000, and $3000 are received at $t = 3$, 4, and 5, respectively. What five equal payments occurring at $t = 1$, 2, 3, 4, and 5, respectively, are equivalent if $i = 10\%$ compounded per period? (3.7)

59. What single deposit of size X into a fund paying 10% compounded annually is required at $t = 0$ in order to make withdrawals of $500 each at $t = 4$, 5, 6, and 7 and a single withdrawal of $1000 at $t = 20$?
If the above withdrawals are immediately placed into another fund paying 6% compounded annually, what amount will be accumulated in this fund at $t = 20$? (3.7)

60. A college student borrows $4000 and repays the loan with four quarterly payments of $400 during the first year and four quarterly payments of $1000 during the second year after receiving the $4000 loan. Determine the effective interest rate for the loan transaction. (3.7)

61. Quarterly deposits of $500 are made at $t = 1$, 2, 3, 4, 5, 6, 7. Then withdrawals of size A are made at $t = 12$, 13, 14, 15 and the fund is depleted with the last withdrawal. If the fund pays 6% compounded quarterly, what is the value of A? (3.7)

62. Determine the value of X so that the following cash flow series are equivalent at 8% interest. (3.7)

EOY	CF(A)	CF(B)
0	−$8,000	−$15,000
1	6,000	4,000
2	5,000	$3,000 + X$
3	4,000	$2,000 + 2X$
4	5,000	$3,000 + 3X$
5	6,000	$4,000 + 4X$
6	5,000	$3,000 + 5X$

63. Given the cash flow diagram shown below and an interest rate of 8% per period, solve for the value of an equivalent amount at (a) $t = 5$, (b) $t = 12$, and (c) $t = 15$. (*Note.* Upward arrows represent cash inflows and downward arrows represent cash outflows.)

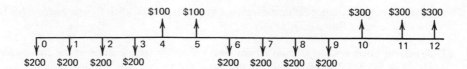

64. Given the cash flow profiles shown below, determine the value of X so that the two cash flow profiles are equivalent at a 10% interest rate (3.7)

EOY	CF(A)	CF(B)
1	−$12,000	−$10,000
2	1,000	7,000
3	3,000	$6,000 + 0.5X$
4	5,000	$5,000 + 1.0X$
5	7,000	$4,000 + 1.5X$
6	9,000	

65. What single sum of money at $t = 4$ is equivalent to the cash flow profile shown below? Use a 6% interest rate in your analysis. (3.7)

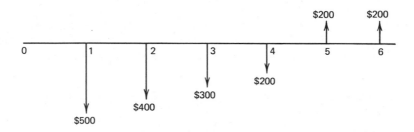

66. Given the two cash flow profiles shown below, for what value of X are the two series equivalent using an interest rate of 10%? Show all work. (3.7)

EOY	CF(A)	CF(B)
0	−$200,000	−$140,000
1	24,000	16,000
2	32,000	16,000
3	40,000	$16,000 + X$
4	48,000	$16,000 + 2X$
5	56,000	$16,000 + 3X$
6	64,000	$16,000 + 4X$
7	72,000	$16,000 + 5X$
8	80,000	$16,000 + 6X$

67. At what interest rate is $1000 today equivalent to $1967 in 10 years? (3.7)

68. Given the cash flow profiles shown below, determine the value of X such that the two cash flow profiles are equivalent at 8% compounded annually. (3.7)

EOY	CF(A)	CF(B)
1	−$12,000	$ − X
2	1,000	7,000
3	4,000	9,000
4	6,000	10,000
5	7,000	10,000
6	5,000	7,000

69. If a machine costs $10,000 and lasts for 10 years, at which time it is sold for $2000, what equal annual cost over its life is equivalent to these two cash flows with a 10% annual interest rate? (3.7)

70. Given the two cash flow profiles shown below, for what value of X are the two series equivalent using an interest rate of 10%? (3.7)

EOY	CF(A)	CF(B)
0	−$100,000	−$70,000
1	12,000	8,000
2	16,000	8,000
3	20,000	8,000
4	24,000	$8,000 + X$
5	28,000	$8,000 + 2X$
6	32,000	$8,000 + 3X$
7	36,000	$8,000 + 4X$
8	40,000	$8,000 + 5X$

71. Given the three cash flow profiles shown below, determine the values of X and Y so that all three cash flow profiles are equivalent at an annual interest rate of 8%. Show all work.

EOY	CF(A)	CF(B)	CF(C)
0	−$1000	−$2500	$ Y
1	X	3000	Y
2	1.5X	2500	Y
3	2.0X	2000	2Y
4	2.5X	1500	2Y
5	3.0X	1000	2Y

72. An investment opportunity offered by commercial banks is certificates of deposit. For example, a certificate of deposit (C.D.) having a four-year maturity date may have a stated interest rate of "7 1/2%, payable quarterly." This statement means that every three months, the issuing bank would pay the C.D. holder an amount of $(0.075)(1/4)$ times the face value (purchase price) of the C.D. Assuming a $5000, four-year, 7 1/2% C.D. were purchased at $t = 0$, quarterly payments of $[(0.075)(0.25)(\$5000)] = \$93.75$ would be received at $t = 1, 2, 3, \ldots , 16$ and, at $t = 16$, the C.D. would be redeemed by the bank for the original purchase price, or $5000. (This payment series is an application of simple interest on a quarterly basis).

It is also possible for the purchaser of the C.D. to arrange with the bank for each interest payment to be deposited into a regular savings account that pays, say, 5 1/2% compounded quarterly. Let it be assumed that this is done; the first deposit of $93.75 occurs at $t = 1$, and the last deposit is made at $t = 16$.

(a) For the initial investment of $5000, what total sum of money will be received by the purchaser at the end of the fourth year (C.D. redemption value plus savings account balance)?

(b) Now consider only the initial investment, $P = \$5000$, and the F value calculated from (a). What effective annual compound interest rate relates the two single sums of money?

(c) For P and F from (b) above, what annual simple interest rate relates the two single sums of money? (3.7)

73. Assume the following two investment plans:
Plan A—Purchase for $4386 and receive the following:
      (1) $400 at the end of each six months for four years.
      (2) A single payment of $2000 at the end of the fourth year.
Plan B—Purchase for $3302 and receive the following:
      (1) $900 at the beginning of each year for four years.
      (2) A single payment of X at the end of the fourth year.
If money is worth 6% compounded semiannually, what is the value of X such that the two plans are equivalent? (3.7)

74. A person borrows $5000 at 6% compounded annually. Five equal annual payments are used to repay the loan, with the first payment occurring one year after receiving the $5000. Determine the interest amount included in each payment. (3.8.1)

75. An individual borrows $8000 at 7% compounded annually. Six equal annual payments are used to repay the loan, with the first payment occurring two years after receiving the $8000. Determine the payment on principal included in each payment. (3.8.1)

76. A man borrows $10,000 at 8% compounded monthly. He is to pay off the loan with 60 monthly payments. One month after making the thirtieth payment, he elects to pay off the unpaid balance on the note. How much should he repay? (3.8.1)

77. An individual makes five annual deposits of $1000 in a savings account that pays interest at a rate of 4% compounded annually. One year after making the last deposit, the interest rate changes to 5% compounded annually. Five years after the last deposit the accumulated money is withdrawn from the account. How much is withdrawn? (3.8.2)

78. A person deposits $10,000 in a savings account paying 5% compounded annually for the first two years and 6% compounded annually for the next two years. Four annual withdrawals are made from the savings account. The size of the withdrawal increases by $1000 per year, with the first withdrawal occurring one year after the deposit. Determine the size of the last withdrawal, which depletes the balance of the account. (3.8.2)

79. Assume that six equal deposits of $500 are made at $t = 1$, 2, 3, 4, 5, and 6 (three-month periods) into a fund paying *6% compounded quarterly*. This interest applies until $t = 16$. The accumulated sum is withdrawn at this time and immediately deposited into a fund paying *8% compounded continuously*. Beginning with $t = 17$, determine the amount of three equal payments, A, that may be withdrawn such that the fund is depleted at $t = 19$. (3.8.2)

80. A man borrows $10,000 and repays the loan with four equal annual payments. The interest rate for the first two years of the loan is 5% compounded continuously, and for the third and fourth years of the loan it is 6% compounded continuously. Determine the size of the annual payment. (3.8.2)

81. A person deposits $5000 in a savings account. One year after the initial deposit, $1000 is withdrawn. Two years after the first withdrawal, $4000 is deposited in the account. Three years after the second deposit, $2000 is withdrawn from the account. Four years after the second withdrawal, all funds are withdrawn from the account. During the period of time the savings account was in use, the bank paid 4% compounded continuously for the first two years, 5% compounded annually for the next four years, and 6% compounded quarterly for the remainder of the time the account was in use. Determine the amount of the final withdrawal. (3.8.2)

82. A man borrows $5000 and repays the loan with three equal annual payments. The interest rate for the first year of the loan is 5% compounded annually, for the second year of the loan is 6% compounded annually, and for the third year of the loan is 7% compounded annually. Determine the size of the equal annual payment. (3.8.2)

83. A person deposits $1000 in a fund that pays interest at a continuous compound rate of r_t for the tth year after the initial deposit, where

$$r_t = \begin{cases} 0.04 + 0.005t, & t = 1, 2, 3, 4, 5, 6 \\ 0.07 & t = 7, 8, 9, 10, \ldots \end{cases}$$

Thus, a continuous compound rate of 4 1/2% is earned during the first year, 5% is earned during the second year, 5 1/2% is earned during the third year, etc. A maximum of 7% compounded continuously is earned during the sixth and each successive year. If the person withdraws $500 two years after the initial deposit, how much money will be in the fund six years after the initial deposit? (3.8.2)

84. A person deposits $2500 in a savings account that pays interest at a rate of 5% compounded annually. Two years after the deposit, the savings account begins paying interest at a rate of 5% compounded continuously. Five years after the deposit, the savings account begins paying interest at a rate of 6% compounded semiannually.
   (a) How much money should be in the savings account 10 years after the initial deposit?
   (b) What annual compound interest rate is equivalent to the interest pattern of the savings account over the 10-year period? (3.8.2)

85. A man deposits $1000 in a fund each year for a 10-year period. The fund initially pays 4% compounded annually. Immediately after the man makes his sixth deposit, the fund begins paying 5% compounded annually. The man removes his money from the fund three years after his last deposit. How much should he be able to withdraw at that time? (3.8.2)

86. Based on discrete cash flows and discrete compounding, compute the present worth and annual worth for the following situations: (3.8.2)

EOP t	CF	Interest Rate During Period
0	−$10,000	0
1	2,000	0.05
2	4,000	0.06
3	6,000	0.07
4	8,000	0.08
5	10,000	0.09

87. Solve problem 86 for the case of continuous compounding with the interest rates shown interpreted as nominal rates. (3.8.2)

88. A person deposits $2000 in a savings account that pays 7% compounded annually; two years after the deposit, the interest rate increases to 8% compounded annually. A second deposit of $2000 is made immediately after the interest rate changes to 8%. How much will be in the fund five years after the second deposit? (3.8.2)

89. An individual makes monthly deposits of $100 in a savings account that pays interest at a rate equivalent to 6% compounded quarterly. How much money should be in the account immediately after the sixtieth deposit? If no interest is earned on money deposited during a quarter and the first deposit coincides with the beginning of a quarter, what will be the account balance immediately after the sixtieth deposit? (3.8.3)

90. An individual makes semiannual deposits of $500 into an account that pays interest equivalent to 7% compounded quarterly. Determine the account balance immediately *before* the tenth deposit. (3.8.3)

91. A rental contract consists of annual charges payable in advance. The first charge is $1000 at $t = 0$ and then decreases by $200 each year. If five annual payments are made and money is worth 15% compounded annually, what is the present worth at $t = 0$ of all payments by using the uniform gradient series formula? (3.8.3)

92. If a fund pays 6% compounded annually, what deposit is required today such that $1000 can be withdrawn every five years forever? (Ignore any tax considerations.) (3.8.4)

93. Maintenance on a reservoir is cyclic with the following *costs* occurring over a five-year period: $3000, $2000, $5000, $0, and $1000. It is anticipated that the sequence will repeat itself every five years forever. Determine the capitalized cost for the maintenance costs based on a time value of money of 8%. (3.8.4)

94. A bond is purchased for $900 and kept for 10 years, at which time it matures at a face value of $1000. During the 10-year period $60 is received every six months (i.e., 20 receipts of $60 each). What is the rate of return for the investment? (3.8.5)

95. A person buys a $2000 bond for $1800. The bond has a bond rate of 8% with bond premiums paid annually. The bond has a life of 10 years. Determine the equivalent annual return (rate of return) for the bond investment. (3.8.5)

96. A person is considering purchasing a bond having a face value of $2500 and a bond rate of 8% payable semiannually. The bond has a life of 15 years. How much should be paid for the bond in order to earn a rate of return of 10% compounded semiannually? (3.8.5)

97. A person wishes to sell a bond that has a face value of $2000. The bond has a bond rate of 8% with bond premiums paid annually. Four years ago, $1800 was paid for the bond. At least a 10% return on investment is desired. What must be the minimum selling price for the bond in order to make the desired return on investment? (3.8.5)

98. A $5000, 10-year, 8% semiannual bond is purchased at $t = 0$ by Mr. Rich for par value. After receiving the twelfth dividend, Mr. Rich sells the bond at a price to yield a 6% nominal rate of return on his original purchase price.
(a) What was the selling price?
(b) If Mr. Rich keeps the bond until maturity and redeems it for par value, what approximate nominal rate of return will he have earned? (3.8.5)

99. The following labor *costs* are anticipated over a five-year period: $7000, $8000, $10,000, $12,000, and $15,000. It is estimated that a 6% inflation rate will apply over the time period in question. The labor costs given above are expressed in then-current dollars. The time value of money, excluding inflation, is estimated to be 5%. Determine the present worth equivalent for labor cost. (3.8.6)

100. Labor costs over a four-year period have been forecast in then-current dollars as follows: $10,000, $12,000, $15,000, and $17,500. The inflation rates for the four years are forecast to be 8, 9, 10, and 10%; the interest rate (excluding inflation) is anticipated to be 6, 6, 5, and 5% over the four-year period. Determine the present worth equivalent for labor cost. (3.8.6)

101. Determine the capital recovery cost for an investment of $100,000 over a 10 year period, with a salvage value of $20,000. Use an interest rate of 20%. (3.8.7)

102. Determine the capital recovery cost for an investment of $75,000 over a 5 year period, with a salvage value of $-\$25,000$. Use an interest rate of 10%. (3.8.7)

CHAPTER FOUR
COMPARISON OF ALTERNATIVES

4.1 INTRODUCTION

Chapter One recommended that engineers and systems analysts solve problems by formulating and analyzing the problem, generating a number of feasible solutions (alternatives) to the problem, comparing the investment alternatives, selecting the preferred solution, and implementing the solution. The process of evaluating the alternatives and selecting the preferred solution is the subject of the remainder of this book.

In this chapter we apply the time value of money concept to the comparison of economic investment alternatives. Although multiple objectives are often involved in performing a comparison of alternatives, for now we concentrate on the comparison of *mutually exclusive* alternatives on the basis of monetary considerations alone. Mutually exclusive alternatives means that no more than one alternative can be chosen. The adage of "not being able to have one's cake and eat it too" illustrates the notion of mutually exclusive alternatives.

A systematic approach that can be used in economic investment alternatives is summarized as follows:

1. Define the set of feasible, mutually exclusive economic investment alternatives to be compared.

2. Define the planning horizon to be used in the economic study.

3. Develop the cash flow profiles for each alternative.

4. Specify the time value of money to be used.

5. Specify the measure(s) of merit or effectiveness to be used.

6. Compare the alternatives using the measure(s) of merit or effectiveness.

7. Perform supplementary analyses.

8. Select the preferred alternative.

The procedures for comparing investment alternatives outlined in this chapter are intended to aid in making better measurements of the *quantitative aspects* of capital investment alternatives. *It cannot be too strongly emphasized that no economic evaluation can replace the sound judgment of experienced managers concerned with both the quantitative and nonquantitative aspects of investment alternatives.* Typical of the aspects of alternatives not considered in this chapter are safety, personnel considerations, product quality, environmental effects, and engineering and construction capability. Such factors are relevant and often control decisions on capital expenditures. However, our concern is with the monetary aspects of the alternatives.

We interpret our role to be the development of logical approaches to be used in *analyzing* investment alternatives and *recommending* action to be taken, based on economic considerations alone. The process of *deciding* which alternative to choose for implementation involves the consideration of monetary and nonmonetary factors. Approaches that can be used to assimilate multiple objectives are treated in Chapter Eight.

4.2 DEFINING MUTUALLY EXCLUSIVE ALTERNATIVES

An individual alternative selected from a set of mutually exclusive alternatives can be made up of several *investment proposals*. Investment proposals are distinguished from investment alternatives by Thuesen, et al. [38] by noting that investment alternatives are decision options; investment proposals are single projects or undertakings that are being considered as investment possibilities.

Example 4.1 —————————————————————————————————

As an illustration of the distinction between investment proposals and investment alternatives, consider a distribution center that receives pallet loads of product, stores the product, and ships pallet loads of product to various customer locations. A new distribution center is to be constructed, and the following proposals have been made:

1. Method of moving materials from receiving to storage and from storage to shipping.
   a. Conventional lift trucks for operating in 12-foot aisles.
   b. Narrow aisle lift trucks for operating in 5-foot aisles.
   c. Driverless tractor system.
   d. Towline conveyor system.
   e. Pallet conveyor system.
2. Method of placing materials in and removing materials from storage.
   a. Conventional lift trucks for operating in 12-foot aisles.
   b. Narrow aisle lift trucks for operating in 5-foot aisles.
   c. Narrow aisle, operator driven, rail guided storage/retrieval vehicle.
   d. Narrow aisle, automated, rail guided storage/retrieval vehicle.
3. Method of storing materials.
   a. Stacking pallet loads of material (8 feet high, 12-foot aisles).
   b. Conventional pallet rack (20 feet high, 12-foot aisles).
   c. Flow rack (20 feet high, 12-foot aisles).
   d. Narrow aisle, pallet rack (20 feet high, 5-foot aisles).
   e. Flow rack (20 feet high, 5-foot aisles).
   f. Medium height, pallet rack (35 feet high, 5-foot aisles).
   g. High rise, pallet rack (70 feet high, 5-foot aisles).

Given the set of proposals, alternative designs for the material handling system can be obtained by combining a proposed method of moving materials from receiving to storage, a proposed method of placing materials in storage, a proposed method of storage, a proposed method of removing materials from storage, and a proposed method of transporting materials from storage to shipping. Some of the combinations of proposals will be eliminated because of their incompatibility. For example, lift trucks requiring 12-foot aisles cannot be used to place materials in and remove materials from storage when 5-foot aisles are used. Other combinations might be eliminated because of budget limitations; a desire to minimize the variation in types of equipment due to maintainability, availability, reliability, flexibility, and operability considerations; ceiling height limitations; physical characteristics of the product (crushable product might require the use of storage racks); and a host of other considerations. Characteristically, experience and judgment are used to trim the list of possible combinations to a manageable number.

Example 4.2

To illustrate the formation of mutually exclusive investment alternatives from a set of investment proposals, consider a situation involving m investment

TABLE 4.1 Developing Mutually Exclusive Investment
Alternatives from Investment Proposals

Alternative	x_1	x_2	x_3	Explanation
1	0	0	0	Do nothing (proposals 1, 2, and 3 not included
2	0	0	1	Accept proposal 3 only
3	0	1	0	Accept proposal 2 only
4	0	1	1	Accept proposals 2 and 3 only
5	1	0	0	Accept proposal 1 only
6	1	0	1	Accept proposals 1 and 3 only
7	1	1	0	Accept proposals 1 and 2 only
8	1	1	1	Accept all three proposals

proposals, let x_j be defined to be 0 if proposal j is not included in an alternative and let x_j be defined to be 1 if proposal j is included in an alternative. Using the binary variable x_j we can form 2^m mutually exclusive alternatives. Thus, if there are three investment proposals, we can form eight mutually exclusive investment alternatives, as depicted in Table 4.1.

As pointed out in the discussion of the design of a material handling system for the distribution center, among the alternatives formed, some might not be feasible, depending on the restrictions or constraints placed on the problem. To illustrate, there might be a budget limitation that precludes the possibility of combining all three proposals; thus, Alternative 8 would be eliminated. Additionally, some of the proposals might be *mutually exclusive proposals*. For example, Proposals 1 and 2 might be alternative computer designs and only one is to be selected; in this case Alternative 7 would be eliminated from consideration. Other proposals might be *contingent proposals* so that one proposal cannot be selected unless another proposal is also selected. As an illustration of a contingent proposal, Proposal 3 might involve the procurement of computer terminals, which depend on the selection of the computer design associated with Proposal 2. In such a situation, Alternatives 2 and 6 would be infeasible. Thus, depending on the restrictions present, the number of feasible mutually exclusive alternatives that result can be considerably less than 2^m.

In many organizations there is a rather formalized hierarchy for determining how the organization will invest its funds. Typically, the entry point in this hierarchy involves an individual analyst or engineer who is given an assignment to solve a problem; the problem may be one requiring the design of a new product, the improvement of an existing manufacturing process, or the development of an improved system for performing a service. The individual performs the steps involved in the problem-solving procedure and recommends the preferred solution to the problem. In arriving at the preferred solution, a number of alternative solutions are normally compared; hopefully, the eight-step approach for comparing economic investment alternatives was followed!

The preferred solution is usually forwarded to the next level of the hierarchy for approval. In fact, one would expect many preferred solutions to various problems to be forwarded to the second level of the hierarchy for approval. Each preferred solution becomes an investment *proposal*, the resulting set of mutually exclusive investment alternatives are formed, and the process of comparing economic investment alternatives is repeated. This sequence of operations is usually performed in various forms at each level of the hierarchy until, ultimately, the preferred solution by the individual analyst or engineer is accepted or rejected. In this textbook we concentrate on the process of comparing investment alternatives at the first level of the hierarchy; however, the need for such comparisons at many levels of the organization should be kept in mind.

4.3 DEFINING THE PLANNING HORIZON

In comparing investment alternatives, it is important to compare them over a common period of time. We define that period of time to be the *planning horizon*. In the case of investments in, say, equipment to perform a required service, the period of time over which the service is required might be used as the planning horizon. Likewise, in one-shot investment alternatives the period of time over which receipts continue to occur might define the planning horizon.

In a sense, the planning horizon defines the width of a "window" that is used to view the cash flows generated by an alternative. In order to make an objective evaluation, the same window must be used in viewing each alternative.

In some cases the planning horizon is easily determined; in other cases the duration of one or more projects is sufficiently uncertain to cause concern over the time period to use. Some commonly used methods for determining the planning horizon to use in economy studies include:

1. Least common multiple of lives for the set of feasible, mutually exclusive alternatives, denoted $\hat{T}$.
2. Shortest project life among the alternatives, denoted T_s.
3. Longest project life among the alternatives, denoted T_l.
4. Some other period of time less than $\hat{T}$.
5. Some other period of time greater than $\hat{T}$.

In the economic analysis literature the most commonly used method of selecting the planning horizon appears to be the least common multiple of lives approach. In most cases, such a selection is made implicitly, not explicitly. Using such a procedure when three alternatives are being considered and the individual lives are six years, seven years, and eight years yields a planning horizon of $\hat{T} = 168$ years. If the lives had been either six years, six years, and eight years or six years, eight years, and eight years, $\hat{T} = 24$ years. Clearly, strict reliance on $\hat{T}$ as the planning horizon is not advisable.

If the shortest project life is used to define the planning horizon, estimates are required for the values of the unused portions of the lives of the remaining alternatives. Thus, for the situation considered above, with $T_s = 6$ years, the salvage or residual values at the end of six years' use must be assessed for the other two alternatives.

If the longest project life, T_l, is used in determining the planning horizon, some difficult decisions must be made concerning the period of time between T_s and T_l. If the alternative selected is to provide a necessary service, that service must continue throughout the planning horizon, regardless of the alternative selected. Consequently, when the shortest life alternative reaches the end of its project life, it must be replaced with some other asset capable of performing the required service. However, since technological developments will probably take place during the period of time T_s, new and improved candidates will be available for selection at time T_s. Thus, the specification of the cash flows for the shortest life alternative during the period of time from T_s to T_l is a difficult undertaking. As a result, T_l is seldom used as the planning horizon.

A number of organizations have adopted a standard planning horizon for all economic alternatives. Letting T denote the planning horizon specified by the organization, different approaches are recommended, depending on whether $T < \hat{T}$ or $T \geq \hat{T}$. If the planning horizon selected is less than the least common multiple of lives, the cash flows for each alternative must be provided for a period of time equal to the planning horizon. When the planning horizon is greater than or equal to the least common multiple of lives, it is recommended that the economic analysis be based on a period of time equal to the least common multiple of lives. The reason for the latter recommendation is that at time $\hat{T}$ a new economic analysis can be performed based on the alternatives available at that time. After $\hat{T}$ years new alternatives might be available; furthermore, one can more accurately estimate the values of cash flows occurring after $\hat{T}$ if one waits until nearer time $\hat{T}$ to make the estimates.

Example 4.3 ────────────────────────────────

To illustrate the difficulties associated with the selection of the planning horizon, consider the two cash flow profiles given in Table 4.2. Alternatives A

TABLE 4.2 Cash Flow Profiles for Two Mutually Exclusive Investment Alternatives Having Unequal Lives

Alternative A		Alternative B	
EOY	CF(A)	EOY	CF(B)
0	−$5000	0	−$6000
1–4	− 3000	1–6	− 2000

and B have anticipated lives of four years and six years, respectively.

Using a least common multiple of lives approach, a planning horizon of $\hat{T} = 12$ years would be used. Using a 12-year planning horizon requires answers to the following questions. What cash flows are anticipated for years 5 to 12 if Alternative A is selected. What will be the cash flows for Alternative B for years 7 to 12?

As shown in Table 4.3, the traditional approach is to assume that Alternative A will be repeated twice and Alternative B will be repeated once and that identical cash flows occur during these repeating life cycles. Inflation effects, as well as the possibility of technological improvements, tend to invalidate such assumptions.

TABLE 4.3 Cash Flow Profiles for Various Planning Horizons

$T = \hat{T} = 12$			$T = T_s = 4$			$T = 10$		
EOY	CF(A)	CF(B)	EOY	CF(A)	CF(B)	EOY	CF(A)	CF(B)
0	−$5000	−$6000	0	−$5000	−$6000	0	−$5000	−$6000
1	− 3000	− 2000	1	− 3000	− 2000	1	− 3000	− 2000
2	− 3000	− 2000	2	− 3000	− 2000	2	− 3000	− 2000
3	− 3000	− 2000	3	− 3000	− 2000	3	− 3000	− 2000
4	− 8000	− 2000	4	− 3000	− 2000	4	− 8000	− 2000
5	− 3000	− 2000			+ 2000*	5	− 3000	− 2000
6	− 3000	− 8000		$T = T_1 = 6$		6	− 3000	− 8000
7	− 3000	− 2000				7	− 3000	− 2000
8	− 8000	− 2000	EOY	CF(A)	CF(B)	8	− 8000	− 2000
9	− 3000	− 2000				9	− 3000	− 2000
10	− 3000	− 2000	0	−$5000	−$6000	10	− 3000	− 2000
11	− 3000	− 2000	1	− 3000	− 2000		+ 2500*	+ 2000*
12	− 3000	− 2000	2	− 3000	− 2000			
			3	− 3000	− 2000			
			4	− 8000	− 2000			
			5	− 3000	− 2000			
			6	− 3000	− 2000			
				+ 2500*				

If the shortest life approach is used, a planning horizon of four years would be used. In such a case, an estimate of the salvage value of Alternative B should be indicated at the end of year 4, as denoted in Table 4.3 by an asterisk.

Using the longest life approach yields a six-year planning horizon. In this instance a decision must be made concerning the cash flows in years 5 and 6 for Alternative A. If an initial investment must be made to provide the required service for years 5 and 6, it would occur at the end of year 4. The assumption made in Table 4.3 is that Alternative A will be repeated with identical cash flows, and a terminal salvage value of $2500 will apply after two years of use.

Suppose a standard planning horizon of 10 years must be used. The same questions that arose in the cases of $\hat{T}$, T_s, or T_l being the planning horizon apply when a 10-year planning horizon is used. As depicted in Table 4.3, one might assume that identical life cycles will be repeated until the end of the planning horizon and provide estimates of terminal salvage values at that time.

Although the use of a standard planning horizon has the benefit of a consistent approach in comparing investment alternatives, there are also some dangers that should be recognized. In some cases the major benefits associated with an alternative might occur in the later stages of its project life. If the planning horizon is less than the project life, such alternatives would seldom be accepted. Just such a practice caused one major textile firm to lose its strong position in the industry. A major modernization of the processing departments had been proposed, but its benefits would not be realized until the bugs had been worked out of the new system, all personnel were trained under the new system, and the marketing people had regained the lost customers. Unfortunately, the planning horizon specified by the firm was too short in duration and the modernization plan was not approved.

For the example, a decision concerning the planning horizon to use should depend on the particular situation instead of on the durations of the individual alternatives. If, for instance, Alternatives A and B are two different lift truck designs and it is anticipated that the material handling function to be performed by the lift truck will continue for at least 12 years, the least common multiple of lives approach might make sense. Likewise, since the lift truck industry is quite dynamic and technological improvements are quite likely to occur in the future, we might prefer to use the shortest life as the planning horizon.

Example 4.4

As a second illustration of the planning horizon selection process, consider the two cash flow diagrams given in Figure 4.1. The two alternatives are

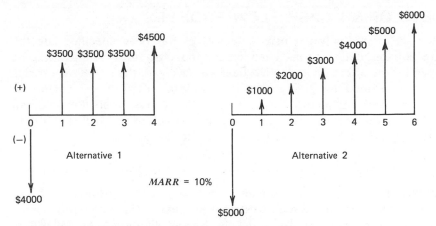

FIGURE 4.1. Cash flow diagrams for the example problem.

mutually exclusive, one-shot investment alternatives. We are unable to predict what investment alternatives will be available in the future, but we do anticipate that recovered capital can be reinvested and earn a 10% return.

For this type situation, a six year planning horizon is suggested, with zero cash flows occurring in years 5 and 6 with Alternative 1. At the end of six years the net future worths for the two alternatives will be

$$FW_1(10\%) = \$4500(F|P\ 10,2) + \$3500(P|A\ 10,3)(F|P\ 10,6)$$
$$-\$4000(F|P\ 10,6)$$
$$= \$13,778.87$$

$$FW_2(10\%) = \$1000(F|A\ 10,6) + \$1000(A|G\ 10,6)(F|A\ 10,6)$$
$$-\$5000(F|P\ 10,6)$$
$$= \$16,014.01$$

Thus, we would recommend Alternative 2.

If one did not give careful thought to the situation involved and blindly assumed a least common multiple of lives planning horizon, with identical cash flows in repeating life cycles, then Alternative 1 would be recommended. Hence, it is important to consider the particular situation involved and specify the planning horizon instead of employing a rule of thumb for establishing planning horizons that does not consider the nature of the investments.

It appears that the preferred approach would be to have a "flexible" standard planning horizon. All of the "routine" economic analyses would be based on the standard planning horizon of, say, 5 to 10 years; nonroutine economic analyses would be based on a planning horizon that was appropriate for the situation.

4.4 DEVELOPING CASH FLOW PROFILES

Once the set of mutually exclusive alternatives has been specified and the planning horizon decision has been made, the cash flow profiles can be developed for the alternatives. As has been emphasized, the cash flow profiles should be developed by giving careful consideration to *future* conditions instead of relying completely on *past* cash flows. The cash flows for an investment alternative are obtained by aggregating the cash flows for all investment proposals included in the investment alternative.

Example 4.5 —————————————————————————————————

To illustrate the approach to be taken, suppose a planning horizon of five years is used and there are three investment proposals. Cash flow profiles for the proposals are given in Table 4.4. A budget limitation of $50,000 is available for investment among the proposals. Proposal 2 is contingent on Proposal 1, and Proposals 1 and 3 are mutually exclusive. Based on the restrictions associated with the combinations of proposals, only four investment alternatives are to be considered. Alternative A is the do nothing alternative; Alternative B involves Proposal 1 alone; Alternative C involves a combination of Proposals 1 and 2; and Alternative D involves Proposal 3 alone. The cash flow profiles for the four alternatives are given in Table 4.5.

TABLE 4.4 Cash Flow Profiles for Three
Investment Proposals

EOY	CF(1)	CF(2)	CF(3)
0	−$20,000	−$30,000	−$50,000
1	− 4,000	+ 4,000	− 5,000
2	+ 2,000	+ 6,000	+ 10,000
3	+ 8,000	+ 8,000	+ 25,000
4	+ 14,000	+ 10,000	+ 40,000
5	+ 25,000	+ 20,000	+ 10,000

TABLE 4.5 Cash Flow Profiles for Four Mutually
Exclusive Investment Alternatives

EOY	CF(A)	CF(B)	CF(C)	CF(D)
0	0	−$20,000	−$50,000	−$50,000
1	0	− 4,000	0	− 5,000
2	0	+ 2,000	+ 8,000	+ 10,000
3	0	+ 8,000	+ 16,000	+ 25,000
4	0	+ 14,000	+ 24,000	+ 40,000
5	0	+ 25,000	+ 45,000	+ 10,000

The do nothing alternative is the status quo condition and serves as the base against which other alternatives are considered. In some cases the do nothing alternative is not feasible (e.g., failing to comply with pollution standards). Moreover, the do nothing alternative does not necessarily have zero cash flows associated with it. In principle, one should forecast the cash flows that will result if the present method is continued and compare the cash flows with those associated with other alternatives.

In most, if not all, economic evaluations, it is not necessary to develop a detailed forecast of all items of cost, revenue, and investment associated with an alternative. Costs and revenues that will be the same regardless of the alternative selected can be omitted. If a cost reduction alternative will not affect sales revenues, no forecast of such revenues need be developed. Attention is focused on the items of cost and revenue that will be affected by the alternative selected.

Since economic analyses are to be performed to judge the merits of investment alternatives for the future, the data required differ from those normally provided by the accounting system. As indicated in Chapter Two, the fundamental and historical purpose of accounting is to maintain a consistent *historical* record of the financial results of the operations of the organization. Accounting figures are based on definitions derived consistent with this objective. Accounting methods are not designed to determine the economic worth of alternative courses of action.

4.5 SPECIFYING THE TIME VALUE OF MONEY

An important step in evaluating investment alternatives involves the specification of the interest or discount rate to be used. Even though a project may be financed entirely from internal sources of funds, an interest rate is recommended in evaluating investment alternatives. One reason for doing so is to reflect the cost of investing money in a particular project instead of investing it elsewhere and earning a return on the investment. The cost of foregoing other investment opportunities is referred to as the *opportunity cost*, as discussed in Chapter Two.

Except where other intangible benefits are involved, the discount rate should be greater than the cost of securing additional capital. Indeed, it should be greater than the cost of capital by an amount that will cover unprofitable investments that a firm must make for nonmonetary reasons. Examples of the latter would include investments in antipollution equipment, safety devices, and recreational facilities for employees.

The discount rate that is specified establishes the firm's *minimum attractive rate of return* (MARR) in order for an investment alternative to be justified. If the present worth for an investment alternative were negative, indicating that a negative cash flow was equivalent to the investment alternative, it would not be recommended for adoption.

Some firms establish a standard discount rate or minimum attractive rate of return to be used in all economy studies; others maintain a flexible posture. As one firm described it,

> The XYZ Company return on investment (ratio of earnings to gross investment) has been, on the average over the past five years, approximately equal to a rate-of-return of 15% per year. Accordingly, 15% per year is being established as a tentative *minimum requirement for investment alternatives* whose results are primarily measurable in quantitative dollar terms. The principle being applied is that such alternatives should be expected to maintain or to improve overall return-on-investment performance for the XYZ Company. The minimum requirement is based on overall XYZ financial results (as opposed to results for various parts of the Company) in order to avoid situations in which investment alternatives of a given level of attractiveness are unknowingly undertaken in one part of the Company and rejected in another.
>
> The 15% minimum attractive rate-of-return standard is intended as a guide, rather than as a hard-and-fast decision rule. Furthermore, it is intended to apply to alternatives having risks of the kind usually associated with investments which are primarily in plant and facilities. For alternatives involving expenditures with substantially lower risks, such as those solely for inventories or those in buy-or-lease alternatives, a lower minimum requirement may apply.

Other approaches that are used to establish the *MARR* include:

1. Add a fixed percentage to the firm's cost of capital (see the discussion in Chapter Two).
2. Average rate of return over the past five years is used as this year's *MARR*.
3. Use different *MARR* for different planning horizons.
4. Use different *MARR* for different magnitudes of initial investment.
5. Use different *MARR* for new ventures than for cost improvement projects.
6. Use as a management tool to stimulate or discourage capital investments, depending on the overall economic condition of the firm.
7. Use the average stockholder's return on investment for all companies in the same industry group.

There are a number of different approaches used by companies in establishing the discount rate to be used in performing economic analyses. The issue is not a simple one.

The proper determination of the discount rate has been the subject of considerable controversy in the economic analysis literature for many years.

In some ways we are no closer to an agreement today than we were 20 years ago. For this reason, among others, step 7 (perform supplementary analyses), is incorporated in the eight-step procedure for comparing economic alternatives. In many cases, a particular alternative will be preferred over a range of possible discount rates; in other cases, the alternative preferred will be quite sensitive to the discount rate used. Hence, depending on the situation under study, it might not be necessary to specify a particular value for the discount rate—a range of possible values might suffice. We examine this situation in more depth in Chapter Seven.

Subsequently, it will be convenient to refer to the interest rate or discount rate used as the minimum attractive rate of return (*MARR*) and to interpret its value using the opportunity cost concept. The argument will be made that money should not be invested in an alternative if it cannot earn a return at least as great as the *MARR*, since it is reasoned that other opportunities for investment exist that will yield returns equal to the *MARR*.

In the case of the public sector, a different interpretation is required in determining the discount rate to use. Since this chapter emphasizes economic analysis in the private sector, Chapter Six discusses establishing the discount rate for the public sector.

4.6 SPECIFYING THE MEASURE OF MERIT

As noted in Chapter Three in discussing equivalence, investment alternatives can be compared in a number of ways. Present worth (*PW*) and annual worth (*AW*) comparisons are two commonly used approaches. Among the several methods of comparing investment alternatives are:

1. Present worth method.
2. Annual worth method.
3. Future worth method.
4. Payback period method.
5. Rate-of-return method.
6. Savings/investment ratio method.

Each of the above measures of merit or measures of effectiveness has been used numerous times in comparing real-world investment alternatives. They may be described briefly as follows:

1. Present worth method converts all cash flows to a single sum equivalent at time zero.
2. Annual worth method converts all cash flows to an equivalent uniform annual series of cash flows over the planning horizon.

3. Future worth method converts all cash flows to a single sum equivalent at the end of the planning horizon.

4. Payback period method determines how long at a zero interest rate it will take to recover the initial investment.

5. Rate of return method determines the interest rate that yields a future worth of zero.

6. Savings/investment ratio method determines the ratio of the present worth of savings to the present worth of the investment.

With the exception of the payback period method, all of the measures of merit listed are equivalent methods of comparing investment alternatives. Hence, applying each of the measures of merit to the same set of investment alternatives will yield the same recommendation (with the possible exception of the payback period method).

Since the present worth, annual worth, future worth, rate-of-return, and savings/investment ratio methods are equivalent, why does more than one of the methods exist? The primary reason for having different, but equivalent, measures of effectiveness for economic alternatives appears to be the differences in preferences among managers. Some individuals (and firms) prefer to express the net economic worth of an investment alternative as a single sum amount; hence, either the present worth method or the future worth method is used. Other individuals prefer to see the net economic worth spread out uniformly over the planning horizon, so the annual worth method is used by them. Yet another group of individuals wishes to express the net economic worth as a rate or percentage; consequently, the rate-of-return method would be preferred. Finally, some individuals prefer to see the net economic worth expressed as a percentage of the investment required; the savings/investment ratio is one method of providing such information.

Since many organizations have established procedures for performing economic analyses, it seems worthwhile to consider in this chapter the more popular measures of merit that are used. Among those listed, it appears that the present worth, rate-of-return, and payback period methods are currently the most popular. However, the U.S. Postal Service and a number of governmental agencies have recently adopted some version of the savings/investment (or benefit/cost) ratio method for purposes of comparing investment alternatives; hence, it is gaining in popularity.

Depending on the organization, specially designed forms are often provided for aiding the analyst in conducting the analysis. Sample forms used by one major industrial organization are provided in Figures 4.2 to 4.4. This particular company uses the rate-of-return method; they refer to it as the discounted cash flow rate-of-return method.

NET CASH FLOW SCHEDULE

ALTERNATE **A** OR **B** (CIRCLE ONE)

TITLE (PROJECT) _____

TITLE (THIS ALTERNATE) _____

PROJECT NO. _____   DATE _____

LINE	EXPENSES *	0	1	2	3	4	PERIOD 5	6	7	8	TOTAL
1	NET BOOK VALUE (BEGINNING OF YEAR)	✕									
2	DEPRECIATION (___ YEARS LIFE)										
3	RENTAL / LEASE COST										
4	IN-PLANT LABOR WITH FRINGE BENEFITS @ ___ %										
5	PURCHASED LABOR										
6	REWORK / SCRAP										
7	TOOLING / ANCILLARY EQUIPMENT										
8	MAINTENANCE & REPAIR										
9	POWER & FUEL / UTILITIES EXPENSE										
10	MATERIALS & SUPPLIES										
11	PRODUCT SHIPPING COSTS										
12	PERSONAL PROPERTY TAX										
13	REAL ESTATE TAX										
14	IMPLEMENTING COST										
15											
16											
17	TOTAL CASH EXPENSES (LINES 2 THRU 16)										
18	AFTER TAX ANNUAL COST ** (50% x LINE 17)										
19	INVESTMENT/EXISTING ASSET VALUE										
20	INVESTMENT TAX CREDIT/REBATE										
21	SALVAGE										
22											
23	NET CASH FLOW (ALGEBRAIC SUM OF LINES 18 THRU 22 AND ADD BACK DEPRECIATION)										

* DO NOT SHOW COSTS WHICH ARE IDENTICAL IN BOTH ALTERNATIVES.
** SHOW ALGEBRAIC SIGNS ON LINES 18 THRU 23.

FIGURE 4.2

DISCOUNTED CASH FLOW SUMMARY

PERIOD	"A" NCF	"B" NCF	"A"-"B" △NCF 0% INTEREST	CUMULATIVE CASH FLOW BACK AMOUNT	%	15% INTEREST FACTOR	PW	25% INTEREST FACTOR	PW	40% INTEREST FACTOR	PW	60% INTEREST FACTOR	PW
-1						1.150		1.250		1.400		1.600	
0						1.000		1.000		1.000		1.000	
TOTAL "X" (DISBURSEMENTS)													
1						.870		.800		.714		.625	
2						.756		.640		.510		.391	
3						.658		.512		.364		.244	
4						.572		.409		.260		.153	
5						.497		.328		.186		.095	
6						.432		.262		.133		.060	
7						.376		.210		.095		.037	
8						.327		.168		.068		.023	
9						.284		.134		.048		.015	
10						.247		.107		.035		.009	
11						.215		.086		.025		.006	
12						.187		.069		.018		.004	

TOTAL "Y" (RECEIPTS)

RATIO "X"/"Y"

PROPOSED INVESTMENT _____

DCF / ROR _____ %

PAYOUT @ 0% _____ YRS.

DIVISION _____

PROJECT _____

PROJECT NO. _____ J.O. NO. _____

PREPARED BY _____ DATE _____

APPROVED BY _____ DATE _____

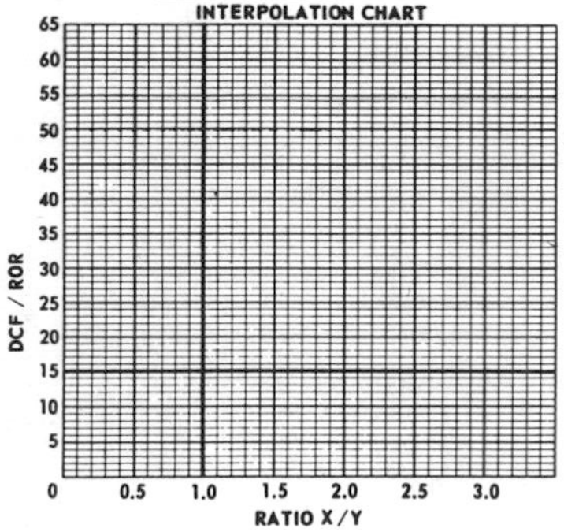

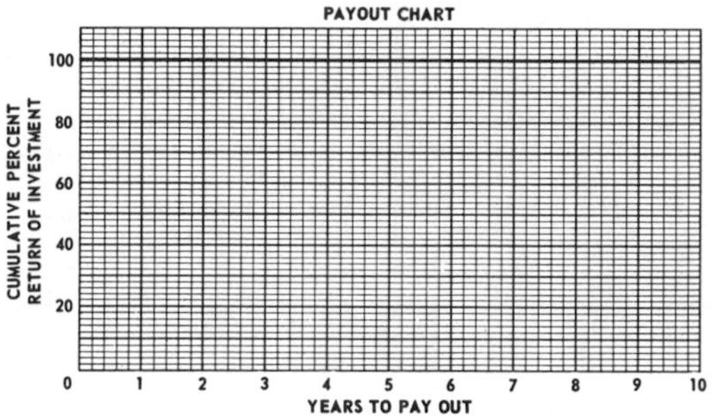

PAYOUT CHART

CUMULATIVE PERCENT RETURN OF INVESTMENT

YEARS TO PAY OUT

INTERPOLATION CHART

DCF / ROR

RATIO X/Y

FIGURE 4.3

DCF / ROR DATA SHEET

PROJECT TITLE_____ PROJECT NO._____ DATE_____

PROJECT LIFE IS _____ YEARS, DETERMINED BY: _____

_____

_____

OTHER ALTERNATIVES CONSIDERED AND REASONS FOR REJECTION: _____

_____

_____

_____

EXPLANATIONS AND CALCULATIONS	LINE REFERENCE	ALTERNATE A TITLE _____ METHOD DESCRIPTION _____	LINE REFERENCE	ALTERNATE B TITLE _____ METHOD DESCRIPTION _____
DEPRECIATION DESCRIBE EXISTING/PROPOSED EQUIPMENT AND EXPLAIN DEPRECIATION	2		2	
EXPENSES (LINES 3–16) NOTE ZERO YEAR EXPENSES SUCH AS IMPLEMENTATION AND TRAINING; MAINTENANCE/MAJOR OVERHAUL; OR ANY UNUSUAL EXPENSES. NOTE IF ONLY DELTA COSTS OF ONE ALTERNATIVE ARE BEING SHOWN.				
INVESTMENT SHOW DERIVATION OF EXISTING ASSET VALUE AND PROPOSED CAPITAL INVESTMENT. IDENTIFY ANY GOVT. EQUIP. STATE IF OLD EQUIP TO BE KEPT.	19		19	
INVESTMENT TAX CREDIT OR REBATE SHOW PERCENT USED.	20		20	
SALVAGE CALCULATE RESIDUAL AFTER TAX VALUE OF ASSETS	21		21	
OTHER CONSIDERATIONS INCLUDES INTANGIBLES, RISK AND LIKELIHOOD, OTHER ANALYSES AND RECOMMENDATION.				

FIGURE 4.4

4.6.1 Present Worth Method

The present worth of Alternative j can be represented as

$$PW_j(i) = \sum_{t=0}^{n} A_{jt}(1 + i)^{-t} \qquad (4.1)$$

with

$PW_j(t)$ = present worth of Alternative j using $MARR$ of $i\%$

n = planning horizon

A_{jt} = cash flow for Alternative j at the end of period t

$i = MARR$

The alternative having the greatest present worth is the alternative recommended using the present worth method.

Example 4.6

A pressure vessel is purchased for $10,000, kept for five years, and sold for $2000. Annual operating and maintenance costs were $2500. Using a 10% minimum attractive rate of return, what was the present worth for the investment?

$$PW_1(10\%) = -\$10,000 - \$2500(P|A\ 10,5) + \$2000(P|F\ 10,5)$$
$$= -\$18,235.20$$

Thus, a single expenditure of $18,235.20 at time zero is equivalent to the cash flow profile for the investment alternative.

4.6.2 Annual Worth Method

The annual worth of Alternative j can be computed as

$$AW_j(i) = \left[\sum_{t=0}^{n} A_{jt}(P|F\,i,t)\right](A|P\ i,n)$$

or

$$AW_j(i) = PW_j(i)(A|P\ i,n) \qquad (4.2)$$

where $AW_j(i)$ denotes the annual worth of Alternative j using $i = MARR$. The alternative having the greatest annual worth is selected using the annual worth method.

Example 4.7

Determine the annual worth for an anticipated investment of $10,000 in an analog computer that will last for eight years and have a zero salvage value at the time. Operating and maintenance costs are projected to be $1000 the first four years and $1500 the last four years. The minimum attractive rate of return is specified to be 10%. One method of determining the annual worth is:

$$AW_1(10\%) = -\$10,000(A|P\ 10,8) - \$1000 - \$500(F|A\ 10,4)(A|F\ 10,8)$$
$$= -\$3077/\text{year}$$

4.6.3 Future Worth Method

The future worth of Alternative j can be determined using the relationship

$$FW_j(i) = \sum_{t=0}^{n} A_{jt}(1+i)^{n-t} \qquad (4.3)$$

where $FW_j(i)$ is defined as the future worth of Alternative j using a MARR of $i\%$. The future worth method is equivalent to the present worth method and the annual worth method, since the ratio of $FW_j(i)$ and $PW_j(i)$ equals a constant, $(F|P\ i,n)$, and the ratio of $FW_j(i)$ and $AW_j(i)$ equals a constant, $(F|A\ i,n)$. The alternative having the greatest future worth is the preferred alternative when the future worth method is used.

Example 4.8

For the previous example, the future worth is given by

$$FW_1(10\%) = -\$10,000(F|P\ 10,8) - \$1000(F|A\ 10,8)$$
$$-\$500(F|A\ 10,4)$$
$$= -\$35,192.40$$

or

$$FW_1(10\%) = AW_1(10\%)(F|A\ 10,8)$$
$$= -\$3077(11.4359)$$
$$= -\$35,188.26$$

Note. the difference in $35,192.40 and $35,188.26 is due to round-off errors.

4.6.4 Payback Period Method

The payback period method involves the determination of the length of time required to recover the initial investment based on a zero interest rate. Letting

C_{0j} denote the initial investment for Alternative j and, R_{jt} denote the net revenue received from Alternative j during period t, if we assume no other negative cash flows occur, then the smallest value of m_j such that

$$\sum_{t=1}^{m_j} R_{jt} \geq C_{0j}$$

defines the payback period for Alternative j. The alternative having the smallest payback period is the preferred alternative using the payback period method.

Example 4.9 ─────────────────────────────────────

Based on the data given in Table 4.6,

$$\sum_{t=1}^{3} R_t = \$2000 + \$2500 + \$3130 < C_0 = \$10,000$$

and

$$\sum_{t=1}^{4} R_t = \$2000 + \$2500 + \$3130 + \$3510 > C_0 = \$10,000$$

Consequently, m equals 4, indicating that four years are required to pay back the original investment.

───

A number of variations of the payback or payout method have been used by different organizations. However, the basic deficiencies of the payback method we have described are present in the variations with which we are familiar; the *timing* of cash flows and the *duration* of the project are ignored.

To illustrate the deficiencies of the payback period, consider the alternatives depicted in Figure 4.5. In the first case, if the payback period is used to compare Alternatives A and B, then Alternative B would be preferred. In the second case, Alternatives C and D would be equally preferred using the payback period method.

Despite its obvious deficiencies, the payback period method continues to be one of the most popular methods of judging the desirability of investing in a project. The reasons for its popularity include the following:

1. No interest rate calculations are required.
2. No decision is required concerning the discount rate (*MARR*) to use.
3. Easily explained and understood.
4. Reflects a manager's attitudes when investment capital is limited.

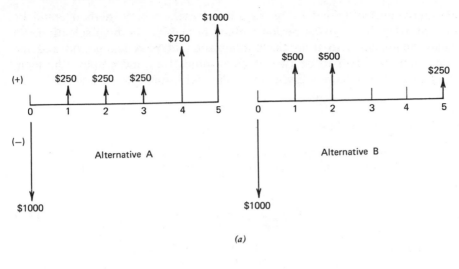

(a)

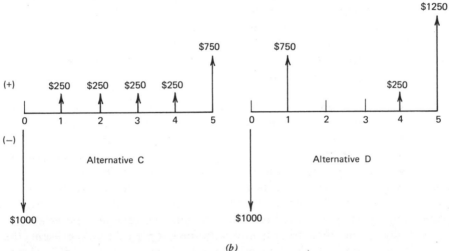

(b)

Figure 4.5. Cash flow diagrams for two mutually exclusive situations, with each involving two mutually exclusive investment alternatives.

5. Hedges against uncertainty of future cash flows.

6. Provides a rough measure of the liquidity of an investment.

The payback period method is recommended as a supplementary method of evaluating investment alternatives. In particular, it is suggested that the payback period method be used after the set of feasible alternatives has been evaluated using the "time value of money"-based methods. The top two or three alternatives obtained using, say, the present worth method could be compared using the payback period method in making the final selection. An alternate approach is to treat the minimization of payback period and the maximization of worth to be two of many objectives and employ the technique presented in Chapter Eight for dealing with multiple objectives.

4.6.5 Rate-of-Return Method

The rate of return for an alternative can be defined as the interest rate that equates the future worth to zero. Letting i_j^* denote the rate of return for Alternative j,

$$0 = \sum_{t=0}^{n} A_{jt}(1 + i_j^*)^{n-t} \qquad (4.4)$$

This method of defining the rate of return is referred to in the economic analysis literature as the discounted *cash flow rate of return, internal rate of return*, and the *true rate of return*. We prefer the term *internal rate of return (IRR)*.

Note that the present worth and the annual worth can be obtained by multiplying both sides of Equation 4.3 by the appropriate interest factor [i.e., $(P|Fi_j^*,n)$ and $(A|Fi_j^*,n)$]. Hence, the internal rate of return can also be defined to be the interest rate that yields either a present worth or an annual worth of zero. Depending on the form of a particular cash flow profile, it might be more convenient to use a present worth or an annual worth formulation to determine the internal rate of return for an alternative.

It is important to understand the definition of rate of return inherent in the use of the *IRR* method. In particular, the internal rate of return on an investment can be defined as *the rate of interest earned on the unrecovered balance of an investment*. This concept was demonstrated in Chapter Three in discussing the amount of a loan payment that was principal. It is illustrated again in Table 4.6, where $10,000 is invested to obtain the receipts shown over a six-year period. A_t denotes the cash flow at the end of period t, B_t represents the unrecovered balance at the *beginning* of period t, E_t is the unrecovered balance at the *end* of period t, and I_t is defined as the interest on the unrecovered balance during period t. The following relationships exist.

$$E_0 = A_0$$
$$B_t = E_{t-1} \qquad t = 1, \ldots, n$$
$$I_t = B_t i \qquad t = 1, \ldots, n$$
$$E_t = A_t + B_t + I_t \qquad t = 1, \ldots, n$$

If i is the internal rate of return, then E_n will equal zero. As indicated in Table 4.6, if i is 20%, E_n is approximately zero. Consequently, i^* is ap-

**TABLE 4.6 Data Illustrating the Meaning of the
Internal Rate of Return**

t	A_t	B_t	I_t^a	E_t
0	−$10,000	—	—	−$10,000
1	+ 2,000	−$10,000	−$2,000	− 10,000
2	+ 2,500	− 10,000	− 2,000	− 9,500
3	+ 3,130	− 9,500	− 1,900	− 8,270
4	+ 3,510	− 8,270	− 1,654	− 6,414
5	+ 4,030	− 6,414	− 1,283	− 3,667
6	+ 4,400	− 3,667	− 733	0

^{a} Based on $i = 0.20$.

proximately 20%. The equivalence of E_n being zero and the future worth being zero is easily understood by recognizing that E_n is actually the future worth of the cash flow profile. To see why this is true, notice that

$$E_n = A_n + B_n + I_n$$

Employing the definition of I_n,

$$E_n = A_n + B_n(1 + i)$$

By the relationship between B_n and E_{n-1}, it is seen that

$$E_n = A_n + E_{n-1}(1 + i)$$

Since a similar relationship exists between E_{n-1} and E_{n-2}, we note that

$$E_n = A_n + A_{n-1}(1 + i) + E_{n-2}(1 + i)^2$$

Generalizing, the recursive relationship between E_t and E_{t-1} gives

$$E_n = A_n + A_{n-1}(1 + i) + A_{n-2}(1 + i)^2 + \cdots + A_0(1 + i)^n$$

Hence, we see that

$$E_n = FW(i\%)$$

as anticipated.

The above example illustrates that the time value of money operations involved in the *IRR* method are equivalent to assuming that all monies received are reinvested and earn interest at a rate equal to the internal rate of return. In particular, if the net cash flow in period t is negative, it is denoted by C_t; if the net cash flow in period t is positive, it is denoted R_t. Letting r_t be the reinvestment rate for positive cash flows occurring in period t and i' be the rate of return for negative cash flows, then the following relationship can be defined.

$$\sum_{t=0}^{n} R_t(1 + r_t)^{n-t} = \sum_{t=0}^{n} C_t(1 + i')^{n-t} \qquad (4.5)$$

The future worth of reinvested monies received must equal the future worth of investments.

If r_t equals i', Equation 4.4 becomes

$$0 = \sum_{t=0}^{n} (R_t - C_t)(1 + i')^{n-t} \qquad (4.6)$$

Letting A_t equal $R_t - C_t$ defines the *IRR* method given by Equation 4.4. Hence, we see that the rate of return obtained using the *IRR* method can be interpreted as the reinvestment rate for all recovered funds.

Since the determination of the rate of return involves solving Equation 4.4 for i_j^*, it is seen that (for a given alternative j), it is necessary to determine the values of x that satisfy the following n-degree polynomial $0 = A_0 x^n + A_1 x^{n-1} + \cdots + A_{n-1} x + A_n$ where $x = (1 + i^*)$. In general, there can exist n distinct roots (values of x) for an n-degree polynomial; however, most cash flow profiles encountered in practice will have a unique root (rate of return).

The number of real positive roots of an n-degree polynomial with real coefficients is less than or equal to the number of changes of sign in the sequence of cash flows, $A_0, A_1, \ldots, A_{n-1}, A_n$. Since the typical cash flow pattern begins with a negative cash flow, followed by positive cash flows, a unique root will normally exist.

Example 4.10 ───

As an illustration of a cash flow profile having multiple roots, consider the data given in Table 4.7. The future worth of the cash flow series will be zero using either a 20 or 50% interest rate.

$$FW(20\%) = -\$1000(1.2)^3 + \$4200(1.2)^2 - \$5850(1.2) + \$2700 = 0$$
$$FW(50\%) = -\$1000(1.5)^3 + \$4200(1.5)^2 - \$5850(1.5) + \$2700 = 0$$

A plot of the future worth for this example is given in Figure 4.6. The future worth polynomial is a third-degree polynomial and there are three changes of sign in the ordered sequence of cash flows (−, +, −, +); however, there are only two unique roots, corresponding to $i = 0.20$ and $i = 0.50$. In this case, there is a repeated root corresponding to $i = 0.50$, since the future worth

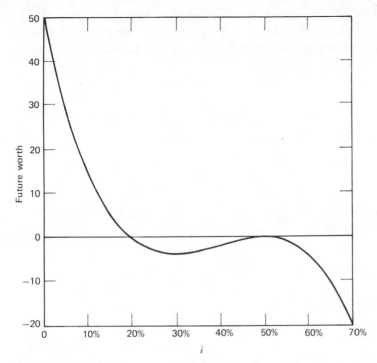

FIGURE 4.6. Plot of future worth for the example problem.

polynomial can be written as

TABLE 4.7 Cash Flow Profile

EOY	CF
0	−$1000
1	4200
2	− 5850
3	2700

$$FW(i\%) = \$1000(1.2 - x)(1.5 - x)^2$$

where $x = (1 + i)$. For additional discussion of the subject of multiple roots in internal rate of return calculations, as well as their interpretation, see Bernhard [4, 5].

The possibility of multiple roots occurring in the internal rate of return calculation, coupled with the reinvestment assumptions concerning recovered funds, has led to the development of an alternative rate-of-return method, called the *external rate of return method*. The external rate of return (*ERR*) method consists of the determination of the value of i' that satisfies Equation 4.7.

$$\sum_{t=0}^{n} R_t(1+r_t)^{n-t} = \sum_{t=0}^{n} C_t(1+i')^{n-t} \qquad (4.7)$$

where, as defined above,

A_t = net cash flow in period t

$$C_t = \begin{cases} A_t, & \text{if } A_t < 0 \\ 0, & \text{otherwise} \end{cases}$$

$$R_t = \begin{cases} A_t, & \text{if } A_t \geq 0 \\ 0, & \text{otherwise} \end{cases}$$

r_t = reinvestment rate for funds recovered in period t

i' = external rate of return

Normally, r_t equals the minimum attractive rate of return, since the *MARR* reflects the opportunity cost for money available for investment. The preferred alternative is that alternative with the greatest investment such that each increment of investment has a return at least equal to the *MARR*.

Example 4.11 ————————————————————————

For the data provided in Table 4.6, suppose money received from the initial investment is reinvested and earns 10% interest. At the end of the sixth year the reinvested funds total

$2000(F|P\ 10,5) + \$2500(F|P\ 10,4) + \$3130(F|P\ 10,3) + \$3510(F|P\ 10,2)$
$+ \$4030(F|P\ 10,1) + \$4400 = \$24,127.38$

Consequently, the *external rate of return* is defined as the interest rate such that the future worth of the $10,000 investment equals $24,127.38. Thus,

$$\$10,000(1+i')^6 = \$24,127.38$$

Taking the logarithm and solving for i yields a value of 15.8% as the rate of return.

As another example, if recovered funds are reinvested at 20%, the future worth will be

$2000(F|P\ 20,5) + \$2500(F|P\ 20,4) + \$3130(F|P\ 20,3) + \$3510(F|P\ 20,2)$
$+ \$4030(F|P\ 20,1) + \$4400 = \$25,859.64$

Setting the future worth of the $10,000 investment equal to $25,859.64 yields

$$\$10,000(1 + i')^6 = \$29,859.64$$

Solving for i gives a value of 20% as the rate of return, which was anticipated.

4.6.6 Savings/Investment Ratio Method

The *savings/investment ratio method* can be defined in various ways; two typical approaches will be described. First, the *savings/investment ratio* can be given as

$$SIR_j(i) = \frac{\displaystyle\sum_{t=1}^{n} A_{jt}(1 + i)^{-t}}{C_{0j} - F_j(1 + i)^{-n}} \tag{4.8}$$

where $SIR_j(i)$ is the savings investment ratio for Alternative j based on an MARR of $i\%$, A_{jt} is the cash flow for Alternative j for period t, C_{0j} is the initial investment for Alternative j, and F_j is the terminal salvage value for Alternative j. An alternative formulation of the savings/investment ratio will be employed in this chapter; the SIR is defined as follows:

$$SIR_j(i) = \frac{\displaystyle\sum_{t=0}^{n} R_{jt}(1 + i)^{-t}}{\displaystyle\sum_{t=0}^{n} C_{jt}(1 + i)^{-t}} \tag{4.9}$$

where R_{jt} denotes the magnitude of the net positive cash flow for Alternative j occurring in period t and C_{jt} denotes the magnitude of the net negative cash flow for Alternative j occurring in period t.

Example 4.12

Consider the cash flows given in Table 4.6 and let the MARR be 10%. The present worth of the cash flows during periods 1 to 6 totals

$$\$2000(P|F\ 10,1) + \$2500(P|F\ 10,2) + \$3130(P|F\ 10,3)$$
$$+ \$3510(P|F\ 10,4) + \$4030(P|F\ 10,5) + \$4400(P|F\ 10,6) = \$13,619.13$$

Thus, the savings/investment ratio is given by

$$SIR_1(10\%) = \frac{\$13,619.13}{\$10,000.00} = 1.361913$$

Any savings/investment ratio greater than one indicates the alternative is economically desirable. An alternative label for the savings/investment ratio is the *benefit-cost ratio*. Savings are interpreted as the benefits derived from a venture and the difference in the initial investment and the present worth of the salvage value is denoted as the cost of the venture. We explore benefit-cost analysis more fully in Chapter Six.

4.7 COMPARING THE INVESTMENT ALTERNATIVES

A comparison of investment alternatives, to be complete, requires a knowledge of all differences in cash flows among the investment alternatives. In a number of public sector applications the quantification of benefits in economic units is not a trivial undertaking. Consequently, we will postpone until Chapter Six a discussion of the comparison of alternatives involving benefits which are not normally measured in dollars and cents.

Additionally, our discussion in this chapter will be based on an assumption that either the cash flows are after-tax cash flows or a before-tax study is desired. In Chapter Five we address the subject of income taxes and their effect on the preference among investment alternatives.

Example 4.13

To illustrate the use of the various methods of comparing investment alternatives, consider the three mutually exclusive alternatives having the cash flow profiles given in Table 4.8 for a planning horizon of five years. A minimum attractive rate of return of 10% is to be used in the analysis.

TABLE 4.8 Cash Flow Profiles for
Three Mutually Exclusive Investment
Alternatives

T	A_{1t}	A_{2t}	A_{3t}
0	0	−$10,000	−$15,000
1	$4,000	7,000	7,000
2	4,000	7,000	8,000
3	4,000	7,000	9,000
4	4,000	7,000	10,000
5	4,000	7,000	11,000

4.7.1 Ranking Approaches

In comparing mutually exclusive investment alternatives using either present worth, annual worth, future worth, or payback period as the measure of merit, one can simply compute the value of the measure of merit for each alternative and rank the alternatives on the basis of the value obtained. The use of the ranking approach will be demonstrated for each of the four measures of merit by comparing the alternatives depicted in Table 4.8.

Present Worth Method. Using the present worth method gives

$$PW_1(10\%) = \$4000(P|A\ 10,5)$$
$$= \$15,163.20$$
$$PW_2(10\%) = -\$10,000 + \$7000(P|A\ 10,5)$$
$$= \$16,535.60$$
$$PW_3(10\%) = -\$15,000 + \$7000(P|A\ 10,5) + \$1000(P|G\ 10,5)$$
$$= \$18,397.33$$

Since Alternative 3 has the greatest present worth, it is recommended.

Annual Worth Method. The annual worths for the three alternatives are:

$$AW_1(10\%) = \$4000$$
$$AW_2(10\%) = -\$10,000(A|P\ 10,5) + \$7000$$
$$= \$4362$$
$$AW_3(10\%) = -\$15,000(A|P\ 10,5) + \$7000 + \$1000(A|G\ 10,5)$$
$$= \$4853$$

(Alternatively, the annual worth values could have been obtained directly from the present worth values.) Since Alternative 3 has the greatest annual worth, it is recommended.

Future Worth Method. The future worths for the three alternatives are computed as follows:

$$FW_1(10\%) = \$4000(F|A\ 10,5)$$
$$= \$24,420.40$$
$$FW_2(10\%) = -\$10,000(F|P\ 10,5) + \$7000(F|A\ 10,5)$$
$$= \$26,630.70$$
$$FW_3(10\%) = -\$15,000(F|P\ 10,5) + \$7000(F|A\ 10,5)$$
$$+\$1000(A|G\ 10,5)(F|A\ 10,5)$$
$$= \$29,926.04$$

(Alternatively, the future worth values could have been obtained directly from the present worth and from the annual worth values.) Since Alternative 3 has the greatest future worth, it is recommended.

Payback Period Method. The payback period is computed based on the difference in the cash flows for Alternatives 2 and 3 and the cash flows for Alternative 1 (do nothing alternative). By investing $10,000, annual savings of

$3000 are provided by Alternative 2; by investing $15,000 savings of $3000, $4000, $5000, $6000 and $7000 are produced for years 1 to 5, respectively. For Alternative 2 the payback period is found to be four years, since

$$\sum_{t=1}^{3} R_{2t} = \$3000 + \$3000 + \$3000 < \$10,000 = C_{02}$$

and

$$\sum_{t=1}^{4} R_{2t} = \$3000 + \$3000 + \$3000 + \$3000 > \$10,000 = C_{02}$$

For Alternative 3 the payback period is also found to be four years, since

$$\sum_{t=1}^{3} R_{3t} = \$3000 + \$4000 + \$5000 < \$15,000 = C_{03}$$

and

$$\sum_{t=1}^{4} R_{3t} = \$3000 + \$4000 + \$5000 + \$6000 > \$15,000 = C_{03}$$

Using the payback period method, Alternatives 2 and 3 are equally preferred.

4.7.2 Incremental Approaches

In comparing mutually exclusive investment alternatives using either the rate of return or the savings/investment method, an incremental approach must be taken. The comparisons are based on the *differences* in the cash flows for combinations of investment alternatives.

It is also possible to use incremental methods to compare alternatives using present worth, annual worth, and future worth methods. A step-by-step representation of the incremental procedure for present worth, annual worth, and future worth comparisons is given below:

Step 1. Order the feasible alternatives according to the size of the initial investment. Go to step 2.

Step 2. Compute the value of the measure of merit for the feasible alternative having the smallest initial investment. Go to step 3.

Step 3. Obtain the cash flow profile for the differences in cash flows for the two remaining feasible alternatives that require the smallest investments. Go to step 4.

Step 4. Compute the value of the measure of merit for the cash flow profile obtained in step 3. If the value of the measure of merit is positive (negative), eliminate from further consideration the alternative that has the smallest (largest) initial investment. Go to step 5.

Step 5. If only one alternative remains, it is the preferred alternative; if more than one alternative remains, go to step 3.

In order to illustrate the step-by-step procedure, an annual worth comparison will be performed for the example problem.

Step 1. Order 1, 2, 3.

Step 2. $AW_1(10\%) = \$4000$.

Step 3.

t	$A_{2t} - A_{1t}$
0	-$10,000
1	3,000
2	3,000
3	3,000
4	3,000
5	3,000

Step 4. $AW_{2-1}(10\%) = -\$10,000(A|P\ 10,5) + \$3000$
$$= \$362$$

$AW_{2-1}(10\%) > 0$, therefore, eliminate Alternative 1.

Step 5. More than one alternative (2, 3) remains; therefore, go to step 3.

Step 3.

t	$A_{3t} - A_{2t}$
0	-$5000
1	0
2	1000
3	2000
4	3000
5	4000

Step 4. $AW_{3-2}(10\%) = -\$5000(A|P\ 10,5) + \$1000(A|G\ 10,5)$
$$= \$491$$

$AW_{3-2}(10\%) > 0$; therefore, eliminate Alternative 2.

Step 5. Only one alternative remains; therefore, Alternative 3 is the preferred alternative.

As exercises, you may wish to solve the example problem using the incremental procedure in conjunction with the present worth and future worth methods. Additionally, you may wish to develop an incremental procedure for the payback period method. Similar, but slightly different, incremental procedures are employed when either the rate-of-return or the savings/investment ratio methods are used.

Rate-of-Return Method. The rate-of-return method includes the same sequence of comparisons used in comparing the differences in alternatives using present worth, annual worth, and future worth methods. The steps to be performed using the rate-of-return method in comparing investment alternatives are:

Step 1. Order the alternatives according to the size of the initial investment. Go to step 2.

Step 2. Compute the rate of return for the alternative having the smallest investment. Go to step 3.

Step 3. If the rate of return obtained in step 2 is less than the minimum attractive rate of return, eliminate the alternative from further consideration and repeat step 1. If the rate of return obtained in step 2 is greater than or equal to the *MARR*, go to step 4.

Step 4. Compute the rate of return for the difference in cash flows for the two alternatives having the smallest initial investments. Go to step 5.

Step 5. If the rate of return based on incremental investment obtained in step 4 is less than (greater than or equal to) the *MARR*, eliminate from further consideration the alternative having the largest (smaller) investment. Go to step 6.

Step 6. If only one alternative remains, it is the preferred alternative; if more than one alternative remains, go to step 4.

Applying the sequence of steps to the example problem using the internal rate-of-return method yields the following results:

Step 1. Order 1, 2, 3.

Step 2. $i_1^* = \infty$, since no "initial investment" is required to obtain income of $4000/year.

Step 3. $i_1^* > MARR$, go to step 4.

Step 4. Compute rate of return on difference in cash flows for Alternatives 1 and 2. Using a present worth formulation yields
$-\$10,000 + \$3000(P|A\ i,5) = 0$, and
solving for i yields $i_{2-1}^* = 15.06\%$.

Step 5. $i_{2-1}^* > MARR = 10\%$; eliminate Alternative 1 from further consideration.

Step 6. More than one alternative remains; go to step 4.

Step 4. Compute rate of return on difference in cash flows for Alternatives 2 and 3. That is,
$-\$5000 + \$1000(P|G\ i,5) = 0$; and
solving for i yields $i_{3-2}^* = 19.48\%$.

Step 5. $i_{3-2}^* > MARR$; eliminate Alternative 2 from further consideration.

Step 6. Only one alternative remains; thus, Alternative 3 is preferred.

It should be noted that the procedure for the *IRR* method given above depends on at least one of the alternatives having a rate of return greater than the *MARR*. However, in a number of investment situations, it is not uncommon to find that all of the alternatives involve only negative cash flows. In such a situation steps 2 and 3 are eliminated from the procedure.

Applying the *ERR* method in comparing the investment alternatives yields the following results.

Step 1. Order 1, 2, 3.

Step 2. $i_1' = \infty$.

Step 3. $i_1' > MARR$; go to step 4.

Step 4. $-\$10,000(F|P\ i',5) + \$3000(F|A\ 10,5) = 0$, and
$i_{2-1}' = 12.87\%$.

Step 5. $i_{2-1}' > MARR$; eliminate Alternative 1 from further consideration.

Step 6. More than one alternative remains; go to step 4.

Step 4. $-\$5000(F|P\ i',5) + \$1000(A|G\ 10,5)(F|A\ 10,5) = 0$, and
$i_{3-2}' = 17.19\%$.

Step 5. $i_{3-2}' > MARR$; eliminate Alternative 2 from further consideration.

Step 6. Only one alternative remains; thus, Alternative 3 is preferred.

Savings/Investment Ratio Method. The savings/investment ratio is computed by comparing the cash flows of Alternatives 1 and 2. Investing $10,000 yields annual savings of $3000/year. The present worth of the savings equals $3000(P|A\ 10,5)$, or $11,372.40. Consequently, the *SIR* value is

$$SIR_{2-1}(10\%) = \frac{\$11,372.40}{\$10,000.00} = 1.137$$

Since $SIR_{2-1}(10\%) > 1.00$, Alternative 2 is preferred to Alternative 1.

Next we compare Alternatives 2 and 3. An incremental investment of $5000 in Alternative 3 yields savings of $1000, $2000, $3000, and $4000 in years 2 to 5. The present worth of the savings equals $6861.73. Thus, the *SIR* value is

$$SIR_{3-2}(10\%) = \frac{\$6861.73}{\$5000.00} = 1.372$$

Since $SIR_{3-2}(10\%) > 1.00$, Alternative 3 is preferred to Alternative 2.

4.8 PERFORMING SUPPLEMENTARY ANALYSES

The sixth step in performing an economic evaluation of investment alternatives is the performance of subsequent analyses. A sensitivity analysis consists of an exploration of the behavior of the measure of merit to changes in the values of the parameters for an investment alternative. The parameters subject to change might include the planning horizon, the discount rate, and any or all of the cash flows. This step recognizes that the process of providing estimates of cash flows and decisions concerning the planning horizon and discount rate is not a precise, errorless process. An examination of break-even, sensitivity, and risk analyses as they relate to economic analyses is reserved for Chapter Seven. For now, we assume that the values assigned to the parameters are neither inaccurate nor subject to change.

4.9 SELECTING THE PREFERRED ALTERNATIVE

The final step in performing a comparison of investment alternatives is the selection of the preferred alternative. Our discussion in this chapter has concentrated solely on the economic factor; we have been concerned with determining the most economical alternative. The final decision may be based on a host of criteria instead of on the single criterion of economics. In Chapter Eight we examine the decision process in the face of multiple objectives.

The selection process is complicated not only by the presence of multiple objectives, but also by the risks and uncertainties associated with the future. The analysis of Chapter Seven is extended in Chapter Eight to include decision making in the face of risk or uncertainty.

As pointed out in Chapter One, the selection or rejection of the recommended solution is heavily dependent on the sales ability of the individual presenting the recommendation to management. Since the corporate decision makers are normally presented with many more investment alternatives than can be funded, it is important to communicate effectively in order to compete favorably for the company's limited capital. Toward this end, Klausner [23] provides four specific suggestions for the engineer or systems analyst.

1. He should recognize that the decision-makers' perspective is broad and develop his proposal accordingly. The project's capital requirements should be related to previous and estimated future capital requirements on similar investments. The project's Discounted Cash Flow return or Net Present Value should be compared to other projects with which the decision-makers are familiar. The proposal should be shown to fit in with long-range corporate plans and support short-range objectives. Comparison should be made with similar investments by competition and competitive advantage (if any) shown. Ancillary marketing, public relations and/or political benefits that the company will derive from the investment should be pointed out. The effect that the investment will have on other functional activities of the company should be noted and overall benefits stressed. In other words, the investment proposal should be related to the well-being of the total enterprise.

2. He should recognize that this investment proposal is only one of several that the decision-makers are reviewing and that not all proposals will be accepted.

The engineer should know how the decision-makers classify investments and what competition there is for available capital resources. He should know the relative strengths and weaknesses of his investment proposal vis-a-vis competing investment proposals and deal with each in the presentation. If the proposed investment is relatively risk-free, that point should be made strongly to possibly offset less desirable aspects of the proposal (e.g., low return on investment).

3. He should know the decision-makers and tailor the investment proposal accordingly. The engineer should, for example, know and use the measure of merit they prefer, and support it with any other measures of merit that may be necessary or helpful. For example, if the decision-makers favor Discounted Cash Flow return on investment and the proposal has one that barely meets minimum return standards but does have a large Net Present Value, the engineer must make sure that the potential contribution to profit is clearly pointed out. If it is known that the decision-makers still calculate the payout period of each investment proposal, the investment's payout period should be stated—either to support the proposal or to point out its irrelevance in hopes of minimizing its impact on the ultimate investment decision. The engineer should always use technical and economical terms that the decision-makers will understand and relate to."

The engineer should become familiar with the values that the decision-makers attach to different aspects of investment proposals—particularly economic uncertainty. Low risk should be emphasized in proposals to decision-makers who tend to be risk-avoiders; potential economic gain should be highlighted for the risk-taking decision-maker. This, of course, does not suggest that the engineer should be less than honest in his proposal, but only makes the point that the communication of the investment proposal should be developed with the decision-maker in mind and emphasis varied accordingly.

4. He should not oversell the technical engineering aspects of the investment proposal. It should be remembered that decision-makers are primarily interested in the economic aspects of the proposal. The engineer must resist the temptation to overstate the complicated and sophisticated technology that might underlie a proposal. The less decision-makers understand about the technical/engineering aspects of a proposal, the more uneasy they become. This apprehension becomes part of the uncertainty which the decision-makers subjectively assign to the proposal and the result could be its rejection in favor of a proposal with which they are more familiar, or at least, more comfortable. The engineer should keep in mind the background and interests of the decision-makers and use technical and commercial terms with which they are familiar."

4.10 ANALYZING ALTERNATIVES WITH NO POSITIVE CASH FLOWS

The previous analysis of an investment decision involved three mutually exclusive alternatives, including the do nothing alternative, which involved positive valued cash flows. However, there exist situations in which no positive valued cash flows are present.

Example 4.14

Consider a cost reduction case in which two cost reduction alternatives have been proposed. The present method, which we refer to as the do nothing alternative, is also a feasible alternative. Thus, three mutually exclusive alternatives are considered. Cash flow profiles for the alternatives are given in Table 4.9.

TABLE 4.9 Cash Flow Profiles for Three Mutually Exclusive Investment Alternatives with No Positive Cash Flows

t	A_{1t}	A_{2t}	A_{3t}
0	0	-$10,000	-$15,000
1	-$12,000	- 9,000	- 9,000
2	- 12,000	- 9,000	- 8,000
3	- 12,000	- 9,000	- 7,000
4	- 12,000	- 9,000	- 6,000
5	- 12,000	- 9,000	- 5,000

Alternative 1 is the do nothing alternative in which the present expenditure of $12,000/year is continued. Alternative 2 involves an initial investment of $10,000 in order to reduce the annual expenditures by $3000 over the five-year period. Alternative 3 requires an initial investment of $15,000 in order to obtain decreasing annual expenditures. The minimum attractive rate of return is specified to be 10%.

The procedures we employed in analyzing alternatives having positive-valued cash flows can be applied to the present situation as well. As an illustration, consider the use of the incremental approach in conjunction with the annual worth procedure.

Step 1. Order 1, 2, 3.

Step 2. $AW_1(10\%) = -\$12,000$.

Step 3.

t	$A_{2t} - A_{1t}$
0	-$10,000
1	3,000
2	3,000
3	3,000
4	3,000
5	3,000

Step 4. $AW_{2-1}(10\%) = \$362 > 0$; therefore, eliminate Alternative 1.

Step 5. Alternatives 2 and 3 remain.

Step 3.

t	$A_{35} - A_{2t}$
0	-$5000
1	0
2	1000
3	2000
4	3000
5	4000

Step 4. Consider differences in cash flows for Alternatives 2 and 3.
$AW_{3-2}(10\%) = \$491 > 0$; therefore, eliminate Alternative 2.

Step 5. Alternative 3 is the only remaining alternative; hence, it is the preferred alternative.

You may wish to verify on your own that the present worth, future worth, payback period, and savings/investment methods can be applied in the case of nonpositive valued cash flows. We find it instructive, however, to consider the rate-of-return method explicitly.

Since the sum of the cash flows (ignoring the time value of money) is negative for each alternative, positive-valued rates of return among the three alternatives do not exist. Consequently, you may ask, "How can the rate-of-return method be used to compare alternatives that do not have rates of return?" Admittedly, Alternative 2 does not have a rate of return by itself, but in comparison with Alternative 1, it is clear that the investment of the $10,000 yields annual savings of $3000 in annual expenditures. Thus, on an incremental basis, the $10,000 incremental investment produces positive valued cash flows of $3000/year for each of the five years.

The return on the incremental investment, using the internal rate-of-return method, is found to be approximately 15.06%. Since we would do better to invest the $10,000 in Alternative 2 and earn 15.06% than to choose Alternative 1 and only earn the minimum attractive rate of return on the remaining money available for investment, Alternative 2 is preferred to Alternative 1.

Of course, we have $15,000 to invest (otherwise Alternative 3 would be unfeasible). Thus, at this point we are willing to invest $10,000 in Alternative 2 and earn 15.06% and invest the remaining $5000 in some other opportunity and earn the minimum attractive rate of return of 10%. The question now considered is, "Should we use the additional $5000 and pool it with the $10,000 to invest in Alternative 3?" The rate of return on the incremental investment required to obtain Alternative 3 can be calculated to be approximately 19.48%.

At this point, the question is "Should we invest $10,000 in Alternative 2 and invest $5000 elsewhere at' the *MARR*, or should we invest $15,000 in Alternative 3?" Investing in Alternative 3 yields a return of 15.06% on the $10,000 increment and 19.48% on the $5000 increment. Consequently, Alternative 3 would be preferred to the investment of $10,000 in Alternative 2 and investing the remaining $5000 to earn only 10%.

The rate of return philosophy is to continue investing so long as each increment of investment is justified. Since the application of this philosophy in evaluating investment alternatives is often difficult to grasp, we will consider yet another example to illustrate this important concept.

Example 4.15 ──

Consider the four alternatives given in Table 4.10. A six-year planning horizon is used, along with a 10% minimum attractive rate of return. The alternatives are ranked in order of increasing investment. Following the step-by-step procedure outlined previously for the rate-of-return method yields the following results:

TABLE 4.10 Data for a Rate of Return Example Problem

t	A_{1t}	A_{2t}	A_{3t}	A_{4t}
0	−$5,000	−$8,000	−$10,000	−$14,000
1−6	1,504	2,026	2,720	3,585

Step 1. Order 1, 2, 3, 4.

Step 2. Compute rate of return for Alternative 1:
$$0 = -\$5000(A|P\ i_1,6) + \$1504$$
$$(A|P\ i_1,6) = 0.3008$$
$$i_1^* = 20\%$$

Step 3. $i_1^* > MARR.$

Step 4. Compute rate of return on the $3000 incremental investment required to go from Alternative 1 to Alternative 2.
$$0 = -\$3000(A|P\ i,6) + \$522$$
$$(A|P\ i,6) = 0.1740$$
$$i_{2-1}^* = 1.25\%$$

Step 5. $i_{2-1}^* < MARR$, eliminate Alternative 2.

Step 6. Alternatives 1, 3, and 4 remain.

Step 4. Compute rate of return on the $5000 incremental investment required to go from Alternative 1 to Alternative 3,
$$0 = -\$5000(A|P\ i,6) + \$1216$$
$$(A|P\ i,6) = 0.2432$$
$$i_{3-1}^* = 12\%$$

Step 5. $i_{3-1}^* > MARR$, eliminate Alternative 1.

Step 6. Alternatives 3 and 4 remain.

Step 4. Compute rate of return on the $4000 incremental investment required to go from Alternative 3 to Alternative 4.

$0 = \$4000(A|P\ i,6) + \$865$

$(A|P\ i,6) = 0.21625$

$i_{4-3}^* \doteq 8\%$

Step 5. $i_{4-3}^* < MARR$, eliminate Alternative 4.

Step 6. Alternative 3 is preferred.

The conclusion we draw is that Alternative 3 yields a return of 20% on the first $5000 increment and 12% on the next $5000 increment. Unused monies earn 10%. Letting the rates of return on individual alternatives be designated as *gross* rates of return, the following gross rates of return are obtained.

$$i_1 = 20\%, \qquad i_2 = 13.5\%, \qquad i_3 = 16.1\%, \qquad i_4 = 13.85\%$$

Ranking alternatives on the basis of gross rate of return indicates that Alternative 1 is preferred. However, using the incremental rate-of-return approach, Alternative 3 is preferred even though it does not have the highest gross rate of return. Again, the reason for choosing Alternative 3 over Alternative 1 is that the second increment of $5000 will earn 12% if Alternative 3 is chosen; investing in Alternative 1 caused the remaining $5000 to be invested and earn *only* the *MARR* of 10%.

As long as the *IRR* method is used to evaluate increments of investment instead of ranking alternatives on the basis of gross rate of return, the recommended alternative will be the same as obtained using the annual worth, present worth, future worth, and savings/investment ratio methods. Likewise, if the *ERR* method employs a reinvestment rate equal to the minimum attractive rate of return, then the recommendation will be consistent with that obtained by using the *IRR* method.

Example 4.16

To illustrate the use of the external rate-of-return method, suppose a 10% reinvestment rate is used in the previous illustration. Following the step-by-step procedure gives the following results:

Step 1. Order 1, 2, 3, 4.

Step 2. $\$1504(F|A\ 10,6) = \$5000(F|P\ i_1',6)$

$(F|P\ i_1',6) = 2.32085$

$i_1' = 0.1506(15.06\%)$

Step 3. $i_1' > MARR$.

Step 4. $522(F|A\ 10,6) = \$3000(F|P\ i_{2-1}',6)$
$(F|P\ i_{2-1}',6) = 1.34251$
$i_{2-1}' = 0.0503(5.03\%)$

Step 6. Alternatives 1, 3, 4 remain.

Step 4. $1216(F|A\ 10,6) = \$5000(F|P\ i_{3-1}',6)$
$(F|P\ i_{3-1}',6) = 0.1106(11.06\%)$

Step 5. $i_{3-1}' > MARR.$

Step 6. Alternatives 3 and 4 remain.

Step 4. $865(F|A\ 10,6) = \$4000(F|P\ i_{4-3}',6)$
$(F|P\ i_{4-3}',6) = 1.6685$
$i_{4-3}' = 0.0891(8.91\%)$

Step 5. $i_{4-3}' < MARR.$

Step 6. Alternative 3 is preferred.

4.11 CLASSICAL METHOD OF DEALING WITH UNEQUAL LIVES

The method that we presented of dealing with investment alternatives having unequal lives is not a widely used approach in the economic analysis literature. Recall that it was recommended that a planning horizon be specified, cash flows over the planning horizon be given explicitly, and the evaluation be performed over the common planning horizon time frame. The classical method of dealing with unequal lives is to use a least common multiple of lives planning horizon and assume identical cash flow profiles in repeating life cycles. It is not uncommon to see alternatives having unequal lives compared on the basis of the annual worths for individual life cycles.

Example 4.17 ————————————————————————

To illustrate the classical approach, consider two mutually exclusive alternatives having cash flow profiles, as depicted in Figure 4.7. Using a minimum attractive rate of return of 15%, the following annual worths are obtained:

$$AW_1(15\%) = -\$5000(A|P\ 15,5) - \$2000 + \$1000(A|F\ 15,5)$$
$$= -\$3343.20$$
$$AW_2(15\%) = -\$7000(A|P\ 15,6) - \$1500 + \$1500(A|F\ 15,6)$$
$$= -\$3178.10$$

Implicit in the comparison of the alternatives using annual worths based on individual life cycles is the assumption that a 30-year planning horizon is

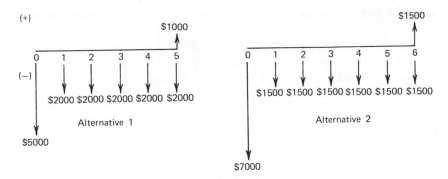

FIGURE 4.7. Cash flow diagrams for two mutually exclusive alternatives.

being used. As depicted in Table 4.11, if the present worths are computed for a 30-year period, the following values will be obtained:

$$PW_1(15\%) = -\$21,952.10$$
$$PW_2(15\%) = -\$20,868.30$$

TABLE 4.11 Cash Flows for Two Mutually Exclusive Alternatives, with a 30-year Planning Horizon

EOY	CF(1)	CF(2)
0	−$5000	−$7000
1–4	− 2000	− 1500
5	− 6000	− 1500
6	− 2000	− 7000
7–9	− 2000	− 1500
10	− 6000	− 1500
11	− 2000	− 1500
12	− 2000	− 7000
13–14	− 2000	− 1500
15	− 6000	− 1500
16–17	− 2000	− 1500
18	− 2000	− 7000
19	− 2000	− 1500
20	− 6000	− 1500
21–23	− 2000	− 1500
24	− 2000	− 7000
25	− 6000	− 1500
26–29	− 2000	− 1500
30	− 1000	0
Present worth	−$21,952.10	−$20,868.30

Converting the present worths to annual worths yields the values obtained using individual life cycles.

An alternative assumption that could be made is that a five-year planning horizon is being used and the salvage value for Alternative 2 is such that an annual worth of $-\$3178.10$ will still be obtained. Hence, S_5, the salvage value at the end of the fifth year, for Alternative 2 must be

$$AW_2(15\%) = -\$7000(A|P\ 15,5) - \$1500 - S_5(A|F\ 15,5)$$
$$-\$3178.10 = -\$3588.10 - 0.1483S_5$$

or

$$S_5 = \$2,764.67$$

(It is interesting to note that the salvage value obtained in this case is the book value at the end of five years of service using the sinking fund depreciation method described in Chapter Five.)

Example 4.18

To illustrate the shortcoming associated with an assumption that the salvage value for unused portions of an asset's life will be such that the annual worth will be unchanged, consider the alternatives depicted in Figure 4.8. The annual worths for individual life cycles are found to be:

$$AW_1(15\%) = -\$5000(A|P\ 15,5) - \$3000 + \$1000(A|F\ 15,5)$$
$$= -\$4343.20$$
$$AW_2(15\%) = -\$6000(A|P\ 15,6) - \$1000 - \$1000(A|G\ 15,6)$$
$$0.1142$$
$$+\$1000(A|F\ 15,6)$$
$$= -\$4568.20$$

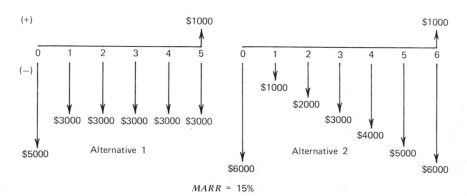

FIGURE 4.8. Cash flow diagrams for two mutually exclusive alternatives.

If a five-year planning horizon is used, to obtain an annual worth of $-\$4568.20$ for Alternative 2 requires that the salvage value at the end of the fifth year, S_5, be such that

$$-\$4568.20 = -\$6000(A|P\ 15,5) - \$1000 - \$1000(A|G\ 15,5)$$
$$0.1483$$
$$+ S_5(A|F\ 15,5)$$

or S_5 equals $-\$374.92$. Thus, instead of having a $1000 salvage value at the end of the sixth year, a salvage value of $-\$374.92$ must exist to yield the annual worth obtained by treating individual life cycles.

We do not recommend that alternatives having unequal lives be blindly compared on the basis of the annual worths for individual life cycles. Such an approach assumes *implicitly* that either a least common multiple of lives planning horizon is appropriate or the salvage values for unused portions of an asset's life are such that the annual worth is unchanged. We prefer to make explicit all assumptions concerning the planning horizon and the salvage values.

4.12 REPLACEMENT ANALYSIS

Among the various types of investment alternatives that are encountered in the real world, the *replacement* alternative deserves special consideration. Typically, in replacement analyses, one of the feasible alternatives involves maintaining the status quo; the remaining alternatives provide various replacement options that are available.

Whether you are considering replacing your automobile or your television set, there are definitely economic considerations involved. The need for considering replacement can be due to a number of factors, including inadequacy, excessive maintenance, declining efficiency, obsolescence (both functional and economic), physical impairment, and rental or lease possibilities. For example, if you are confronted with the need for a major overhaul on your car, you begin to consider replacing the car.

Changing demands can render a present asset inadequate to meet the production levels required. As an asset ages, it tends to deteriorate gradually over time and either operate less efficiently or experience increasing demands for maintenance. A decrease in demand might result in the present asset being obsolete functionally, since new assets having lower capacity are available to accommodate the new demand level. New and improved designs can render an existing asset obsolete economically. The rapid development of new models of pocket calculators is an illustration of the possibility of an asset becoming obsolete economically. When an asset fails and the service it renders is still required, a major repair is made or a replacement is acquired. The development of equipment rental firms has caused some homeowners to change from an ownership role to a leasor role with regard to some expensive, but infrequently used, tools; the same is true for businesses.

4.12.1 Cash Flow Approach

The replacement decision can be analyzed using the approaches outlined previously in this chapter. Also, we continue to recommend you use a cash flow approach and ask of each alternative, "How much money will be spent and received if I adopt this alternative?" Past costs should be viewed from the proper perspective; as pointed out in Chapter Two, unrecoverable past costs are *sunk costs* and are not to be included in economy studies that deal with the future, except as these sunk costs may affect income taxes if a present asset is disposed. The past is beneficial in that it allows the identification of estimation errors previously made and assists in making better estimates of the future.

We will be concerned with defining the planning horizon based on the remaining life of the asset in current use and the lives of the alternative candidates for replacement. Likewise, if we replace the current asset we will want to know how much money will be received for it (i.e., its salvage value). Another consideration involves income taxes, which are treated subsequently in Chapter Five.

Example 4.19 ————————————————————————————

To illustrate a replacement alternative, consider a situation involving a chemical plant that owns a filter press that was purchased three years ago for $10,000. Actual operating and maintenance (O&M) expenses (excluding labor) for the press have been $2000, $2500, and $3000 each of the past three years, as depicted in Table 4.12. It is anticipated that the filter press can be used for five more years and salvaged for $1000 at that time. The current market value for the used filter press is $4000. If the old filter press is retained, annual operating and maintenance costs are anticipated to be as shown in Table 4.12.

A new filter press is available and can be purchased for $18,000. The new filter press is expected to have annual operating and maintenance costs as depicted in Table 4.12.

The new filter press has an anticipated useful life of 10 years. Based on historical data concerning salvage values of filter presses, estimated salvage values (S_t) for the new press are given in Table 4.12.

Since the "old" press can only be used for five more years, a planning horizon of five years is specified; the salvage value for the "new" press is estimated to be $6000 in five years.

TABLE 4.12 Data for a Replacement
Alternative

| | Alternative 1 | | Alternative 2 | |
t	O & M	t	O & M	S_t
-3	—	0	—	$18,000
-2	-$2,000	1	—	15,000
-1	- 2,500	2	-$ 500	12,300
0	- 3,000	3	- 1,000	9,900
1	- 3,500	4	- 1,500	7,800
2	- 4,000	5	- 2,000	6,000
3	- 4,500	6	- 2,500	4,500
4	- 5,000	7	- 3,000	3,300
5	- 5,500	8	- 3,500	2,400
		9	- 4,000	1,800
		10	- 4,500	1,500

Alternative 1 is defined to be "keep the old press," Alternative 2 is defined to be "replace the old press with the new press." Cash flows for each alternative are given in Table 4.13.

TABLE 4.13 Cash Flows for a
Replacement Alternative

t	A_{1t}	A_{2t}
0	0	-$18,000 + 4,000
1	-$3,500	0
2	- 4,000	- 500
3	- 4,500	- 1,000
4	- 5,000	- 1,500
5	- 5,500 + 1,000	- 2,000 + 6,000

Computing the annual worths for each alternative using a minimum attractive rate of return of 10% yields the following results:

$$AW_1(10\%) = -\$3500 - \$500(A|G\ 10,5) + \$1000(A|F\ 10,5)$$
$$= -\$4241.25$$
$$AW_2(10\%) = -\$14,000(A|P\ 10,5) - \$500(A|G\ 10,5) + \$6000(A|F\ 10,5)$$
$$= -\$3,615.45$$

On the basis of the annual worth comparison, it is recommended that the new filter press be purchased and the old press be sold.

Note that, in carrying out the replacement analysis when the cash flows were listed for each alternative in Table 4.13, we ignored the operating and maintenance costs that will occur after the fifth year if the new filter press is purchased. The argument against including the operating and maintenance costs is based on the specification of a planning horizon of five years. The five-year planning horizon was based on the maximum useful life for the old filter press. If the old press is retained, then it will have to be replaced in five years (if not before, because of the possible development of attractive replacement alternatives in the future). Consequently, in five years we might have available an alternative that will yield even greater operating and maintenance savings than the filter press currently being considered. Thus, it is not fair to include savings that might occur after five years for one alternative without including such anticipated savings for the other alternative.

The above interpretation defines the planning horizon as a window through which only the cash flows can be seen that occur during the planning horizon, *with the exception of the terminal value for the alternative.* At the end of the planning horizon, an estimate is provided for the terminal value for each alternative, even though the alternative might not be physically replaced at that time. The end of the planning horizon defines a point in time at which another replacement study is planned. At that time, the future savings and costs will be compared against other available replacement candidates.

Example 4.20 ───

If, in Example 4.19, a 10-year planning horizon is desired, we recommend that consideration be given to the replacement of the "old" filter press in five years. Based on a forecast of the growth of filter press

TABLE 4.14 Cash Flows for a Replacement Alternative Using
a 10-Year Planning Horizon

t	A_{1t}	A_{2t}	$A_{2t} - A_{1t}$
0	0	−$18,000 + $4,000	−$14,000
1	−$3,500	0	3,500
2	− 4,000	− 500	3,500
3	− 4,500	− 1,000	3,500
4	− 5,000	− 1,500	3,500
5	−$5,500 + 1,000 − 15,000	− 2,000	3,500 + 14,000
6	0	− 2,500	− 2,500
7	− 500	− 3,000	− 2,500
8	− 1,000	− 3,500	− 2,500
9	− 1,500	− 4,000	− 2,500
10	− 2,000 + 7,500	− 4,500 + 1,500	− 2,500 − 6,000

technology, suppose we anticipate that at the end of five years a filter press will be available at a cost of $20,000; net operating and maintenance costs, and a terminal salvage value are anticipated to be as depicted in Table 4.14. A computation establishes that the difference in annual worths for the two alternatives is $38.54, with Alternative 1 being most economic. Thus, on the basis of technological forecasts of filter press alternatives in five years, it would appear advantageous to postpone the replacement. However, the degree of uncertainty in our forecast of the cash flows for a projected future replacement candidate would cause us to question the merits of postponing the replacement because of a difference of $38.54/year. These kinds of considerations are explored more fully in Chapters Seven and Eight.

4.12.2 Classical Approach

As with the treatment of unequal lives the classical approach in analyzing replacement decisions differs from the cash flow approach. The classical approach considers the salvage value of the old asset (defender) to be the investment cost for the defender if it is retained in service. Such an approach is consistent with the opportunity cost concept described in Chapter 2. Since the retention of the defender is equivalent to a decision to forego the receipt of its salvage value, then an opportunity cost is assigned to the defender.

Example 4.21

In order to illustrate the classical approach, consider Example 4.19. Applying the classical approach, the "cash flows" would be as depicted in Table 4.15. As can be seen the differences in the alternatives are the same; hence, the cash flow approach and classical approach are equivalent approaches.

While the concept underlying the classical approach is sound, there are some potential pitfalls in its application which should be avoided. Such pitfalls arise when either the defender has multiple trade-in values or unequal lives exist. Both of these situations will be illustrated with example problems.

TABLE 4.15 "Cash Flows" Using the Classical Approach

T	A_{1t}	A_{2t}	$A_{2t} - A_{1t}$
0	−$4,000	−$18,000	−$14,000
1	− 3,500	0	3,500
2	− 4,000	− 500	3,500
3	− 4,500	− 1,000	3,500
4	− 5,000	− 1,500	3,500
5	−5,500 + 1,000	− 2,000 + 6,000	8,500

Example 4.22

In the previous example, suppose a second replacement alternative is available. A filter press which sells for $20,000 is being considered; annual operating and maintenance costs are expected to equal $500 the first year and increase by 20% thereafter. A salvage value of $10,000 is anticipated in 5 years. If the new filter press is purchased then a $5,000 trade-in allowance will be provided for the old filter press. A cash flow approach yields the data presented in Table 4.16.

TABLE 4.16 Cash Flows for Example 4.21

T	A_{1t}	A_{2t}	A_{3t}
0	0	$-\$18,000 + 4,000$	$-\$20,000 + 5,000$
1	$-3,500$	0	$-\quad 500$
2	$-4,000$	$-\quad 500$	$-\quad 600$
3	$-4,500$	$-\quad 1,000$	$-\quad 720$
4	$-5,000$	$-\quad 1,500$	$-\quad 864$
5	$-5,500 + 1,000$	$-\quad 2,000 + 6,000$	$-\quad 1,037 + 5,000$

If the classical approach is used then a decision must be made concerning the appropriate investment cost for the defender. Should it be $4,000 or $5,000? If $4,000 is used then the cash flows at $t = 0$ will be $-\$4,000$, $-\$18,000$, and $-\$19,000$ for the three alternatives; if $5,000 is used then the cash flows at $t = 0$ will be $-\$5,000$, $-\$19,000$, and $-\$20,000$, respectively. Since the classical approach is based on the opportunity cost concept, it seems appropriate for a $5,000 investment cost to be assigned to the defender. However, regardless of which investment cost is used, the differences in the alternatives will be the same.

Example 4.23

As in Example 4.20 suppose a planning horizon of 10 years is used. If the cash flow approach is used, then, as was shown in Table 4.14, an explicit decision must be made concerning the replacement of the defender after 5 years; whereas, the classical approach can result in such a decision being suppressed. In particular, with the classical approach the equivalent uniform annual cost of the defender is often compared with the equivalent uniform annual cost of the challenger. The dangers of such an approach were elaborated on in Section 4.11; in the case of replacement decisions the dangers are magnified.

Using the classical approach, the *EUAC* for the defender and the challenger can be determined to be

$$EUAC_1(10\%) = \$4,000(A|P\ 10,5) + \$3,500 + \$500(A|G\ 10,5)$$
$$-\$1,000(A|F\ 10,5) = \$5,296.45$$

$$EUAC_2(10\%) = \$18,000(A|P\ 10,5) + \$500(A|G\ 10,5)$$
$$-\$6,000(A|F\ 10,5) = \$4,670.65$$

However, in order for the equivalent uniform annual costs to be accurate the defender will have to be replaced with an asset identical in cash flow profile to that of the defender. Such an assumption does not seem reasonable.

Recall, the cash flow approach requires that the cash flows be stated explicitly for each and every year during the planning horizon. Consequently, unequal trade-in values and unequal lives pose no difficulties. In this case, the treatment of the trade-in value as an investment cost can still be accommodated by letting A_{1t} equal $-\$4,000$ for $t = 0$ and A_{2t} equal $-\$18,000$ for $t = 0$ in Table 4.14.

Although logical arguments can be given for treating a replacement decision as just another economic investment alternative, many firms fail to subject their existing equipment to careful scrutiny on a periodic basis to determine if replacement is required. Despite the fact that replacement studies can yield significant reductions in costs, it still remains that many firms postpone replacing assets beyond the "optimum" time for replacement, perhaps because a decision to replace an asset involves a change, and resistance to change is inherent in most individuals. For example, an engineer who two years ago successfully argued that compressor Z should be replaced by compressor Y may now find that compressor X is more economical than compressor Y. If Y is now championed it may be viewed by management as an admission that the wrong compressor was selected as a replacement for Z.

Some reasons for delaying the replacement of assets beyond the economic replacement time are:

1. The firm is making a profit with its present equipment.
2. The present equipment is operational and is producing an acceptable quality product.
3. There is risk or uncertainty associated with predicting the expenses of a new machine, whereas one is relatively certain about the expenses of the current machine.
4. A decision to replace equipment is a stronger commitment for a period of time into the future than keeping the existing equipment.
5. Management tends to be conservative in decisions regarding the replacement of costly equipment.

6. There may be a limitation on funds available for purchasing new equipment, but no limitation on funds for maintaining existing equipment.

7. There may be considerable uncertainty concerning the future demand for the services of the equipment in question.

8. Sunk costs psychologically affect decisions to replace equipment.

9. An anticipation that technological improvements in the future might render obsolete equipment available currently; a wait-and-see attitude prevails.

10. Reluctance to be a pioneer in adopting new technology; instead of replacing now, wait for the competition to act.

As a unit of equipment ages, operating and maintenance costs increase. At the same time, the capital recovery cost decreases with prolonged use of the equipment. The combination of decreasing capital recovery costs and increasing annual operating and maintenance cost results in the equivalent uniform annual cost taking on a form similar to that depicted in Figure 4.9.

By forecasting the operating and maintenance costs for each year of service, as well as the anticipated salvage values for various replacement ages, one can determine the replacement interval for equipment.

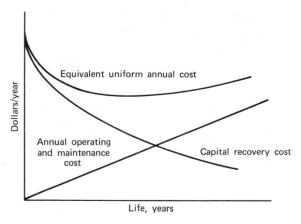

FIGURE 4.9. Portrayal of components of equivalent annual cost.

Example 4.18

Suppose a small compressor can be purchased for $1000; the salvage value for the compressor is assumed to be negligible, regardless of the replacement interval. Annual operating and maintenance costs are expected to increase by $75/year, with the first year's cost anticipated to be $150. Using a minimum attractive rate of return of 8%, the following equivalent uniform annual costs are obtained:

$$EUAC(n = 1) = \$1000(A|P\ 8,1) + \$150 + \$75(A|G\ 8,1)$$
$$= \$1230$$

$$EUAC(n = 2) = \$1000(A|P\ 8,2) + \$150 + \$75(A|G\ 8,2)$$
$$= \$746.52$$

$$EUAC(n = 3) = \$1000(A|P\ 8,3) + \$150 + \$75(A|G\ 8,3)$$
$$= \$608.25$$

$$EUAC(n = 4) = \$1000(A|P\ 8,4) + \$150 + \$75(A|G\ 8,4)$$
$$= \$555.49$$

$$EUAC(n = 5) = \$1000(A|P\ 8,5) + \$150 + \$75(A|G\ 8,5)$$
$$= \$536.26$$

$$EUAC(n = 6) = \$1000(A|P\ 8,6) + \$150 + \$75(A|G\ 8,6)$$
$$= \$533.07$$

$$EUAC(n = 7) = \$1000(A|P\ 8,7) + \$150 + \$75(A|G\ 8,7)$$
$$= \$538.72$$

$$EUAC(n = 8) = \$1000(A|P\ 8,8) + \$150 + \$75(A|G\ 8,8)$$
$$= \$549.34$$

$$EUAC(n = 9) = \$1000(A|P\ 8,9) + \$150 + \$75(A|G\ 8,9)$$
$$= \$563.03$$

$$EUAC(n = 10) = \$1000(A|P\ 8,10) + \$150 + \$75(A|G\ 8,10)$$
$$= \$578.41$$

As n increases beyond six years, the equivalent uniform annual cost increases. Hence, for this example, a replacement interval of six years is indicated.

The assumptions inherent in the determination of the optimum replacement interval should be considered closely. In particular, the analysis is based on a planning horizon that is an integer multiple of the replacement interval selected. (Recall our discussion of the use of the annual worth based on one life cycle and a planning horizon of the least common multiple of lives.) Second, we have assumed that each time the compressor is replaced it will be replaced with a compressor having an identical cash flow profile. If neither assumption is valid, then the approach we used is not valid.

Example 4.19

Suppose in the previous case a nine-year planning horizon is appropriate. For simplicity, we will continue to assume that replacements will have identical cash flow profiles. If the original compressor is kept for nine years, the present worth equivalent will be

$$PW(n = 9) = -\$563.03(P|A\ 8,9)$$
$$= -\$3517.19$$

If the compressor is to be replaced at some intermediate point during the planning horizon, say after k years, and the replacement is to be kept until the end of the planning horizon, the following present worth calculations result for $k = 5$, 6, or 7:

$$PW(k = 5) = -\$536.26(P|A\ 8,5) - \$555.49(P|A\ 8,4)(P|F\ 8,5)$$
$$= -\$3393.32$$

$$PW(k = 6) = -\$533.07(P|A\ 8,6) - \$608.25(P|A\ 8,3)(P|F\ 8,6)$$
$$= -\$3452.18$$

$$PW(k = 7) = -\$538.72(P|A\ 8,7) - \$746.52(P|A\ 8,2)(P|F\ 8,7)$$
$$= -\$3581.59$$

Thus, the compressor should be replaced after five years and the replacement should be kept until the end of the planning horizon. [Why is it unnecessary to consider the remaining values of k (i.e., $k = 4$, 3, 2, and 1)? In order to answer this, consider the case of $k = 4$ and compare the cash flow profiles over the nine-year planning horizon using $k = 4$ and $k = 5$. As shown in Figure 4.10, a consideration of the time value of money eliminates the possibility of $k = 4$

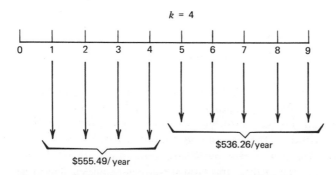

FIGURE 4.10. Comparison of replacement strategies involving replacement at the end of year 4 versus the end of year 5.

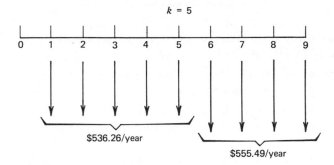

$k = 5$

$536.26/year

$555.49/year

FIGURE 4.10. (*cont'd*)

having a lower present worth than $k = 5$. Similar comparisons can be used to eliminate the cases of $k = 1, 2,$ and 3.]

For this particular example failure to replace the equipment at precisely the optimum time will not result in a significantly lower annual *worth* or present *worth*. This is often the case with many replacement situations; the resulting measure of effectiveness is relatively insensitive to deviations from the optimum strategy. Hence, the firm might establish an operating policy of reviewing the *actual* operating and maintenance costs for the equipment at the anticipated "optimum" time of replacement. Then, perform a replacement study at that time, using the eight-step procedure for comparing investment alternatives.

The subject of replacement analysis is popular among engineering economists. In fact, sufficient literature is available on the subject to devote an entire book to it. In fact, such books do exist, e.g. [37]. For further treatment of replacement analysis, you may wish to consult the *Engineering Economist* publication, Canada [9], Morris [28], and Terborgh [37], among others.

4.13 SUMMARY

The process of comparing investment alternatives was described in this chapter. A cash flow approach was recommended, as was a planning horizon approach. Replacement problems were treated as alternative comparison problems. Measures of merit examined included the present worth, annual worth, payback period, rate of return, and savings/investment ratio. The discussion of the comparison of investment alternatives did not consider either the effects of income taxes, the public sector requirements for measuring benefits in economic units, or the effects of risk and uncertainty; such considerations are reserved for the following chapters.

BIBLIOGRAPHY

1. Adler, M., "The True Rate of Return and the Reinvestment Rate," *The Engineering Economist, 15* (3), 1970, pp. 185–187.

2. Baldwin, R. H., "How to Assess Investment Proposals," *Harvard Business Review, 37* (3), 1959, pp. 98–99.

3. Beenhakker, H. L., "Discounting Indices Proposed for Capital Investment Evaluation: A Further Examination," *The Engineering Economist, 18* (3), 1973, pp. 149–168.

4. Bernhard, R. H., "Discount Methods for Expenditure Evaluation—A Clarification of Their Assumptions," *The Journal of Industrial Engineering, 13* (1), 1962, pp. 19–27.

5. Bernhard, R. H., "On the Inconsistency of the Soper and Sturm-Kaplan Conditions for Uniqueness of the Internal Rate of Return," *The Journal of Industrial Engineering, 18* (8), 1967, pp. 498–500.

6. Bernhard, R. H., "Mathematical Programming Models for Capital Budgeting—A Survey, Generalization, and Critique," *The Journal of Financial and Quantitative Analysis, 4* (2), 1969, pp. 111–158.

7. Bernhard, R. H., "A Comprehensive Comparison and Critique of Discounting Indices Proposed for Capital Investment Evaluation," *The Engineering Economist, 16* (3), 1971, pp. 157–186.

8. Bullinger, C. E., "Comments on 'A Critical Evaluation of the Field of Engineering Economy,'" *The Journal of Industrial Engineering, 15* (6), 1964, pp. 316–318.

9. Canada, J. R., *Intermediate Economic Analysis for Management and Engineers*, Prentice-Hall, 1971.

10. Canada, J. R., "Rate of Return: A Comparison Between the Discounted Cash Flow Model and a Model Which Assumes an Explicit Reinvestment Rate for the Uniform Income Flow Case," *The Engineering Economist, 9* (3), 1964, pp. 1–15.

11. de Faro, C., "On the Internal Rate of Return Criterion," *The Engineering Economist, 19* (3), 1974, pp. 165–194.

12. DeGarmo, E. P., and Canada, J. R., *Engineering Economy*, Fifth Edition, Macmillan, 1973.

13. Fleischer, G. A., "Two Major Issues Associated with the Rate of Return Method for Capital Allocation: The 'Ranking Error' and 'Preliminary Selection,'" *The Journal of Industrial Engineering, 17* (4), 1966, pp. 202–208.

14. Grant, E. G., Ireson, W. G., and Leavenworth, R. S., *Principles of Engineering Economy*, Sixth Edition, Ronald Press, 1976.

15. Grant, E. L., "Reinvestment of Cash Flow Controversy—A Review of: FINANCIAL ANALYSIS IN CAPITAL BUDGETING by Pearson Hunt," *The Engineering Economist*, *11* (3), 1966, pp. 23–29.

16. Heebink, D. V., "Rate of Return, Reinvestment and the Evaluation of Capital Expenditures," *The Journal of Industrial Engineering*, *13* (1), 1962, pp. 48–49.

17. Hirshleifer, J., "On the Theory of Optimal Investment Decision," *Journal of Political Economy*, *66* (5), 1958, pp. 329–352.

18. Horowitz, I., "Engineering Economy: An Economist's Perspective," *AIIE Transactions*, forthcoming.

19. Jeynes, P. H., "Minimum Acceptable Return," *The Engineering Economist*, *9* (4), 1964, pp. 9–25.

20. Jeynes, P. H., "Comment on Ralph O. Swalm's Letter," *The Engineering Economist*, *10* (3), 1965, pp. 38–42.

21. Jeynes, P. H., "The Significance of Reinvestment Rate," *The Engineering Economist*, *11* (1), 1965, pp. 1–9.

22. Kirshenbaum, P. S., "A Resolution of the Multiple Rate-of-Return Paradox," *The Engineering Economist*, *10* (1), 1964, pp. 11–16.

23. Klausner, R. F., "Communicating Investment Proposals to Corporate Decision-Makers," *The Engineering Economist*, *17* (1), 1971, pp. 45–55.

24. Lutz, R. P., "How to Justify the Purchase of Material Handling Equipment to Management," *Proceedings of the 1976 MHI Material Handling Seminar*, The Material Handling Institute, Pittsburgh, Pa., 1976.

25. Mao, J.C.T., "The Internal Rate of Return as a Ranking Criterion," *The Engineering Economist*, *11* (4), 1966, pp. 1–13.

26. Mao, J.C.T., "An Analysis of Criteria for Investment and Financing Decisions Under Certainty: A Comment," *Management Science*, *13* (3), 1966, pp. 289–291.

27. Mao, J.C.T., *Quantitative Analysis of Financial Decisions*, Macmillan, 1969.

28. Morris, W. T., *Engineering Economic Analysis*, Reston Publishing Company, 1976.

29. Oakford, R. V., *Capital Budgeting: A Quantitative Evaluation of Investment*, Ronald Press, 1970.

30. Radnor, M., "A Critical Evaluation of the Field of Engineering Economy," *The Journal of Industrial Engineering*, *15* (3), 1964, pp. 133–140.

31. Renshaw, E., "A Note on the Arithmetic of Capital Budgeting Decisions," *The Journal of Business*, *30* (3), 1957, pp. 193–204.

32. Solomon, E., "The Arithmetic of Capital Budgeting Decisions," *The Journal of Business, 29* (2), 1956, pp. 124–129.

33. Swalm, R. O., "On Calculating the Rate of Return on an Investment," *The Journal of Industrial Engineering, 9* (2), 1958, pp. 99–103.

34. Swalm, R. O., "Comment on 'Minimum Acceptable Return' by Paul H. Jeynes," *The Engineering Economist, 10* (3) 1965, pp. 35–38.

35. Teichroew, D., Robicheck, A. A., and Montalbano, M., "Mathematical Analysis of Rates of Return Under Certainty," *Management Science, 11* (3), 1965, pp. 395–403.

36. Teichroew, D., Robicheck, A. A., and Montalbano, M., "An Analysis of Criteria for Investment and Financing Decisions Under Certainty," *Management Science, 12* (3), 1965, pp. 151–179.

37. Terborgh, G., *Business Investment Management*, Machinery and Allied Products Institute, 1958.

38. Thuesen, H. G., Fabrycky, W. J., and Thuesen, G. J., *Engineering Economy*, Fifth Edition, Prentice-Hall, 1977.

39. Weaver, J. B., "False and Multiple Solutions by the Discounted Cash Flow Method for Determining Interest Rate of Return," *The Engineering Economist, 3* (4), 1958, pp. 1–31.

40. White, J. A., Case, K. E., and Agee, M. H., "Rate of Return: An Explicit Reinvestment Rate Approach," *Proceedings of the 1976 AIIE Conference*, American Institute of Industrial Engineers, Norcross, Ga., 1976.

41. *Economic Analysis Handbook*, Publication P-442, Department of the Navy, Naval Facilities Engineering Command, Washington, D.C., 1971.

PROBLEMS

1. Three investment proposals have been collected by the capital budget committee. Proposals A and C are mutually exclusive; Proposal A is contingent on Proposal B; and Proposal B is contingent on Proposal C. A budget limitation of $100,000 exists. The cash flow profiles for the three investment proposals are given below for the five-year planning horizon.

EOY	CF(A)	CF(B)	CF(C)
0	−$60,000	−$40,000	−$80,000
1–5	20,000	10,000	15,000
5	—	20,000	20,000

Using a minimum attractive rate of return of 5%, compare the set of feasible, mutually exclusive alternatives and specify the preferred alternative. Use all measures of merit described in the text. Assume a payback period of 3 year applies for the payback period method. (4.2, 4.7)

2. Four investment proposals, A, B, C, and D, are available. Proposals A and B are mutually exclusive and Proposals A and C are mutually exclusive. Proposal D is contingent on either Proposal A or B. Funds available for investment are limited to $50,000. Using a *MARR* of 15% and a present worth analysis, determine the recommended investment program. Justify the elimination of any mutually exclusive alternatives. (4.2, 4.7)

EOY	CF(A)	CF(B)	CF(C)	CF(D)
0	-$20,000	-$30,000	-25,000	-20,000
1-4	7,000	9,000	8,000	8,500

3. Two mutually exclusive proposals, each with a life of five years, are under consideration. Each proposal has the following cash flow profile:

EOY	CF(A)	CF(B)
0	-$20,000	-$34,000
1	4,878	4,762
2	4,878	4,762
3	4,878	4,762
4	4,878	4,762
5	4,878	4,762

(a) Specify clearly the mutually exclusive *alternatives* available to the decision maker. (4.2)
(b) Using a rate of return analysis, which *alternative* should the decision maker select? Assume *MARR* = 10%. Use both the internal rate-of-return method and the external rate-of-return method. (4.7)

4. A firm has available four proposals, A, B, C, & D. Proposal A is contingent on acceptance of either Proposal C or Proposal D. The firm has a budget limitation of $200,000. In addition, Proposal C is contingent on Proposal D, while Proposal D is contingent on either Proposal A or Proposal B. Using a *MARR* of 10%, determine the preferred alternative using the savings/investment ratio method. (4.2, 4.7)

EOY	CF(A)	CF(B)	CF(C)	CF(D)
0	-$100,000	-$140,000	-$20,000	-$15,000
1	10,000	7,500	3,000	1,000
2	10,000	8,500	3,000	1,500
3	10,000	9,500	3,000	2,000
4	10,000	10,500	3,000	2,500
5	20,000	30,000	3,000	3,000

5. A firm has available three investment proposals, A, B, and C, having the cash flow profiles shown below. Proposals B and C are mutually exclusive and Proposal C is contingent on Proposal A being chosen. The firm has a *MARR* of 20%.

	CF(A)	CF(B)	CF(C)
Initial investment	$200,000	$300,000	$150,000
Life	8 years	12 years	8 years
Annual receipts	$160,000	$190,000	$200,000
Annual disbursements	$115,000	$120,000	$150,000
Salvage value	$ 50,000	$100,000	$ 50,000

The firm is willing to use a planning horizon of 24 years and accepts the assumption of identical cash flow profiles for successive life cycles of a proposal.
   (a) Determine the set of feasible alternatives and give the cash flow profile for each over the planning horizon. (4.2, 4)
   (b) Determine the preferred alternative, based on an annual worth analysis. (4.7)

6. Four investment proposals W, X, Y, and Z, are being considered by the Ajax Corporation. Proposals X and Z are mutually exclusive. Proposal Y is contingent on either X or Z. Proposals W and Y are mutually exclusive. A budget limitation of $200,000 exists. Either Proposal X or Proposal Z must be included in the alternative selected. Using a *MARR* of 20%, determine the annual worth for each *feasible investment alternative.* (4.2, 4.7)

EOY	CF(W)	CF(X)	CF(Y)	CF(Z)
0	−$100,000	−$125,000	−$ 90,000	−$100,000
1–8	30,000	50,000	25,000	20,000
8	40,000	10,000	120,000	100,000

7. A firm is faced with three investment proposals A, B, and C, having the cash flow profiles shown below over the planning horizon of five years. Proposals B and C are mutually exclusive, and Proposal C is contingent on Proposal A being selected. A budgetary limitation of $125,000 exists on the amount that can be invested at the end-of-year (EOY) zero. A *MARR* of 10% is to be used in performing a present worth analysis.
   (a) Specify the cash flow profiles for all mutually exclusive, feasible investment alternatives. (4.4)
   (b) Determine which investment alternative yields the maximum present worth. (4.6)

EOY	CF(A)	CF(B)	CF(C)
0	−$100,000	−$50,000	−$20,000
1	20,000	20,000	5,000
2	30,000	20,000	5,000
3	40,000	20,000	5,000
4	50,000	20,000	5,000
5	60,000	20,000	15,000

8. Ms. Brown invested $64,000 in a business venture with the following cash flow results:

EOY	CF	EOY	CF	EOY	CF
0	−$64,000	3	$ 9,000	6	$9,000
1	5,000	4	11,000	7	7,000
2	7,000	5	11,000	8	5,000

To the nearest percent, what was her internal rate of return on the venture? (4.6)

9. Mr. Smith invests $10,000 in an investment fund. One year after making the investment he receives $1000 and continues to receive $1000 annually until 10 such amounts are received. He receives nothing further until 15 years after the initial investment, at which time he receives $10,000. Over the 15-year period, what was the man's internal rate of return? (4.6)

10. A firm purchases a pressure vessel for $20,000. Half of the purchase price is borrowed from a bank at 6% compounded annually. The loan is to be paid back with equal annual payments over a five-year period. The pressure vessel is expected to last 10 years, at which time it will have a salvage value of $4000. Over the 10-year period the operating and maintenance costs are anticipated to equal $4000/year. The firm expects to earn 10% on its investments. What is the equivalent uniform annual cost for the investment? (4.6)

11. A firm purchases an analog computer for $10,000; the computer is used for five years and is then sold for $2000. Annual disbursement for operating and maintenance costs equaled $4000/year over the five-year period. On the basis of a 20% MARR, what was the equivalent annual worth for the investment? (4.6)

12. A firm purchases a gravity settling tank by borrowing the $10,000 purchase price. The loan is to be repaid with six equal annual payments at an annual compound rate of 7%. It is anticipated that the tank will be used for nine years and then be sold for $1000. Annual operating and maintenance expenses are estimated to be $5000/year. The firm uses a MARR of 15% for its economic analyses. Determine the equivalent uniform annual cost for the piece of equipment. (4.6)

13. A firm borrows $50,000 to purchase a new numerically controlled milling machine and pays the loan back over a five-year period with equal payments. Interest on the loan is 10% compounded annually. The machine is estimated to have annual operating and maintenance costs of $12,000/year and have a life of 10 years. Salvage value is estimated to be $15,000. The firm has a *MARR* of 8%. Determine the equivalent uniform annual cost over the 10-year period. (4.6)

14. The Belt Company is considering four investment proposals, A, B, C, and D. Proposals A and B are mutually exclusive, and Proposal C is contingent on Proposal D. Proposals A and D are mutually exclusive. The cash flow data for the investments over a 10-year planning horizon are given below. The Belt Company has a budget limit of $750,000 for investments of the type being considered currently.

	CF(A)	CF(B)	CF(C)	CF(D)
Initial investment	$450,000	$600,000	$350,000	$300,000
Life	10 years	10 years	10 years	10 years
Salvage value	$ 50,000	$100,000	$ 50,000	$ 50,000
Annual receipts	$300,000	$450,000	$200,000	$350,000
Annual disbursements	$100,000	$200,000	$ 50,000	$200,000
MARR = 25%				

(a) State all mutually exclusive investment alternatives available. (4.2)
(b) Develop the cash flow profiles for the mutually exclusive alternatives. (4.4)
(c) Compare the alternatives using the present worth method. (4.7)

15. The Acme Manufacturing Company purchased an automatic transfer machine for $15,000 installed. At the end of 10 years the company paid $2000 to have the machine removed and junked. The operating and maintenance cost history was as follows.

EOY	O&M Cost
1	$1000
2	1000
3	1000
4	1000
5	1500
6	1500
7	1500
8	1500
9	2000
10	2000

The company has a 10% *MARR*. Supply the values of U, V, W, X, and Y in the following equation for computing the equivalent uniform annual cost for the machine. (4.6)

$$(\$15,000 - U)(A/F\ 10\%, 10) + V(0.10) + W +$$
$$\$500[(F/A\ 10\%, 6) + (F/A\ 10\%, X)](Y, 10\%, 10)$$

16. A distillation column is purchased for \$200,000. Operating and maintenance costs for the first year are \$20,000. Thereafter, operating and maintenance costs increase by 10%/year over the previous year's costs. At the end of 10 years the column is sold for \$50,000. During the life of investment revenue was produced that could be related directly to the investment in the column. The revenue the first year was \$50,000. Thereafter, revenue increased by \$4000 over the previous year's revenue. Using a *MARR* of 15%, determine the equivalent present worth for the investment. (4.6)

17. Machine A initially costs \$40,000. Its resale value at the end of the kth year of service equals $\$40,000 - \$3000k$ for $k = 1, 2, \ldots, 12$. Operating and maintenance expenses equal \$8000/year. Machine B initially costs \$18,000. Its resale value at the end of the kth year, S_k, equals $\$18,000(0.80)^k$ for $k = 1, 2, \ldots, 8$. Operating and maintenance expenses equal \$10,000/year. What is the equivalent uniform annual cost for each machine, based on a planning horizon of (a) 24 years? (b) 5 years? Use a *MARR* of 10%. Assume the recommended machine will be kept until the end of the planning horizon. (4.6)

18. The Ajax Manufacturing Company wishes to choose one of the following machines.

	Machine A	Machine B	Machine C
First cost	\$10,000	\$14,000	\$18,000
Planning horizon	8 years	8 years	8 years
Salvage value	—	\$2,000	\$3,000
Operating and maintenance cost for year k, $k = 1, \ldots, n$	$\$600(1.10)^{k-1}$	$\$400(1.08)^{k-1}$	$\$200 + 100k$

What is the equivalent uniform annual cost for each machine, based on a *MARR* of (a) 0%, (b) 8%, (c) 10%? Assume machines are not replaced during the planning horizon (4.6)

19. Mr. Shrewd invested \$100,000 in a business venture with the following cash flow results. Determine his rate of return using the internal rate of return method. (4.6)

EOY	CF	EOY	CF	EOY	CF
0	−$100,000	7	$14,000	14	$7,000
1	20,000	8	13,000	15	6,000
2	19,000	9	12,000	16	5,000
3	18,000	10	11,000	17	4,000
4	17,000	11	10,000	18	3,000
5	16,000	12	9,000	19	2,000
6	15,000	13	8,000	20	1,000

20. The Ajax Manufacturing Company wishes to choose one of the following machines

	Machine A	Machine B
First cost	$10,000	$16,000
Planning horizon	6 years	6 years
Salvage value	$1,000	$1,000
Operating and maintenance cost for year k	$500 + 80k$	$200 + 100k$

What are the equivalent uniform annual costs for each machine, based on a 10% *MARR*? (4.6)

21. Owners of a nationwide motel chain are considering building a new 200-unit motel in Fairfax, New Mexico. The initial cost of building the motel is $2,250,000; the firm estimates furnishings for the motel will cost an additional $700,000 and will require replacement every five years. Annual operating and maintenance costs for the facility are estimated to be $60,000. The average rate for a unit is anticipated to be $14/day. A 15-year planning horizon is used by the firm in evaluating new ventures of this type; a terminal salvage value of 15% of the original building cost is anticipated; furnishings are estimated to have no salvage value at the end of each five-year replacement interval. Assuming an average daily occupancy percentage of 70%, 80%, and 90%, a *MARR* of 20%, and 365 operating days/year, should the motel be built? Ignore the cost of the land. (4.7)

22. A consulting engineer is designing an irrigation system and is considering two pumps to meet a pumping demand of 15,000 gallons/minute at 15 feet total dynamic head. The specific gravity of the fertilizer mixture being pumped is 1.25. Pump A operates at 78% efficiency and costs $9,450; Pump B operates at 86% efficiency and costs $15,250. Power costs $0.015/kilowatt-hour. Continuous pumping for 200 days/year is required (i.e., 24 hours/day). Using a *MARR* of 10%, a planning horizon of five years, and equal salvage values for the two pumps, which should be selected? (*Note.* Dynamic head × gallons/minute × specific gravity × 0.746 ÷ 3960 = kilowatts.) (4.7)

23. The motor on a numerically controlled machine tool must be replaced. Two different 20-horsepower electric motors are being considered. Motor U sells for $850 and has an efficiency rating of 90%; Motor V sells for $625 and has a rating of 84%. The cost of electricity is $0.032/kilowatt-hour. An eight-year planning horizon is used, and zero salvage values are assumed for both motors. Annual usage of the motor averages approximately 4000 hours. A *MARR* of 20% is to be used. Assume the motor selected will be loaded to capacity. Compare the alternatives using present worth, savings/investment ratio, internal rate-of-return, and external rate-of-return methods. (*Note.* 0.746 kilowatt = 1 horsepower). (4.7)

24. A chemical plant is considering the installation of a storage tank for water. The tank is estimated to have an initial cost of $142,000; annual costs for maintenance are estimated to be $1200/year. As an alternative, a holding pond can be provided some distance away at an initial cost of $90,000 for the pond, plus $15,000 for pumps and piping; annual operating and maintenance costs for the pumps and holding pond are estimated to be $4500. Based on a 20-year planning horizon, zero salvage values, and a *MARR* of 10%, which alternative is preferred on the basis of a present worth alternative? (4.7)

25. An investment of $16,000 for a new condenser is being considered. Estimated salvage value of the condenser is $5000 at the end of an estimated life of eight years. Annual gross income each year for the eight years is $4500. Annual operating expenses are $1500. Assume money is worth 25% compounded annually.
    (a) Using the annual worth method, should the investment in the condenser be made?
    (b) Is the internal rate of return on this investment greater than 25% or less than 25%? (4.7)

26. An individual is considering two mutually exclusive investment alternatives, each requiring an initial investment of $1000. Alternative 1 returns $200 after one year and $1200 after two years; Alternative 2 returns $150/year for the first three years and $1150 after four years. The individual has established a *MARR* of 8%. Which alternative should be selected using internal rate-of-return, external rate-of-return, present worth, and annual worth methods? (4.7)

27. Peachtree Creek in Atlanta has a history of flooding during heavy rainfalls. Two improvement alternatives are under consideration by the city engineer.
    A Leave the existing 36-inch corrugated steel culvert in place and install another of the same size alongside at an installed cost of $4200.
    B Remove the culvert in place currently and replace with a single 60-inch culvert at an installed cost of $6400.
    If Alternative A is adopted, the existing 36-inch culvert will need to be replaced in 10 years. It is assumed that a replacement at that time will cost $5000. A newly installed culvert will last for 20 years; replacement prior to the end of 20 years will result in a zero salvage value. Use a planning horizon of 20 years and a 10% *MARR* to determine the preferred alternative. (4.7)

28. Insulation is to be installed on a steam pipe. Either 1-inch or 2-inch insulation is to be selected. The annual heat loss from the pipe at present is estimated to be

$2.25/foot of pipe. The 1-inch insulation costs $0.60/foot and will reduce the heat loss by 86%; the 2-inch insulation costs $1.15/foot and will reduce the heat loss by 90%. If insulation lasts 10 years with no salvage value and an 8% *MARR* is used, which size (if any) should be selected? (4.7)

29. The state civil defense organization wishes to establish a communications network to cover the state. It is desired to maintain a specified minimum signal strength at all points in the state. Two alternatives have been selected for detailed consideration. Design I involves the installation of 10 transmitting stations of low power. The investment at each installation is estimated to be $40,000 in structure and $25,000 in equipment. Design II involves the installation of two transmitting stations of much higher power. Investment in structure will be approximately $50,000/installation; equipment at each installation will cost $200,000. Annual operating and maintenance costs per installation are anticipated to be $15,000 for Design I and $18,000 for Design II. Structures are anticipated to last for 20 years; equipment life is estimated to be 10 years, with replacements assumed to be identical in costs and performance. Using a planning horizon of 20 years and a *MARR* of 10%, which design should be recommended on the basis of a present worth analysis? (4.7)

30. Two compressors are being considered by the Ajax Company. Compressor A can be purchased for $5500; annual operating and maintenance costs are estimated to be $1950. Alternatively, Compressor B can be purchased for $4150; annual operating and maintenance costs are estimated to be $2325. An eight-year planning horizon is to be used; salvage values are estimated to be 15% of the original purchase price; a *MARR* of 25% is to be used. Compare the alternatives using both internal and external rate-of-return methods, as well as the savings/investment ratio. (4.7)

31. The manager of the distribution center for the southeastern territory of a major pharmaceutical manufacturer is contemplating installing an improved material handling system linking receiving and storage, as well as storage and shipping. Two designs are being considered. The first consists of a conveyor system that is tied into an automated storage/retrieval system. Such a system is estimated to cost $750,000 initially, have annual operating and maintenance costs of $48,000, and a salvage value of $75,000 at the end of the 15-year planning horizon.
The second design consists of manually operated narrow aisle, high stacking lift trucks. To provide service comparable to that provided by the alternative design, an initial investment of $325,000 is required. Annual operating and maintenance costs of $103,000 are anticipated. An estimated salvage value of $20,000 is expected at the end of the planning horizon.
Using a *MARR* of 20%, compare the alternatives using annual worth, savings/investment ratio, and rate-of-return methods. (4.7)

32. A manufacturing plant in Minnesota has been contracting snow removal at a cost of $250/day. The past four years have yielded unusually heavy snowfalls, resulting in the cost of snow removal being of concern to the plant manager. The plant engineer has found that a snow-removal machine can be purchased for $38,000; it is estimated to have a useful life of 10 years, and a zero salvage value at that time.

Annual costs for operating and maintaining the equipment are estimated to be $12,000. Based on a *MARR* of (a) 0%, (b) 10%, (c) 20%, and (d) 30% and an estimated demand for the equipment equal to (a) 40, (b) 80, (c) 100, and (d) 150 days per year, determine the preferred alternative using an annual worth analysis. (4.7)

33. Given the following data, which pump is more economical, using the annual cost comparison? Assume a 10-year planning horizon. (4.7)

	Pump A	Pump B
First cost	$6,000	$10,000
Salvage value	0	2,000
Annual operating cost	500	750
Life	5 years	10 years

34. A firm is considering two compressors X and Y. One must be chosen. Based on the following data, compare the present worths of each alternative and recommend the compressor having the largest present worth. Use a *MARR* = 10% and a planning horizon of 42 years. (4.7)

	Compressor X	Compressor Y
Initial investment	$10,000	$20,000
Life	6 years	7 years
Salvage value	$ 1,000	$ 5,000
Annual disbursements	$ 5,000	$ 3,000

35. A firm has available two mutually exclusive investment proposals, A and B. Based on the cash flow profiles shown below and using a savings/investment ratio analysis, recommend the preferred *alternative*. Use a *MARR* of 10%. (4.7)

EOY	CF(A)	CF(B)
0	-$40,000	-$30,000
1	6,000	6,500
2	6,000	6,000
3	6,000	5,500
4	6,000	5,000
5	6,000	4,500
6	6,000	4,000
7	6,000	3,500
8	6,000	3,000
9	6,000	2,500
10	6,000	2,000

36. Two mutually exclusive alternatives are to be evaluated and the least cost alternative specified. The cash flow profiles for the alternatives are shown below.

EOY	CF(A)	CF(B)
0	−$10,000	−$15,000
1	1,000	1,500
2	1,500	2,000
3	—	500
4	2,500	3,000
5	3,500	4,000
6	—	500
7	4,500	5,000
8	6,000	11,500

Using a 12% before-tax *MARR* and assuming one of the alternatives must be chosen, compare the alternatives by employing an internal rate-of-return approach. (4.7)

37. Consider the following investment decision.

	Machine A	Machine B
First cost	$15,000	$20,000
Estimated life	5 years	10 years
Estimated annual revenues	$ 9,000	$11,066
Estimated annual operative costs	$ 6,000	$ 8,000
Estimated salvage value at end of life	$ 1,500	$ 3,000

(a) If the *MARR* is 10% and a 10-year planning horizon is assumed (two cycles of A), which of the machines should be purchased, if either? Solve by using the annual worth method.
(b) What are the approximate internal and external rates of return for the incremental investment in Machine B?
(c) What would the salvage value of Machine A have to be in order for this investment to yield an internal and an external rate of return equal to 10%? (4.7)

38. Using the rate-of-return method, compare the three mutually exclusive investment alternatives having the cash flow profiles shown below. Base your analysis on a 30% *MARR*. Use both future worth and savings/investment methods. (4.7)

EOY	CF(A)	CF(B)	CF(C)
0	−$8000	−$5000	0
1	3400	1000	0
2	3400	2000	0
3	3400	3000	0
4	3400	4000	0
5	3400	5000	0

39. Two numerically controlled drill presses are being considered by the production department of a major corporation; one must be selected. Both machines meet the quality and safety standards of the firm. Comparative data are as follows.

	Drill Press X	Drill Press Y
Initial investment	$20,000	$30,000
Estimated life	10 years	10 years
Salvage value	$ 5,000	$ 7,000
Annual operating cost	$12,000	$ 6,000
Annual maintenance cost	$ 2,000	$ 4,000

Using a 10% interest rate and a present worth comparison, which machine is preferred? (4.7)

40. A firm is considering either leasing or buying a small computer system. If purchased, the initial cost will be $200,000; annual operating and maintenance costs will be $80,000/year. Based on a five-year planning horizon, it is anticipated the computer will have a salvage value of $50,000 at that time. If the computer is leased, annual operating and maintenance costs in excess of the annual lease payment will be $60,000/year. Based on an interest rate of 10%, what annual end-of-year lease payment will make the firm be indifferent between leasing and buying, based on economic considerations alone? (4.7)

41. Company W is considering investing $10,000 in a heat exchanger. The heat exchanger will last eight years, at which time it will be sold for $2000. Maintenance costs for the exchanger are estimated to increase by $200/year over its life. The maintenance cost for the first year is estimated to be $1000. As an alternative, the company may lease the equipment for $X/year, including maintenance. For what value of X should the company lease the heat exchanger? The company expects to earn 10% on its investments. Assume end-of-year lease payments. (4.7)

42. A firm is faced with four investment proposals, A, B, C, and D, having the cash flow profiles shown below. Proposals A and C are mutually exclusive, and Proposal D is contingent on Proposal B being chosen. Currently, $400,000 is available for investment, and the firm has stipulated a MARR of 20%. Determine the preferred alternative. (4.7)

	CF(A)	CF(B)	CF(C)	CF(D)
Initial investment	$200,000	$200,000	$300,000	$150,000
Planning horizon	10 years	10 years	10 years	10 years
Annual receipts	$140,000	$160,000	$200,000	$200,000
Annual disbursements	$110,000	$125,000	$120,000	$150,000
Salvage value	$ 50,000	$ 50,000	$100,000	$ 50,000

43. A firm will either lease an office copier at an end-of-year cost of $10,000 for a 10-year period, or they will purchase the copier at an initial cost of $66,117. If purchased, the copier will have a zero salvage value at the end of its 10-year life. No other costs are to be considered. Using rate-of-return methods, should the firm lease or buy if their *MARR* is (a) 6%, or (b) 10%? Justify your answer on the basis of an internal rate-of-return analysis. (4.7)

44. A company can construct a new warehouse for $700,000, or it can lease an equivalent building for $75,000/year for 25 years with the option of purchasing the building for $100,000 at the end of the 25-year period. Lease payments are due at the *beginning* of each year. The company can earn 12%/year before taxes on its invested capital. Indicate the preferred alternative. (4.7)

45. In Problem 4.44, for what (a) beginning-of-year (b) end-of-year lease payment will the company be indifferent between leasing and buying?

46. The XYZ Company must decide if they should purchase a small computer or lease the computer. The computer costs $30,000 initially and will last five years, having a $5000 salvage value at that time. If the computer is purchased, all maintenance costs must be paid by the XYZ Company. Maintenance costs are $2000/year over the life of the equipment. The XYZ Company uses an interest rate of 10% in evaluating investment alternatives. For what end-of-year annual leasing charge is the firm indifferent between purchasing and leasing over the five-year period? (4.7)

47. A firm is considering replacing its material handling system and either purchasing or leasing a new system. The old system has an annual operating and maintenance cost of $12,000, a remaining life of 10 years, and an estimated salvage value of $5000 in 10 years.
A new system can be purchased for $100,000; it will be worth $20,000 in 10 years; and it will have annual operating and maintenance costs of $8000/year. If the new system is purchased, the old system can be sold for $20,000.
Leasing a new system will cost $5000/year, payable at the beginning of the year, plus operating costs of $5000/year, payable at the end of the year. If the new system is leased, the old system will be scrapped at no value.
Use a *MARR* of 10% to compare the annual worths of keeping the old system, buying a new system, and leasing a new system. (4.12)

48. A firm is considering replacing a compressor that was purchased four years ago for $50,000. Currently, the compressor has a book value of $30,000, based on straight line depreciation. If the compressor is retained it will probably be used for four years more, at which time it is estimated to have a salvage value of $10,000. If

the old compressor is retained for eight years more, it is estimated that its salvage value will be negligible at that time. Operating and maintenance costs for the compressor have been increasing at a rate of $1000/year, with the cost during the past year being $9000.

A new compressor can be purchased for $60,000. It is estimated to have uniform annual operating and maintenance costs of $8000/year. The salvage value for the compressor is estimated to be $30,000 after four years and $15,000 after eight years. If a new compressor is purchased, the old compressor will be traded in for $20,000.

Using a before-tax analysis, a *MARR* of 6%, and an annual worth comparison, determine the preferred alternative using a planning horizon of (a) four years, and (b) eight years. (4.12)

49. A firm is contemplating replacing a computer they purchased three years ago for $400,000. Operating and maintenance costs have been $75,000/year. Currently the computer has a trade-in value of $250,000 toward a new computer that costs $600,000 and has a life of five years, with a value of $250,000 at that time. The new computer will have annual operating and maintenance costs of $80,000.

If the current computer is retained, another small computer will have to be purchased in order to provide the required computing capacity. The smaller computer will cost $300,000, has a value of $50,000 in five years, and have annual operating and maintenance costs of $50,000.

Using an annual worth comparison before taxes, with a *MARR* of 30%, determine the preferred course of action. (4.12)

50. A firm owns a pump that it is contemplating replacing. The old pump has annual operating and maintenance costs of $5000/year, it can be kept for five years more and will have a zero salvage value at that time.

The old pump can be traded in on a new pump. The trade-in value is $2500, with the purchase price for the new pump being $12,000. The new pump will have a value of $5000 in five years and will have annual operating and maintenance costs of $2000/year.

Using a *MARR* of 20%, evaluate the investment alternative using the present worth method. (4.12)

51. A company owns a five-year old turret lathe that has a book value of $10,000. The present market value for the lathe is $8000. The expected decline in market value is $1000/year to a minimum market value of $1000. Maintenance plus operating costs for the lathe equal $2200/year.

A new turret lathe can be purchased for $20,000 and will have an expected life of 12 years. The market value for the turret lathe is expected to equal $20,000(0.80)^k$ at the end of year k. Annual maintenance and operating cost is expected to equal $600.

Based on a 10% before-tax *MARR*, should the old lathe be replaced now? Use an equivalent uniform annual cost comparison. (4.12)

52. The ABC Company has an overhead crane that has an estimated remaining life of 10 years. The crane can be sold for $6000. If the crane is kept in service it must be overhauled immediately at a cost of $3000. Operating and maintenance costs will

be $2000/year after the crane is overhauled. After overhauling it, the crane will have a zero salvage value at the end of the 10-year period. A new crane will cost $16,000, will last for 10 years, and will have a $3000 salvage value at that time. Operating and maintenance costs are $1000 for the new crane. The company uses an interest rate of 10% in evaluating investment alternatives. Should the company buy the new crane? (4.12)

53. The Ajax Specialty Items Corporation has received a five-year contract to produce a new product. To do the necessary machining operations, the company is considering two alternatives.
Alternative A involves continued use of the currently owned lathe. The lathe was purchased five years ago for $12,000. Today the lathe is worth $5000 on the used machinery market. If this lathe is to be used, special attachments must be purchased at a cost of $2000. At the end of the five-year contract, the lathe (with attachments) can be sold for $1000. Operating and maintenance costs will be $4000/year if the old lathe is used.
Alternative B is to sell the currently owned lathe and buy a new lathe at a cost of $15,000. At the end of the five-year contract, the new lathe will have a salvage value of $8000. Operating and maintenance costs will be $2500/year for the new lathe.
Using a present worth analysis, should the firm use the currently owned lathe or buy a new lathe? Base your analysis on a minimum attractive rate of return of 20%. (4.12)

54. The Telephone Company of America purchased a numerically controlled production machine five years ago for $450,000. The machine currently has a trade-in value of $100,000. If the machine is continued in use, another machine, X, must be purchased to supplement the old machine. Machine X costs $300,000, has annual operating and maintenance costs of $50,000, and will have a salvage value of $50,000 in 10 years. If the old machine is retained, it will have annual operating and maintenance costs of $80,000 and will have a salvage value of $20,000 in 10 years.
As an alternative to retaining the old machine, it can be replaced with Machine Y. Machine Y costs $600,000, has anticipated annual operating and maintenance costs of $100,000, and has a salvage value of $200,000 in 10 years.
Using a MARR of 20% and an annual worth comparison, determine the preferred economic alternative. (4.12)

55. A small foundry is considering the replacement of a No. 1 Whiting cupola furnace that is capable of melting gray iron only with a reverberatory-type furnace that has gray iron and nonferrous metals melting capability. Both furnaces have approximately the same melting rates for gray iron in pounds per hour. The foundry company plans to use the reverberatory furnace, if purchased, primarily for melting gray iron, and the total quantity melted is estimated to be about the same with either furnace. Annual raw material costs would therefore be about the same for each furnace. Available information and cost estimates for each furnace is given below.
Cupola furnace. Purchased used and installed eight years ago for a cost of $5000. The present market value is determined to be $2000. Estimated remaining life is

somewhat uncertain but, with repairs, the furnace should remain functional for seven years more. If kept seven years more, the salvage value is estimated as $500 and average annual expenses expected are:

Fuel	$10,000
Labor (including maintenance)	$12,000
Payroll taxes	10% of direct Labor costs
Taxes and insurance on furnace	1% of purchase price
Other	$ 6,000

Reverberatory furnace. This furnace costs $8000. Expenses to remove the cupola and install the reverberatory furnace are about $600. The new furnace has an estimated salvage value of $800 after seven years of use and annual expenses are estimated as:

Fuel	$7500
Labor (operating)	$9000
Payroll taxes	10% of direct labor costs
Taxes and insurance on furnace	1% of purchase price
Other	$6000

In addition, the furnace must be relined every two years at a cost of $1000/occurrence. If the foundry presently earns an average of 20% on invested capital before income taxes, should the cupola furnace be replaced by the reverberatory furnace? (4.12)

56. A firm has an automatic chemical mixer that it has been using for the past four years. The mixer originally cost $18,000. Today the mixer can be sold for $9000. The mixer can be used for 10 years more and will have a $2000 salvage value at that time. The annual operating and maintenance costs for the mixer equal $5000/year.

Because of an increase in business, a new mixer must be purchased. If the old mixer is retained, a new mixer will be purchased at a cost of $16,000 and have a $2000 salvage value in 10 years. This new mixer will have annual operating and maintenance costs equal to $4000/year.

The old mixer can be sold and a new mixer of larger capacity purchased for $28,000. This mixer will have a $3000 salvage value in 10 years and will have annual operating and maintenance costs equal to $8000/year.

Based on a *MARR* of 15%, what do you recommend? (4.12)

57. A firm is presently using a machine that has a market value of $8000 to do a specialized production job. The requirement for this operation is expected to last only six years more, after which it will no longer be done. The predicted costs and salvage values for the present machine are:

Year	1	2	3	4	5	6
Operating cost	$1000	$1200	$1400	$1800	$2300	$3000
Salvage value	$5000	$4500	$4000	$3300	$2500	$1400

A new machine has been developed that can be purchased for $12,000 and has the following predicted cost performance.

Year	1	2	3	4	5	6
Operating cost	$ 500	$ 700	$ 900	$1,200	$1,500	$1,900
Salvage value	$11,000	$10,500	$10,000	$9,500	$8,500	$7,500

If interest is at 0%, when should the new machine be purchased? (4.12)

58. A firm has received a production contract for a new product. The contract lasts for five years. To do the necessary machining operations, the firm can use one of its own lathes, which was purchased four years ago at a cost of $11,000. Today the lathe can be sold for $5000. In five years the lathe will have a zero salvage value. Annual operating and maintenance costs for the lathe are $4000/year. If the firm uses its own lathe it must also purchase an additional lathe at a cost of $7000; its value in five years will be $1000. The new lathe will have annual operating and maintenance costs of $2500/year.

As an alternative the presently owned lathe can be sold and a new lathe of larger capacity purchased for a cost of $16,000; its value in five years is estimated to be $5000, and its annual operating and maintenance costs will be $5000/year.

An additional alternative is to sell the presently owned lathe and subcontract the work to another firm. Company X has agreed to do the work for the five-year period at an annual cost of $8000/end-of-year.

Using a 10% interest rate, determine the least cost alternative for performing the required production operations. (4.12)

59. A machine was purchased five years ago for $8000. At that time, its estimated life was 10 years with an estimated end-of-life salvage value of $800. The average annual operating and maintenance costs have been $12,000 and are expected to continue at this rate for the next five years. However, average annual revenues have been and are expected to be $15,000. Now, the firm can trade in the old machine for a new machine for $3000. The new machine has a list price of $10,000, an estimated life of 10 years, annual operating plus maintenance costs of $5000, annual revenues of $9000, and salvage values at the end of the jth year according to

$$S_j = \$10,000 - \$1000\, j, \qquad \text{for } j = 0, 1, 2, 3, 4, 5, 6, 7, 8, 9, 10$$

Determine whether to replace or not by the annual equivalent method using a *MARR* equal to 10% compounded annually. (Use a five-year planning horizon.) (4.12)

60. An automatic car wash has been experiencing difficulties in keeping its equipment operational. The owner is faced with the alternative of overhauling the present equipment or replacing it with new equipment. The cost of overhauling the present equipment is $5000. The present equipment has annual operating and maintenance costs of $5000. If it is overhauled, the present equipment will last for five years

more and be scrapped at zero value. If it is not overhauled, it has a trade-in value of $2000 toward the new equipment.

New equipment can be purchased for $20,000. At the end of five years the new equipment will have a resale value of $8000. Annual operating and maintenance costs for the new equipment will be $2000.

Using a *MARR* of 8%, what is your recommendation to the owner of the car wash. Base your recommendation on a present worth comparison. (4.12)

61. A highway contractor must decide whether to overhaul a tractor and scraper or replace it. The old equipment was purchased five years ago for $70,000; it had a projected life of 15 years, with a $5000 salvage value at that time. If traded in on a new tractor and scraper, it can be sold for $30,000. Overhauling the old equipment will cost $10,000. If overhauled, operating and maintenance costs will be $8000/year, and the overhauled equipment will have a projected salvage value of $4000 in 10 years.

A new tractor and scraper can be purchased for $80,000; it will have annual operating and maintenance costs of $5000, and will have a salvage value of $40,000 in 10 years.

Using an annual worth comparison with a *MARR* of 10%, should the equipment be overhauled or replaced? (4.12)

62. A highway construction firm purchased an item of earth-moving equipment three years ago for $50,000. The salvage value at the end of 10 years was estimated to be 34% of first cost. The firm earns an average annual gross revenue of $45,000 with the equipment and the average annual operating costs have been and are expected to be $26,000.

The firm now has the opportunity to sell the equipment for $38,000 and sub-contract the work normally done by the equipment over the next seven years. If the subcontracting is done, the average annual gross revenue will remain $45,000, but the subcontractor charges $35,000/end-of-year for these services.

If a 25% rate of return before taxes is desired, determine whether the firm should subcontract or not by the annual worth method. (4.12)

63. A building supplies distributor purchased a gasoline-powered forklift truck five years ago for $8000. At that time, the estimated useful life was 10 years with a salvage value of $800 at the end of this time. The truck can now be sold for $2500. For this truck, average annual operating expenses for year j have been

$$C_j = \$2000 + \$200(j-1)$$

Now the distributor is considering the purchase of a smaller battery-powered truck for $6500. The estimated life is 10 years, with the salvage value decreasing by $600 each year. Average annual operating expenses are expected to be $1600. If a *MARR* = 10% is assumed and a five-year planning horizon is adopted, should the replacement be made now? (4.12)

64. A particular unit of production equipment has been used by a firm for a period of time sufficient to establish very accurate estimates of its operating and maintenance costs. Replacements can be expected to have identical cash flow profiles

in successive life cycles if constant worth dollar estimates are used. The appropriate discount rate is 30%. Operating and maintenance costs for a unit of equipment in its tth year of service, denoted by C_t, are as follow.

t	C_t	t	C_t
1	$2000	6	$5000
2	2500	7	5750
3	3050	8	6550
4	3650	9	7400
5	4300	10	8300

Each unit of equipment costs $15,000 initially. Because of its special design, the unit of equipment cannot be disposed of at a positive salvage value following its purchase; hence, a zero salvage value exists, regardless of the replacement interval used.

(a) Determine the optimum replacement interval assuming an infinite planning horizon. (Maximum feasible interval = 10 years.)

(b) Determine the optimum replacement interval assuming a planning horizon of 15 years.

(c) Solve parts a and b using a discount rate of 0%.

(d) Based on the results obtained, what can you conclude concerning the effect the discount rate has on the optimum replacement interval? (4.12)

65. Given an infinite planning horizon, identical cash flow profiles for successive life cycles, and the following functional relationships for C_t, the operating and maintenance cost for the tth year of service for the unit of equipment in current use, and F_n, the salvage value at the end of n years of service:

$$C_t = \$1000(1.25)^t \qquad t = 1, 2, \ldots, 12$$
$$F_n = \$11,000(0.75)^n \qquad n = 0, 1, 2, \ldots, 12$$

Determine the optimum replacement interval assuming a *MARR* of (a) 0%, (b) 10%. (Maximum life = 12 years.) (4.12)

66. Solve problem 65 given the following functional relationships. (4.12)

$$C_t = \$1000(1.10)^t \qquad t = 1, 2, \ldots, 12$$
$$F_n = \$11,000(0.50)^n \qquad n = 0, 1, 2, \ldots, 12$$

CHAPTER FIVE
DEPRECIATION AND INCOME TAX CONSIDERATIONS

5.1 INTRODUCTION

Depreciation and taxes are particularly important in engineering economy analyses. Although depreciation allowances are not actually cash flows, their magnitude and timing do affect taxes. Tax dollars are cash flows and are therefore just as green as dollars spent on wages, utilities, and raw materials. The care taken to use the most favorable depreciation and tax methods allowed by law can result in the saving of millions of dollars per year by large companies.

This chapter presents the basic depreciation methods commonly used and shows their effects on corporate after-tax cash flow profiles. In addition, the importance of the depreciation write-off period, borrowed funds, capital gains and losses, and tax credits will be illustrated. No attempt will be made to discuss the minute aspects of depreciation or tax law. Experts in these areas devote entire careers to keeping abreast of the latest laws and judgments. Instead, the objective here will be to motivate sufficiently the importance of depreciation and taxes such that you will perform economic analyses on an after-tax basis and seek corporate legal assistance as required. All tax codes or regulations quoted are those pertaining to U.S. federal corporate income tax.

5.2 THE MEANING OF DEPRECIATION

Depreciation is based on a recognition that most property decreases in value with use and time. The U.S. Supreme Court has ruled that the depreciation allowance represents reduction in value through the wear and tear of depreciable assets. The amount of the allowance is the sum that should be set aside for the taxable year, in order that the total of the sums set aside throughout the asset's useful life will, with salvage value, equal the original cost [1]. The components used to determine the depreciation allowance are the investment cost basis, salvage value, and useful life of the asset, the latter two being estimates.

The federal income tax is based on net income that results by deducting certain items such as expenses from gross income. For tax purposes, an investment is treated as a prepaid expense, and the depreciation allowance allocates that expense over time. It is important to note that depreciation is not an actual cash flow, but is merely treatable as an expense for income tax purposes. A larger depreciation allowance in a year decreases net taxable income and, hence, income taxes, making more money available for reinvestment by the firm. Because of the time value of money, it is generally desirable to take larger depreciation allowances in the early years and lesser allowances in the latter years of an asset's life. This is true if the firm is, in fact, earning money from which to take the depreciation deduction. Also, it must be done within limits prescribed by the Treasury Department and discussed subsequently.

5.3 FACTORS USED TO DETERMINE DEPRECIATION

Depreciation can be taken on any property that is used in the taxpayer's trade or business or held for the production of income. Depreciation cannot be taken on inventories held for sale to customers, nor on land. A useful life and a cost basis (investment) must be ascertained before depreciation can be taken. The *cost basis* (P) is essentially the taxpayer's investment. In most cases, this is the cost of the property plus the cost of capital additions to that property, including installation cost. The *useful life* (n) depends on the use that the taxpayer intends for the asset. As pointed out in Chapter Two, this life may have little relationship to the inherent physical life of the asset. For example, obsolescence because of technological improvements and foreseeable economic changes may render an asset useless long before it is physically worthless. The IRS has published *Depreciation Guidelines and Rules* to assist in setting a useful life; however, the IRS guidelines need not be followed if the user can justify a different life. *Salvage value* (F) is an estimate of the market value at the end of an asset's useful life. The salvage value enters explicitly into the calculation of some allowable depreciation methods and not at all in others. In no case may total depreciation exceed the cost basis less the estimated salvage value.

5.4 METHODS OF DEPRECIATION

The Treasury Department has stated that several types of depreciation procedures are acceptable. The straight line, declining-balance, and sum of the years' digits methods are specifically mentioned. They and others will be considered throughout this chapter.

5.4.1 Straight Line Depreciation

The *straight line method* provides for the uniform write-off of an asset. The depreciation allowed at the end of each year (D_t) is equal throughout the asset's useful life and is given by

$$D_t = \frac{P - F}{n} \qquad (5.1)$$

The undepreciated or book value at the end of each year (B_t) is given by

$$B_t = P - \left(\frac{P - F}{n}\right)t \qquad (5.2)$$

Example 5.1 ───

We have just purchased a minicomputer at a cost of $10,500 with an estimated salvage value of $500 and a projected useful life of six years. The depreciation and book value for each year are given in Table 5.1.

TABLE 5.1 Straight Line Depreciation and Book Value

End of Year, t	Depreciation, D_t	Book Value, B_t
0		$10,500.00
1	$1,666.67	8,833.33
2	1,666.67	7,166.67
3	1,666.67	5,500.00
4	1,666.67	3,833.33
5	1,666.67	2,166.67
6	1,666.67	500.00

The straight line method must be used in all cases where the taxpayer has not adopted a different method explicitly acceptable to or approved by the IRS. In addition to being the most commonly used method of depreciation in its own right, one may switch to the straight line method in the latter years of life of an asset initially depreciated under an accelerated method.

5.4.2 Sum of the Years' Digits Depreciation

The *sum of the years' digits method* of depreciation is known for its accelerated write-off of assets. That is, it provides relatively high depreciation allowances in the early years and lower allowances throughout the rest of an asset's useful life. The name "sum of the years' digits" comes from the fact that the sum

$$1 + 2 + \cdots + n - 1 + n = \frac{n(n+1)}{2}$$

is used directly in the calculation of depreciation. The depreciation allowance during any year t is expressed as

$$D_t = \frac{n - (t-1)}{n(n+1)/2}(P - F) \tag{5.3}$$

The book value at the end of each year t is given by

$$B_t = P - \sum_{j=1}^{t} \frac{n - (j-1)}{n(n+1)/2}(P - F)$$

which reduces to

$$B_t = (P - F)\frac{(n - t)(n - t + 1)}{n(n+1)} + F \tag{5.4}$$

Example 5.2

Returning to our minicomputer in Example 5.1 where $P = \$10,500$, $F = \$500$, and $n = 6$, sum of the years' digits depreciation results in the allowances and book values summarized in Table 5.2.

Sum of the years' digits depreciation may be used for federal income tax purposes only in the case of tangible property having a useful life of three years or more (special rules apply to depreciable realty) where the original use (any use) of the property began with the taxpayer.

TABLE 5.2 Sum of the Years' Digits, Depreciation and Book Value

End of Year, t	Value of $\dfrac{n-(t-1)}{n(n+1)/2}$	Depreciation, D_t	Book Value, B_t
0			$10,500.00
1	6/21	$2,857.14	7,642.86
2	5/21	2,380.95	5,261.90
3	4/21	1,904.76	3,357.14
4	3/21	1,428.57	1,928.57
5	2/21	952.38	976.19
6	1/21	476.19	500.00

5.4.3 Declining Balance Depreciation

The *declining balance method*, like sum of the years' digits depreciation, is known for its accelerated write-off of assets. In this method, the depreciation allowed at the end of each year t is a constant fraction (p) of the book value at the end of the previous year. That is,

$$D_t = pB_{t-1} \qquad (5.5)$$

The book value at the end of each year t is given by

$$B_t = P(1-p)^t \qquad (5.6)$$

Substituting Equation 5.6 into Equation 5.5 allows us to calculate the year t depreciation directly as

$$D_t = pP(1-p)^{t-1} \qquad (5.7)$$

Note that in the declining balance method of depreciation the estimated salvage value need not come into play in figuring the deduction. Twice the straight line rate, or $2/n$, is the maximum constant fraction permissible under law, and then only under certain conditions. When $p = 2/n$, as it most frequently is, this method of depreciation is known as the *double declining balance method*.

Example 5.3

Assuming that double declining balance depreciation is acceptable, let us again calculate depreciation and book value for the minicomputer example. The rate $p = 2/6 = 0.333$. The results are tabulated in Table 5.3.

TABLE 5.3 Double Declining Balance, Depreciation and Book Value

End of Year, t	Depreciation, D_t	Book Value, B_t
0		$10,500.00
1	$3,500.00	7,000.00
2	2,333.33	4,666.67
3	1,555.56	3,111.11
4	1,037.04	2,074.07
5	691.36	1,382.72
6	460.91	921.81

In this example, the book value at the end of year 6 was $921.81. Since the declining balance rate does not require the estimate of a salvage value, the book value in the last year need not be the same as for the other methods. In the event that the resale value is different from the book value, compensating adjustments of a capital or ordinary income (these terms are discussed subsequently in this chapter) nature must be made at the time of asset disposal.

The IRS also allows switching from the double declining balance method to straight line depreciation.[1] This may be desirable in order to present a greater depreciation charge, resulting in lower taxes in the current year, deferring taxes until later years, and thus providing a present worth tax advantage. In this case, the switch should take place whenever straight line depreciation on the undepreciated portion of the asset exceeds the double declining balance allowance. That is, we should switch to straight line at the first year for which

$$\boxed{\frac{B_{t-1} - F}{n - (t-1)} > pB_{t-1}} \tag{5.8}$$

The estimated salvage value will be used in determining the straight line depreciation component, even though it is neglected in the double declining

---

[1] Switching from sum of the years digits to straight line depreciation is also allowed, but normally cannot be justified economically.

balance method. Switching to straight line depreciation will never be desirable if the estimated salvage value F exceeds the double declining balance book value for the last year B_n.

Example 5.4

Reconsidering the double declining balance results in Example 5.3, we see that the last year's book value of $B_6 = \$921.81$ exceeds the estimated salvage value $F = \$500$. We therefore have reason to believe that switching to straight line depreciation would be desirable. First let us consider year 4. The straight line depreciation for the last three years would be

$$\frac{\$3111.11 - \$500}{3} = \$870.37$$

This would not exceed the double declining balance depreciation of $\$1037.04$. Trying year 5, the straight line depreciation for the last two years would be

$$\frac{\$2074.07 - \$500}{2} = \$787.04$$

This does exceed the $\$691.36$ allowance under double declining balance depreciation and, consequently, in year 5 we will switch to straight line. The results are illustrated in Table 5.4.

TABLE 5.4 Double Declining Balance Switching to Straight Line, Depreciation and Book Value

End of Year, t	Double Declining Balance Depreciation D_t	Straight Line Depreciation on Remaining Life, D_t	Book Value, B_t
0			$10,500.00
1	$3,500.00[a]	$1,666.67	7,000.00
2	2,333.33[a]	1,300.00	4,666.67
3	1,555.56[a]	1,041.67	3,111.11
4	1,037.04[a]	870.37	2,074.07
5	691.36	787.04[a]	1,287.03
6	460.91	787.04[a]	500.00

[a] Indicates depreciation allowance actually used. Switching to straight line occurred in year 5.

Other changes in depreciation methods are permissible; however, since a change in accounting method is involved, IRS approval is generally required [2].

Another approach to declining balance depreciation is to select the value of the rate p for which the last year's book value equals an estimated salvage value. This is the only time that estimated salvage is used explicitly in declining balance calculations. The value of p required to make the last year's book value equal to the salvage value can be found by setting $B_n = F$ and solving for p in the book value expression $B_t = P(1-p)^t$, where $t = n$. That is,

$$B_n = F = P(1-p)^n \qquad (5.9)$$

Consequently,

$$p = 1 - \sqrt[n]{\frac{F}{P}} \qquad (5.10)$$

Example 5.5 ─────────────────────────────

In the examples used to this point in the chapter, $P = \$10,500$, $F = \$500$, and $n = 6$. We now want to calculate the fixed rate p that will result in a book value of $\$500$ after six years. Solving for p,

$$p = 1 - \sqrt[6]{\frac{\$500}{\$10,500}} = 0.39795$$

Unfortunately, this value of p exceeds the maximum allowable rate, $2/n = 0.333$, and we revert to the solutions tabulated previously for the double declining balance method. The same would be true any time p exceeds $2/n$, such as when $F = 0$.

───

Example 5.6 ─────────────────────────────

Suppose in our example that $F = \$1000$ instead of $\$500$. Now calculate the depreciation values and the book value using a value of p that will result in a book value of $\$1000$ after six years. Solving for p,

$$p = 1 - \sqrt[6]{\frac{\$1000}{\$10,500}} = 0.324$$

The results using a fixed declining balance rate of $p = 0.324$ are illustrated in Table 5.5.

───

Declining balance depreciation in excess of 1.5 times the straight line rate may be used only in the case of tangible property having a useful life of three years or more (special rules apply to depreciable realty) where the original use (any use) of the property began with the taxpayer. Used tangible personal property and certain depreciable real property may be limited to 1.5 or 1.25 times the straight line rate.

TABLE 5.5 Declining Balance with Calculated p, Depreciation and Book Value

End of Year, t	Depreciation, D_t	Book Value, B_t
0		$10,500.00
1	$3,404.37	7,095.63
2	2,300.59	4,795.05
3	1,554.68	3,240.37
4	1,050.61	2,189.76
5	709.98	1,479.78
6	479.78	1,000.00

5.4.4 Sinking Fund Depreciation

Sinking fund depreciation was used in certain industries a few decades ago and in several cities (nontaxpaying institutions) until a few years ago. It is now of lessened importance in accounting practice; however, it is still referred to on Professional Engineering registration and certification examinations; hence, our discussion of the sinking fund method. The sinking fund model assumes that the asset depreciates at an increasing rate. If sum of the years' digits and declining balance depreciation are considered to be accelerated methods, sinking fund depreciation is a decelerated method.

Sinking fund depreciation can be determined by first imagining a bank account (call it a sinking fund) into which we deposit an equal amount at the end of each year. The sinking fund pays interest at a rate of $i\%$ and will have a balance equal to the total amount to be depreciated, $P - F$, after n years. From Chapter Three, we know that the equal annual deposit must be

$$A = (P - F)(A|F\,i,n) \qquad (5.11)$$

The depreciation allowance for any year t is then considered to be the sum of the deposit, A, plus the interest earned on the account. That is, the first year's depreciation is A, the second year's is $A(1 + i)$, and the tth year's is $A(1 + i)^{t-1}$, which equals $A(F|P\,i,t - 1)$. Sinking fund depreciation for any year t may then be expressed as

$$D_t = (P - F)(A|F\,i,n)(F|P\,i,t - 1) \qquad (5.12)$$

Since total depreciation taken at any time t is just the amount in our imaginary bank account, the book value at the end of year t equals first cost less the size of the sinking fund at that time. That is,

$$B_t = P - A(F|A\ i,t)$$

or

$$B_t = P - (P - F)(A|F\ i,n)(F|A\ i,t) \qquad (5.13)$$

Sinking fund depreciation has an interesting property in that the depreciation allowance for each year (D_t) plus interest charged on the undepreciated balance at the beginning of year (iB_{t-1}) is equal during each of the n years. This is sometimes called *capital recovered* (through depreciation) *plus return* (interest) *on the unrecovered capital* (book value). It is calculated for sinking fund depreciation as

$$D_t + iB_{t-1} = A(F|P\ i,t-1) + Pi - Ai(F|A\ i,t-1)$$
$$= A(1+i)^{t-1} + Pi - A[(1+i)^{t-1} - 1]$$

Hence,

$$D_t + iB_{t-1} = Pi + A \qquad (5.14)$$

which does not depend on the year t in question. It is interesting to note that this is equal to the equivalent uniform annual cost (*EUAC*) of an asset, regardless of its method of depreciation. This is seen as follows:

$$EUAC = Pi + A$$
$$= Pi + (P - F)(A|F\ i,n)$$
$$= Pi + (P - F)\frac{i}{(1+i)^n - 1}$$
$$= \frac{Pi(1+i)^n - Fi}{(1+i)^n - 1}$$
$$= \frac{Pi(1+i)^n}{(1+i)^n - 1} - \frac{Fi}{(1+i)^n - 1}$$

or

$$EUAC = P(A|P\ i,n) - F(A|F\ i,n) \qquad (5.15)$$

Example 5.7

Reconsidering our minicomputer example, where $P = \$10,500$, $F = \$500$, and $n = 6$, let us calculate not only depreciation and book value, but also illustrate that capital recovery plus return on the unrecovered capital is equal each year. Letting $i = 10\%$, the results are illustrated in Table 5.6.

TABLE 5.6 Sinking Fund Depreciation, Book Value, and Capital Recovery Plus Return

End of Year, t	Depreciation (Capital Recovered), D_t	Book Value (Capital Unrecovered), B_t	Return on Capital Unrecovered, iB_{t-1}	Capital Recovered, Plus Return $D_t + iB_{t-1}$
0		$10,500.00		
1	$1,296.07	9,203.93	$1,050.00	$2,346.07
2	1,425.68	7,778.25	920.39	2,346.07
3	1,568.25	6,210.00	777.82	2,346.07
4	1,725.07	4,484.92	621.00	2,346.07
5	1,897.58	2,487.34	448.49	2,346.07
6	2,087.34	500.00	258.73	2,346.07

We can see how the depreciation allowances increase as time progresses, contrary to the depreciation methods seen previously. Also, we see that the capital recovery plus return remains constant during each year as the item is depreciated.

Now we calculate the uniform annual cost of an asset where $P = \$10,500$, $F = \$500$, $n = 6$, and $i = 10\%$. Having performed similar calculations in previous chapters, we know that

$$EUAC = P(A|P\ i,n) - F(A|F\ i,n)$$
$$= \$10,500(0.22961) - \$500(0.12961)$$
$$= \$2346.07$$

The annual cost equals the capital recovery plus return for each year of the sinking fund depreciation method. Since $EUAC$ is calculated here without respect to depreciation, it should be apparent that the uniform annual cost of an asset before taxes remains the same, regardless of the depreciation method used.

5.5 COMPARISON OF DEPRECIATION METHODS

Straight line (A), sum of the years' digits (B), double declining balance (C), double declining balance switching to straight line (D), and sinking fund (E) depreciation may be compared using value-time curves, as in Figure 5.1. The data displayed are those from several of the previous examples. Note that the method of depreciation determines the path of book values.

We can see that the book value, particularly in the intermediate years, depends largely on the method of depreciation chosen. This will have a strong bearing on taxes paid during a given year and on the resulting after-tax cash flow profile.

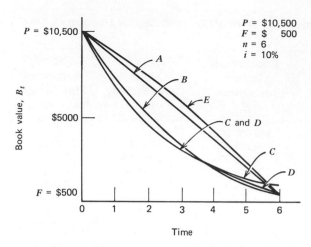

$P = \$10,500$
$F = \$\ \ 500$
$n = 6$
$i = 10\%$

FIGURE 5.1. Value time curves of five different depreciation methods.

5.6 ADDITIONAL FIRST-YEAR DEPRECIATION

Taxpayers can elect to take an additional deduction in the first year of an asset's life equal to 20% of the cost of qualified new or used tangible personal property, such as equipment or machinery, having a useful life of at least six years. This 20% applies only to $10,000 of investment, and the limit applies to each taxpayer, not to each asset being considered. In addition, the company's investment cost basis in the asset (P) must be reduced by the additional allowance. Regular depreciation is then calculated using the adjusted cost basis. We can see that the maximum additional first-year depreciation for a company in any year is $2000, and this will have its greatest effect in assisting small businesses.

Example 5.8 ────────────────────────────────

In our earlier examples we had a minicomputer with investment cost basis $P = \$10,500$, $F = \$500$, and $n =$ six years. If earlier in the tax year we have already claimed the 20% first-year depreciation on a $4000 asset, only $6000 of our $10,500 investment qualifies for the additional first-year allowance. That allowance (D_f) is

$$D_f = 0.20(\$6000)$$
$$= \$1200$$

Adjusting the investment cost basis, we have

$$P = P_{old} - D_f$$
$$= \$10,500 - \$1200$$
$$= \$9300$$

A depreciation schedule would then be set up on the basis of $P = \$9300$, $F = \$500$, and $n = 6$ years.

Additional first-year depreciation will not be illustrated in the remaining examples of the chapter. It is an easily used allowance; however, its use may mask some of the points to be made by subsequent examples. Therefore, the assumption is made that our company has already taken advantage of the additional depreciation allowance on other assets newly acquired in this tax year. Because of the turbulent history of additional first-year depreciation, it would be wise to seek legal assistance to determine the current allowance should this be a factor in an economic analysis.

5.7 OTHER METHODS OF DEPRECIATION

We are not limited to the methods of depreciation treated to this point. Any other consistent method may be used to determine the depreciation allowance for property similar to that for which the declining balance method is applicable. The major limitation is that total allowances at the end of each year do not exceed, during the first two thirds of the property's useful life, the total allowances that would result if the declining balance method were used. Some recognized methods are briefly presented below.

5.7.1 Units of Production Method

This procedure allows equal depreciation per each unit of output, regardless of the lapse of time involved. The allowance for year t is equal to the total depreciable amount $(P - F)$ times the ratio of units produced during the year (U_t) to the total units that may be produced during the useful life of the asset (U). That is,

$$D_t = (P - F)\frac{U_t}{U} \qquad (5.16)$$

5.7.2 Operating Day Method

This is similar to the previous method in that year t depreciation is based on the ratio of days used during the year (Q_t) to total days expected in a useful life (Q). Depreciation is expressed as

$$D_t = (P - F)\frac{Q_t}{Q} \qquad (5.17)$$

5.7.3 Income Forecast Method

This method is applicable to depreciate the cost of rented property such as moving picture films. The ratio of year t rental income (R_t) to the total useful life income (R) is multiplied by the total lifetime depreciation, or

$$D_t = (P - F)\frac{R_t}{R} \qquad (5.18)$$

5.7.4 Multiple-Asset Accounts

Until now each asset has been considered separately, having its own depreciation method, useful life, and salvage value. With multiple-asset accounts, a number of items can be combined and a single depreciation scheme applied to the entire account.

Through statistical analysis and past history, survivor curves can be determined that indicate the proportion of items in the group that will survive to any age. It should not be surprising that this is possible for property—life insurance companies base their entire business future on survivor curves for human beings. Depreciation is charged against the entire group of assets instead of against just a single item, using any reasonable method. If the useful life is taken as the average life of the grouped assets, it is logical to expect some items to last less time and some to last more. To retire some items early is considered "normal," and accounting for the capital loss on short-lived items is eliminated.

Three types of accounts often used are *group accounts, composite accounts*, and *classified accounts*. A group account contains similar assets with nearly the same useful lives, such as typewriters, lathes, passenger cars, or file cabinets. Composite accounts include assets of dissimilar character and useful lives, such as machinery, office equipment, *and* furniture in the same account. Classified accounts have items of homogeneous character without regard to useful life, such as transportation equipment, office equipment, *or* machinery. Classified accounts are used frequently in the manufacturing industries with item accounts used mainly for large special assets, buildings, and structures.

5.8 TAX CONCEPTS

The taxes paid by a corporation represent a real cost of doing business and, consequently, affect the cash flow profile. For this reason, it is wise to perform economic analyses on an *after-tax* basis. After-tax analysis procedures are identical to the before-tax evaluation procedures studied already; however, the cash flows are adjusted for taxes paid or saved.

There are numerous kinds of taxes including *ad valorem* (property), *sales*, *excise* (a tax or duty on the manufacture, sale, or consumption of various commodities), and *income taxes*. Income taxes are usually the only significant taxes to be considered in an economic analysis. Property taxes, if considered, are normally treated as annual disbursements. Income taxes are assessed on gross income less certain allowable deductions, incurred both in the normal course of business as well as on capital gains resulting from the disposal of property.

Federal and state income tax regulations are not only detailed and intricate, but they are subject to change over time. During periods of recession and inflation, there is a tendency for the tax laws to be changed in order to improve the state of the economy. For this reason, only the general concepts and procedures for calculating after-tax cash flow profiles and performing after-tax analyses are emphasized here. Furthermore, only federal income tax will be considered because of the diversity of state laws. Practitioners involved in an actual analysis should seek the assistance of corporate legal counsel regarding any uncertainties about tax laws in effect at that time.

5.9 CORPORATE INCOME TAX—ORDINARY INCOME

Corporate income tax is not limited to business organizations that have actually incorporated. Lawyers, doctors, and other professionals may be treated as corporations if they have formally organized as professional associations. Associations, joint-stock companies, insurance firms, and certain limited partnerships are also taxed as corporations.

Ordinary federal income tax imposed on these corporations is composed of two components—a *normal tax* and a *surtax*. Effective in 1975 the corporate tax rate included a normal tax of 20% on the first $25,000 of taxable income plus 22% on taxable income over $25,000, and a surtax of 26% on taxable income over $50,000. Prior to 1975, the normal tax was 22% on taxable income and the surtax was 26% on taxable income over $25,000 [2]. This means that prior to 1975 every dollar of taxable income over $25,000 was taxed at the rate of 48%, whereas in 1975 every dollar in excess of $50,000 was taxed at the rate of 48%. For illustration and consistency throughout this chapter, an applicable tax rate of 50% will be used. This will not only provide a workable tax rate, but will be assumed to account partially for state income taxes as well. This assumption is used so long as the firm has sufficient taxable income to qualify for the surtax rate, regardless of the outcome of the analysis in question. Also, the pre-1975 tax structure will be used throughout.

Taxable income must first be determined before any tax rate can be applied. Basically, *taxable income is gross income less allowable deductions. Gross income* is income in a general sense less any monies specifically exempt from tax liability. Corporate deductions are subtracted from gross income and commonly include items such as salaries, wages, repairs, rent, bad debts, taxes (other than income), charitable contributions, casualty losses, interest, and depreciation. Interest and depreciation are of particular interest, since we can control them to some extent through financing arrangements and accounting procedures.

Taxable income is represented pictorially in Figure 5.2, which shows that taxable income for any year is what is left after deductions, including interest on borrowed money and depreciation, are subtracted from gross income. These components are not all cash flows, since depreciation is simply treated as an expense in determining taxable income.

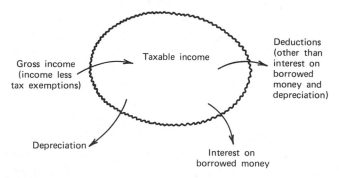

FIGURE 5.2. Pictorial representation of taxable income.

5.10 AFTER-TAX CASH FLOW

We have now looked at the basic elements needed to calculate after-tax cash flows. These elements are summarized in Figure 5.3. This shows that the after-tax cash flow is the amount remaining after income taxes and deductions, including interest but excluding depreciation, are subtracted from gross income.

In many of the tables to follow, we simplify our terminology by speaking of before-tax cash flows. The term *before-tax cash flow* is used when no borrowed money is involved, and it equals gross income less deductions, not including depreciation. *Before-tax and loan cash flow* is used when borrowed

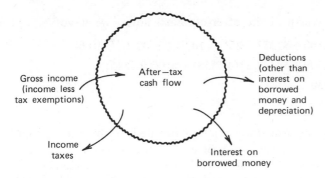

Gross income
(income less
tax exemptions)

After-tax
cash flow

Deductions
(other than
interest on
borrowed
money and
depreciation)

Income
taxes

Interest on
borrowed money

FIGURE 5.3. Pictorial representation of after-tax cash flow.

money is involved, and it equals gross income less deductions, not including either depreciation or principal or interest on the loan.

Example 5.9

Recall the earlier minicomputer examples for which $P = \$10,500$, $F = \$500$, and $n = 6$. We expect it to be responsible for reducing operating expenses by $3000/year. Therefore, before-tax and interest cash flow will increase by $3000 during each year of the computer's useful life. We plan to depreciate it by the straight line method and desire to know the after-tax cash flows. We will then apply the present worth method as an effectiveness measure to see whether the alternative to purchase the minicomputer yields at least a 10% after-tax rate of return. The cash flow calculations are given in Table 5.7.

TABLE 5.7 After-Tax Cash Flow Profile Using Straight Line Depreciation

End of Year, A	Before-Tax Cash Flow, B	Depreciation, C	Taxable Income $B - C$, D	Tax $D \times 0.50$, E	After-Tax Cash Flow $B - E$, F
0	-$10,500				-$10,500.00
1	3,000	$1,666.67	$1,333.33	$666.67	2,333.33
2	3,000	1,666.67	1,333.33	666.67	2,333.33
3	3,000	1,666.67	1,333.33	666.67	2,333.33
4	3,000	1,666.67	1,333.33	666.67	2,333.33
5	3,000	1,666.67	1,333.33	666.67	2,333.33
6	3,000	1,666.67	1,333.33	666.67	2,333.33
6	500 (salvage)				500.00

Calculating the present worth of the after-tax cash flows, we have

$$PW(10) = -\$10,500 + \$2333.33(P|A\ 10,6) + \$500(P|F\ 10,6)$$
$$= -\$10,500 + \$2333.33(4.3553) + \$500(0.5645)$$
$$= -\$55.38$$

This example shows that the after-tax cash flow profile is nothing more than lumps of money flowing in or out at different points in time over a planning horizon. Although the present worth effectiveness measure was used in the example, a rate of return, future worth, or other equivalent analysis could have been applied. Since the present worth of the discounted cash flows in our example was negative (−$55.38), we are able to conclude that the alternative as described does not yield a 10% after-tax rate of return.

When we look at the taxable income and after-tax cash flow representations in Figures 5.2 and 5.3, it is natural to wonder about the tax effects of depreciation method and useful life and the tax effect of interest on borrowed money. We will show that these factors have a substantial effect on taxes and thus on cash flow profiles.

5.11 EFFECT OF DEPRECIATION METHOD

We have seen how the accelerated depreciation methods provide for a higher depreciation allowance during the early years of an asset's life and a correspondingly lower allowance in later years. This will place a lower tax burden on the asset during the early years followed by a higher yearly burden. In most cases the total undiscounted ordinary income tax dollars paid will be the same, regardless of the depreciation method used. This is always true when the tax rate remains constant and the total depreciation allowance for each depreciation method is the same. We may therefore expect the after-tax cash flow profile under accelerated depreciation to be superior to that using a nonaccelerated method such as straight line depreciation.

Example 5.10 ————————————————————————————————

Let us repeat the previous example, now using sum of the years' digits depreciation. The actual depreciation values were calculated earlier, in Example 5.2 and the after-tax results of this method are as given in Table 5.8. Note that the after-tax cash flows for years 1 to 6 result in a decreasing gradient series when

TABLE 5.8 After-Tax Cash Flow Profile Using Sum of the Years' Digits
Depreciation

End of Year, A	Before-Tax Cash Flow, B	Depreciation, C	Taxable Income B − C, D	Tax D × 0.50, E	After-Tax Cash Flow B − E, F
0	−$10,500				−$10,500
1	3,000	$2,857.14	$ 142.86	$ 71.43	2,928.57 _238.10
2	3,000	2,380.95	619.05	309.53	2,690.47 _238.10
3	3,000	1,904.76	1,095.24	547.62	2,452.38 _238.10
4	3,000	1,428.57	1,571.41	785.71	2,214.29 ''
5	3,000	952.38	2,047.62	1,023.81	1,976.19 ''
6	3,000	476.19	2,523.81	1,261.91	1,738.09 ''
6	+ 500 (salvage)			500	

we have a uniform series of before-tax cash flows and sum of the years' digits
depreciation is used.

Calculating the present worth of the after-tax cash flows at $i = 10\%$:

$$PW(10) = -\$10,500 + [\$2928.57 - \$238.10(A|G\ 10,6)](P|A\ 10,6)$$
$$+\$500(P|F\ 10,6)$$
$$= -\$10,500 + [\$2928.57 - \$238.10(2.2236)](4.3553)$$
$$+\$500(0.5645)$$
$$= \$231.18$$

Comparing Tables 5.8 and 5.7, we can see that the accelerated depreciation
method caused cash flows to be higher in years 1, 2, and 3 and lower in years
4, 5, and 6. This effect on the present worth of after-tax cash flows was also
noticeable. Merely changing depreciation methods caused our effectiveness
measure to jump from −$55.38 to $231.18 when cash flows were discounted at
the *MARR* of 10%.

Example 5.11 ───

Let us see how double declining balance switching to straight line depreciation affects the cash flow profile for our minicomputer. Note that we
previously calculated depreciation for this asset in Example 5.4; the after-tax
results are shown in Table 5.9.

TABLE 5.9 After-Tax Cash Flow Profile Using Double Declining Balance Switching to Straight Line Depreciation

End of Year, A	Before-Tax Cash Flow, B	Depreciation, C	Taxable Income B − C, D	Tax D × 0.50, E	After-Tax Cash Flow B − E, F
0	−$10,500				−$10,500
1	3,000	$3,500.00	−$ 500.00	−$ 250.00	3,250.00
2	3,000	2,333.33	666.67	333.33	2,666.67
3	3,000	1,555.56	1,444.44	722.22	2,277.78
4	3,000	1,037.04	1,962.96	981.48	2,018.52
5	3,000	787.04	2,212,96	1,106.48	1,893.52
6	3,000	787.04	2,212.96	1,106.48	1,893.52
6		500 (salvage			500

The present worth of the after-tax cash flows when discounted at 10% is:

$$PW(10) = -\$10,500 + \$3250(P|F\ 10,1) + \$2666.67(P|F\ 10,2)$$
$$+\$2277.78(P|F\ 10,3) + \$2018.52(P|F\ 10,4)$$
$$+\$1893.52(P|A\ 10,2)(P|F\ 10,4) + \$500(P|F\ 10,6)$$
$$= -\$10,500 + \$3250(0.9091) + \$2666.67(0.8265)$$
$$+\$2277.78(0.7513) + \$2018.52(0.6830)$$
$$+\$1893.52(1.7355)(0.6830) + \$500(0.5645)$$
$$= \$275.25$$

This accelerated depreciation method also improved upon straight line depreciation and, for the specific example used above, double declining balance switching to straight line was slightly better than sum of the years' digits. In the first year, depreciation was quite high. In fact, the allowance exceeded before-tax and interest cash flow and yielded a negative taxable income resulting in a negative tax. This simply implies that the company has other positive taxable income and tax obligations in year 1 that may be offset by the negative taxable income and tax for this alternative. If there were no positive taxable income to offset, the taxable income and tax entries for year 1 would have been zero, and losses could then be carried backward or forward in time to be applied against positive taxable incomes in other years. Carry-back and carry-forward provisions are discussed later in this chapter.

The three previous examples illustrate the important point that the method of depreciation does affect cash flow and, therefore, the economic desirability of a project. Can we make any generalized statements *from these examples* about the preferability of depreciation methods with respect to effects on taxes and after-tax cash flows? No, because these examples are worked only for a particular set of conditions. Reisman [3], however, shows mathematically that certain methods of depreciation are superior to others in that they provide a higher present worth of tax savings, assuming that the effective tax rate remains the same from year to year. For example, the sum of the years' digits method is always superior to the straight line method. The declining balance method without switching is also preferable to straight line when using a rate of $p = 1.5/n$ for high values of the *MARR* and high values of the salvage to cost basis ratio, F/P. The double declining balance method without switching is preferable to sum of the years' digits for high values of the *MARR* and high values of F/P [3]. Reisman goes on to show that we always desire to switch from double declining balance depreciation to straight line if the imputed salvage value (book value at the end of the last year of useful life) is not as low as the estimated salvage value. When switching from double declining balance to straight line depreciation is used, it is always preferable to straight line depreciation, yet still not always better than the sum of the years' digits method.

5.12 EFFECT OF ESTIMATED USEFUL LIFE

The useful life of an asset as declared by the taxpayer can also affect the after-tax cash flow analysis of an alternative under consideration because the annual depreciation allowances increase as the estimate of useful life decreases. If all incomes, expenses, and interest on borrowed money remain status quo, and with a constant tax rate over time, it will be favorable to declare a shorter useful life for tax purposes. The life used for an asset, however, may be questioned by the IRS, in which case the burden of proof falls on the taxpayer.

Example 5.12 ───

To illustrate the effect of a shorter estimated useful life as applied to taxes, reconsider Example 5.11, now using a tax life of four years but maintaining the physically useful life of our minicomputer over the full six years. It will be beneficial first to calculate the double declining balance switching to straight line depreciation. A declining balance rate of $p = 2/4 = 0.5$ will be used. The depreciation results over four years are summarized in Table 5.10, and the cash flow results over six years are given in Table 5.11.

TABLE 5.10 Double Declining Balance Switching to Straight
Line, Depreciation and Book Value

End of Year, t	Double Declining Balance Depreciation, D_t	Straight Line Depreciation on Remaining Life, D_t	Book Value, B_t
0			$10,500
1	$5,250[a]	$2,500	5,250
2	2,625[a]	1,583.33	2,625
3	1,312.50[a]	1,062.50	1,312.50
4	656.25	812.50[a]	500

[a] Indicates depreciation allowance actually used. Switching to straight line
occurred in year 4.

TABLE 5.11 After-Tax Cash Flow Profile Using Double Declining Balance
Switching to Straight Line Depreciation and a Four-Year Estimated Useful Tax Life

End of Year, A	Before-Tax Cash Flow, B	Depreciation, C	Taxable Income $B - C$, D	Tax $D \times 0.50$, E	After-Tax Cash Flow $B - E$, F
0	−$10,500				−$10,500
1	3,000	$5,250	−2,250	−$1,125	4,125
2	3,000	2,625	375	187,50	2,812.50
3	3,000	1,312.50	1,687.50	843.75	2,156.25
4	3 000	812.50	2,187.50	1,093.75	1,906.25
5	3,000	0	3,000	1,500	1,500
6	3,000	0	3,000	1,500	1,500
6	500 (salvage)				500

$$PW(10) = -\$10,500 + \$4125(P|F\ 10,1) + \$2812.50(P|F\ 10,2)$$
$$+\$2156.25(P|F\ 10,3) + \$1906.25(P|F\ 10,4)$$
$$+\$1500(P|A\ 10,2)(P|F\ 10,4) + \$500(P|F\ 10,6)$$
$$= -\$10,500 + \$4125(0.9091) + \$2812.50(0.8265)$$
$$+\$2156.25(0.7513) + \$1906.25(0.6830) + \$1500(1.7355)(0.6830)$$
$$+\$500(0.5645)$$
$$= \$556.80$$

We see that the shorter life increased the discounted present worth of cash
flows from $275.25 to $556.80. This increase may or may not be permissible,

however, depending on our ability to justify using a tax life of only four years in the analysis.

5.13 EFFECT OF INTEREST ON BORROWED MONEY

Investment alternatives may be financed using equity (owner's) funds or debt (borrowed) funds. Until now we have implicitly assumed all financing to be through equity, although many companies use a mix of debt and equity for financing plant and equipment. Borrowed funds must be repaid, including both principal and interest. The interest repaid each year affects taxable income and, consequently, taxes. Both the principal and interest payments affect after-tax cash flows.

There are four common (and many less common) ways in which money can be repaid. First is the periodic payment of interest over the stipulated repayment period with the entire principal being repaid at the end of that time. The second requires an annual payment that uniformly repays the principal and also covers the annual interest. In this method, the annual payments decrease as the interest on the unrepaid principal decreases. Third is the method requiring a uniform annual payment for the sum of principal plus interest. In each payment the proportion of principal gradually increases as the proportion of interest decreases. The fourth method repays nothing, neither interest nor principal, until the end of a specified period.

Example 5.13

Let us illustrate these four basic plans for repaying principal and interest on borrowed money. Assume that a business borrows $5000 to be used in financing an alternative, and the interest rate on this loan is 8% compounded annually. The stipulated repayment period is six years.

A summary of all relevant components of our example is presented in Table 5.12. In method 1, the interest equals $5000(0.08) = \$400$/year. Only the interest is paid, and only the principal of $5000 is owed after each year's payment. Method 2 repays the principal in equal amounts of $5000/6 = \$833.33$ as well as the interest for year t, which is given by $[\$5000 - \$833.33(t - 1)]0.08$. Clearly, the interest payment, total payment, and total money owed after yearly payments are decreasing gradient series. Method 3 requires equal annual total payments. This annual payment is equal to $\$5000(A|P\ 8,6) = \$1081.58$. The principal component of this annual payment for year t can be found quickly as $\$1081.58(P|F\ 8,6 - t + 1)$, using the method given in Chapter Three. It is interesting to note that the principal payment is an 8% increasing compound series. Also, the interest payment and the total money owed after yearly payment decrease each year, and the amount by which they decrease is

TABLE 5.12 Illustration of Four Common Methods of Principal and Interest Repayment

End of Year	Interest Accrued During Year	Total Money Owed Before Yearly Payment	Interest Payment	Principal Payment	Total Payment	Total Money Owed After Yearly Payment
Method 1						
0						$5000
1	$400	$5400	$400	$ 0	$ 400	5000
2	400	5400	400	0	400	5000
3	400	5400	400	0	400	5000
4	400	5400	400	0	400	5000
5	400	5400	400	0	400	5000
6	400	5400	400	5000	5400	0
Method 2						
0						$5000
1	$400	$5400	$ 400	$ 833.33	$1233.33	4166.67
2	333.33	4500	333.33	833.33	1166.67	3333.33
3	266.67	3600	266.67	888.33	1100.00	2500
4	200	2700	200	833.33	1033.33	1666.67
5	133.33	1800	133.33	833.33	966.67	833.33
6	66.67	900	66.67	833.33	900.00	0
Method 3						
0						$5000
1	$400	$5400	$ 400	$ 681.58	$1081.58	4318.42
2	345.47	4663.90	345.47	736.10	1081.58	3582.32
3	286.59	3868.91	286.59	794.99	1081.58	2787.33
4	222.99	3010.31	222.99	858.59	1081.58	1928.74
5	154.30	2083.04	154.30	927.28	1081.58	1001.46
6	80.12	1081.58	80.12	1001.46	1081.58	0
Method 4						
0						$5000
1	$400	$5400	$ 0	$ 0	$ 0	5400
2	432	5832	0	0	0	5832
3	466.56	6298.56	0	0	0	6298.56
4	503.88	6802.44	0	0	0	6802.44
5	544.20	7346.64	0	0	0	7346.64
6	587.73	7934.37	2934.37	5000	7934.37	0

an 8% increasing compound series. In method 4, the interest accrued during each year is added to the principal such that the total amount owed after t years is $\$5000(1.08)^t$. When payment is made at the end of year 6, everything over $5000 is considered interest.

We have seen that the interest on borrowed money is deductible for tax purposes, whereas the principal repayment does not enter into taxable income. In addition, both the interest and principal portions of a payment are real and must be taken into account when calculating cash flows.

Example 5.14 ————————————————————————————————

Let us illustrate the effect of borrowed money by assuming that $5000 of the $10,500 paid for our minicomputer is through debt funding. The loan is to be repaid in equal annual installments (method 3) at 8% over six years. The remaining $5500 will be equity money. Applicable depreciation and loan activity (method 3) have previously been calculated in Example 5.13. The resulting after-tax cash flow profile is detailed in Table 5.13.

$$PW(10) = -\$5500 + \$2368.42(P\,|F\ 10,1) + \$1757.83(P\,|F\ 10,2)$$
$$+\$1339.50(P\,|F\ 10,3) + \$1048.44(P\,|F\ 10,4)$$
$$+\$889.09(P\,|F\ 10,5) + \$852(P\,|F\ 10,6) + \$500(P\,|F\ 10,6)$$
$$= -\$5500 + \$2368.42(0.9091) + \$1757.83(0.8265)$$
$$+\$1339.50(0.7513) + \$1048.44(0.6830)$$
$$+\$889.09(0.6209) + \$852(0.5645) + \$500(0.5645)$$
$$= \$1143.67$$

The value of our effectiveness measure jumped to $1143.67 for this particular example. Note that the present worth calculation on cash flows was made using a discount rate equal to our 10% MARR. The 8% loan rate was used only in sizing the loan repayments. If we actually needed to borrow the $5000 to implement the project, our estimates indicate that a handsome monetary return will be received on the $5500 equity investment. If, on the other hand, we had at least the other $5000 available, borrowing allowed us to invest that money as equity capital in another alternative that will earn a return at least equal to the MARR. This, of course, depends on the availability of investments that will yield a rate at least equal to the MARR.

We cannot conclude from the example that borrowing money is always favorable. The desirability of borrowed funds depends on the terms of the loan, including method of repayment, interest, and repayment period. Furthermore, collateral is frequently required, which may be lost (after legal action) if the principal and interest cannot be paid on schedule. In summary, each alternative investment and financing strategy should be compared on its own merits.

We have used the same basic example to illustrate the implications of depreciation and financing strategies thus far throughout the chapter. The present worth of after tax cash flows was seen to increase from −$55.38 to $1143.67 simply by going from the commonly used equity financing and straight line depreciation to the more sophisticated mix of debt and equity financing and using double declining balance switching to straight line depreciation. Nothing about the basic incomes or costs relating to the asset were changed. By now you should have a good idea how to calculate and assess after-tax cash flows under differing depreciation schedules and financing arrangements, and you should be motivated to apply these techniques in the analysis of alternative investments.

5.14 CARRY-BACK AND CARRY-FORWARD RULES

In three of the previous examples we saw that an alternative's taxable income and tax can be negative for a particular year. It was implicitly assumed that those values were used to offset positive tax liabilities from other corporate activities. When there are insufficient profits to be offset, the taxable income losses can be carried back three years or forward five years in order to reduce positive taxable income. That is, if a loss occurs in 1978, we can reopen books from 1975, 1976, and 1977, or apply the loss against positive taxable income in 1979, 1980, 1981, 1982, and 1983. We will continue to assume that negative taxable incomes are used to offset positive values for the year in question.

5.15 CAPITAL GAINS AND LOSSES

Many tax experts agree that there are no provisions in the internal revenue laws more difficult to understand and apply than those pertaining to capital gains and losses. Assets receiving capital treatment are important for tax purposes because they receive preferential tax rates, being less severely taxed than ordinary income. Congress originally desired to lessen the blow of ordinary income tax on gains caused by appreciation of items held in a risk situation over a substantial period of time. Over the years, however, taxpayers have maneuvered to assure that certain items are classified as eligible for capital gains treatment. As a result, unforeseen loopholes have been closed by the IRS, increasing the complexity of the tax provisions. Many of the common capital gain and loss situations faced in economic analysis will be presented in this chapter; however, any uncertainty should be satisfied by legal counsel.

TABLE 5.13 After-Tax Cash Flow Profile Using Double Declining Balance Switching to Straight Line Depreciation Using $5000 Borrowed Money at 8%

End of Year, A	Before-Tax and Loan Cash Flow, B	Loan Principal Payment, C	Loan Interest Payment, D	Depreciation Charges, E	Taxable Income B − D − E, F	Taxes F × 0.50, G	After-Tax Cash Flow B − C − D − G, H
0	−$10,500	−$5,000					−$5,500
1	3,000	681.58	$400	$3,500	−$ 900	−$ 450	2,368.42
2	3,000	736.10	345.47	2,333.33	321.20	160.60	1,757.83
3	3,000	794.99	286.59	1,555.56	1,157.85	578.93	1,339.50
4	3,000	858.59	222.99	1,037.04	1,739.97	869.99	1,048.44
5	3,000	927.28	154.30	787.04	2,058.66	1,029.33	889.09
6	3,000	1,001.46	80.12	787.04	2,132.84	1,066.42	852
6	500 (salvage)						500

227

A capital asset is any property held by a taxpayer *except*:

Stock in trade or other property properly included in inventory. Property held *primarily* for sale to customers in the ordinary course of the taxpayer's trade or business.

Depreciable property used in a trade or business.

Real property used in a trade or business.

Copyrights; literary, musical or artistic compositions; a letter or memorandum; or similar property; held by a taxpayer whose personal efforts created the property; or a taxpayer for whom a letter, memorandum or similar property was prepared or produced; or one receiving property as a gift from the person who created it.

Accounts or notes receivable acquired in the ordinary course of trade or business for services rendered, or from the sale of stock in trade, inventory or property held for sale to customers in the ordinary course of trade or business.

Certain short-term government obligations issued on a discount basis [2].

Shares of stock, bonds, notes, debentures, and similar securities are considered capital assets unless they fall under one of the exceptions listed. Real property and property held for the production of income, not used in a trade or business, are capital assets. Patents held for investment, life estates, inherited jewelry, bank accounts, and cotton acreage allotments, for example, are also considered capital assets.

Eligibility is the key difficulty in complying with IRS capital gains codes. For example, the courts have used purpose of acquisition and disposal, frequency and extent of sales, nature of the taxpayer's business, the taxpayer's sales efforts, and so forth, to determine whether or not capital treatment is allowable. Any items sold to customers in the ordinary course of trade or business may not be considered capital assets by the seller, and they are therefore subject to ordinary gains or losses.

Before applying capital treatment, the extent of the capital gain or loss must be determined. This requires a balancing of long- and short-term gains and losses. A capital gain or loss may occur when there is a sale or exchange of a capital asset. The excess of selling price over book value is a *capital gain*. If the selling price is less than the book value, then a capital loss has occurred. If the period during which the taxpayer holds the asset is less than or equal to six months, the gain or loss is referred to as a *short-term gain or loss*. If the asset is held longer than six months, a *long-term gain or loss* has occurred. A net short-term gain (loss) is the sum of short-term gains minus the sum of short-term losses. Similarly, a net long-term gain (loss) is the sum of long-term gains minus long-term losses.

Example 5.15

Our corporation has the following capital gains and losses during a tax year. The net short- and long-term gains and losses are summarized as follows:

Short-term capital gains	$19,300
Short-term capital losses	$27,600
Net short-term capital loss	($ 8,300)
Long-term capital gains	$77,500
Long-term capital losses	$18,750
Net long-term capital gain	$58,750

5.16 TAX TREATMENT OF CAPITAL ASSETS

Short- and long-term capital gains may be consolidated for tax purposes. The result may be a net capital gain, or a net capital loss. The consolidated result dictates whether the tax treatment will be as capital gains, ordinary income, or a loss to be carried back or over. The consolidation rules and tax treatments are summarized in Table 5.14.

TABLE 5.14 Results and Corporate Tax Treatment of Consolidated Long- and Short-Term Capital Gains and Losses

		Net Long-Term Capital Gains and Losses		
		None	Long-Term Gain	Long-Term Loss
Net Short-Term Capital Gains and Losses	None	No capital treatment	Net capital gain taxed as capital gain	Net capital loss may be carried over
	Short-Term Gain	Net capital gain taxed as ordinary income	Net capital gain. Long-term gain taxed as capital gain *and* short-term gain taxed as ordinary income	If a net capital loss, may be carried back or over. If a net capital gain, taxed as ordinary income
	Short-Term Loss	Net capital loss may be carried back or over	If a net capital gain, taxed as capital gain. If a net capital loss, may be carried back or over	Net capital loss may be carried back or over

Note. Capital losses carried back and over are treated as short-term capital losses.

Example 5.16 ──

In Example 5.15 we determined that our corporation has a net short-term capital loss of $8300 and a net long-term capital gain of $58,750. Consolidating these in accordance with Table 5.14, we have a net capital gain of $58,750 – $8300 = $50,450, which may be taxed as a capital gain.

──

There are two basic ways of computing taxes when capital gains and losses are involved. First is the method whereby all taxable income including capital gains are treated as ordinary income. An alternative procedure excludes the excess of net long-term capital gains over net short-term capital losses. The ordinary income is then taxed as usual, but the excluded net capital gain is taxed at the rate of 30%.

Example 5.17 ──

Now, suppose our corporation has $170,450 taxable income of which we found $50,450 to be a net capital gain taxable as a capital gain. Therefore $120,000 would be ordinary income. Computing the federal taxes by both the regular and alternative methods, we have:

Regular method
Normal tax at 22% of $170,450   = $37,499
Surtax at 26% of $170,450 – $25,000 = $37,817
                    Total tax   = $75,316

Alternative method
Note. Exclude $50,450 capital gains
         income from $120,000 ordinary income.
Normal tax at 22% of $120,000   = $26,400
Surtax at 26% of $120,000 – $25,000 = $24,700
Capital gains tax at 30% of $50,450   = $15,135
                    Total tax   = $66,235

The alternative method for capital gains would save over $9000 in taxes for our company.

──

Example 5.18 ──

A small business has taxable income of $22,000, including $5000 of net long-term capital gains and $3000 of net short-term capital gains. Table 5.14 shows that we have a net capital gain, and the long-term gains may be taxed as capital gains and the short-term gains as ordinary income. Computing the taxes using the regular method, we have:

> Regular method
> Normal tax at 22% of $22,000 = $4840

> There is no surtax, since total taxable
> income does not exceed $25,000.

We can see that a company would not elect to use the alternate procedure here, since only the normal 22% tax rate is applied to its taxable income.

When capital losses are subtracted from capital gains and the result is negative, Table 5.14 shows that the net loss may be carried back or over. The carry-back period is three years and the carry-forward period is five years, during which time the capital loss is treated as a short-term capital loss and can be used to offset capital gains.

5.17 DEPRECIATION RECAPTURE

Depreciable property and real property used in a trade or business may, under certain conditions, receive capital treatment after disposal. Taxpayers were quick to recognize the monetary advantages of capital gains tax as compared to ordinary income tax. A favorite loophole was then to declare a low salvage value on a depreciable asset, depreciating it far below the market value. The excess depreciation was used to offset ordinary income, thereby saving taxes at the rate of 48%. After disposing of the asset, the lower capital gains tax rate was applied to the difference between fair market value and book value. In 1964, Congress ruled that gains realized at the time of sale, to a limited extent, would be taxed as ordinary income [2]. This is known as *depreciation recapture*; in effect, it closed the loophole of overdepreciating as a means of converting ordinary income into capital gains.

Depreciation recapture can best be explained by considering a depreciable asset, purchased after 1963, and depreciated by the straight line method. If the selling price is below the original cost and above the book value, the gain is considered a recapture of previously charged depreciation and is taxed as ordinary income. If the item actually appreciates and sells for more than the original cost, the difference between original cost and book value represents depreciation recapture taxed as an ordinary income gain, while the difference between selling price and original cost is a capital gain. Finally, if the selling price is below the book value, the loss is treated as a capital loss. The rules are more complex for assets purchased before 1963 or depreciated using accelerated methods. In these cases, expert assistance should be sought.

Example 5.19 ─────────────────────────────

In Example 5.9, straight line depreciation was used to depreciate a minicomputer from $10,500 down to $500 over a six-year horizon. The computer was estimated to provide an additional before-tax cash flow of $3000/year. The asset, we will now assume, was sold for $2000, resulting in a $1500 gain. Since this gain is considered to be depreciation recapture, it is taxed at the ordinary income rate (we will again use 50%). That is,

$$\text{Tax on gain} = 0.50(\$2000 - \$500) = \$750$$

Table 5.15 illustrates the after-tax cash flow profile for this example when depreciation recapture is involved.

TABLE 5.15 After-Tax Cash Flow Profile with Depreciation Recapture

End of Year, A	Before-Tax Cash Flow, B	Depreciation, C	Taxable Income $B - C$, D	Tax $D \times 0.50$, E	After-Tax Cash Flow $B - E$, F
0	−$10,500				−$10,500
1	3,000	$1,666.67	$1,333.33	$666.67	2,333.33
2	3,000	1,666.67	1,333.33	666.67	2,333.33
3	3,000	1,666.67	1,333.33	666.67	2,333.33
4	3,000	1,666.67	1,333.33	666.67	2,333.33
5	3,000	1,666.67	1,333.33	666.67	2,333.33
6	3,000	1,666.67	1,333.33	666.67	2,333.33
6	2,000 (salvage)				2,000
6		− 1,500	1,500	750	−750

Note that three after-tax cash flows are shown for year 6. The first, $2333.33, is for regular year 6 operations. The second, $2000, is the salvage value received after six years. The third, −$750, represents the ordinary income tax liability on the $1500 of excess depreciation recaptured. The present worth of the after-tax cash flows may now be calculated using a *MARR* of 10%.

$$PW(10) = -\$10,500 + \$2333.33(P|A\ 10,6) + \$2000(P|F\ 10,6) - \$750(P|F\ 10,6)$$
$$= -\$10,500 + \$2333.33(4.3553) + \$2000(0.5645) - \$750(0.5645)$$
$$= \$367.98$$

5.18 INVESTMENT TAX CREDIT

As of this writing, taxpayers can claim a tax credit for investments in certain depreciable property. This means that they can deduct from their tax liability an amount up to a percentage of their investment in qualified property. The credit applies to depreciable tangible personal property and to depreciable real property, except buildings and their structural components. Typical depreciable tangible personal property includes office equipment, production machinery, outside gasoline pumps, and neon signs. Real property would include oil and gas pipelines, fractionating towers, microwave towers, and blast furnaces. To qualify, real property must be used in manufacturing, production, extraction, the furnishing of utilitylike services, or meet other stated criteria.

The history of the tax credit has been somewhat sporadic, being established by the 1962 Revenue Act at the 7% level. It was suspended briefly between 1966 and 1967 and terminated in 1969. It was restored to 7% by the 1971 Revenue Act, only to be temporarily increased by the 1975 Tax Reduction Act to 10% for qualified property acquired and placed in service after January 21, 1975 and before January 1, 1977 [2]. The exact deduction allowed depends not only on the current tax credit percentages but also on the useful life as determined for depreciation purposes as well as the tax liability of the company.

Qualified property must be intended for use over at least a seven-year period to qualify for the full percentage. Table 5.16 indicates the effect of useful life on investment tax credit [2].

The credit may be used to offset all tax liability up to the first $25,000 of tax and up to 50% of the tax liability in excess of $25,000. If more credit is available than can be used, it may be carried back for three years or carried forward for seven years, in accordance with rather specific rules. It is important to note that the tax credit, since 1964, does not reduce the amount of the investment to be depreciated, even though it is essentially

TABLE 5.16 Effect of Useful Life on Investment Tax Credit

Declared Useful Life	Investment Tax Credit
Less than three years	None
Three or four years	1/3 of 10%[a]
Five or six years	2/3 of 10%
Seven or more years	10%

[a] Tax credit rate of 10% is used for illustration. Consult Internal Revenue Codes for current rate.

a cash rebate. In other words, the tax credit is in addition to, not in lieu of, normal depreciation.

Example 5.20

A firm invests $500,000 in an asset expected to be used for four years, and $1,200,000 in another property to be used for 12 years. Both qualify for the investment tax credit. Tax liability for the tax year 1976 is $200,000. Let us determine the maximum possible tax credit as well as the tax credit used in 1976.

$$\text{Maximum tax credit} = \$500,000 \times 1/3 \times 0.10 + \$1,200,000 \times 0.10$$
$$= \$136,666.67$$

$$\text{Tax credit used} = \$25,000 + (\$200,000 - \$25,000)0.50$$
$$= \$112,500$$

Note that the excess tax credit of $136,666.67 - $112,500 = $24,166.67 can be carried backward or forward to apply against taxes in other years.

Example 5.21

Let us again consider Example 5.9 in which straight line depreciation is used to depreciate a minicomputer from $10,500 down to $500 over six years. The computer was estimated to provide an additional before-tax cash flow of $3000/year. Now, however, we will assume that the investment tax credit applies.

The maximum permissible tax-credit will be as follows:

$$\text{Maximum tax credit} = \$10,500 \times 2/3 \times 0.10$$
$$= \$700$$

Assuming that the purchase was made at the end of fiscal year 0, the $700 tax credit is applied against year 0 taxes, and the net present worth of the after-tax cash flows is increased by $700, from $-$55.38$ up to $644.62. If, however, the purchase occurs at the beginning of fiscal year 1, the $700 tax credit applies against the first year tax liability. This is illustrated in Table 5.17.

The present worth of the after-tax cash flows may now be calculated.

$$PW(10) = -\$10,500 + \$3033.33(P|F\ 10,1)$$
$$+\$2333.33(P|A\ 10,5)(P|F\ 10,1) + \$500(P|F\ 10,6)$$
$$= -\$10,500 + \$3033.33(0.9091)$$
$$+\$2333.33(3.7908)(0.9091) + \$500(0.5645)$$
$$= \$581.01$$

TABLE 5.17 After-Tax Cash Flow Profile with Investment Tax Credit

End of Year, A	Before-Tax Cash Flow, B	Depreciation, C	Taxable Income B − C, D	Tax D × 0.50, E	Investment Tax Credit, F	After-Tax Cash Flow, B − E + F, G
0	−$10,500					−$10,500
1	3,000	$1,666.67	$1,333.33	$666.67	$700.00	3,033.33
2	3,000	1,666.67	1,333.33	666.67	0	2,333.33
3	3,000	1,666.67	1,333.33	666.67	0	2,333.33
4	3,000	1,666.67	1,333.33	666.67	0	2,333.33
5	3,000	1,666.67	1,333.33	666.67	0	2,333.33
6	3,000	1,666.67	1,333.33	666.67	0	2,333.33
6	500 (salvage)					500

The investment tax credit increased the after-tax cash flow present worth from −$55.38 to $581.01. Obviously, the investment tax credit is a significant factor in an economic analysis. In the problems at the end of this chapter, *the tax credit, if applicable, will be applied at the end of year 0.*

5.19 LEASE-BUY CONSIDERATIONS

It is often attractive to lease property as opposed to owning it. When an asset is leased, a schedule of payments over time is agreed on. Lease charges, being expenses, apply directly against taxable income during the year in which they occur. Thus, each dollar spent on a leased item has the effect of costing only $0.50 (assuming a 0.50 tax rate). This is not true for dollars spent on a purchased item, because this cost must be written off over the life of the asset. Thus, the effective cost per dollar spent on a purchased item is more than $0.50.

Example 5.22 ———————————————————————————————

Our minicomputer has a $10,500 first cost, salvage value of $500, and life of six years. The minicomputer may be purchased and depreciated using double declining balance switching to straight line depreciation. This case was illustrated in Table 5.9, and had a $275.25 present worth of after-tax cash flows.

A lessor offers to buy the computer and lease it to us for $3037.55 − $100 $(t − 1)$ during the tth year. This decreasing gradient is equivalent to 120% of the pretax equivalent uniform annual cost of the computer. The after-tax cash flows for this lease alternative are presented in Table 5.18.

TABLE 5.18 After-Tax Cash Flow for Lease Alternative

End of Year, A	Reduced Expenses Due to Asset, B	Cost to Lease Asset, C	Taxable Income B − C, D	Tax D × 0.50, E	After-Tax Cash Flow B − C − E, F
0					
1	$3000	$3037.55	−$ 37.55	−$ 18.78	−$ 18.77
2	3000	2937.55	62.45	31.23	31.22
3	3000	2837.55	162.45	81.23	81.22
4	3000	2737.55	262.45	131.23	131.22
5	3000	2637.55	362.45	181.23	181.22
6	3000	2537.55	462.45	231.23	231.22

The present worth of the after-tax cash flows at a *MARR* of 10% is as follows:

$$PW(10) = [-\$18.77 + \$50(A|G\ 10,6)](P|A\ 10,6)$$
$$= [-\$18.77 + \$50(2.2236)](4.3553)$$
$$= \$402.47$$

Thus, leasing appears to be more favorable than purchasing. However, if we are eligible for an investment tax credit, the decision will change. That is, assuming the tax credit is applied to the purchase alternative in year 0, the present worth of the after-tax cash flows will be $275.25 + $700, or $975.25.

5.20 DEPLETION

Depletion is a gradual reduction of minerals, gas and oil, timber, and natural deposits. In a sense, depletion is closely akin to depreciation. The difference is that while a depleting asset is losing value by actually being removed and sold, a depreciable asset is losing value through wear, tear, and obsolescence in the manufacture of goods to be sold. Money recovered through the depletion allowance is likely to be used in the exploration and development of depletable assets, just as depreciation reserves are reinvested for new equipment. As in most tax-related matters, laws relating to lessors, lessees, royalties, and sales are complex and will probably require expert assistance. However, the importance of depletion and the alternative methods that can be used for figuring the allowance will be illustrated here.

The *cost method* is the general rule for calculating a depletion deduction. It is expressed as

$$\text{Depletion deduction} = \frac{\text{Cost of property} \times \text{Number of units sold during year}}{\text{Number of units in property}}$$

The calculated deduction may or may not be used, depending on whether or not the alternative depletion allowance method provides a larger deduction.

Percentage depletion provides an allowance equal to a percentage of the gross income from the property. The applicable percentage depends on the type of property depleted. Table 5.19 gives some recent depletion percentages; however, those for oil and gas particularly are likely to be modified substantially over the next decade.

The depletion allowance may not be less than that calculated under the cost method, but may not otherwise exceed 50% of the taxable income before allowance for depletion.

TABLE 5.19 Depletion Percentages for Some Minerals and Similar Resources

Type	Depletion Percentage
Oil and gas (applies only to certain production from gas wells and to independent producers and royalty owners)	22
Sulphur, uranium, clay, asbestos, bauxite, chromite, graphite, mica, quartz crystals, cadmium, cobalt, lead, mercury, nickel, tin, and zinc	22
Gold, silver, oil shale, copper, and iron ore	15
Various clays, diatonaceous earth, granite, marble, etc., plus metals if not allowed in the 22% group	14
Coal, lignite, perlite, and sodium chloride	10
Clay and shale used for sewer pipe and brick	7 1/2
Gravel, peat, pumice, and sand	5

Example 5.23 ─────────────────────────────────

We own a small oil company and have purchased the rights, drilled, and developed a 500,000-barrel oil well in Baldhill, Oklahoma, for $325,000. Operating expenses, based on past experience, are equal to $1 + $0.2(t − 1)$/barrel where t is the year in which the oil is removed and sold. This increase in cost per barrel over time indicates the increased difficulty of recovering the oil as the field is depleted. Our geologist expects the field to last six years, yielding 150,000, 125,000, 100,000, 60,000, 40,000, and 25,000 barrels of oil over that time. Each barrel will be worth $3 to our company. Using the cost method, $325,000/500,000 barrels or $0.65/barrel will be our depletion allowance. From Table 5.20, our depletion percentage is seen to be 22%, assuming we meet all the qualifications that allow a small producer to claim a percentage depletion. In Table 5.20 we calculate each year's estimated depletion, tax, and after-tax cash flow.

Note that in years 5 and 6 we chose the cost method because the percentage depletion was too high, being over 50% of net income. When working with depletion problems, it is always advisable to calculate the allowance both ways in order to assure the greatest tax advantage.

5.21 SUMMARY

This chapter has presented the most important elements of depreciation and income tax law as they pertain to economic analyses. It is clear that depreciation (or depletion) method, useful life, financing, and tax credits can

TABLE 5.20 Depletion Allowance, Tax, and After-Tax Cash Flow for Oil Well Example

End of Year, A	Barrels Sold, B	Gross Income, C	Operating Cost, D	Before Tax Cash Flow, C − D, E	50% of Net Income, E × 0.5, F	Cost Depletion at $0.65/barrel B × 0.65, G	Percentage Depletion at 22% of Gross Income C × 0.22, H	Taxable Income E − max{G, [min (F,H)]}, I	Income Tax 0.5 × I, J	After-Tax Cash Flow E − J, K
0		−$325,000								−$325,000
1	150,000	450,000	$150,000	$300,000	$150,000	$97,500	$99,000[a]	$201,000	$100,500	199,500
2	125,000	375,000	150,000	225,000	112,500	81,250	82,500[a]	142,500	71,250	153,750
3	100,000	300,000	140,000	160,000	80,000	65,000	66,000[a]	94,000	47,000	113,000
4	60,000	180,000	96,000	84,000	42,000	39,000	39,600[a]	44,400	22,200	61,800
5	40,000	120,000	72,000	48,000	24,000	26,000[a]	26,400	22,000	11,000	37,000
6	25,000	75,000	50,000	25,000	12,400	16,250[a]	16,500	8,750	4,375	20,625

[a] Indicates depletion rate used.

have significant effects on the desirability of making an investment. Many of these factors are law related, and changes are being made daily. Therefore, in cases of uncertainty regarding depreciation and tax treatment, it is wise to seek competent legal advice.

Capital gains and losses are often not important in engineering economic analyses. However, it is wise to know when a capital gain or loss has occurred and when to apply capital gains treatment versus depreciation recapture.

Finally, unless there are overriding considerations, lease versus buy alternatives should be analyzed closely. Quite often, a lease alternative will compare with surprising favor against an equivalent purchase alternative.

BIBLIOGRAPHY

1. Holzman, Robert S. *Dun and Bradstreet's Handbook of Executive Tax Management.* Thomas Y. Crowell, 1974.

2. *Prentice-Hall Federal Tax Handbook.* Prentice-Hall, 1976.

3. Reisman, Arnold. *Managerial and Engineering Economics.* Allyn and Bacon, 1971.

PROBLEMS

1. An electrostatic precipitator is purchased for $130,000; it has an anticipated life of 10 years, and a terminal salvage value of $20,000. Determine the book value for the precipitator at the end of the ninth year of its life and the depreciation charge for the tenth year of its life, using:
   (a) Straight line depreciation.
   (b) Sum of the years' digits depreciation.
   (c) Sinking fund depreciation at $i = 10\%$. (5.4)

2. Rework problem 1 using the following methods of depreciation.
   (a) Double declining balance switching to straight line depreciation.
   (b) Declining balance depreciation using the rate that will yield a book value of $20,000 at the end of year 10.
   (c) Double declining balance depreciation. (5.4)

3. A firm purchases a computer for $100,000. It has a life of nine years and a terminal salvage value of $10,000 at that time. Determine the depreciation charge for year 6 and the book value at the *beginning* of year 6, using:
   (a) Straight line depreciation.
   (b) Sum of the years' digits depreciation.
   (c) Double declining balance depreciation.
   (d) Sinking fund depreciation at $i = 10\%$. (5.4)

4. A furnace has a first cost of $50,000 and a salvage value after four years of $0. The method of depreciation to be used is double declining balance switching to straight

line at the most attractive time. What will be the depreciation charge each year? (5.4)

5. A compressor is purchased for $4000 and has an estimated salvage value of $1000 after a useful life of three years. Interest is 10%.
   (a) Determine capital recovered, return on the capital unrecovered, and capital recovered plus return for each year using straight line depreciation.
   (b) Determine capital recovered, return on the capital unrecovered, and capital recovered plus return for each year using sum of the years' digits depreciation.
   (c) Determine the annual equivalent capital recovered plus return for each of parts a and b.
   (d) Determine the annual equivalent before-tax cost of the asset using $(P - F)(A|P\ i,n) + Fi$. (5.4)

6. For the compressor in problem 5, what would be the capital recovered plus return for the second year using sinking fund depreciation? (5.4)

7. A numerically controlled lathe is purchased by a small machine shop for $8000. It is to be used for six years, will have a salvage value of $400, and is to be depreciated using the straight line method. Assuming it qualifies for "additional first-year depreciation," prepare a depreciation schedule for each year of its life. (5.6)

8. An automatic control mechanism is estimated to provide 3000 hours of service during its life. The mechanism costs $4800 and has a salvage value of $300 after 3000 hours of use. Its use is projected over a four-year period as follows:

Year	Hours of Use
1	1500
2	800
3	400
4	300

Calculate the depreciation charge for each year using a method similar to the "operating day" method, but based on operating hours. (5.7)

9. A utility trailer costs $1000 and is rented out by the hour, day, or week. It is expected to depreciate to zero by the time it has been rented out for a total of $5000 in gross income. If its forecasted annual revenues are $1800, $1500, $900, and $800, what will be the appropriate annual depreciation charges? (5.7)

10. What is the federal income tax for each of the following corporate taxable incomes? Assume a tax year prior to 1975.
    (a) $12,800.
    (b) $25,000.
    (c) $50,000.
    (d) $1 million.
    Plot a graph of federal income tax versus taxable income. Be sure to label critical values on both scales. (5.9)

11. A dust collector with a first cost of $30,000 will be depreciated to zero by the straight line method. The collector will be used for five years, and will yield an annual gross income less operating expenses of $15,000. The effective tax rate is 50%. Determine the after-tax rate of return. (5.10)

12. An automatic soldering machine is purchased for $18,000. Installation cost is $3000 because of the extreme care and provisions needed. It is estimated that the asset can be sold for $2500 after a useful life of 10,000 hours of operation. However, before selling, it must be removed at a cost of $500 to Company A. Inspection and maintenance costs $6/operating hour. It requires one half hour to solder one unit on the machine, and a total of 10,000 units are soldered per year. The after-tax *MARR* is 15%, and the effective tax rate is 50%. What will be the cost of soldering per unit assuming depreciation is based on hours of service? (5.10)

13. An investment proposal is described by the following tabulation:

	Year		
	1	2	3
Gross income during year	$ 8,000	$16,000	$20,000
Investment at *beginning* of year	24,000	0	0
Operating cost	2,000	3,000	4,000
Depreciation charge (straight line)	8,000	8,000	8,000

The company considering this proposal is profitable in its other activities. Determine the taxable income, tax, and after tax cash flow for years 0, 1, 2, and 3. Assume an effective tax rate of 50%. (5.10)

14. An investment opportunity has the following financial outlook over a five year period.

EOY	Before-Tax Cash Flow	Depreciation Charges	Taxes
0	−$30,000		
1	10,000	$10,000	$ 0
2	10,000	8,000	800
3	10,000	6,000	1,600
4	10,000	4,000	2,400
5	10,000	2,000	3,200

If our firm requires a 12% rate of return after taxes, should we invest? (5.10)

15. A firm purchases a centrifugal separator for $8000; it is estimated to have a life of five years. Operating and maintenance costs are anticipated to increase by

$500/year, with the cost for the first year estimated to be $500. A $500 salvage value is anticipated. Using straight line depreciation, a 50% income tax rate, and a 10% after-tax *MARR*, determine the annual equivalent cost of the separator. (5.10)

16. A firm may either invest $10,000 in a numerically controlled lathe, that will last for five years and have a zero salvage value at that time, or invest $X in a methods design study. Both investment alternatives yield an increase in income of $4000/year for six years. With straight line depreciation used for all investments in capital assets and with a 50% tax rate, for what value of X will the firm be indifferent between the two investment alternatives? Assume an after-tax *MARR* of 10%. (5.10)

17. An investment proposal has the following estimated net cash flow before taxes.

EOY	0	1	2	3	4	
	−$70,000	42,000	14,000	8,000	3,000 + 20,000	salvage value

The effective tax rate is 50% and the after-tax *MARR* is 10%. Assume the company considering this proposal is profitable in its other activities. Find the after-tax cash flows and the present worth of those cash flows if sum of the years' digits depreciation is used. (5.11)

18. An earth-moving company is contemplating an investment of $35,000 in equipment that will have a useful life of five years with a $5000 salvage value at that time. Increases in the firm's income from cut and fill contracts due to the investment will total $10,600/year for five years. The company pays taxes at a rate of 50%. Using sum of the years digits' depreciation and an after-tax *MARR* of 10%, should the firm undertake the investment? (5.11)

19. An automatic copier is being considered that will cost $8000, have a life of four years, and no prospective salvage value at the end of that time. It is estimated that because of this venture, there will be a yearly gross income of $8000. The operating cost during the first year will be $2000, increasing by $100 each year thereafter. The company contemplating the purchase of this asset has an effective tax rate of 50%. Determine the after-tax cash flow for each year. First use straight line depreciation, and then use double declining balance switching to straight line depreciation. (5.11)

20. Two mutually exclusive alternatives, A and B, are available. Alternative A requires an original investment of $80,000, has a life of five years, zero annual operating costs, and a salvage value of $20,000. Alternative B requires an original investment of $120,000, has a life of seven years, annual operating costs of $2000, and a salvage value of $36,000. The after-tax *MARR* is 10%. A 50% tax rate is to be used. Use an annual worth comparison and recommend the least cost al-

ternative, using a planning horizon of
(a) five years.
(b) seven years.
(c) thirty-five years.
For tax purposes, as well as predicting salvage values for incomplete life cycles, use sum of the years digits' depreciation. If replacements are required, assume that they have identical cash flow profiles. (5.11)

21. Suppose a back-hoe is purchased for $12,000 and has an estimated salvage value of $2000 at the end of five years. Annual revenues and annual operating costs, excluding depreciation, are $5000 and $1000, respectively. If $i = 10\%$ and the firm's tax rate is considered 50%,
(a) Determine the present worth of "taxes paid" for the five-year period (1) if straight line depreciation is used and, (2) if the sum of the years' digits depreciation method is used.
(b) Determine the present worth of the depreciation charges for the five year period (1) if straight line depreciation is used and, (2) if the sum of the years' digits depreciation method is used.
Do your answers from parts a and b suggest that the present worth of taxes is minimized if the present worth of depreciation charges is maximized? (5.11)

22. The ABC Company is considering the purchase of a computer-controlled printing press that will cost $10,000, have a life of six years, and an estimated salvage value of zero. The press will develop a gross income of $4000/year. The annual cost connected with the press (exclusive of depreciation) will amount to $500/year. The company will use straight line depreciation and the effective tax rate is 50%. If the MARR is 15% after taxes, determine the present worth of the after-tax cash flows. Now, assuming that the press is depreciated over four years, although it is used for six years as estimated, determine the present worth of the after-tax cash flows. (5.12)

23. A company has the opportunity to invest in a water purification system requiring $100,000 capital. The firm has only $60,000 available and must borrow the additional $40,000 at an interest rate of 10%/year. The system has a life of four years and an estimated salvage value at the end of its life of $0. The before-tax and loan cash flow for each of years 1 to 4 is $25,000. Only the loan interest is paid each year and the entire principle is repaid at the end of year 4. If depreciation charges are based on the sum of the years' digits method, the tax rate is 50%, and the firm is able to sell the system for $10,000 at the end of year 4 (note depreciation recapture), construct a table showing the following for each of the four years:
(a) Loan principal payment.
(b) Loan interest payment.
(c) Depreciation charges.
(d) Taxable income.
(e) Taxes.
(f) After-tax cash flow. (5.13)

24. Rework problem 23 assuming that the loan is paid back as follows:
(a) In equal annual amounts.
(b) The principal in equal annual amounts plus yearly interest. (5.13)

25. Complete the partial table below.

EOY	Before-Tax and Loan Cash Flow	Loan Principal Payment	Loan Interest Payment	Depreciation Charges	Taxable Income	Taxes (Rate = 0.50)	After-Tax Cash Flow
0	−$100,000	−$40,000					−$60,000
1	20,000		$4,000	$18,000	?	?	?
2	20,000		4,000	17,000	?	?	?
3	?		?	15,000	?	$ 5,500	?
4	?		?	?	?	3,000	?
5	40,000		4,000	?	?	13,000	?
6	20,000		4,000	?	?	4,000	?
7	?		4,000	7,000	$19,000	?	?
8	?		4,000	5,000	31,000	?	?
9	45,000		4,000	3,000	?	?	?
10	30,000		4,000	2,000	?	?	?
10		40,000					− 40,000

Note 1. There is no uniformity to the before-tax and loan cash flows.
Note 2. Depreciation does not follow any well-known pattern, but totals $100,000.
Note 3. The effective income tax rate is 50%. (5.13)

26. A firm purchases a heat exchanger by borrowing the $10,000 purchase price. The loan is to be repaid with six equal annual payments at an annual compound rate of 7%. It is anticipated that the exchanger will be used for nine years and then be sold for $1000. Annual operating and maintenance expenses are estimated to be $5000/year. Assume sum of years' digits depreciation, a 50% tax rate, and an after-tax *MARR* of 8%. Compute the after-tax, equivalent uniform annual cost for the heat exchanger. (5.13)

27. A firm borrows $10,000, paying back $2000/year on the principal plus 10% on the unpaid balance at the end of each year. The $10,000 is used to purchase a digital frequency meter that lasts five years and has a $1000 salvage value at that time. The firm uses sum of the years' digits depreciation; it requires a 15% after-tax *MARR* in such analyses, and has a 50% tax rate on taxable income. Determine the after-tax equivalent uniform annual cost for the meter. (5.13)

28. An electronic multiaxis wheel balancer was purchased for $10,000 three years ago. At that time, the estimated life was eight years, with an estimated salvage value of $1000. Today, it appears that the balancer will only last two years more with an estimated salvage value of $2000. The depreciation schedule will *not* be changed. The effective tax rate is 0.50, the capital gains tax rate is 0.30, and the sum of the years digits' depreciation method is being used. Operating expenses are $4000/year and gross income is $8000/year. Assume that total capital gains are always greater than total capital losses for the company. How much will be paid in taxes for *each of the next two years*? Do not neglect the effect of the capital loss. (5.16)

29. A company had the following capital gains and losses and ordinary taxable income in 1974:

Short-term capital gains	$ 25,000
Short-term capital losses	4,000
Long-term capital gains	103,000
Long-term capital losses	8,000
Ordinary taxable income	80,000

If the normal, surtax, and capital gains taxes were 22, 26, and 30%, how much tax was paid that year? (5.16)

30. A firm purchases a crane for $100,000. Originally, it was estimated that the crane would be retained for nine years and sold for $10,000. However, the crane is used for five years and sold for $40,000. Income less operating and maintenance costs for the crane have been $30,000/year. Using a 50% income tax rate, a 30% capital gains tax, and straight line depreciation, compute the after-tax cash flows for the five-year period. Do not neglect depreciation recapture or capital loss treatment as required. The company annually has a large amount of long term capital gains. (5.17)

31. A $39,000 investment in an automatic control panel is proposed. It is anticipated that this investment will cause a reduction in net annual operating disbursements of $12,000/year for eight years. The investment will be depreciated for income tax purposes by the double declining balance switching to straight line method, assuming an eight year life and a $3000 salvage value. The effective tax rate is 50%. A 10% investment tax credit applies at the end of year 0. With a required after-tax return of 10%, what is the equivalent present worth cost for the investment? (5.18)

32. A sum of $10,000 is borrowed at 10% compounded annually and paid back with equal annual payments over a two-year period. The $10,000 is combined with $15,000 of equity funds to purchase a transformer that has a service life of eight years and a terminal salvage value of $3400. Annual operating and maintenance costs are anticipated to be $5000/year.
A 10% investment tax credit applies in year 0, along with an income tax rate of 50%. Sum of the years' digits depreciation is used. The after-tax *MARR* is 20%. Determine the present worth of costs for this machine. (5.18)

33. Rework problem 32 assuming that the transformer qualifies for additional first year depreciation. (5.18)

34. A $50,000 heat recovery incinerator is expected to cause a reduction in net out-of-pocket costs of $11,000/year for 10 years. The incinerator will be depreciated for income tax purposes using straight line depreciation with a life of 10 years and a zero salvage value. A 10% investment tax credit is allowed in year 0. If the tax rate is 50%, and the after-tax *MARR* is 10%, determine whether or not the incinerator should be purchased. (5.18)

35. A firm is considering investing $765,000 in a material handling system that will reduce annual operating costs by $150,000/year over a 15-year planning horizon. Perform a before-tax analysis using a *MARR* of 20%. Is the investment justified? Suppose further analysis indicates $540,000 of the investment is for equipment; the remaining $225,000 is for expense items. A 50% tax rate, 10% investment tax credit at year 0, eight-year equipment write-off period, sum of the years' digits depreciation method, and 15% after-tax *MARR* are to be used. Is the investment justified on an after-tax basis? (5.18)

36. A tractor originally costs $10,000. Operating costs for year j are equal to $2000 + $200 $(j-1)$. A 50% tax rate, 10% investment tax credit in year 0, straight line depreciation, zero salvage value at all times, 10% minimum attractive rate of return after taxes, and indefinite planning horizon is to be assumed. Determine the optimum replacement interval. (5.18)

37. A firm is contemplating either purchasing or renting a computer. If purchased, the computer will cost $400,000 and have annual operating and maintenance costs of $75,000. Because of the obsolescence rate on computing equipment, a five-year study period is chosen. At the end of five years, it is anticipated the computer will have a value of $150,000. The same computer can be leased, with *beginning of year* lease charges of $100,000, which includes maintenance. If leased, the firm will have end of year operating costs of $50,000. Using straight line depreciation, a 50% tax rate, and an after-tax *MARR* of 20%, perform an annual worth comparison to determine if the computer should be leased. (5.19)

38. A firm is considering purchasing an analog computer for $200,000. The computer is estimated to have a life of nine years and a salvage value of $20,000. Operating and maintenance costs are estimated to be $25,000/year. Sum of the years' digits depreciation is used and a 10% investment tax credit applies in year 0. Alternatively, the computer can be leased for X at the *beginning* of each year for a nine-year period. If leased, the annual operating and maintenance cost to be paid by the firm reduces to $15,000. Using a 50% tax rate, an after-tax *MARR* of 10%, and an annual worth comparison, determine the value of X that will yield equivalent annual costs between buying and leasing. (5.19)

39. A firm is considering purchasing a digital computer for $500,000. The computer is estimated to have a life of five years and a salvage value of $50,000. End of year operating and maintenance costs are estimated to be $60,000/year. Sum of the years' digits depreciation is used. A 10% investment tax credit applied in year 0. Alternatively, the computer can be leased for X/year for the five-year period with the lease payment due at the *beginning* of the year. If leased, the annual end of year operating and maintenance costs are estimated to be $40,000. Using a 50% tax rate, an after-tax *MARR* of 10%, and an annual worth comparison, determine the value of X that will yield equivalent annual costs between buying and leasing. (5.19)

40. A highway construction firm purchased an item of earth-moving equipment three years ago for $50,000. It has been company policy to depreciate such equipment in 10 years by the sum of the years' digits method of depreciation and use a salvage value of 34% of first cost at $t = 10$. However, the company switches to the straight

line method of depreciation for the fourth and subsequent years. The firm earns an average annual gross revenue of $45,000 with the equipment, and the average annual operating costs have been and are estimated to be $26,000.

Now ($t = 3$), the firm has the opportunity to sell the equipment for book value and subcontract the work normally done by the equipment over the next seven years. The firm will not have this opportunity after the present time. If the subcontracting is done, the average annual gross revenue remains $45,000 but the subcontractor charges $35,000/year for these services.

If the firm's income tax rate is 50% and a 25% rate of return after taxes is desired, determine whether the firm should subcontract or not by the *annual worth method*. (5.19)

41. The XYZ Construction Company has been subcontracting a particular job at a cost of $30,000 annually, payable at the end of the year. Annual revenues associated with this particular job have averaged $40,000. The agreement with the subcontractor has now expired but could be renewed for another five years. The same revenues and subcontract charges are expected to continue.

The XYZ Company is also considering the purchase of equipment to handle this same job and terminate the subcontracting. The initial cost of the equipment plus installation expenses is $50,000. The economic life of the equipment is established as 10 years with a salvage value of $5,000 at this time. However, the company is only interested in a planning horizon of five years at which time a salvage value of $20,000 is anticipated. With the equipment, annual revenues are expected to remain $40,000 but average annual operating expenses expected are $20,000.

The Company uses the straight line depreciation method of accounting. If the company's annual tax rate is 50%, a 10% after-tax rate of return is desired, and a five-year planning horizon is assumed, should the equipment be purchased or the subcontract agreement be renewed? (5.19)

42. A gold mine that is expected to produce 30,000 ounces of gold is purchased for $600,000. The gold can be sold for $130/ounce; however, it costs $75/ounce for mining and processing costs. If 3500 ounces are produced this year, what will be the depletion allowance for:

(a) Unit depletion.

(b) Percentage depletion where the fixed percentage for gold is 15%? (5.20)

43. A West Virginia coal mine having 6 million tons of coal has a first cost of $10 million. The gross income for this coal is $18/ton. Operating costs are $14.80/ton. If, during the first two years of operation, the mine yields 150,000 tons and 200,000 tons, respectively, determine the after-tax cash flow for each year, assuming that this mine is the owner's only venture. Use the better method of depletion, and assume that these are tax years prior to 1975. (5.20)

CHAPTER SIX

ECONOMIC ANALYSIS OF PROJECTS IN THE PUBLIC SECTOR

6.1 INTRODUCTION

Knowing how to evaluate and select projects to be approved, paid for, and operated by the government is at least as important to today's engineer as a similar knowledge relating to the private sector. Fortunately, analysis methods for public and private projects are very similar, even though there are some basic, significant differences between the two. The methods considered in this chapter are those most frequently used in evaluating government (national, state, or local) projects—*benefit-cost* and *cost-effectiveness* analysis. More emphasis will be spent on benefit-cost methods, which require that benefits and costs be evaluated on a monetary basis. Cost effectiveness requires a numerical measure of effectiveness; however, that measure need not be in terms of money.

6.2 THE NATURE OF PUBLIC PROJECTS

There are many types of government projects and many agencies involved. Four classes reasonably cover the spectrum of projects entered into by government. They include cultural development, protection, economic services, and natural resources. *Cultural development* is enhanced through education, recreation, historic, and similar institutions or preservations. *Protection* is achieved through military services, police and fire protection, and the judicial system. *Economic services* include transportation, power generation, and housing loan programs. *Natural resource* projects might entail wildland management, pollution control, and flood control. Although these are obviously incomplete project lists in each class, it is not so obvious that some projects belong in more than one area. For example, flood control is certainly a form of protection for some, as well as being related to natural resources.

Government projects have a number of interesting characteristics that set them apart from projects in the private sector. Many government projects are huge, having first costs of tens of millions of dollars. They tend to have extremely long lives, such as 50 years for a bridge or a dam. The multiple-use concept is common, as in wildland management projects, where economic (timber), wildlife preservation (deer, squirrel), and recreation projects (camping, hiking) are each considered uses of import for the land. The benefits or enjoyment of government projects are often completely out of proportion to the financial support of individuals or groups. Also, there are almost always multiple government agencies that have an interest in a project. Also public sector projects are not easily evaluated, since it may be many years before their benefits are realized. Finally, there is usually not a clear-cut measure of success or failure in the private sector such as the rate of return or net present worth criteria.

6.3 OBJECTIVES IN PROJECT EVALUATION

If large, complex, lengthy, multiple-use projects of interest to several groups are to be evaluated for their desirability, the criteria for the evaluation must first be agreed on. The setting for modern evaluation of government projects dates back to the River and Harbor Act of 1902, which "required a board of engineers to report on the desirability of Army Corps of Engineers' river and harbor projects, taking into account the amount of commerce *benefited* and the *cost*." Even more applicable to today is the criterion specified in the Flood Control Act of June 22, 1936, which stated ". . . that the Federal Government should improve or participate . . . if the *benefits to whomsoever they may accrue* are in excess of the *estimated costs* . . ." [13]. Obviously, the idea of benefit-cost analysis has been around for a long time, even though only in recent years has there been much written about it.

Prest and Turvey [10] give a short, reasonable definition of benefit-cost analysis: It is "a practical way of assessing the desirability of projects where it is important to take a long view (in the sense of looking at repercussions in the further, as well as the nearer, future) and a wide view (in the sense of allowing for side-effects of many kinds on many persons, industries, regions, etc.) i.e., it implies the enumeration and evaluation of all the relevant costs and benefits." The "long view" is nothing more than considering the entire planning horizon which, granted, is generally far longer for projects in the public sector than for those in the private sector. The "wide view" and the notion of evaluating all "relevant costs and benefits" probably spell out the greatest single difference between government and private economic evaluation. That is, government projects often affect many individuals, groups, and things, either directly or indirectly, for better or for worse. In evaluating these projects, the analyst tries to capture these effects on the public, quantify them and, where possible, make the measures in monetary terms. Positive effects are referred to as *benefits*, while negative effects are *disbenefits*. In contrast, the private sector "effects" of primary importance are those that relate to income being returned to the organization. *Costs* of construction, financing, operation, and maintenance are estimated in much the same way in both the public and private sectors.

6.4 BENEFIT-COST ANALYSIS

The notion of benefit-cost analysis is simple in principle. It follows the same systematic approach used in selecting between economic investment alternatives, including the following eight steps adopted from Chapter Four.

1. Define the set of feasible, mutually exclusive, public sector alternatives to be compared.
2. Define the planning horizon to be used in the benefit-cost study.
3. Develop the cost-savings and benefit-disbenefit profiles in monetary terms for each alternative.
4. Specify the interest rate to be used.
5. Specify the measure(s) of merit or effectiveness to be used.
6. Compare the alternatives using the measure(s) of merit on effectiveness.
7. Perform supplementary analyses.
8. Select the preferred alternative.

These will be discussed from the viewpoint of a benefit-cost analysis.

The heart of many engineering problems is in identifying all of the *alternatives* available to achieve a particular goal. Some alternatives can be excluded from further consideration immediately. Each remaining alternative should be described thoroughly. This involves specifying all of the good and bad effects that a project will have on the public. This includes effects directly on people, land values, and the environment. In addition, all aspects of project development, operation, maintenance, and eventual salvage must be stated.

Defining the *planning horizon* is essential in order to define the period over which the best project(s) is to be selected. If the planning horizon is longer than the life of a nonrenewable public project, there may be several years during which no benefits or costs are considered. If the planning horizon is shorter than the life of a project, residual values (same concept as salvage values) may be estimated and used.

Quantifying all *costs and savings* refers to governmental expenditures and incomes received relating to a project over the planning horizon. Disbursements include all first and continuing costs of a project, while income may result from tolls, fees, or other charges to the user public. Residual or salvage values may exist if the project is believed to still be in operation at the end of the planning horizon, and these are treated as savings or negative costs at that time. Costs, unlike benefits, are quantified for public projects in a way similar to private projects.

Each *benefit or disbenefit* during the planning horizon must also be quantified monetarily. Benefits, or the positive effects of a government investment, refer to desirable consequences on the public instead of on any governmental body. Disbenefits are the negative effects on the public instead of on any government organization. Unfortunately, placing dollar values on benefits received by a diverse public is often not an easy task. Neither is deciding on the interest rate to be used.

A base measure such as annual equivalent benefits and costs or present worths of benefits and costs is then established, and the *measure of merit* is chosen. Benefit-cost analyses frequently use the benefit-cost ratio (B/C) or, to a lesser extent, a measure of benefits less costs ($B - C$). If

B_{jt} = public benefits associated with project j during year t, $t = 1, 2, \ldots, n$
C_{jt} = governmental costs associated with project j during year t, $t = 0, 1, 2, \ldots, n$

and

i = appropriate interest rate

then the B/C criterion may be expressed mathematically, using a present worth base measure, as

$$B/C_j(i) = \frac{\sum_{t=1}^{n} B_{jt}(1+i)^{-t}}{\sum_{t=0}^{n} C_{jt}(1+i)^{-t}} \tag{6.1}$$

Note that the B/C ratio is an alternative name for the savings/investment ratio treated in Chapter Four. The $B - C$ criterion is expressed as

$$(B - C)_j(i) = \sum_{t=0}^{n} (B_{jt} - C_{jt})(1 + i)^{-t} \qquad (6.2)$$

which is similar to the present worth method described in Chapter Four.

When two or more project *alternatives are being compared* using a B/C ratio, the analysis should be done on an *incremental basis*. That is, first the alternatives should be ordered from lowest to highest cost (present worth, annual equivalent, etc.). Then, the incremental benefits of the second alternative over the first, $\Delta B_{2-1}(i)$, are divided by the incremental costs of the second over the first, $\Delta C_{2-1}(i)$. That is,

$$\Delta B/C_{2-1}(i) = \frac{\Delta B_{2-1}(i)}{\Delta C_{2-1}(i)} = \frac{\displaystyle\sum_{t=1}^{n} (B_{2t} - B_{1t})(1 + i)^{-t}}{\displaystyle\sum_{t=0}^{n} (C_{2t} - C_{1t})(1 + i)^{-t}} \qquad (6.3)$$

Note that if the first alternative is "do nothing," the incremental B/C ratio is also the straight B/C ratio for the second alternative. As long as $\Delta B/C_{2-1}(i)$ exceeds 1.0, Alternative 2 is preferable to Alternative 1. Otherwise, Alternative 1 is preferred to Alternative 2. The winner of these is then compared on an incremental basis with the next most costly alternative. These pairwise comparisons continue until all alternatives have been exhausted, and only one "best" project remains. The procedure used is very similar to that specified for the rate-of-return method in Chapter Four.

With the $B - C$ criterion, an incremental basis may be used following the some rules as for the B/C ratio, but preferring Alternative 2 to Alternative 1 as long as the following condition holds:

$$\Delta(B - C)_{2-1}(i) = \Delta B_{2-1}(i) - \Delta C_{2-1}(i)$$

$$= \sum_{t=1}^{n} [B_{2t}(i) - B_{1t}(i)](1 + i)^{-t} - \sum_{t=0}^{n} [C_{2t}(i) - C_{1t}(i)](1 + i)^{-t} \geq 0$$

$$(6.4)$$

Where benefits and costs are known directly, the value of $(B - C)_j$ for each Alternative j may be calculated and the maximum value selected.

Apparently straightforward, there are a number of potential pitfalls in the evaluation of government projects. That is, economic analyses of public projects can be easily biased unknowingly by a project evaluator. The more significant of these pitfalls are discussed in subsequent sections.

Next, *supplementary analyses* in the form of risk analysis, sensitivity analysis, and break-even analysis may be performed. These are often useful in determining how critical various inputs can be and how close various alternatives are economically, and they otherwise quantify much other information that would likely remain intangible.

The final step is *selecting the preferred alternative*. Not only is it to be selected, but all quantitative and qualitative supporting considerations should be recorded in detail. This is particularly true in the case of public sector projects.

These seven major steps in conducting a benefit-cost analysis, as well as the B/C and $B - C$ relationship to the present worth, annual equivalent, and rate of return criteria, are illustrated in the remainder of this section.

Example 6.1

We are given the task of deciding between three highway alternatives to replace a winding, old, dangerous road. As the crow drives, the length of the current route is 26 miles. Alternative A is to overhaul and resurface the old road at a cost of $2,200,000. Resurfacing will then be required at a cost of $2 million at the end of each 10-year period. Annual maintenance for Alternative A will cost $8000/mile. Alternative B is to cut a new road following the terrain; it will be only 22 miles long. Its first cost will be $8,800,000, and surface renovation will be required every 10 years at a total cost of $1,800,000. Annual maintenance will be $8000/mile. Alternative C also involves a new highway which, for practical considerations, will be built along a 20.5 mile straight line. Its first cost, however, will be $17,300,000, because of the extensive additional excavating necessary along this route. It, too, will require resurfacing every 10 years at a cost of $1,800,000. Annual maintenance will be $15,500/mile. This increase over route B is due to the additional roadside bank retention efforts that will be required.

Our task is to select one of these alternatives, considering a planning horizon of 30 years with negligible residual value for each of the highways at that time. One of these alternatives is required, since the old road has deteriorated below acceptable standards. We can calculate the annual equivalent first cost and maintenance cost of each alternative using an interest rate of 6%.

Construction and resurfacing cost:

Route A: $[\$2,200,000 + \$2,000,000(P|F\ 6,10)$
$\quad + \$2,000,000(P|F\ 6,20)](A|P\ 6,30)$
$\quad = [\$2,200,000 + \$2,000,000(0.5584) + \$2,000,000(0.3118)](0.0726)$
$\quad = \$286,269/\text{year}$

Route B: [$8,800,000 + $1,800,000(P|F 6,10)
        + $1,800,000(P|F 6,20)](A|P 6,30)
        = [$8,800,000 + $1,800,000(0.5584) + $1,800,000(0.3118)(0.0726)
        = $753,115/year

Route C: [$17,300,000 + $1,800,000(P|F 6,10)
        + $1,800,000)(P|F 6,20)](A|P 6,30)
        = [$17,300,000 + $1,800,000(0.5584) + $1,800,000(0.3118)](0.0726)
        = $1,370,640/year

Maintenance cost:

Route A: $\left(\$8000\ \dfrac{\$}{\text{mile}-\text{year}}\right)(26\ \text{miles}) = \$208,000/\text{year}$

Route B: ($8000)(22) = $176,000/year

Route C: ($15,500)(20.5) = $317,750/year

Clearly, route A costs less than route B, which itself costs less than route C. Can we now conclude that route A should be selected based on cost to the government? Absolutely not, according to the criterion that seeks to maximize benefits to the public as a whole, minus costs. In fact, we have analyzed only one side of the problem. Now, we must attempt to quantify the benefits along each of the routes.

Traffic density along each of the three routes will fluctuate widely from day to day, but will average 4000 vehicles/day throughout the year. This volume is composed of 350 light commercial trucks, 250 heavy trucks, 80 motorcycles, and the remainder are automobiles. The average cost per mile of operation for these vehicles is $0.20, $0.40, $0.05, and $0.12, respectively.

There will be a time savings because of the different distances along each of the routes, as well as the different speeds which each of the routes will sustain. Route A will allow heavy trucks to average 35 miles/hour, while other traffic can maintain an average speed of 45 miles/hour. Routes B and C will allow heavy trucks to average 40 miles/hour, and the rest of the vehicles can average 50 miles/hour. The cost of time for all commercial traffic is valued at $9/vehicle/hour, and for noncommercial traffic, $3/vehicle/hour. Twenty-five percent of the automobiles and all of the trucks are considered commercial.

Finally, there is a significant safety factor that should be included. Along the old winding road, there has been an excessive number of accidents per year. Route A will reduce the number of vehicles involved in accidents to 105, and routes B and C are expected to involve only 75 and 70 vehicles in accidents, respectively, per year. The average cost per vehicle in an accident is estimated to be $4400, considering actual physical property damages, lost wages because of injury, medical expenses, and other relevant costs.

We now set about to analyze the various benefits in monetary terms. We have considered savings in vehicle operation, time, and accident prevention. The costs incurred by the public for these items are calculated in the following steps.

Operational costs

Route A: $\left(350\dfrac{\text{light trucks}}{\text{day}}\right)\left(26\dfrac{\text{miles}}{\text{light truck}}\right)\left(0.20\dfrac{\$}{\text{mile}}\right)\left(365\dfrac{\text{days}}{\text{year}}\right)$

$+\left(250\dfrac{\text{heavy trucks}}{\text{day}}\right)\left(26\dfrac{\text{miles}}{\text{heavy truck}}\right)\left(0.40\dfrac{\$}{\text{mile}}\right)\left(365\dfrac{\text{days}}{\text{year}}\right)$

$+\left(80\dfrac{\text{motorcycles}}{\text{day}}\right)\left(26\dfrac{\text{miles}}{\text{motorcycle}}\right)\left(0.05\dfrac{\$}{\text{mile}}\right)\left(365\dfrac{\text{days}}{\text{year}}\right)$

$+\left(3320\dfrac{\text{automobiles}}{\text{day}}\right)\left(26\dfrac{\text{miles}}{\text{automobile}}\right)\left(0.12\dfrac{\$}{\text{mile}}\right)\left(365\dfrac{\text{days}}{\text{year}}\right)$

$= \$5,432,076/\text{year}$

Route B: $[350(\$0.20)+250(\$0.40)+80(\$0.05)+3320(\$0.12)](22)(365)$
$= \$4,596,372/\text{year}$

Route C: $[350(\$0.20)+250(\$0.40)+80(\$0.05)+3320(\$0.12)](20.5)(365)$
$= \$4,282,983/\text{year}$

Time costs

Route A: $\left(350\dfrac{\text{light trucks}}{\text{day}}\right)\left(26\dfrac{\text{miles}}{\text{light trucks}}\right)\left(\dfrac{1\text{ hour}}{45\text{ miles}}\right)\left(365\dfrac{\text{days}}{\text{year}}\right)\left(9\dfrac{\$}{\text{hour}}\right)$

$+\left(250\dfrac{\text{heavy trucks}}{\text{day}}\right)\left(26\dfrac{\text{miles}}{\text{heavy truck}}\right)\left(\dfrac{1\text{ hour}}{35\text{ miles}}\right)\left(365\dfrac{\text{days}}{\text{year}}\right)\left(9\dfrac{\$}{\text{hour}}\right)$

$+\left(80\dfrac{\text{motorcycles}}{\text{day}}\right)\left(26\dfrac{\text{miles}}{\text{motorcycle}}\right)\left(\dfrac{1\text{ hour}}{45\text{ miles}}\right)\left(365\dfrac{\text{days}}{\text{year}}\right)\left(3\dfrac{\$}{\text{hour}}\right)$

$+\left(3320\dfrac{\text{automobiles}}{\text{day}}\right)\left(26\dfrac{\text{miles}}{\text{automobile}}\right)\left(\dfrac{1\text{ hour}}{45\text{ miles}}\right)\left(365\dfrac{\text{days}}{\text{year}}\right)$

$\times\left(0.25\times9\dfrac{\$}{\text{hour}}+0.75\times3\dfrac{\$}{\text{hour}}\right)$

$= \$4,475,665/\text{year}$

Route B: $\left[\dfrac{350}{50}(\$9)+\dfrac{250}{40}(\$9)+\dfrac{80}{50}(\$3)+\dfrac{3320}{50}(0.25\times\$9+0.75\times\$3)\right](22)(365)$

$= \$3,395,486/\text{year}$

Route C: $\left[\dfrac{350}{50}(\$9) + \dfrac{250}{40}(\$9) + \dfrac{80}{50}(\$3)\right.$

$\left. + \dfrac{3320}{50}(0.25 \times \$9 + 0.75 \times \$3)\right](20.5)(365)$

$= \$3,163,975/\text{year}$

Safety costs

Route A: $\left(105\ \dfrac{\text{vehicles}}{\text{year}}\right)\left(4400\ \dfrac{\$}{\text{vehicle}}\right) = \$462,000/\text{year}$

Route B: $(75)(\$4400) = \$330,000/\text{year}$

Route C: $(70)(\$4400) = \$308,000/\text{year}$

We can summarize all relevant government and public costs as in Table 6.1.

TABLE 6.1 Summary of Annual Equivalent Government and Public Costs

	Route A	Route B	Route C
Government first cost of highway	$ 286,269/year	$ 753,115/year	$1,370,640/year
Government cost of highway maintenance	208,000/year	176,000/year	317,750/year
Public operational costs	5,432,076/year	4,596,372/year	4,282,983/year
Public time costs	4,475,665/year	3,395,486/year	3,163,975/year
Public safety costs	462,000/year	330,000/year	308,000/year
Total government costs	494,269/year	929,115/year	1,688,390/year
Total public costs	10,369,741/year	8,321,858/year	7,754,958/year

We can desire to compare these three alternative routes using benefit-cost criteria. Let our first criterion be the popular benefit-cost ratio. Since one of these alternatives must be selected, we will assume the lowest government cost alternative, route A, will be selected unless the extra expenditures for routes B or C prove more worthy. Since we have not defined "benefits" per se, user benefits will be taken as the incremental reduction in user costs from the less expensive to the more expensive alternatives. Since we are looking at incremental benefits, it makes sense to compare these against the respective incremental costs needed to achieve these additional benefits.

The incremental benefits and costs for route B as compared to route A for $i = 6\%$ are given as follows:

$$\Delta B_{B-A}(6) = \text{public costs}_A(6) - \text{public costs}_B(6)$$
$$= \$10,369,741 - \$8,321,858 = \$2,047,883/\text{year}$$

$$\Delta C_{B-A}(6) = \text{government costs}_B(6) - \text{government costs}_A(6)$$
$$= \$929,115 - \$494,269 = \$434,846/\text{year}$$

That is, for an incremental expenditure of \$434,846/year, the government can provide added benefits of \$2,047,883/year for the public. The appropriate benefit cost ratio is then

$$\Delta B/C_{B-A}(6) = \frac{\Delta B_{B-A}(6)}{\Delta C_{B-A}(6)} = \frac{\$2,047,883}{\$434,846} = 4.71$$

This clearly indicates that the additional funds for route B are worthwhile, and we desire route B over route A.

Using a similar analysis, we now calculate the benefits, costs, and $\Delta B/C$ ratio to determine whether or not route C is preferable to route B.

$$\Delta B_{C-B}(6) = \$8,321,858 - \$7,754,958 = \$566,900/\text{year}$$
$$\Delta C_{C-B}(6) = \$1,688,390 - \$929,115 = \$759,275/\text{year}$$
$$\Delta B/C_{C-B}(6) = \frac{\$566,900}{\$759,275} = 0.75$$

This benefit-cost ratio, being less than 1.00, indicates that the additional expenditure of \$759,275/year to build and maintain route C would not provide commensurate benefits to the public. In fact, the user savings would be only \$566,900/year. Of the three alternative routes, route B is preferred.

The next benefit-cost criterion takes advantage of the fact that if $\Delta B/C > 1$, then $\Delta(B - C) > 0$. That is, the difference in incremental benefits and costs may be used in place of the incremental benefit-cost ratio.

Example 6.2 ────────────

Applying this measure to the routes in the previous example results in the following calculations for comparing routes A and B. We know that

$$\Delta B_{B-A}(6) = \$2,047,883/\text{year}$$
$$\Delta C_{B-A}(6) = \$\ 434,846/\text{year}$$

This results in

$$\Delta(B - C)_{B-A}(6) = \Delta B_{B-A}(6) - \Delta C_{B-A}(6) = \$2,047,883 - \$434,846$$
$$= \$1,613,037/\text{year}$$

leaving us again to conclude that route B is preferred to route A. Similarly,

$$\Delta(B - C)_{C-B}(6) = \Delta B_{C-B}(6) - \Delta C_{C-B}(6) = \$566{,}900 - \$759{,}275 = -\$192{,}375/\text{year}$$

which indicates that route C is not worthy of the additional expenditure required and that route B should be constructed.

The benefit-cost criteria are consistent with the methods of alternative evaluation presented in Chapter Four. For example, suppose it is desired to base a decision on the present worth criterion, minimizing the sum of government construction and maintenance costs as well as public user costs.

Example 6.3

The present worth of all costs for each highway alternative is given by the following formula:

$$PW(\text{total}) = PW_{\text{total government costs}}(i) + PW_{\text{total public costs}}(i)$$

Then,

Route A: $PW_A(6) = \$494{,}269(P|A\ 6{,}30) + \$10{,}369{,}741(P|A\ 6{,}30)$
$$= \$494{,}269(13.7648) + \$10{,}369{,}741(13.7648)$$
$$= \$149{,}541{,}263$$

Route B: $PW_B(6) = \$929{,}115(P|A\ 6{,}30) + \$8{,}321{,}858)(P|A\ 6{,}30)$
$$= \$929{,}115(13.7648) + \$8{,}321{,}858(13.7648)$$
$$= \$127{,}338{,}081$$

Route C: $PW_C(6) = \$1{,}688{,}390(P|A\ 6{,}30) + \$7{,}754{,}958(P|A\ 6{,}30)$
$$= \$1{,}688{,}390(13.7648) + \$7{,}754{,}958(13.7648)$$
$$= \$129{,}986{,}091$$

Since the present worth of all costs for route B is smallest, we again see that this alternative is preferred.

The annual equivalent cost is calculated by simply multiplying a constant times the present worth. Thus, the benefit-cost criteria are also comparable to the annual worth measure of merit.

Example 6.4

Calculating the equivalent uniform annual cost using

$$AW_{\text{total}}(i) = AW_{\text{total government costs}}(i) + AW_{\text{total public costs}}(i)$$

we have

Route A: $AW_A(6) = \$494,269 + \$10,369,741 = \$10,864,010/\text{year}$

Route B: $AW_B(6) = \$929,115 + \$8,321,858 = \$9,250,973/\text{year}$

Route C: $AW_C(6) = \$1,688,390 + \$7,754,958 = \$9,443,348/\text{year}$

Again, route B is preferred.

It is interesting and logical to note that the differences in equivalent uniform annual costs for routes A,B and B,C are the same as the differences in incremental benefits minus incremental costs for routes A,B and B,C. That is,

$$AW_A(6) - AW_B(6) = \$10,864,010 - \$9,250,973 = \$1,613,037/\text{year}$$
$$\Delta B_{B-A}(6) - \Delta C_{B-A}(6) = \$2,047,883 - \$434,846 = \$1,613,037/\text{year}$$

and

$$AW_B(6) - AW_C(6) = \$9,250,973 - \$9,443,348 = -\$192,375/\text{year}$$
$$\Delta B_{C-B}(6) - \Delta C_{C-B}(6) = \$566,900 - \$759,275 = -\$192,375/\text{year}$$

Finally, the benefit-cost criteria can even be related to the internal rate-of-return approach discussed in Chapter Four. That is, the interest rate is found that causes the annual equivalent incremental benefits less incremental costs to equal zero.

Example 6.5 ─────────────────────────

The internal rate of return approach may be used by finding the value of the interest rate i that equates the incremental benefits of Alternative k over Alternative j to their incremental costs. That is,

$$AW_{k-j}(i) = 0 = \Delta B_{k-j}(i) - \Delta C_{k-j}(i)$$

Using this expression to consider route B versus route A, we have

$$\Delta B_{B-A}(i) = \$10,369,741 - \$8,321,858 = \$2,047,883/\text{year}$$
$$\Delta C_{B-A}(i) = [\$8,800,000 + \$1,800,000(P|Fi,10)$$
$$+\$1,800,000(P|Fi,20)](A|P\,i,30) + \$176,000$$
$$-\{[\$2,200,000 + \$2,000,000(P|Fi,10) + \$2,000,000(P|Fi,20)]$$
$$[A|P\,i,30] + \$208,000\}$$

Searching for the value of i yielding $AW_{B-A}(i) = \Delta B_{B-A}(i) - \Delta C_{B-A}(i) = 0$, we find

$$AW_{B-A}(30) = \$103,827/\text{year}$$
$$AW_{B-A}(35) = -\$226,726/\text{year}$$

The interest rate we are seeking is between 30 and 35%, or approximately 32%. In any event, i easily exceeds 6% and we prefer route B to route A.

Using the rate-of-return approach, we can also compare routes B and C.

$$\Delta B_{C-B}(i) = \$8,321,858 - \$7,754,958 = \$566,900$$

$$\Delta C_{C-B}(i) = [\$17,300,000 + \$1,800,000(P|Fi,10) + \$1,800,000(P|Fi,20)]$$
$$(A|P\,i,30) + \$317,750$$
$$-\{[\$8,800,000 + \$1,800,000(P|Fi,10) + \$1,800,000(P|Fi,20)]$$
$$(A|P\,i,30) + \$176,000\}$$

The rate of interest causing $AW_{C-B}(i) = \Delta B_{C-B}(i) - \Delta C_{C-B}(i) = 0$ is between 2.5 and 3%, since

$$AW_{C-B}(2.5) = \$19,020/\text{year}$$
$$AW_{C-B}(3) = -\$8,520/\text{year}$$

Since i is less than 6%, we prefer route B to route C.

Several different approaches to the same problem have been illustrated and shown to be consistent project evaluators. Typically, where government projects are involved, one of the benefit-cost criteria (B/C or $B - C$) are used. More often than not, the criterion is the benefit-cost ratio. This is unfortunate because, just as in rate of return analyses in the private sector, the benefit-cost ratio is easy to misuse and misinterpret, and it is very sensitive to the classification of problem elements as "benefits" or "costs." These problems will be discussed subsequently.

The present worth, annual worth, and rate-of-return methods are seldom used in government analyses. Even though the mechanics underlying these approaches are perfectly suitable, their underlying philosophy is more attuned to return on investment, discounted net profits, and minimum attractive rate of return, implying measures of return *to investors* based on capital investment *by the same investors*. Government investments, however, do not necessarily result in *any* monetary incomes to the government and are seldom evaluated solely on the basis of monetary return. Also, government employees and politicians prefer benefit-cost analyses to enable them to speak intelligently of the "benefits" derived by the public as a result of government's expenditures of the public's tax, assessment, and other money.

6.5 IMPORTANT CONSIDERATIONS IN EVALUATING PUBLIC PROJECTS

In the section on benefit-cost analysis, it was stated that there are a number of pitfalls that can affect the analysis of government projects. Actually, opportunities for error pervade benefit-cost analyses, from the very initial philosophy, through to the interpretation of a B/C ratio. It is important to talk about the more significant of these, both to help prevent analysts from erring, and to help those who may be reviewing a biased evaluation.

The major topics to be considered include the following:

1. Point of view (national, state, local, individual).
2. Selection of the interest rate.
3. Assessing benefit-cost factors.
4. Overcounting.
5. Unequal lives.
6. Tolls and fees.
7. Multiple-use projects.
8. Problems with the B/C ratio.

6.5.1 Point of View

The stance taken by the engineer in analyzing a public venture can have an extensive effect on the economic "facts." The analyst may take any of several viewpoints, including those of:

1. An individual who will benefit or lose.
2. A particular governmental organization.
3. A local area such as a city or county.
4. A regional area such as a state.
5. The entire nation.

The first of these viewpoints is not particularly interesting from the standpoint of economic analysis. Nonetheless, all too frequently, an isolated road is paved, a remote stretch of water or sewer line is extended under exceptional circumstances, or a seemingly ideal location for a public works facility is suddenly eliminated from consideration. In these cases, the "benefit-cost analysis," its review, and the implementation decision are usually made by a small, select group.

The other four viewpoints are, however, of considerable interest to those involved in public works evaluation. Analyzing projects or project components from viewpoint 2, that of a particular government agency, is analogous to economic comparisons in private enterprise. That is, only the gains and losses to the organization involved are considered. This viewpoint, which seems contrary to benefit-cost optimization to the public as a whole, may be appropriate under certain circumstances.

Example 6.6

Consider a Corps of Engineers construction project in which the water table must be lowered in the immediate area so work can proceed. Any of several water cutoff or dewatering systems may be employed. Water cutoff techniques include driving a sheet pile diaphram or using a bentonite slurry trench to cut off the flow of water to the construction area. Dewatering methods include deep well turbines, an educator system, or wellpoints for lowering the water level. It is sometimes appropriate for the Corps to evaluate these different techniques from an "organization" point of view, since each of the feasible methods provides the same service or outcome—a dry construction site. Therefore, the most economical decision from the Corps' point of view is also correct from the view of the public as a whole, since the benefits or contributions to the project are the same regardless of the method chosen.

The third point of view, that of a locality such as a city or county, is popular among local government employees and elected officials. Unfortunately, seemingly localized projects often impact a much wider range of the citizenry than is apparent.

Example 6.7

County officials are to decide whether or not future refuse service should be county owned and operated, or whether a private contractor should be employed. The job requires front-end loader compaction trucks as well as roll on-off container capability. Primarily rural roads are traveled, and from 1 container (a roadside picnic area) to 50 containers (a large rurally located industrial plant) must be collected at each stop. Front end loader containers range from 2 to 8 cubic yards, while roll on-off containers are sized from 15 to 45 cubic yards. Several trucks and drivers will be required, including a base for operations and maintenance.

The cost in dollars per ton of refuse collected, removed, and disposed of, is given below:

Personnel services	$ 6.08/ton
Materials, supplies, utilities	5.13/ton
Maintenance and repair	5.62/ton
Overhead	3.41/ton
Depreciation	3.61/ton
Five percent interest on half financed by bonds	1.80/ton
Total county cost	$25.65/ton

Federal taxes foregone	$ 1.87/ton
State taxes foregone	0.20/ton
Property taxes foregone	1.54/ton
Eight percent return on half financed by tax money	2.88/ton
Not necessarily paid by county	$ 6.49/ton

County cost to provide refuse service will be $25.65/ton. The county is not, however, required to pay the additional $6.49/ton for federal and state taxes, *ad valorem* taxes, or a return on appropriated money, as would a private firm.

It is obvious from the example that a "local" county decision can affect a much wider public. Suppose, based on $25.65/ton, the county decided to own and operate the needed refuse service. Federal taxes of $1.87/ton that would have been paid by a private contractor will not be paid. Since the federal government will still have the same revenue requirements, the difference will be made up by passing on an infinitesimally small burden to the national public as a whole. Although this is easily rationalized at the local level—to spread a portion of the cost of county refuse service over the entire country—consider the result if every town, city, and county took this attitude.

An analogous argument follows for state income taxes foregone; however, now the burden is being spread over the people of the state. Even though this is much smaller than the national population, the burden per person to make up the lost state tax income is still very small. Again, providing refuse service at less cost to the local populace, at the expense of the state, is tempting from a parochial point of view.

If the county were to plan on not having to pay the *ad valorem* tax, the slack would be taken up by increasing the property tax rates in the county. Although this approach increases the burden on county property owners, that burden may be entirely disproportionate when compared to the refuse service each requires.

The last two points of view include a regional (e.g., a state) or national perspective. Ideally, a national outlook is preferable for local, regional, and national public works projects. Experience indicates, however, that the primary concern of public works officials and politicians is their particular constituency.

Perhaps the best advice for evaluators and decision makers in the public realm is to examine multiple viewpoints. That is, project evaluators are often not in a position to decide on a single specific point of view. They should instead present a thorough analysis clearly indicating any benefits or costs that depend on the perspective taken. Similarly, decision makers should require multiple points of view so they can be aware of the kind and degree of repercussions resulting from their actions.

6.5.2 Selection of the Interest Rate

The interest rate, discount rate, or minimum attractive rate of return is another premier factor to be decided upon when evaluating public works projects. In the route selection examples, the interest rate was taken at 6% with no question of the appropriateness of such a figure. Clearly, however, the interest rate has a significant effect on the net present worth or annual worth of cash flows in the private sector. Similarly, there is a significant effect on the net present worth of benefits minus costs or the benefit-cost ratio in public sector analyses.

Example 6.8

Let us reconsider the highway route selection Examples 6.1 to 6.5 again, this time using a minimum attractive rate of return of 2.5%. Since route B was preferable to route A at $i = 6\%$, that decision continues to hold. However, route C was not preferred to route B using a *MARR* of 6%. In the rate-of-return approach, it was determined that the incremental "rate of return" of route C over route B exceeded 2.5% but fell short of 3%. Now, however, using an interest rate of 2.5%, we find that spending the substantial additional funds for route C is worthwhile.

One may argue with the above example, saying that $i = 2.5\%$ is lower than would ever be used. Such an argument is probably true for private industry. In public works, however, various knowledgeable individuals have explicitly proposed values on i ranging from 0 to 13%, and rather vague support has been seen ranging from 0 to over 20%. Before considering various schools of thought on the appropriate interest rate, it is helpful to know how public activities are financed.

Financing of Government Projects. There are several different ways that units of government finance public sector projects. The most obvious way is, of course, through taxation such as income tax, property tax, sales tax, and road user tax. Another popular approach is through the issuance of bonds for either specific projects or general use. Other forms of borrowing, such as notes, may be thought of in the same category as bonds. A third type of fund raising includes income generating activities such as a municipally owned power plant, a tool road, or other activity where a charge is made to cover (or partially offset) the cost of the service performed. Although these are the primary sources of government funds, there are a number of ways in which this money may be passed from one government authority to another by way of direct payments, loans, subsidies, and grants.

Federal funds are raised through tax money and federal borrowing. Federal projects may then be financed through direct payment. In this case, no monetary return is expected by the government; however, the "return" is expressed through the benefits incurred by the public. Direct payment financing may be total as in the case of many corps of engineers projects, or partial, for example 90%, as in cost sharing with states for interstate highways.

Financing for projects of national interest and impact may also be available through no-interest or low-interest rate loans. Both are available for long periods of time, say up to 40 years, with terms obviously more favorable than could normally be expected from conventional sources of money. Such loans are available for financing large projects such as the Tennessee Valley Authority where revenues resulting from the projects are used to pay back the loans plus interest, if any. Certain university buildings such as dormitories may also be financed in this way. There are also occasions when principal payment deferment is permitted during the early years of the project.

Other forms of federal financing include subsidies and federal loan insurance. Subsidies are used to encourage projects or services believed to be in the public's best interest, such as in the area of transportation. Loan insurance is used to eliminate the private lending institution's risk, allowing lower interest loans over longer periods than conventionally available. Insured loans began with the Federal Housing Administration.

State and local public projects are financed from taxes or bonds. There are, however, constraints on bond financing. First, bond issues must be approved by the voters. Often, there must be a 60% or even two thirds vote in favor of the bond. Second, in order to prevent excessive borrowing, states have limited the amount of bond debt that may be undertaken. This is often a fraction of the property valuation assessment in the local area. Finally, there are also restrictions that govern the payback requirements and lives of the bonds. Considering these restrictions, the temperament of the public regarding bond issues, and the future needs that are likely to require bond financing, care is obviously needed in selecting public works projects to be implemented or put before the people.

Considerations in the Selection of the Interest Rate. Many arguments over the correct philosophy to use in selecting the interest rate have surfaced over time. For practical purposes, most of these philosophies are somewhat aligned with one of the following positions:

1. A zero interest rate is appropriate when tax monies are used for financing.

2. The interest value need only reflect society's time preference rate.

3. The interest rate should match that paid by government for borrowed money.

4. The appropriate interest rate is dictated by the opportunity cost of those

investments foregone by private investors who pay taxes or purchase bonds.

5. The appropriate interest rate is dictated by the opportunity cost of those investments foregone by government agencies due to budget constraints.

Advocates of a zero interest rate when tax money is used argue that current taxes require no principal or interest payment at all. Hence, current tax monies (e.g., highway user taxes) should be considered "free" money, and a no-interest or discount rate applied. Counterarguments to this stance point out that a zero (or even low) interest rate will allow very marginal projects or marginal "add-on" project enhancements to achieve a B/C ratio greater than 1. This, in turn, takes money away from other projects that are truly deserving. If, in fact, it is not true that more deserving projects are precluded while very marginal projects (say 1 or 2% "return") are being approved, perhaps there is an excess of money available and taxes should be lowered. A final position against the zero interest rate advocates is that if government does not invest funds in high-benefit or "profitable" projects, the people should be able to retain and invest their own money to provide benefits more economically.

The "societal time preference rate" advocates contend that i is merely "a planning parameter reflecting society's feelings about providing for the future as opposed to current consumption" says Howe [6]. Or Henderson [5] says i is the rate that "reflects the government's judgment about the relative value which the community as a whole is believed to assign, or which the government feels it ought to assign, to present as opposed to future consumption at the margin." As such, Howe [6] makes clear that the societal time preference rate "need bear no relation to the rates of return in the private sector, interest rates, or any other measurable market phenomena." Estimated societal time preference rates tend to be quite low (e.g., 2 to 4%).

Many people back the use of an interest rate that matches that paid by government for borrowed money. This seems reasonable in that government bonds are in direct competition with other investment opportunities available in the private sector. Of course, many government bond coupon rates tend to be lower because of their tax exempt status (the bond interest is not taxable) than their private industry counterparts. One way of obtaining an interest figure is to determine the rate on "safe" long-term federal bonds [6]. Another possibility is to use the rate paid on borrowings by the particular government unit in question. That is, if a project will be financed by borrowing, the rate of interest to be paid is used. If nonborrowed funds are to be used, the appropriate rate is the average rate of interest being paid on long-term (over 15 years) borrowing.

There are also good arguments that the cost of government money may be too low an interest rate. For example, the opportunities foregone by other government agencies or investors in the private sector may have provided a far higher "return" than investments approved using the cost of government money. In addition, using a rate equal to the cost of government borrowings includes no provision for risk, nor does it include the subsidizing effect of the tax exemption. Finally, it is argued that plain good judgment dictates that projects worthy of approval under higher interest rates are more justifiable in cases where benefits and disbenefits received by the public are very nonuniform.

The next philosophy calls for an opportunity cost approach, taking into account many of the factors not considered in the pure cost of government borrowed money. For example, Howe [6] says this philosophy is that "no public project should be undertaken that would generate a rate of return less than the rate of return that would have been experienced on the private uses of funds that would be precluded by the financing of the public project (say, through taxes or bonds)." How are these private use rates determined? Consider a situation in which private investments would yield an average annual return of $i_1\%$. Also consider the rate of return on consumption, which is measured by the rate that consumers are willing to pay to consume now instead of later. This rate is, say, $7\frac{1}{2}$ to 10% for a house, 12 to 24% for a new or used car, and so on. Let this rate of return on consumption average out to $i_2\%$. Now, if a fraction α of government financing precludes private investments, while $1 - \alpha$ precludes consumption, the rate of return foregone to finance public projects is given as i.

$$i = \alpha i_1 + (1 - \alpha)i_2 \qquad 0 \le \alpha \le 1 \qquad\qquad (6.5)$$

An empirical study conducted by Haveman [4] in 1966 showed the appropriate weighted average to be 7.3% *at that time.* With the many economic changes since that time, a rate of perhaps 10 to 13% is more appropriate for the late 1970s.

The last philosophy also requires an opportunity cost approach in which an artificial interest rate reflects the rates of return foregone on government projects by virtue of having insufficient funds. That interest rate is found by continuing to increase the value of i until only the projects remain having $B/C > 1$ or $B - C > 0$ which can be afforded with monies available.

What conclusions can be reached from these philosophical arguments? What guidelines are available for evaluators of public works projects? Clearly, no one answer is universally applicable. But, as a general rule, we recommend using a rate that is at *least* as high as the average effective yield on tax exempt long-term government bonds. That is, projects or enhancements to projects not providing at least this rate of return should not be undertaken. This is not to say, however, just because a project is attractive at the *MARR*, that it should be implemented. Only the most worthy projects, in the environment of limited funds, should be constructed. These projects may be identified by conducting several analyses of prospective projects, continually raising the interest rate ($i = 6, 8, 10, 12, \ldots \%$) until only the projects remain that may be accomplished with available or legally borrowable funds.

6.5.3 Assessing Benefit-Cost Factors

The benefit-cost analyst knows, before starting a study, that placing a monetary figure on certain "societal benefits" may be difficult. Actually, there is even a more fundamental problem—what factors to assess. Some insight is available by considering the following four types of factors discussed by Cohn [2].

1. *"Internal"* effects are those which accrue directly or indirectly to the individual(s) or organization(s) with which the analyst is primarily concerned. These effects are always included in a benefit-cost analysis.

2. *"External"* *technological (or real)* effects are those which cause changes in the physical opportunities for consumption or production. For example, effects on navigation and water sport recreation due to a new hydroelectric plant would fall into this category. These effects should be included in an analysis.

3. *"External"* *pecuniary* effects relate to changes in the distribution of incomes through changes in the prices of goods, services, and production factors. For example, if a firm's expansion is sufficiently large to affect industry prices, an increase in its output is most likely to result in lower prices for output from other firms in the industry as a whole. Many authors agree that these effects can safely be ignored.

4. *Secondary effects* involve changes in the demand for and supply of goods, services, resources, and production factors which *arise from* a particular project. As an example, phosphate mining in Idaho on government lands will bring instant population increases to nearby small towns. Secondary effects include increasing the incomes of various producers such as vehicle repairmen and barbers who provide the vehicle repairmen with haircuts. McKean [9] contends that only incremental income arising from such effects should be included, if reasonable to do so.

Example 6.9

A dam and reservoir are contemplated in an effort to reduce flood damage to homes and crops in a low area of northeastern Oklahoma. Annual damage to property varies from year to year, but averages approximately $230,000/year. The dam and reservoir contemplated should virtually eliminate damage to the area in question. No other benefits (e.g., irrigation, power generation, recreation) will be provided.

In performing a benefit-cost analysis of the flood control project, the engineer notes that the primary benefit to the public will be the $230,000/year damage prevented. However, the engineer argues that this will also cut back on money paid to contractors and servicemen for home and car repair, to health care units, for insurance premiums, and so forth. In other words, the building of a dam, it could be argued, will provide disbenefits to those who would normally receive part of their livelihood from helping flood-damaged families. Should the engineer include in the evaluation only the direct benefits to the flood damage victims, or should the other effects be included as well? That depends on how the analyst chooses to handle secondary effects, as indicated in the following discussion.

The disbenefits to those who would lose income if the dam and reservoir were built are considered *secondary effects*. That is, there would be a decrease in the demand for and supply of post-flood restoration goods and services. It is argued that only the *incremental* incomes or incremental profits (losses or lost profits in this case) should be considered when secondary effects are involved.

Another argument calls upon the "ripple" effect of the economy. That is, every secondary effect disbursement by one person or organization is a receipt to another person or organization. Each receipt then contributes to another disbursement, and so on. If the ripple philosophy were tracked or followed for the secondary effects of a particular alternative, the sum of receipts less disbursements would equal zero, and there would be no economic evaluation.

With either philosophy, the secondary effects of the example's flood control dam and reservoir would be small, if not negligible. This is intuitively reasonable, because the dam's main benefits represent the measure of the direct usefulness of the dam serving its intended purpose, whereas the diseconomies described are, in fact, secondary and diffuse.

Example 6.10 ————————————————————————————

Now, reconsider the previous flood control dam and reservoir of Example 6.9. Suppose the reservoir would cause a loss of agricultural land for grazing and crops. Should this loss be considered in the benefit-cost analysis? Yes, because it is an external real effect causing changes in the physical opportunities for consumption or production.

External technological effects are often well defined and, in practice, are usually included in benefit-cost analyses. External pecuniary effects, however, may not be vividly apparent, as illustrated in the following example.

Example 6.11

A large irrigation project is being considered in the heart of cotton country. The irrigation will provide a significant effect on the quantity and quality of the cotton grown. This additional supply of cotton will, however, depress the price of cotton, lowering the profitability of other cotton growers. Also, the same effect will be felt throughout, say, the clothing industry and manufacturers of cotton substitutable products (products that may be used in place of cotton items) will likely have to reduce prices. At the same time, producers of cotton complementary goods (items that go well with or are used in conjunction with cotton products) will note increased demand, and may increase prices because of an insufficient supply.

Which of these effects would we include in an evaluation of the irrigation project? Of the factors mentioned above, none would be included in the analysis. Each of the effects described relates to changes in the distribution of incomes through changes in the prices of goods, services, and production factors. As such, they are considered external pecuniary effects, which are not "real" benefits or disbenefits and, hence, are not included.

In the above examples, internal and external technological effects were considered to be factors in an analysis, while external pecuniary and secondary effects were not included. There is, of course, no complete agreement on either these classifications or whether or not they should be evaluated. For the practitioner, good practical judgment is probably the best asset in deciding what factors to include.

As a guide, all identifiable effects of a project should be delineated. Some effects will clearly provide direct benefits or disbenefits to the public that should be counted. There may also be some factors that obviously should not be included. The third group—the controversial factors, if any—should be studied in depth and either included for good reason or be used implicitly as supporting information. The reason for including or not considering these controversial factors should be stated in writing and become a part of the evaluation for the record.

6.5.4 Overcounting

A common dysfunction of trying to consider a large variety of effects in a benefit-cost analysis is to overcount, or unknowingly count some factors twice.

Example 6.12

Many years ago, Connecticut's Department of Health calculated the following loss from *preventable* disability, hoping to increase interest in health problems.

Individual income lost

Number of wage earners affected	350,000
Average days per year lost due to preventable accidents	4
Average hourly wage	$2.50
Total individual loss per year	$3,500,000

Industry loss:

It is accepted that the loss to industry in disorganization, idle overhead, and lessened production is $2\frac{1}{2}$ times the wage loss, or

Total industry loss per year	$8,750,000

These figures were totally inadequate because of a classic (but not overly apparent) case of double-counting. If we assume the loss to individuals is, correctly, their foregone earnings, then the loss to industry would be *its* foregone earnings, or its lost profits, and not the entire value of lessened production. As the figures stand above, lost wages are counted once from the viewpoint of the employee, and are again included as part of industry's losses. That is, Weisbrod [11] emphasizes that employee wages are double-counted.

Example 6.13 ─────────────────────────────────────

Reconsider Example 6.11 involving a cotton irrigation system. The increased quantity of cotton will require that additional gin and seed mill hands be employed, removing a significant number of persons from the welfare rolls. The amount of their new wage, equal to the sum of their old welfare payments plus some increase, represents an increase in real output and constitutes a legitimate national benefit of the project. To then add the reduction in welfare payments from the taxpayers to the unemployed would be to double-count welfare payments, once from the standpoint of the recipient, and once from the taxpayer's viewpoint.

6.5.5 Unequal Lives

When comparing one-shot public works projects having unequal lives, the planning horizon will commonly coincide with the longest lived alternative. This selection of a planning horizon was recommended in Chapter Four. When some projects are expected to have a long life while others have a shorter life, changes in the discount rate could change the attractiveness of some projects with respect to the alternatives.

Example 6.14

Two projects each have first costs of $200,000, with annual operating costs of $30,000. The life of Project A is 15 years, while that of Project B is 30 years and benefits accrued to the public are estimated at $60,000/year and $52,808/year, respectively. Each is a one-shot project; hence, Project A will have no benefits or costs after year 15. A 30-year planning horizon will be used. If the *MARR* is set at 5%, which project is more attractive?

Using the $B - C$ measure of merit and a present worth base, we have

$$(B - C)_A(5) = \$60,000(P|A\ 5,15) - \$30,000(P|A\ 5,15) - \$200,000$$
$$= \$60,000(10.380) - \$30,000(10.380) - \$200,000$$
$$= \$111,400$$

$$(B - C)_B(5) = \$52,808(P|A\ 5,30) - \$30,000(P|A\ 5,30) - \$200,000$$
$$= \$52,808(15.372) - \$30,000(15.372) - \$200,000$$
$$= \$150,605$$

Over a 30-year planning horizon, and using an interest rate of 5%, Project B is clearly better. Now suppose that insufficient money is available to implement all of the many projects, B included, that a government agency would like to do. A decision is made to consider all projects using an interest rate that reflects the opportunity cost of investments foregone by government agencies. That is, the interest rate is increased until only those projects that can be afforded remain. Assuming the interest rate is up to 12%, now which project is more desirable?

$$(B - C)_A(12) = \$60,000(P|A\ 12,15) - \$30,000(P|A\ 12,15) - \$200,000$$
$$= \$60,000(6.811) - \$30,000(6.811) - \$200,000$$
$$= \$4330$$

$$(B - C)_B(12) = \$52,808(P|A\ 12,30) - \$30,000(P|A\ 12,30) - \$200,000$$
$$= \$52,808(8.055) - \$30,000(8.055) - \$200,000$$
$$= -\$16,282$$

At the higher interest rate, Project A is more favorable because of the increased emphasis on early year net benefits as opposed to the heavily discounted net benefits during the latter years of the planning horizon.

When a planning horizon shorter than some project durations is selected, a residual value must be estimated for those projects. The residual value is handled in the same way as a salvage value.

Example 6.15

Reconsidering the previous example and selecting a 15-year planning horizon, let us now determine the more favorable alternative, assuming project B has a residual value of 40% of first cost. Let $i = 12\%$.

$$(B - C)_A(12) = \$60,000(P|A\ 12,15) - \$30,000(P|A\ 12,15) - \$200,000$$
$$= \$60,000(6.811) - \$30,000(6.811) - \$200,000$$
$$= \$4330$$

$$(B - C)_B(12) = \$52,808(P|A\ 12,15) + 0.4(\$200,000)(P|F\ 12,15)$$
$$- \$30,000(P|A\ 12,15) - \$200,000$$
$$= \$52,808(6.811) + \$80,000(0.1827) - \$30,000(6.811) - \$200,000$$
$$= -\$30,039$$

6.5.6 Tolls, Fees, and User Charges

Tolls, fees, and user charges have an interesting effect on the fiscal aspects of public projects. If a toll, fee, or user charge is regarded as a payment or partial payment for benefits derived, it can be argued that net benefits received are reduced by the amount of the payment. Similarly, the amount of the payment decreases the cost of the project to the government. Thus, the B/C ratio will change, but the $B - C$ measure of merit will remain constant so long as total user benefits remain constant.

Example 6.16

Suppose 10,000 people/year attend a public facility that has an equivalent uniform annual cost of $20,000. The people, on the average, receive re-creational benefits in the amount of $3 each. The B/C ratio would be

$$B/C = \frac{3(\$10,000)}{\$20,000} = 1.5$$

and the $B - C$ measure of merit is

$$B - C = 3(\$10,000) - \$20,000 = \$10,000/\text{year}$$

Based on either criterion, the public facility appears worthwhile. Now suppose that a fee of $1.5/season is charged. The net benefits are now $3 - \$1.5$, or $1.5/person, and the government cost is reduced by $15,000/year. Thus, the B/C and $B - C$ measures are as follows.

$$B/C = \frac{\$30,000 - \$15,000}{\$20,000 - \$15,000} = 3$$

$$B - C = \$30,000 - \$15,000 - (\$20,000 - \$15,000) = \$10,000/\text{year}$$

Note that in the example B/C changed while $B - C$ did not. This pheno-menon will be discussed in a subsequent section on "Problems With the B/C Ratio."

It might be concluded that tolls, fees, and user charges are irrelevant, at least with respect to the $B - C$ measure of merit. This, however, *is not* true if the number of users or degree of use is linked to the fee charged, as it almost always will be.

Example 6.17 ────────────────────────────────

As an extreme, suppose that the 10,000 users of the public facility in the previous example receive different levels of benefits, but they average out to $3/person. The actual breakout is that 8000 persons perceive $1.45 worth of enjoyment, 1000 persons perceive $3 worth, and 1000 persons expect to derive $15.40 in recreational benefits. With a user fee of $1.50/person, only 2000 will patronize the facility. Thus, the B/C and $B - C$ measures would be:

$$B/C = \frac{1000(\$3) + 1000(\$15.40) - 2000(\$1.50)}{\$20,000 - 2000(\$1.50)} = \frac{\$15,400}{\$17,000} = 0.91$$

$$B - C = [1000(\$3) + 1000(\$15.40) - 2000(\$1.50)] - [\$20,000 - 2000(\$1.5)]$$
$$= -\$1600$$

In this case, the reaction of demand to a fee would cause the costs to exceed realized benefits. Thus, when tolls, fees, and user charges are ex-pected, their effect on user demand, and hence total user benefits, must be determined and accounted for.

6.5.7 *Multiple-Use Projects*

Multiple-use projects receive a great deal of attention, both pro and con. Multiple uses, and hence multiple benefits, are often available at slight incremental costs over single-use projects. Of course, the incremental capital and net operating costs required for an additional use must provide at least a like worth of benefits.

Example 6.18 ────────────────────────────────

A dam and reservoir for irrigation will provide present worth benefits of $25 million over the next 50 years. The present worth cost of construction, operation, and maintenance of the irrigation facility will be $14,500,000. A single purpose flood control dam providing present worth benefits of $6 million would cost a present worth of $9 million. Suitable design modifications can be made to the irrigation dam and reservoir to provide the flood control benefits, too, at a total package present worth cost of $18,500,000. Funds permitting, what should be done?

First, it must be understood that the benefits and costs discounted to the present were discounted at the *MARR* deemed suitable by the decision maker. Assuming this to be true, it is clear that the irrigation project is worthwhile, providing a benefit-cost ratio of

$$B/C_{\text{irrigation}} = \frac{\$25,000,000}{\$14,500,000} = 1.72^1$$

As a single-purpose facility, a flood control dam would not provide benefits commensurate with its costs, yielding a *B/C* ratio of

$$B/C_{\text{flood control}} = \frac{\$6,000,000}{\$9,000,000} = 0.67^1$$

As a multiple-use facility, however, the flood control benefits may be provided at a sufficiently low incremental cost to be justifiable. The incremental *B/C* ratio is

$$\Delta B/\Delta C_{\substack{\text{irrigation plus} \\ \text{flood control}}} = \frac{\$6,000,000}{\$18,500,000 - \$14,500,000} = 1.5$$

Thus, a multiple-use facility should be built.

The example illustrates how multiple uses can draw on each other, providing benefits economically that could never have been provided using a single-purpose facility (e.g., the *B/C* flood control ratio of 0.67). Multiple-purpose projects also have their problems. For example, it is frequently desirable to "allocate" the costs of a project to its various uses.

Example 6.19 ───

A city's refuse is used to fire a power generation facility owned by a municipality. Not only is electrical power supplied to a segment of the city, but burning of the refuse after processing has virtually eliminated the need for an expensive landfill operation. Since this project is self supporting, construction, operation, and maintenance costs must be allocated between the disposal and power benefits provided in order to determine the user charge for refuse disposal and electrical energy. Arguments for cost allocation range from (1) no costs should be allocated to refuse disposal because the refuse is being used in place of fuel oil or coal and, in fact, a credit should be issued, to

---

[1] This is also the incremental *B/C* ratio of irrigation over doing nothing.

(2) refuse disposal should receive sufficient cost allocation to raise rates above those for conventional disposal to include the aesthetic benefits of no unsightly public landfill.[2]

Another problem with multiple-purpose projects includes proponents from and opposition by various interest groups, on the basis of politics, environment, and the like. Yet another problem has to do with coordinating project finances, where some aspects may be federally financed with no payback required, other aspects may be self-supporting, and yet others may require state funds. Finally, there may be conflicts between the various purposes involved. That is, an enhancement of one use may detract from another use. Nevertheless, multiple-use projects are here to stay and will be confronted frequently by engineers in the public sector.

6.5.8 Problems with the *B/C* Ratio

There are two frequent problems with the *B/C* ratio that require an explanation and warning. Either can give misleading results that may cause an otherwise perfect analysis to point toward the wrong project.

First, it is sometimes difficult to decide whether an item is a benefit to the public or a cost savings to the government. Similarly, there is often uncertainty between disbenefits and costs.

Example 6.20

A project provides annual equivalent benefits of $100,000, disbenefits of $60,000/year, and annual costs of $5000. What is the benefit-cost ratio?

Let us first calculate the *B/C* ratio for the problem as stated.

$$B/C = \frac{\$100,000 - \$60,000}{\$5000} = 8$$

Such a high ratio leads one to believe that the project is outstanding.

Now another analyst notes that the government will reimburse those incurring damages from the project in an amount equivalent to $60,000/year. Thus, the analyst concludes that the public disbenefits have been compensated for by the government and calculates a *B/C* ratio of

$$B/C = \frac{\$100,000}{\$5000 + \$60,000} = 1.54$$

which is considerably lower.

---

[2] These extremes in arguments have actually been used by public officials of one major city.

Example 6.20 shows that a wide range of B/C ratios may reasonably be obtained on a single project simply by interpreting certain elements of the problem differently. The resolution of this problem is not difficult. The analyst should simply calculate the net benefits less the net costs.

Example 6.21 ————————————————————————————

Let us reconsider Example 6.20 and calculate the annual net benefits less net costs.

The first analyst would have calculated

$$B - C = (\$100,000 - \$60,000) - (\$5000) = \$35,000/\text{year}$$

and the second analyst would have calculated

$$B - C = (\$100,000) - (\$5000 + \$60,000) = \$35,000/\text{year}$$

Calculating $B - C$ eliminates the inherent bias in the B/C ratio and does not require an incremental approach between alternatives where benefits and costs are known directly for each alternative. That is, if mutually exclusive alternatives are involved, over the same time horizon, the one having the highest $B - C$ value should be selected. Unfortunately, the B/C ratio is by far the more popular criterion of the two.

——

The potentiality for the other B/C problem was illustrated in the dam and reservoir irrigation example. That is, when the B/C ratio is used, it should be based on *incremental benefits* and *incremental costs*.[3] Simply to calculate the B/C ratio of each alternative and take the one with the largest ratio is incorrect and will frequently lead to errors in project selection.

Example 6.22 ————————————————————————————

In the dam and reservoir of Example 6.18 we calculated the B/C irrigation ratio to be 1.72. To compare this value against a total project $B/C_{\text{irrigation+flood control}}$ ratio of

$$\frac{\$25,000,000 + \$6,000,000}{\$18,500,000} = 1.68$$

would cause us to select irrigation only, in error.

——

A related error is to require the incremental B/C ratio to be above that for the previous incremental B/C ratio. In the dam and reservoir example, had

————

[3] Actually, the B/C irrigation ratio in the referenced example is a ratio of incremental benefits to incremental costs where the pairwise comparison is between irrigation and doing nothing.

the incremental B/C ratio, $\Delta B/\Delta C_{\text{irrigation plus flood control}} = 1.5$ been compared against the $B/C_{\text{irrigation}} = 1.72$ ratio, again an incorrect conclusion would have resulted. As long as the incremental B/C ratio exceeds 1, the incremental benefits justify the incremental costs. In this regard, the B/C ratio criterion is closely akin to the rate-of-return criterion discussed in Chapter Four.

6.6 COST-EFFECTIVENESS ANALYSIS

To this point it has been assumed that public project effects were measurable, either directly or indirectly, in monetary terms. There are circumstances, however, when project outputs are not measurable monetarily and must be expressed in physical units appropriate to the project. In these cases, *cost-effectiveness analysis* has proven to be a useful technique for deciding between projects or systems for the accomplishment of certain goals. Although cost-effectiveness is most often associated with the economic evaluation of complex defense and space systems, it has also proven useful in the social and economic sectors. In fact, it has roots dating back to Arthur M. Wellington's *The Economic Theory of Railway Location* in 1887 [12].

Cost-effectiveness analyses require that three conditions be met, according to Kazanowski [7]. They are:

1. Common goals or purposes must be identifiable and attainable.
2. There must be alternative means of meeting the goals.
3. There must be perceptible constraints for bounding the problem.

Common goals are required in order to have a basis for comparison. For example, it would not make sense to compare a submarine with a sophisticated single sideband communication network. Obviously, alternative methods of accomplishing the goals must be available in order to have a comparison. Finally, reasonable bounds for constraining the problem by time, cost, and or effectiveness are necessary to limit and better define the alternatives to be considered.

6.6.1 The Standardized Approach

Kazanowski [7] presents 10 standardized steps that constitute a correct approach to cost-effectiveness analyses. They are, in their usual order, presented below.

Step 1. *Define* the *goals*, purpose, missions, etc., that are to be met. Cost effectiveness analysis will identify the best alternative way of meeting these goals.

Step 2. *State* the *requirements* necessary for attainment of the goals. That is, state any requirements which are essential if the goals are to be attained.

Step 3. *Develop alternatives* for achieving the goals. There must be at least two alternative ways of meeting or exceeding the goals.

Step 4. *Establish* evaluation *measures* which relate capabilities of alternatives to requirements. An excellent list of evaluation criteria is contained in Kazanowski [8]. Typical measures are performance, availability, reliability, maintainability, etc.

Step 5. *Select* the fixed-effectiveness or fixed-cost *approach.* The fixed effectiveness criterion is minimum alternative cost to achieve the specified goals or effectiveness levels. Alternatives failing to achieve these levels may either be eliminated or assessed penalty costs. The fixed cost criterion is the amount of effectiveness achieved at a given cost. "Cost" is usually taken to mean a present worth or annual equivalent of "life cycle cost" which includes research and development, engineering, construction, operation, maintenance, salvage, and other costs incurred throughout the life cycle of the alternative.

Step 6. *Determine capabilities* of the alternatives in terms of the evaluation measures.

Step 7. *Express* the alternatives and their *capabilities* in a suitable manner.

Step 8. *Analyze* the various *alternatives* based upon the effectiveness criteria and cost considerations. Often, some alternatives are clearly dominated by others and should be removed from consideration.

Step 9. *Conduct* a *sensitivity analysis* to see if minor changes in assumptions or conditions cause significant changes in alternative preferences.

Step 10. *Document all* considerations, analyses, and decisions from the above nine steps.

Clearly, one cannot become a cost-effectiveness expert by reading these 10 steps. They do, however give an indication of what is involved in a cost-effectiveness study. Excellent detailed presentations on cost effectiveness are available in English [3] and Blanchard [1].

Example 6.23

Three different propulsion systems are under consideration. The life cycle cost is not to exceed $2.4 million. A single effectiveness measure is decided on, that being reliability. Reliability would be defined as "the probability that the propulsion system will perform without failure under given conditions for a given period of time." Four contractors submit candidate systems, which are evaluated as follows:
Since there is only one effectiveness measure, these results may be expressed graphically as shown in Figure 6.1.

If we look at these systems in pairs, comparing just 1 and 2 is equivalent to a fixed-cost comparison in which we prefer system 1 because of its higher reliability. Similar reasoning leads us to prefer system 3 over system 4. If we

Propulsion System	Life Cycle Cost (Millions)	Reliability
1	2.4	0.99
2	2.4	0.98
3	2.0	0.98
4	2.0	0.97

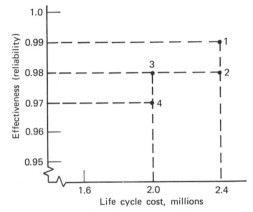

FIGURE 6.1. Effectiveness (reliability versus life cycle cost).

were to compare only systems 2 and 3, this would be a fixed-effectiveness comparison, in which case we prefer system 3 due to its lower cost. Clearly, system 1 dominates 2, 3 dominates 2, and 3 dominates 4. We are left with making the decision between systems 1 and 3. This choice will depend on whether or not an additional percent reliability justifies the expenditure of an additional $400,000.

In this example, it is tempting to go the next step and say that based on the ratios of reliability to cost 0.99/$2.4 million and 0.98/$2.0 million, system 3 should be selected. Unfortunately, apparent as such a decision may seem, the correct decision may depend on many other considerations, not the least of which are the payload (perhaps human life) and the consequences of system failure (perhaps total destruction with no payload recovery).

6.7 SUMMARY

Benefit-cost and cost-effectiveness analyses are accepted facts of life in the public sector. Because of the rapid shifting of the majority of employees from private industry to the public and service sectors, no doubt tomorrow's economic evaluators will be more and more closely involved with these techniques over time.

BIBLIOGRAPHY

1. Blanchard, Ben S., "Cost Effectiveness/Life-Cycle Cost Analysis," unpublished report, Virginia Polytechnic Institute and State University, Blacksburg, Virginia, 24061.
2. Cohn, Elchanan, *Public Expenditure Analysis*, D. C. Heath, 1972.
3. English, J. Morley (Editor), *Cost Effectiveness*, Wiley, 1968.
4. Haveman, R. H., "The opportunity cost of displaced private spending and the social discount rate," *Water Resources Research*, 5 (5), 1969, pp. 947–957.
5. Henderson, P. D., "The Investment Criteria for Public Enterprises," in *Public Enterprise*, edited by Ralph Turvey, Penguin Books, 1968.
6. Howe, Charles W., *Benefit-Cost Analysis for Water System Planning*, Water Resources Monograph 2, American Geophysical Union, Washington, D.C., 1971.
7. Kazanowski, A. D., "A Standardized Approach to Cost-Effectiveness Evaluations," in J. Morley English (Editor), *Cost Effectiveness*, Wiley, 1968.
8. Kazanowski, A. D., "Some Cost-Effectiveness Evaluation Criteria," Appendix B in J. Morley English (Editor), *Cost Effectiveness*, Wiley, 1968.
9. McKean, Roland N., *Efficiency in Government Through Systems Analysis*, Wiley, 1958.
10. Prest, A. R., and Turvey, Ralph, "Cost-Benefit Analysis: A Survey," *Economic Journal*, 75, December 1965, pp. 683–735.
11. Weisbrod, Burton A., *Economics of Public Health*, University of Pennsylvania Press, Philadelphia, 1961, pp. 7–8.
12. Wellington, A. M., *The Economic Theory of Railway Location*, Wiley, 1887.
13. United States Code, 1940 edition, U.S. Government Printing Office, Washington, D.C., p. 2964.

PROBLEMS

1. Identify the benefits, both monetary and intangible, that would accrue to the public for the following projects: (a) a museum, (b) a fence enclosed walkway over a busy

four-lane road connecting two city subdivisions, one having a playground and park, (c) a city sanitary system, and (d) a visiting troupe of performers to appear in a downtown city mall for several days. (6.3)

2. Two four-lane roads intersect and traffic is controlled by a standard green, yellow, red stoplight. From each of the four directions a left turn is permitted from the inner lane; however, this impedes the flow of traffic while the person desiring to turn left waits until a safe turn may be accomplished. The light operates on a cycle allowing 30 seconds of green-yellow, followed by 30 seconds of red light for each direction. Approximately an eighth of the 8000 vehicles using the intersection daily are held up for one full minute cycle of the light, and made to perform one extra stop-start operation, solely because of the left turn bottleneck. These delays are only during the 260 working days/year. A stop-start costs 0.5¢/vehicle. A widening of the intersection to accommodate a left-turn lane plus a new lighting system to provide for a left-turn signal will cost $48,000. The cost of time for commercial traffic is three times that of private traffic, and private traffic accounts for 90% of the vehicles. How much must commercial traffic time be worth to justify the intersection changes if the interest rate is 7% and a lifetime of 10 years is expected? (6.4)

3. Three alternatives are available, A, B, and C. Their respective annual benefits, disbenefits, costs, and savings are as follows:

	A	B	C
Benefits	$200,000	$300,000	$400,000
Disbenefits	37,000	69,000	102,000
Costs	150,000	234,000	312,000
Savings	15,000	31,000	42,000

(a) Calculate the B/C ratios for each project. Can you tell from these calculations which should be selected?
(b) Determine which should be selected using the incremental B/C ratio.
(c) Calculate $B - C$ for each alternative. (6.4)

4. The following costs and benefits have been listed for a dam and reservoir which will be used for both flood control and electrical power generation.
Investment

Dam, including access roads, clearing, and foundation treatment	$31,330,000
Generation equipment and transmission apparatus	16,176,000
Land	2,200,000
Highway relocation	2,770,000
Miscellaneous	180,000

Operating and maintenance costs
   Two percent of investment during the first year and increasing by 5% of itself each subsequent year.
Annual benefits

Flood losses prevented	$ 835,000
Property value enhancement	149,000
Power value	3,190,000

For a planning horizon of 50 years and $i = 5\%$, what is the B/C ratio? (6.4)

5. A city is trying to decide between coal, fuel oil 3, low-sulphur fuel oil, and natural gas to power their electrical generators. Fuel forecasts indicate the following needs for the upcoming year, and the gradient increase during each subsequent year.

Fuel	First Year	Gradient Each Subsequent Year
Coal	460,000 tons	23,000 tons
Fuel oil 3	1,760,000 barrels	88,000 barrels
Low-sulphur fuel oil	1,820,000 barrels	91,000 barrels
Natural gas	11,000 10^6 cubic feet	550 10^6 cubic feet

The cost of the fuel, transportation, and various pollution effects have been estimated as follows,

Fuel	Cost	Transportation and Storage	Health	Crops	Unclean liness
Coal	$4.85/ton	$2.25/ton	$0.70/ton	$1.75/ton	$1.05/ton
Fuel oil 3	1.80/barrel	0.20/barrel	0.10/barrel	0.25/barrel	0.15/barrel
Low-sulphur fuel oil	2.25/barrel	0.20/barrel	0.03/barrel	0.075/barrel	0.045/barrel
Natural gas	0.00048/cubic foot	0.00002/cubic foot	Negligible	Negligible	Negligible

Calculate the annual equivalent benefits and costs of these fuels considering $i = 7\%$ and a life of 30 years. Use an incremental B/C analysis to determine the best fuel to use. (6.4)

6. A highway is to be built connecting Baldhill with Broken Arrow (Baldhill is south of Bixby). Route A follows the old road and costs $4 million initially and $210,000/year thereafter. A new route, B, will cost $7 million initially and $180,000/year thereafter. Route C is simply an enhanced version of route B with wider lanes, shoulders, and so on. It will cost $9 million at first, plus $240,000/year to maintain. Relevant annual user costs considering time, operation, and safety are

$1 million for A, $700,000 for B, and $500,000 for C. Using a *MARR* of 7%, a 15-year study period, and a residual (salvage) value of 50% of first cost, which should be constructed? Use a B/C analysis. Which route is preferred if $i = 0$? (6.4)

7. Solve problem 6 using the $B - C$ criterion. (6.4)

8. A municipal zoo is to be enlarged, and the initial cost of the enlargement for physical facilities will be $225,000. Animals for the addition will cost another $44,000. Maintenance, food, and animal care will run $31,000/year. The zoo is expected to be in operation for an indefinite period; however, a study period of 20 years is to be assumed, with a residual (salvage) value of 50% for all physical facilities. Interest is 7%. An estimated 240,000 persons will visit the zoo each year, and they will receive, on the average, an additional $0.35/person in enjoyment when the new area is complete. Should the new area be built? Use a $B - C$ measure of merit. (6.4)

9. A proposed expressway is under study and is found to have an unfortunately high annual cost of $2,200,000 as compared to benefits of only $1,300,000. However, with only inconsequential changes in design, the highway may be eligible for incorporation into the interstate highway system, in which case 90% of the cost would be paid by the federal government. If so, it is argued that the expressway would cost only $220,000/year, yielding a handsome B/C ratio of 5.91. Is this reasoning sound? Discuss the reasoning from different points of view. (6.5)

10. Seven projects are available as summarized below. It is desired to select only the projects that government can afford on an available budget of $75,000 for first cost. Operating and maintenance costs are no worry. All have a favorable B/C ratio at the cost of money, $i = 6\%$. Raise the interest rate until only those projects remain that continue to have a $B/C > 1$ and that government can afford. Use a 10-year planning horizon. Which projects are selected? What is the opportunity cost of those investments foregone by government? (6.5)

Project	First Cost	Operating and Maintenance	Residual Value After 10 Years	Benefits/Year
A	$33,000	$3,000	$16,000	$14,400
B	27,000	1,500	8,000	7,630
C	41,000	2,500	12,000	10,800
D	38,000	1,600	14,500	8,460
E	30,000	4,200	16,000	8,600
F	34,000	600	8,000	5,400
G	25,000	5,100	3,000	8,980

11. Many benefits resulting from public sector projects may be argued to be at the expense of someone else, thus counterbalancing the supposed direct benefits. For example, in Example 6.1, operating costs saved include the cost of fuel, oil, tires,

wear and tear, and the like. This is money that will *not* be spent at service stations, tire stores, and new car dealers. The safety costs saved include fees that otherwise would go to lawyers, hospitals, doctors, auto repair shops, and so forth. Do you think the fact that the direct savings represent lost revenues to others should void or nullify these types of benefits in a benefit-cost analysis? Why or why not? (6.5)

12. Three projects, each having a first cost of $1 million and annual operating costs of $100,000 are proposed. The lives of Projects A, B, and C are 20, 30, and 40 years, respectively, after which the project will be over, providing no benefits and requiring no costs. The annual benefits provided over the lives of Projects A, B, and C will be $268,000, $250,000, and $243,000, respectively.
Select only one project to be implemented. Use a present worth basis and a $B - C$ measure of merit. First, use an *MARR* of 5%, and a study period of 40 years. Projects A and B will be void during the last 20 and 10 years of the study period, respectively. Second, continually increase the interest rate, again using a study period of 40 years, until only one project remains with a favorable $B - C$ value. Do the two methods point to the same project? Why or why not? (6.5)

13. Rework problem 12 using a value of $i = 12\%$, a planning horizon of 20 years, and residual values of 20% and 40% of first cost for projects B and C, respectively. (6.5)

14. A recreation area suitable for camping, picnicking, hiking, and water sports is to be developed at an annual equivalent cost of $87,600, including initial cost of construction (clearing, water, restrooms, road, etc.), operation, upkeep, and security. An average of 60 families will camp each night throughout the year. In addition, another 100 persons will be admitted for day use of the recreational facilities. The perceived benefits provided to overnight and day users will vary, due to the subjective nature of people. However, the average family camping at the area is estimated to be willing to pay $5 for the privilege, and the average day user $1. What is the B/C ratio of this recreational area? $B - C$? Should it be built? (6.5)

15. In problem 14, assume that 50% of the potential camping families perceive $3.50 in benefits, 25% perceive $5 in benefits, and 25% expect to receive $8 worth of recreation. If a charge of $4 is imposed on campers only, and day users are still permitted free, recalculate B/C and $B - C$. Should the facility be built? (6.5)

16. Reconsider problem 8 and assume that half of the people will derive $0.50 of enjoyment per person, while the other half anticipate only $0.20 in benefits. With this information available, and assuming that an incremental $0.25 entrance fee per person will be charged separately for attendance to the addition, do you recommend building the addition? (6.5)

17. A community must develop a new power generation capability. One alternative is a coal-burning power plant having a first cost of $18 million, a life of 40 years, with no residual value, and operating and maintenance costs of $5,570,000/year, including delivered fuel. If this alternative is selected, rates per kilowatt hour will be set so as to make the plant just self-supporting. Another alternative is to use a dam and reservoir costing $70 million, with a life of 40 years, and having an annual

operating and maintenance cost of $2 million. Electrical energy benefits would be no greater than for the coal-fired plant, and thus the energy charges would be the same per kilowatt hour. However, if the dam and reservoir are used, for another $1,500,000 first cost and $175,000/year, a fine boating and recreation area can be developed as an enhancement to the reservoir. Such a facility would have a first cost of $6 million if constructed by itself. It would provide recreational benefits estimated at $900,000/year for 40 years to the general public. The community has several alternatives from which to choose, including A) a coal-fired power plant, B) a coal-fired power plant plus a separate recreational area, C) a dam and reservoir for power generation only, and D) a dam and reservoir for power generation and recreation. Use a $B - C$ analysis to decide between these alternatives, assuming that the cost of money is 8%. One of these alternatives must be chosen because of the need for new power generation capability. (6.5)

18. Comment on the following analysis if you think it is incorrect. Benefits of $1.75 each are now received by 9000 persons and benefits are perceived to be $3 each by another 9000 users. The annual cost of the recreational facility is $36,000. Commissioners have argued that an entrance fee of $2 should be charged to make the facility self-supporting. They argue that $2 is below the average benefit received per person. They do point out, though, that 9000 persons do not perceive but a $1.75 recreational value, and hence a $0.25 disbenefit per person should be noted. The B/C ratio is then

$$B/C = \frac{\$3(9,000) + \$1.75(9,000) - \$0.25(9,000)}{36,000 - \$2(18,000)} = \infty$$

and the user fee of $2 should definitely be implemented immediately. What do you think? (6.5)

19. Analysts recognize that with a $2 entrance fee as in problem 18, only 9000 persons will utilize the facility, and the benefits derived by 9000 more persons at $1.75 each will be lost. They argue that the *true* benefits are those received only by 9000 persons at $3 each, less the $2 fee, less a disbenefit of $1.75(9000) for the persons who chose not to pay $2 to enter and lose the previous recreational enjoyment from the facility. As such, the B/C ratio is as follows:

$$B/C = \frac{\$3(9,000) - \$2(9,000) - \$1.75(9,000)}{36,000 - \$2(9,000)} = -.38$$

Do you agree with this analysis? Why or why not? (6.5)

20. A government has the following estimates:

Annual benefits	$250,000
Annual disbenefits	200,000
Annual costs	150,000
Annual savings	145,000

(a) Calculate the B/C ratio.
(b) Mistakenly treating disbenefits as costs and savings as benefits, calculate the B/C ratio.

(c) Calculate $B - C$.

(d) Which do you prefer, $B - C$ or B/C? Why? (6.5)

21. Six subsystem designs have been proposed for a critical part of a communications network. Each will be "burned in" to eliminate the infant mortality problem. Hence, reliability of the subsystem will have a constant hazard rate. That is, the failure density function will be exponential and reliability may be expressed as

$$R(t) = e^{-\lambda t}$$

where t is the mission time. Mission time for the subsystem will be one year, as a thorough annual preventative maintenance and recalibration procedure is standard, returning the subsystem to "new" condition. The unit will be on call 24 hours/day, seven days/week. The design data are as follows:

System	Annual Equivalent Initial Cost	Annual Maintenance Cost	λ
A	$ 9,000	$1,700	1.1446886 10^{-5}
B	9,800	1,400	1.0989011 10^{-5}
C	10,200	1,800	1.175824 10^{-5}
D	11,000	1,300	1.070000 10^{-5}
E	11,000	1,500	1.081250 10^{-5}
F	13,100	900	0.998000 10^{-5}

Using a cost-effectiveness analysis, plot reliability versus annual cost. Which subsystems can be eliminated from consideration? How might you decide among the rest? (6.6)

22. Four designs are under consideration for a space shuttle designed to carry a particular type of payload. It is estimated that each will have equal lives; however, the first cost, operating cost, and renovation cost, as well as the salvage value, will differ, as will the size of the payload that may be carried. These characteristics are summarized below for shuttles having lives of three years, and fired once per year at the end of the year.

Shuttle Design	First Cost	Operating Cost/Firing	Renovation Cost/Firing	Salvage Value	Payload in Units
1	$18,000,000	$5,000,000	$1,400,000	$2,600,000	176
2	22,000,000	4,400,000	2,400,000	3,100,000	152
3	24,000,000	5,200,000	1,900,000	3,000,000	134
4	30,000,000	3,600,000	1,200,000	5,000,000	168

Based on $i = 7\%$, and using a cost-effectiveness analysis, which designs may be eliminated and which should be considered further? (6.6)

CHAPTER SEVEN
BREAK-EVEN, SENSITIVITY, AND RISK ANALYSES

7.1 INTRODUCTION

In the previous chapters we have assumed that all of the values of the parameters of the economic models were known with certainty. In particular, correct estimates of the values for the length of the planning horizon, the minimum attractive rate of return, and each of the individual cash flows were assumed to be available. In this chapter we consider the consequences of estimating the parameters incorrectly.

The discussion will concentrate on answering a number of "what if . . ." questions concerning the effects of different parameter values on the measure of economic effectiveness of interest. For example, when we are completely *uncertain* of the possible values a parameter can take on, we will be interested in determining the set of values for which an investment alternative is justified economically and the set of values for which an alternative is not justified; this process is called *break-even analysis*. Alternatively, when we are reasonably sure of the possible values a parameter can take on, but *uncertain* of their chances of occurrence, we will be interested in the sensitivity of the measure of merit to various parameter values; this process is referred to as *sensitivity analysis*. Finally, when probabilities can be assigned to the occurrence of the various values of the parameters, we can make probability statements concerning the values of the measure of merit for the various alternatives; this process is referred to as *risk analysis.*

The conditions that lead to break-even and sensitivity analyses are described in the economic analysis literature as conditions under *uncertainty*, since one is completely uncertain of the chances of a parameter taking on a given value. When the conditions are such that probabilities can be assigned to the various parameter values, the decision environment that results is said to be a decision under *risk*; hence, the term risk analysis is used.

In Chapter Eight we will present *prescriptive* or *normative* models under uncertainty and risk conditions, models will be presented that allow us to *prescribe* the best or optimal course of action using a number of different decision criteria. In this chapter we present *descriptive* models under uncertainty and risk conditions. Instead of attempting to *prescribe* the action to be taken, we attempt to *describe* the behavior of the measure of merit under conditions of uncertainty and risk. Thus, we view break-even, sensitivity, and risk analyses as descriptive processes, not normative processes. Our objective in this chapter will be to gain insight into the behavior of the measure of merit under conditions of uncertainty and risk; the decision of how to respond, given the information from the sensitivity analysis depends on the decision process employed by the decision maker. Hence, the decision models from Chapter Eight may be used to assist the decision maker in making the final selection.

7.2 BREAK-EVEN ANALYSIS

Although we did not label the process as such, in Chapter Three we performed a number of break-even analyses; there we referred to the process as *equivalence.* In particular, in presenting the concept of equivalence, a situation was posed and you were asked to determine the value of a particular parameter in order for two cash flow profiles to be equivalent. Another way of stating the problem could have been "Determine the value of X that will yield a break-even situation between the two alternatives." In this case, X denotes the parameter whose value is to be determined. Additionally, when the cash flow profiles for two alternatives are equivalent, a break-even situation can be said to exist between the two alternatives.

Another application of break-even analysis occurs in the use of the rate-of-return method. Specifically, the rate of return can be interpreted as the break-even value of the reinvestment rate, since such a reinvestment rate will yield a zero future worth for either an individual alternative or the differences in two alternatives.

Break-even analysis is certainly not an unfamiliar concept. Furthermore, the information obtained from a break-even analysis can be of considerable aid to anyone faced with an investment alternative involving a degree of uncertainty concerning the value of some parameter. The term "break-even" is derived from the desire to determine the value of a given parameter that will result in neither a profit nor a loss.

Example 7.1 ————————————————————————————————

Suppose a firm is considering manufacturing a new product and the following data have been provided:

Sales price	$12.50/unit
Equipment cost	$200,000
Overhead cost	$50,000/year
Operating and maintenance cost	$25/operating hour
Production time/1000 units	100 hours
Planning horizon	5 years
Minimum attractive rate of return	15%

Assuming a zero salvage value for all equipment at the end of five years, and letting X denote the annual sales for the product, the annual worth for the investment alternative can be determined as follows:

$$AW(15\%) = -\$200,000(A|P\ 15,5) - \$50,000 - 0.100(\$25)X + \$12.50X$$
$$= -\$109,660 + \$10.00X$$

Solving for X yields a *break-even* sales value of 10,966/year. If it is felt that annual sales of at least 10,966 units can be achieved each year, then the alternative appears to be worthwhile economically. Even though one does not know with certainty how many units of the new product will be sold annually, it is felt that the information provided by the break-even analysis will assist management in deciding whether or not to undertake the new venture.

A graphical representation of the example is given in Figure 7.1. The chart is referred to as a *break-even* chart, since one can determine graphically the *break-even point* by observing the value of X when annual revenue equals annual cost.

Example 7.2 ————————————————————————————————

Consider a contractor who experiences a seasonal pattern of activity for compressors. The manager currently owns eight compressors and suspects that this number will not be adequate to meet the demand. The contractor realizes that there will arise situations when more than eight compressors will be required, and is considering purchasing an additional compressor for use during heavy demand periods.

A local equipment rental firm will rent compressors at a cost of $25/day.

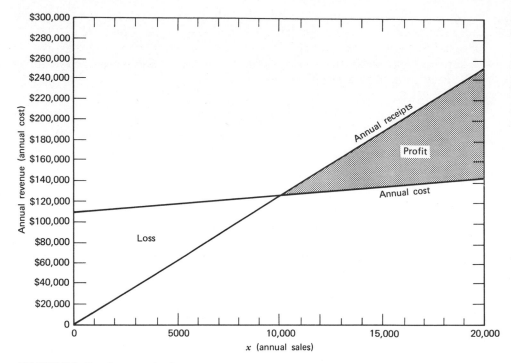

FIGURE 7.1. Break-even chart.

Compressors can be purchased for $3000. The difference in operating and maintenance costs between owned and rented compressors is estimated to be $1500/year.

Letting X denote the number of days a year that more than eight compressors are required, the following break-even analysis is performed. A planning horizon of five years, zero salvage values, and 10% minimum attractive rate of return are assumed.

Annual worth (purchasing compressor)

$$AW_1(10\%) = -\$3000(A|P\ 10,5) - \$1500$$
$$= -\$2291.40$$

Annual worth (renting compressor)

$$AW_2(10\%) = -\$25X$$

Setting the annual worths equal for the two alternatives yields a *break-even value* of $X = 91.656$ days/year.

Hence, if the contractor anticipates that a demand will exist for an additional compressor more than 91 days/year over the next five years, then an additional compressor should be purchased.

292 • BREAK-EVEN, SENSITIVITY, AND RISK ANALYSES

TABLE 7.1 Forecasts of Demand for Compressors

	X Number of Compressors Demanded On a Given Day	f(X) Number of Days per Year Demand Equals X
	≤ 8	140
	9	20
	10	30
	11	30
	12	30
	13	10
	≥14	0

N Number of Com- pressors Owned by the Contractor	Number of Compressor-Rental-Days if N are Owned	Difference in Number of Compressor-Rental- Days if N versus n + 1 Are Owned
8	20(1) + 30(2) + 30(3) + 30(4) + 10(5) = 340	—
9	30(1) + 30(2) + 30(3) + 10(4) = 220	120
10	30(1) + 30(2) + 10(3) = 120	100
11	30(1) + 10(2) = 50	70
12	10(1) = 10	40
13	0	10

Extending the previous example problem to the question of how many compressors should be purchased, it is interesting to note that compressors should continue to be purchased so long as each one reduces the number of compressor-rental-days by at least 91/year over the five-year period. To illustrate, if the contractor provides a forecast of daily demand for compressors as given in Table 7.1, it is seen that two additional compressors should be purchased.

To show that two more compressors should be purchased, note that if only one additional compressor is purchased, the annual worth will be

$$AW(10\%) = -\$3000(A|P\ 10,5) - \$1500 - \$25(220)$$
$$= -\$7791.40$$

Purchasing two more compressors yields an annual worth of

$$AW(10\%) = -\$6000(A|P\ 10,5) - \$3000 - \$25(120)$$
$$= -\$7582.80$$

Purchasing three more compressors yields an annual worth of

$$AW(10\%) = -\$9000(A|P\ 10,5) - \$4500 - \$25(50)$$
$$= -\$8124.20$$

Hence, two additional compressors should be purchased in order to provide a total of 10 compressors to meet annual demand.

Example 7.3

As a third illustration of break-even analysis, consider an office building that has decided to convert from a coal burning furnace to a furnace that burns either natural gas or number two fuel oil. The cost of converting to natural gas is estimated to be $50,000 initially; additionally, annual operating and maintenance costs are estimated to be $4000 less than that experienced currently. Approximately 1000 Btu's are produced per cubic foot of natural gas; it is estimated natural gas will cost $0.0005/cubic foot. The cost of converting to number two fuel oil is estimated to be $70,000 initially; annual operating costs are estimated to be $2500 less than that experienced using the coal furnace. Approximately 140,000 Btu's are produced per gallon of fuel oil; number two fuel oil is anticipated to cost $0.065/gallon. A planning horizon of 20 years is used; zero salvage values and a 10% minimum attractive rate of return are assumed. It is desired to determine the break-even value for annual Btu requirements for the heating system.

Letting X denote the Btu requirement per year, the equivalent uniform annual cost for the natural gas alternative is found to be

$$EUAC_1(10\%) = \$50,000(A|P \ 10,20) - \$4000 + \frac{\$0.0005}{1000} X$$

$$= \$1875 + \$0.5 \times 10^{-6}X$$

Similarly, for the number two fuel oil, the equivalent uniform annual cost will be

$$EUAC_2(10\%) = \$70,000(A|P \ 10,20) - \$2500 + \frac{\$0.065}{140,000} X$$

$$= \$5725 + \$0.4642 \times 10^{-6}X$$

Equating the annual costs and solving for X yields a break-even value of approximately 107.5×10^9 Btu's/year. Since a consulting engineer has estimated that Btu requirements will be considerably less than the break-even value, it is decided to convert to natural gas.

7.3 SENSITIVITY ANALYSIS

Break-even analysis is normally used when an accurate estimate of a parameter cannot be provided, but intelligent judgments can be made as to whether or not the parameter's value is less than or greater than some break-even value. Sensitivity analysis is used to analyze the effects of making errors in estimating parameter values.

Although the analysis techniques employed in break-even analysis and sensitivity analysis are quite similar, there are some subtle differences in the objectives of each. Because of the similarities in the two, it is not uncommon to see the terms used interchangeably.

Example 7.4 ————————————————————————————————

To illustrate what we mean by a sensitivity analysis, consider the investment alternative depicted in Figure 7.2. The alternative is to be compared against the do nothing alternative, which has a zero present worth. If errors are made in estimating the size of the required investment ($10,000), the magnitude of the annual receipt ($3000), the duration of the project (five years), the minimum attractive rate of return (12%), and/or the form of the series of receipts (uniform versus nonuniform), then the economic desirability of the alternative might be affected. It is anticipated that the future states (possible values) for each parameter will be contained within an interval having a range from −40 to +40% of the initial estimate. Hence, in the uncertain environment we have defined an infinite number of future states are to be considered in the continuum from −40 to +40% of the initial estimate.

If it is assumed that all estimates are correct except the estimate of annual receipts, the annual worth for the alternative can be given as

$$AW(12\%) = -\$10,000(A|P\ 12,5) + \$3000(1 + X)$$

where X denotes the percent error in estimating the value for annual receipts. Plotting annual worth as a function of the percent error in estimating the value of annual receipts yields the straight line having positive slope shown in Figure 7.3. Performing similar analyses for the initial investment required, the

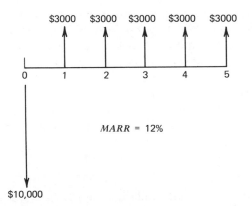

$3000 $3000 $3000 $3000 $3000

0   1   2   3   4   5

MARR = 12%

$10,000

FIGURE 7.2. Cash flow diagram.

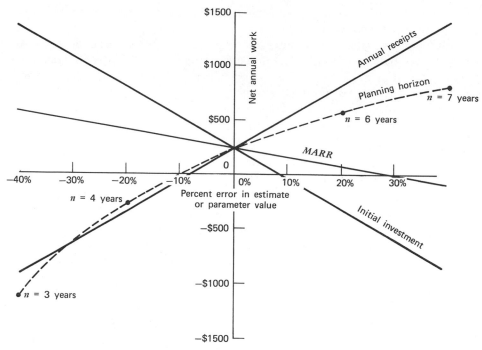

Figure 7.3. Deterministic sensitivity analysis.

duration of the investment (planning horizon) and the minimum attractive rate of return yields the results given in Figure 7.3.

As shown in Figure 7.3, the net annual worth for the investment is affected differently by errors in estimating the values of the various parameters. The net annual worth is relatively insensitive to changes in the minimum attractive rate of return; in fact, as long as the *MARR* is less than approximately 15.25%, the investment will be recommended. A *break-even* situation exists if either the annual receipts decrease by approximately 7.47% to $2774/year or the required investment increases by approximately 8.14% to $10,814. If the project life is four years or less, then the investment will not be profitable.

The analysis depicted in Figure 7.3 examines the sensitivity of individual parameters one at a time. In practice, estimation errors can occur for more than one parameter. In such a situation, instead of a sensitivity curve, a sensitivity surface is needed.

Example 7.5 ───

Consider an investment alternative involving the modernization of a warehousing operation in which automated storage and retrieval equipment is to be installed in a new warehouse facility. The building has a projected life of 30 years and the equipment has a projected life of 15 years. A minimum attractive rate of return of 15% is to be used in the economic analysis. It is anticipated that the new warehouse system will require 70 fewer employees than the present system. Each employee costs approximately $18,000/year. The building is estimated to cost $2,500,000 and the equipment is estimated to cost $3,500,000. Annual operating and maintenance costs for the building and equipment are estimated to be $150,000/year more than the current operation. Existing equipment and buildings not included in the new warehouse have terminal salvage values totaling $600,000. Investing in the new warehouse will negate the need to replace existing equipment in the future; the present worth savings in replacement cost is estimated to be $200,000. The estimated cash flow profile for the investment alternative is given in Table 7.2.

A 30-year planning horizon is to be employed in the analysis. Since equipment life is estimated to be 15 years, it is assumed that identical replacement equipment will be purchased after 15 years. Furthermore, since constant worth dollar estimates are being used in contending with inflationary effects, it is assumed that the replacement equipment will have cash flows that are identical to those that occur during the first 15 years.

TABLE 7.2 Estimated Cash Flows for Warehouse System

EOY	Building	Equipment	Labor Savings	Operating and Maintenance	Salvage[a]	Total
0	−$2,500,000	−$3,500,000			$800,000	−$5,200,000
1–15			$1,260,000	−$150,000		1,110,000
15		− 3,500,000				− 3,500,000
16–30			1,260,000	− 150,000		1,110,000
30	0	0				0

[a] Includes present worth of savings in equipment replacement.

The architectural and engineering estimate of $2,500,000 for the building is believed to be quite accurate as are the estimates of $150,000 for annual operating and maintenance costs, terminal salvage values totaling $600,000, and savings of $200,000 in replacement costs. However, it is felt that the estimate of $3,500,000 for equipment and the labor savings estimate of 70 employees are subject to error. A sensitivity analysis for these two parameters is to be performed on a before tax basis.

Letting x denote the percent error in the estimate of equipment cost and y denote the percent error in estimating the annual labor savings, it can be seen that the warehouse modernization will be justified economically if

$$PW = -\$2,500,000 - \$3,500,000(1 + x)[1 + (P|F\ 15,15)]$$
$$-\$150,000(P|A\ 15,30) + \$18,000(70)(1 + y)(P|A\ 15,30)$$
$$+\$800,000 \geqslant 0$$

or if

$$PW = \$1,658,110 - \$3,930,150x + \$8,273,160y \geqslant 0$$

Solving for y gives

$$y \geqslant -0.2004 + 0.47505x$$

Plotting the equation, as shown in Figure 7.4, indicates the favorable region $(PW > 0)$ lies above the breakeven line and the unfavorable region $(PW < 0)$ lies below the breakeven line.

If no errors are made in estimating the equipment cost, i.e., $x = 0$, then up to a 20.04% reduction in annual labor savings can be tolerated. Likewise, if no errors are made in estimating the annual labor savings, i.e., $y = 0$, then up to a

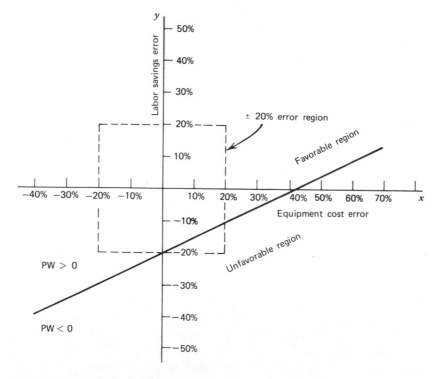

FIGURE 7.4. Multiparameter sensitivity analysis.

298 • BREAK-EVEN, SENSITIVITY, AND RISK ANALYSES

42.185% increase in equipment cost could occur and the warehouse modernization continue to be justified economically.

Since very little of the ±20% estimation error zone results in a negative present worth, it appears that the recommendation to modernize the warehouse is insensitive to errors in estimating either equipment cost, labor savings, or both. However, the decision is more sensitive to errors in estimating labor savings than in estimating equipment costs.

(This particular example illustrates the difficulties in distinguishing between break-even and sensitivity analyses. However, no matter what you call it, the analysis can be quite beneficial in gaining added understanding of the possible outcomes associated with a new venture involving a large capital investment.)

Example 7.6

To illustrate the use of sensitivity analysis in comparing investment alternatives, recall the three mutually exclusive alternatives described in Chapter Four having cash flow profiles as given in Table 7.3. The comparison performed in Chapter Four assumed a minimum attractive rate of return of

TABLE 7.3 Cash Flow Profiles for Three
Mutually Exclusive Investment
Alternatives

t	A_{1t}	A_{2t}	A_{3t}
0	$0	−$10,000	−$15,000
1	−12,000	− 9,000	− 9,000
2	−12,000	− 9,000	− 8,000
3	−12,000	− 9,000	− 7,000
4	−12,000	− 9,000	− 6,000
5	−12,000	− 9,000	− 5,000

10%. Suppose some uncertainty exists concerning the appropriate *MARR* to use in the analysis. As depicted in Figure 7.5, Alternative 3 is preferred if the *MARR* is less than approximately 17%; Alternative 1 is preferred if the *MARR* is greater than 17%; and Alternative 2 is never preferred.

In Chapter Four, when the internal rate of return method was used to compare the alternatives, the differences in Alternatives 1 and 2 were analyzed and a rate of return (break-even value) of 15.02% was obtained; next, Alternatives 2 and 3 were compared and a break-even value of 19.48% was obtained. The combination of Alternatives 1 and 3 was not considered when the rate of return method was used.

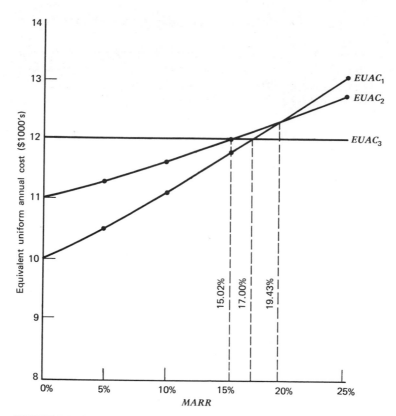

FIGURE 7.5. Sensitivity of equivalent uniform annual cost to the *MARR*.

Example 7.6 illustrates two points made earlier: the rates of return obtained from the internal rate of return method are actually break-even values for the minimum attractive rate of return for the combinations of alternatives considered; and sensitivity analysis and break-even analysis are closely related. The example will be treated again in Chapter Eight.

The final approach we consider in performing sensitivity analyses assumes the possible number of values for the parameters can be reasonably represented by three values for each: pessimistic; most likely; and optimistic estimates will be provided. Given the three estimates of the value of each parameter, the value of the measure of merit is determined for each combination of estimates.

Example 7.7 ────────────────────────────

To illustrate the suggested approach, consider once more the example problem depicted in Figure 7.2 and involving an initial investment of $10,000

TABLE 7.4 Pessimistic, Most Likely, and Optimistic Estimates

	Pessimistic	Most Likely	Optimistic
Investment	-$12,000	-$10,000	-$8,000
Annual receipts	$ 2,500	$ 3,000	$4,000
Discount rate	20%[a]	15%	12%
Planning horizon	4 years	5 years	6 years

[a] Here we interpret the pessimistic value as a value which will lower the present worth; hence, a high discount rate is viewed as a pessimistic value.

followed by annual receipts of $3000 for five years. Realizing that any or all of the parameter estimates could be in error, pessimistic, most likely, and optimistic estimates are provided in Table 7.4 for the parameters. As can be seen, there are 3^m combinations to consider, where m is the number of parameters to be included in the sensitivity analysis. Thus, in this case, 81 combinations are to be considered. Four representative combinations are given below.

$$PW = -\$12,000 + \$4000(P|A\ 15,4) = -\$580$$
$$PW = -\$8000 + \$2500(P|A\ 20,6)9 = \$313.75$$
$$PW = -\$10,000 + \$3000(P|A\ 15,4) = -\$1435$$
$$PW = -\$12,000 + \$3000(P|A\ 12,5) = -\$1185.60$$

The range on present worth can be obtained by evaluating the totally pessimistic case and the totally optimistic case.

$$PW(\text{pessimistic}) = -\$12,000 + \$2500(P|A\ 20,4) = -\$5528.25$$
$$PW(\text{optimistic}) = -\$8000 + \$4000(P|A\ 12,6) = \$8445.60$$

A tabulation of the present worths for the 81 possible combinations of pessimistic (P), most likely (M), and optimistic (O) estimates is provided in Table 7.5. The legend (PMOM) indicates a pessimistic estimate for the investment required, a most likely estimate for annual receipts, an optimistic estimate for the discount rate, and a most likely estimate for the planning horizon were used to determine the present worth of -$1185.60. Plotting the results in the form of a frequency histogram yields the frequencies provided in Table 7.6.

Unfortunately, it is difficult to draw any significant conclusions from the data given in Tables 7.5 and 7.6 concerning the parameter that has the greatest effect on present worth. Instead, the use of the three estimates for each parameter provides us with a feel for the possible values of present worth and their relative frequencies. However, we should not place too much weight on

TABLE 7.5 Present Worth Values for Combination of Estimates

PPPP	−$5528.25	MPPP	−$3528.25	OPPP	−$1528.25
PPPM	− 4523.50	MPPM	− 2523.50	OPPM	− 523.50
PPPO	− 3686.25	MPPO	− 1686.25	OPPO	313.75
PPMP	− 4862.50	MPMP	− 2862.50	OPMP	− 862.50
PPMM	− 3619.50	MPMM	− 1619.50	OPMM	380.80
PPMO	− 2538.75	MPMO	− 538.75	OPMO	1461.25
PPOP	− 4406.75	MPOP	− 2406.75	OPOP	− 406.75
PPOM	− 2988.00	MPOM	− 988.00	OPOM	1012.00
PPOO	− 1721.50	MPOO	278.50	OPOO	2278.50
PMPP	− 4233.90	MMPP	− 2233.90	OMPP	− 233.90
PMPM	− 3028.20	MMPM	− 1028.20	OMPM	971.80
PMPO	− 2023.50	MMPO	− 23.50	OMPO	1976.50
PMMP	− 3435.00	MMMP	− 1435.00	OMMP	565.00
PMMM	− 1943.40	MMMM	56.60	OMMM	2056.60
PMMO	− 646.50	MMMO	1353.50	OMMO	3353.50
PMOP	− 2888.10	MMOP	− 888.10	OMOP	1111.90
PMOM	− 1185.60	MMOM	814.40	OMOM	2814.40
PMOO	334.20	MMOO	2334.20	OMOO	4334.20
POPP	− 1645.20	MOPP	354.80	OOPP	2354.80
POPM	− 37.60	MOPM	1962.40	OOPM	3962.40
POPO	1302.00	MOPO	3302.00	OOPO	5302.00
POMP	− 580.00	MOMP	1420.00	OOMP	3420.00
POMM	1408.80	MOMM	3408.80	OOMM	5408.80
POMO	3138.00	MOMO	5138.00	OOMO	7138.00
POOP	149.20	MOOP	2149.20	OOOP	4149.20
POOM	2419.20	MOOM	4419.20	OOOM	6419.20
POOO	4445.60	MOOO	6445.60	OOOO	8445.60

TABLE 7.6 Frequency Tabulation
of Present Worths

Present Worth	Frequency
−$6000 to −$5000.01	1
− 5000 to − 4000.01	4
− 4000 to − 3000.01	5
− 3000 to − 2000.01	8
− 2000 to − 1000.01	9
− 1000 to − 0.01	11
0 to 999.99	10
1000 to 1999.99	9
2000 to 2999.99	7
3000 to 3999.99	6
4000 to 4999.99	4
5000 to 5999.99	3
6000 to 6999.99	2
7000 to 7999.99	1
8000 to 8999.99	1

the relative frequencies, since we have not verified that each combination of estimates is equally likely to occur. The assignment of probabilities to combinations of parameter estimates lies in the domain of risk analysis and will be treated next.

7.4 RISK ANALYSIS

Risk analysis will be defined as the process of developing probability distributions for some measure of merit for an investment proposal. Typically, probability distributions are developed for either present worth, annual worth, or the rate of return for an individual investment alternative. Consequently, probability distributions are required for random variables such as the cash flows, planning horizon, and the discount rate. The probability distributions are then aggregated analytically or through simulation to obtain the desired probability distribution for the measure of merit.

The cash flow occurring in a given year is often a function of a number of other variables such as selling prices, size of the market, share of the market, market growth rate, investment required, inflation rate, tax rates, operating costs, fixed costs, and salvage values of all assets. The values of a number of these random variables can be correlated with each other, as well as auto-correlated.[1] Consequently, an analytical development of the probability distribution for the measure of merit is not easily achieved in most real-world situations. Thus, simulation is widely used in performing risk analyses.

Risk analysis has gained acceptance in a large number of industries. One industry survey indicated that over 50% of the companies responding to the survey were using risk analysis for operational and/or strategic planning.

By incorporating the concept of utility theory, it is possible to extend risk analysis and obtain a normative or prescriptive model. To be more specific, by combining a manager's utility function and the probability distribution for, say, present worth, it is possible to specify the alternative that maximizes expected utility. Lifson [12] has recommended this approach previously. Hillier [7, 8, 9] incorporates utility theory in applying risk analysis to inter-related investment alternatives.

Our discussion will concentrate on the contributions of risk analysis as a descriptive technique instead as a normative technique. For those interested in extending our discussion to the normative mode as well as the incorporation of utility theory with risk analysis, see Hillier [7, 8, 9].

[1] The term autocorrelation means correlated with itself over time.

7.4.1 Distributions

The risk analysis procedure was developed to take into consideration the imprecision in estimating the values of the inputs required in making economic evaluations. The imprecision is represented in the form of a probability distribution.

Probability distributions for the random variables are usually developed on the basis of subjective probabilities. Typically, the further an event is into the future, the less precise is our estimate of the value of the outcome of the event. Hence, by letting the variance reflect our degree of precision, we would expect the variance of the probability distributions to increase with time.

Among the theoretical probability distributions commonly used in risk analysis are the normal distribution and the beta distribution. Examples of these distributions are depicted in Figures 7.6 and 7.7. For a discussion of a number of distributions and their process generators in the context of simulation, see Schmidt and Taylor [14] and Shannon [15], among others.

In some situations the subjective probability distribution cannot be re-presented accurately using a well-known theoretical distribution. Instead, one

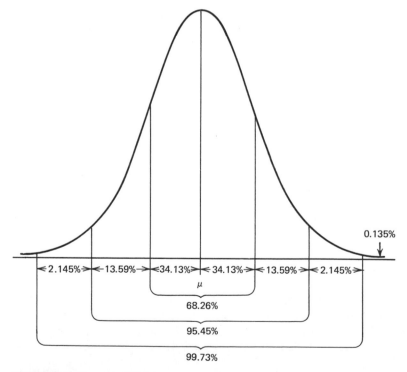

FIGURE 7.6. Normal distribution.

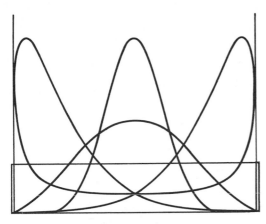

FIGURE 7.7. Sample beta distributions.

must estimate directly the probability distribution for the random variable.

One approach that can be used to estimate the subjective probability distribution is to provide optimistic, pessimistic, and most likely estimates for the random variable. The optimistic and pessimistic values should be ones that you do not anticipate will be exceeded with a significant probability (e.g., 1% chance). Given the practical limits on the range of values anticipated for the random variable, an estimate is provided of the chance that the most likely estimate will not be exceeded. Next, a smooth curve is passed through the three points obtained; the resulting cumulative distribution function for the random variable is divided into an appropriate number of intervals and the individual probabilities estimated.

Example 7.8

To illustrate the process, suppose we wish to develop the probability distribution for, say, the anticipated salvage value of a machine in five years. We estimate the salvage value will range from $0 to $3000, with the most likely value being $1250. We estimate that there is a 40% chance of salvage value being less than $1250. Since we believe the extreme values of $0 and $3000 are not likely to occur, we use an S-shaped curve to represent the cumulative distribution function, as depicted in Figure 7.8a. Using intervals of $500, the cumulative distribution function is transformed into the probability distribution function shown in Figure 7.8b. Letting the midpoints of the intervals represent the probabilities associated with the intervals yields the probability mass function given in Figure 7.8c. Depending on the use to be made of the probabilities obtained, either of the three representations of the probability distribution could be used.

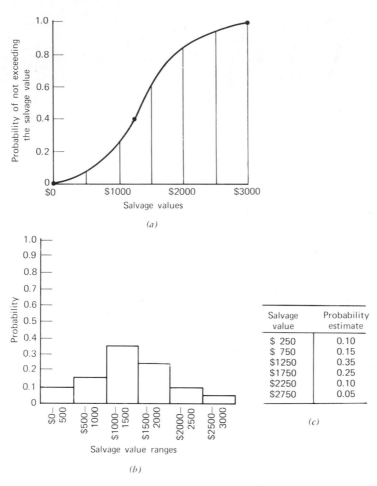

FIGURE 7.8. Developing the subjective probability distribution.

The process described above is quite subjective, since a number of different curves could be used to describe the cumulative distribution function. However, the probability distribution itself is subjectively based. What is sought is a probability distribution that best describes one's beliefs about the outcomes of the random variable. If the probability distribution obtained above does not reflect your beliefs, the process should be repeated until a satisfactory probability distribution is obtained.

The critics of risk analysis cite the degree of subjectivity involved in developing probability distributions. However, it is argued by those who favor the technique that the only alternative to using subjectively based probabilities is to use the traditional single estimate approach, which implies that conditions of certainty exist. Interestingly, the final distribution obtained for the measure of merit is often quite insensitive to deviations in the shapes of the distributions for the parameters.

7.4.2 Risk Aggregation

Given the essential factors and their associated probability distributions, we are in a position to aggregate the distributions and obtain the probability distribution for the measure of merit. Three measures of merit have been mentioned: present worth, annual worth, and rate of return. In practice, a combination of the rate of return and either the present worth or annual worth measures of merit are often used. Each method of comparing investment alternatives has its limitations. For example, if the rate of return is being determined, the explicit reinvestment rate of return method is recommended, since multiple rates of return can arise when using the internal rate of return method.

Risk aggregation is achieved in basically two ways: analytically and by using simulation. Analytic approaches can be used in a number of simple cases. For more complex situations involving a large number of variables, simulation is used.

Analytic Approaches. As an illustration of the use of analytic approaches in developing the probability distribution for present worth, consider the following present worth relation.

$$PW = \sum_{j=0}^{n} C_j(1+i)^{-j} \tag{7.1}$$

Suppose the cash flows, C_j, are *random variables* with expected values $E(C_j)$ and variances $V(C_j)$. Since the *expected value* of a sum of random variables is given by the sum of the expected values of the random variables, the expected present worth is given by

$$E(PW) = \sum_{j=0}^{n} E[C_j(1+i)^{-j}] \tag{7.2}$$

Furthermore, since the expected value of the product of a constant and a random variable is given by the product of the constant and the expected value of the random variable,

$$E[C_j(1+i)^{-j}] = E(C_j)(1+i)^{-j} \tag{7.3}$$

Substituting Equation 7.14 into Equation 7.13 yields

$$E(PW) = \sum_{j=0}^{n} E(C_j)(1+i)^{-j}$$

Hence, we see that the expected present worth of a series of cash flows is found by summing the present worths of the expected values of the individual cash flows.

To determine the variance of the present worth, we first recall that if X_1, X_2, X_3, and Y are random variables related as follows,

$$Y = X_1 + X_2 + X_3 \tag{7.4}$$

then the variance of Y is given by

$$\text{Var}\,(Y) = E(Y^2) - E(Y)^2 \tag{7.5}$$

or

$$\text{Var}\,(Y) = E[(X_1 + X_2 + X_3)^2] - E(X_1 + X_2 + X_3)^2$$

Expanding and collecting terms yields

$$\begin{aligned}
\text{Var}\,(Y) &= E(X_1^2 + X_2^2 + X_3^2 + 2X_1X_2 + 2X_1X_3 + 2X_2X_3) \\
&\quad - [E(X_1) + E(X_2) + E(X_3)]^2 \\
&= E(X_1^2) + E(X_2^2) + E(X_3^2) + 2E(X_1X_2) + 2E(X_1X_3) + 2E(X_2X_3) \\
&\quad - E(X_1)^2 - E(X_2)^2 - E(X_3)^2 - 2E(X_1)E(X_2) - 2E(X_1)E(X_2) \\
&\quad - 2E(X_2)E(X_3) \\
&= E(X_1^2) - E(X_1)^2 + E(X_2^2) - E(X_2)^2 + E(X_3^2) - E(X_3)^2 \\
&\quad + 2[E(X_1X_2) - E(X_1)E(X_2)] + 2[E(X_1X_3) - E(X_1)E(X_3)] \\
&\quad + 2[E(X_2X_3) - E(X_2)E(X_3)]
\end{aligned} \tag{7.6}$$

Since $E(X_k^2) - E(X_k)^2$ defines the variance of the random variable X_k and $E(X_pX_k) - E(X_p)E(X_k)$ defines the covariance of the random variables X_p and X_k, we see that

$$\begin{aligned}
\text{Var}\,(Y) &= \text{Var}\,(X_1) + \text{Var}\,(X_2) + \text{Var}\,(X_3) + 2\,\text{Cov}\,(X_1X_2) \\
&\quad + 2\,\text{Cov}\,(X_1X_3) + 2\,\text{Cov}\,(X_2X_3)
\end{aligned} \tag{7.7}$$

where $\text{Cov}\,(X_pX_k)$ denotes the covariance of X_p and X_k.

Generalizing to the sum of $(n + 1)$ random variables, if

$$Y = \sum_{j=0}^{n} X_j \qquad (7.8)$$

then the expected value and variance of Y are given by

$$E(Y) = \sum_{j=0}^{n} E(X_j)$$

$$\text{Var}\,(Y) = \sum_{j=0}^{n} \text{Var}\,(X_j) + 2 \sum_{j=0}^{n-1} \sum_{k=j+1}^{n} \text{Cov}\,(X_j X_k) \qquad (7.9)$$

Additionally, if a_1 and a_2 are constants and X_1, X_2, and Y are random variables related by

$$Y = a_1 X_1 + a_2 X_2 \qquad (7.10)$$

recall that

$$\text{Var}\,(Y) = a_1{}^2 \, \text{Var}\,(X_1) + a_2{}^2 \, \text{Var}\,(X_2) + 2 a_1 a_2 \, \text{Cov}\,(X_1 X_2) \qquad (7.11)$$

Combining the relationships given in Equations 7.9 and 7.11, the variance of present worth is found to be

$$\text{Var}\,(PW) = \sum_{j=0}^{n} \text{Var}\,(C_j)(1 + i)^{-2j} + 2 \sum_{j=0}^{n-1} \sum_{k=j+1}^{n} \text{Cov}\,(C_j C_k)(i + 1)^{-(j+k)} \qquad 7.12)$$

Hillier [5] argues that it is probably unrealistic to expect investment analysts to develop accurate estimates for covariances. Consequently, he suggests that the net cash flow in any year be divided into those components of cash flow that are reasonably independent from year to year and those that are correlated over time. Specifically, it is assumed that

$$C_j = X_j + Y_{j1} + Y_{j2} + \cdots + Y_{jm} \qquad (7.13)$$

where the X_j values are mutually independent over j but, for a given value of h, $Y_{0h}, Y_{1h}, \ldots, Y_{nh}$ are *perfectly* correlated.

When two random variables X and Y are perfectly correlated, one can be expressed as a linear function of the other. Hence, if

$$Y = a + bX \qquad (7.14)$$

where a and b are constants, then the covariance of X and Y is given by

$$Cov(XY) = E(XY) - E(X)E(Y) \qquad (7.15)$$

However, from Equation 7.14, we note that

$$E(Y) = a + bE(X) \qquad (7.16)$$

and

$$Var(Y) = b^2 Var(X) \qquad (7.17)$$

Substituting Equations 7.14 and 7.16 into Equation 7.15 gives

$$
\begin{aligned}
Cov(XY) &= E(aX + bX^2) - E(X)[a + bE(X)] \\
&= aE(X) + bE(X^2) - aE(X) - bE(X)^2 \\
&= b\,Var(X)
\end{aligned}
\qquad (7.18)
$$

Recalling that the *standard deviation* of a random variable is defined as the square root of the variance of the random variable, we see that Equation 7.18 can be expressed as

$$Cov(XY) = bSD(X)SD(X) \qquad (7.19)$$

where $SD(X)$ denotes the standard deviation of X. However, from Equation 7.17, we see that $SD(Y)$ equals $bSD(X)$. Hence,

$$Cov(XY) = SD(X)SD(Y) \qquad (7.20)$$

where X and Y are perfectly correlated.

The model suggested by Hillier yields the following expressions for the expected present worth and variance of present worth:

$$E(PW) = \sum_{j=0}^{n} E(X_j)(i+1)^{-j} + \sum_{j=0}^{n} \sum_{h=1}^{m} E(Y_{jh})(1+i)^{-j} \qquad (7.21)$$

310 • BREAK-EVEN, SENSITIVITY, AND RISK ANALYSES

and

$$\text{Var}\,(PW) = \sum_{j=0}^{n} \text{Var}\,(X_j)(1+i)^{-2j} + \sum_{h=1}^{m} \left[\sum_{j=0}^{n} SD(Y_{jh})(1+i)^{-j} \right]^2 \qquad (7.22)$$

where $SD(Y_{jh})$ denotes the standard deviation of the random variable Y_{jh}.

A very important theorem from probability theory, the *central limit theorem*, is usually invoked at this point. The central limit theorem establishes under very general conditions that the sum of independently distributed, random variables tends to be distributed normally as the number of terms in the summation approaches infinity. Hence, it is argued that present worth, as defined by Equation 7.1, is normally distributed with mean and variance, as given by Equations 7.21 and 7.22. Of course, this assumes that the C_j are statistically independent. However, the view is usually taken that the normal distribution is a reasonable approximation to the distribution of present worth.

Given that present worth is assumed to be normally distributed, one can compute for each investment alternative the probability of achieving a given aspiration level. For example, one is usually interested in knowing the probability that present worth is less than zero. Some analysts interpret this to be a measure of the risk associated with an investment alternative.

In performing an analysis of the risk associated with an investment alternative, we have treated the simplest situation; only the cash flows were considered to be random variables and the measure of effectiveness employed was present worth, not rate of return. When either the discount rate or the planning horizon are treated as random variables and when the probability distribution for rate of return is desired, simulation approaches are usually employed.

Example 7.9

As an illustration of the analytic approach in developing the probability distribution for PW, consider the following example problem based on one given by Giffin [3].

A flight school operator is considering the alternatives of purchasing a utility category training aircraft versus purchasing an acrobatic version of the same aircraft. Having been in the flight training business for a number of years, the operator has reason to believe that income and expense from a utility category aircraft will be nearly independent from year to year. The pertinent data are summarized in Table 7.7.

Investment in the acrobatic aircraft is a more risky but promising investment. The acrobatic aircraft is also expected to have a life of five years. The operator feels that maintenance costs will be nearly independent year to

TABLE 7.7 Estimated Net Cash Flows for a Utility Model Aircraft[a]

Year	Source	Symbol	Expected Value	Range	Standard Deviation
0	Purchase	C_0	−$11,000	$9,500–12,500	$500
1	Income-expense	C_1	2,200	2,050– 2,350	50
2	Income-expense	C_2	2,200	1,900– 2,500	100
3	Income-expense	C_3	2,200	1,900– 2,500	100
4	Income-expense	C_4	2,000	1,700– 2,300	100
5	Income-expense	C_{51}	1,000	700– 1,300	100
5	Salvage	C_{52}	6,000	4,800– 7,200	400

[a] The data have been modified slightly from those given by Giffin [3].

year with this aircraft. However, since the flight school has never offered an acrobatic course before, there is some uncertainty regarding the demand for time in such an aircraft. It is felt that the net cash flow from the sale of flight time for each of the five years will be perfectly correlated. The pertinent data are summarized in Table 7.8. An interest rate of 10% is to be used in the analysis.

From Equations 7.21 and 7.22, we obtain the following:

$$E(PW_1) = \$190 \qquad E(PW_2) = \$355$$
$$\mathrm{Var}\,(PW_1) = 334{,}741 \qquad \mathrm{Var}\,(PW_2) = 11{,}631{,}930$$

Assuming normally distributed PW, the probability of an equivalent present

TABLE 7.8 Estimated Net Cash Flows for an Acrobatic Model Aircraft[a]

Year	Source	Symbol	Expected Value	Range	Standard Deviation
0	Purchase	X_0	−$14,000	$13,100–14,900	$ 300
1	Expense	X_1	− 10,000	9,100–10,900	300
2	Expense	X_2	− 10,000	9,100–10,900	300
3	Expense	X_3	− 11,000	10,100–11,900	300
4	Expense	X_4	− 12,000	10,800–13,200	400
5	Expense	X_{51}	− 12,000	10,800–13,200	400
1	Income	Y_{11}	12,500	9,500–15,500	1000
2	Income	Y_{21}	13,500	10,500–16,500	1000
3	Income	Y_{31}	13,500	10,800–16,200	900
4	Income	Y_{41}	14,500	12,100–16,900	800
5	Income	Y_{51}	13,500	11,400–15,600	700
5	Salvage	X_{52}	7,500	6,000– 9,000	500

[a] The data have been modified slightly from those given by Giffin [3].

worth less than zero is found to be

$$Pr(PW_1 > 0 | i = 10\%) = 0.37$$

for the utility model aircraft and

$$Pr(PW_2 < 0 | i = 10\%) = 0.46$$

for the acrobatic model aircraft. Here we are faced with a choice between the alternative having the greatest expected value and the alternative having the smallest variance. The choice will depend on the owner's attitudes toward risk. At this point we leave the flight school operator "up in the air" relative to a choice between the investment alternatives. If the decision were yours to make, which model aircraft would you choose? Why?

Simulation Approaches. Even though analytic approaches can be used in some situations to perform the risk aggregation step in risk analysis, simulation approaches are typically required to cope with the complexities of a real-world situation. Simulation, in the general sense, may be thought of as performing experiments on a model. Basically, simulation is an "if . . . , then . . ." device (i.e., *if* a certain input is specified, *then* the output can be determined). Some of the major reasons for using simulation in risk analysis are:

1. Analytic solutions are impossible to obtain without great difficulty.
2. Simulation is useful in selling a system modification to management.
3. Simulation can be used as a verification of analytical solutions.
4. Simulation is very versatile.
5. Less background in mathematical analysis and probability theory is generally required.

Some of the major disadvantages of simulation are:

1. Simulations can be quite expensive.
2. Simulations introduce a source of randomness not present in analytic solutions.
3. Simulations do not reproduce the input distributions exactly (especially the tails of the distribution).
4. Validation is easily overlooked in using simulation.
5. Simulation is so easily applied it is often used when analytic solutions can be easily obtained at considerably less cost.

Example 7.10 ————————————————————————

To illustrate the simulation approach in risk analysis, consider the following situation. An individual is planning on purchasing a used computer terminal for $1000 and performing certain billing and accounting functions for several retail businesses in the neighborhood. It is anticipated that the business will last only four years because of the growth of competition. If income in the third year exceeds expenses by more than $400, operations will continue the fourth year. However, if income is less than or equal to $400 more than expenses in the third year, the computer terminal and software will probably be sold at the end of the third year. For simplicity a probability of 0.70 is assigned to the possibility of selling out at the end of the third year, given income is not greater than $400 above expenses in the third year. Let

$$E_j = \text{expense for year } j$$
$$I_j = \text{income for year } j$$
$$n = \text{life of the investment}$$
$$S_n = \text{salvage value based on an } n\text{-year life}$$

Assuming a zero discount rate for simplicity of calculations, the present worth of the investment is given as

$$PW = -\$1000 + \sum_{j=1}^{n} (I_j - E_j) + S_n$$

The probability distributions assumed to hold for this example are provided in Table 7.9. In practice, a computer would be used to perform the simulation. However, to illustrate the technique, we will manually perform 10 simulations of the investment. A table of two-digit random numbers is given in Table 7.10. Using the worksheet given in Table 7.11, 10 simulations of the investment yielded an average present worth of $805 for the investment. Of course, 10 simulations is not an adequate number of trials to draw strong conclusions concerning the investment. However, the example does illustrate the simulation approach.

To illustrate the approach taken, the first random number selected will provide the simulated value for expenses in the first year. A random number of 90 was obtained from row 1, column 1 in Table 7.10. Consulting Table 7.9, it is seen that a random number of 90 represents an expense of $400; hence $400 is entered appropriately on the worksheet given in Table 7.11. The second random number is selected from row 2, column 1 of Table 7.10 to generate the income for year 1. A random number of 78 is obtained and, from Table 7.9, a simulated income of $600 is obtained for the first year. Continuing through the third year, it is found that income exceeds expenses by $500; hence the business will continue through the fourth year. A random number of 97 is drawn to generate the salvage value of $400 for the investment.

Note that in the second trial it was decided that the business should be discontinued after three years. Furthermore, income was never less than expenses in any year; this illustrates insufficient observations (trials) have

TABLE 7.9 Data for the Risk Analysis Example Problem

E_1	$p(E_1)$	RN	I_1	$p(I_1)$	RN
$200	0.25	00–24	$ 400	0.50	00–49
300	0.50	25–74	600	0.50	50–99
400	0.25	75–99			

E_2	$p(E_2)$	RN	I_2	$p(I_2)$	RN
$300	0.10	00–09	$ 500	0.20	00–19
400	0.40	10–49	750	0.40	20–59
500	0.40	50–89	1,000	0.40	60–99
600	0.10	90–99			

E_3	$p(E_3)$	RN	I_3	$p(I_3)$	RN
$400	0.20	00–19	$ 800	0.30	00–29
500	0.30	20–49	1,000	0.50	30–79
600	0.30	50–79	1,200	0.20	80–99
700	0.20	80–99			

E_4	$p(E_4)$	RN	I_4	$p(I_4)$	RN
$500	0.25	00–24	$ 600	0.25	00–24
600	0.25	25–49	800	0.25	25–49
700	0.25	50–74	1,000	0.25	50–74
800	0.25	75–99	1,200	0.25	75–99

N	$p(N|I_3-E_3\leq \$400)$	RN
3	0.70	00–69
4	0.30	70–99

N	$p(N|I_3-E_3>\$400)$
4	1.00

S_3	$p(S_3)$	RN	S_4	$p(S_4)$	RN
$400	0.50	00–49	$ 300	0.60	00–59
500	0.50	50–99	400	0.40	60–99

been obtained since, in year 2, expense can exceed income with probability

$$Pr(E_2 = 600 \text{ and } I_2 = 500) = 0.10(0.20) = 0.02$$

and in year 4 expense can exceed income with probability

$$Pr(E_4 = 700 \text{ and } I_4 = 600) + Pr(E_4 = 800 \text{ and } I_4 = 600) = 0.125$$

TABLE 7.10 Two-Digit Random Numbers

90	43	78	83	82	99	54	02
78	31	58	98	68	09	87	80
51	81	42	35	21	42	03	62
93	97	15	95	07	56	60	39
27	37	12	63	31	35	66	93
79	39	44	22	83	96	51	00
89	61	73	29	43	84	91	34
29	38	30	84	90	18	00	10
97	64	33	29	17	48	26	04
07	64	15	02	44	32	92	99
82	13	50	83	35	39	50	51
59	83	21	30	86	90	16	09
04	46	19	63	60	53	33	97
96	54	91	43	44	40	09	02
31	27	71	18	03	65	53	62
03	45	70	42	22	16	67	13
08	35	45	92	79	97	46	02
37	60	80	55	05	35	75	57
90	43	63	17	56	21	69	09
22	07	69	85	38	74	02	58
05	33	79	00	69	29	67	08
48	97	91	14	53	00	03	42
94	68	64	58	97	32	27	80
15	39	85	87	82	38	52	16
09	37	81	73	37	01	66	84

As an alternative to using the discrete probability distributions given in Table 7.9, we might employ continuous distributions such as the normal, gamma, or beta distributions to represent income and expenses. If such distributions are to be used, appropriate techniques for generating simulated values of the random variables are required. Since a detailed treatment of simulation is beyond the scope of this text, we refer you to texts devoted to simulation for additional discussion [14], [15].

Risk analysis offers a number of important advantages over traditional deterministic approaches. Klausner [11] summarizes some of the most significant advantages as follows:

1. Uncertainty Made Explicit: The uncertainty which an estimator feels about his estimate of an element value is brought out into the open and incorporated into the investment analysis. The analysis technique permits maximum information utilization by providing a vehicle for the inclusion of "less likely" estimates in the analysis.

TABLE 7.11 Simulation of the Sample Investment Problem

Trial	Year	RN(E)	E	RN(I)	I	I - E	RN(N)	RN(S)	S	PW
1	1	90	$400	78	$ 600	$200				
	2	51	500	93	1000	500				
	3	27	500	79	1000	500	—			
	4	89	800	29	800	0		97	$400	$ 600
2	1	07	200	82	600	400				
	2	59	500	04	500	0				
	3	96	700	31	1000	300	03	08	400	100
	4	—	—	—	—	—				
3	1	37	300	90	600	300				
	2	22	400	05	500	100				
	3	48	500	94	1200	700	—			
	4	15	500	09	600	100		43	300	500
4	1	31	300	81	600	300				
	2	97	600	37	750	150				
	3	39	500	61	1000	500	—			
	4	38	600	64	1000	400		64	400	750
5	1	13	200	83	600	400				
	2	46	400	54	750	350				
	3	27	500	45	1000	500	—			
	4	35	600	60	1000	400		43	300	950
6	1	07	200	33	400	200				
	2	97	600	68	1000	400				
	3	39	500	37	1000	500	—			
	4	78	800	58	1000	200		42	300	600
7	1	15	200	12	400	200				
	2	44	400	73	1000	600				
	3	30	500	33	1000	500	—			
	4	15	500	50	1000	500		21	300	1100
8	1	19	200	91	600	400				
	2	71	500	70	1000	500				
	3	45	500	80	1200	700	—			
	4	63	700	69	1000	300		79	400	1300
9	1	91	400	64	600	200				
	2	85	500	81	1000	500				
	3	83	700	98	1200	500	—			
	4	35	600	95	1200	600		63	400	1200
10	1	22	200	29	400	200				
	2	84	500	29	750	250				
	3	02	400	83	1200	800	—			
	4	30	600	63	1000	400		43	300	950
										8050

2. **More Comprehensive Analysis.** This technique permits a determination of the effect of simultaneous variation of all the element values on the outcome of an investment. This approximates the "real world" conditions under which an actual investment's outcome will be determined. The Probabilistic Cash Flow Simulation generates an overall indication of potential variation in outcome and project risk. This indicator, in the form of a probability distribution, accounts statistically for element interaction.

3. **Variability of Outcome Measured.** One of the most significant advantages of this analysis technique is that it gives a measure of the dispersion around the investment outcome based on the expected cash flow. This dispersion, or variability, is an important consideration in the comparison of alternative investments. Other things being equal, lower variability for the same return is usually desirable. The probability distribution associated with each investment's outcome gives a clear picture of this important evaluation consideration.

4. **Promotes More Reasoned Estimating Procedures.** By requiring that element values be given as probability distributions rather than as single values, more reasoned consideration is given to the estimating procedure. Judgment is applied to the individual element values rather than to the investment's outcome which is jointly determined by all of the elements. Thinking through the uncertainties in a project and recognizing what is known and unknown will go far toward ensuring the best investment decision. Understanding and dealing effectively with uncertainty and risk is the key to rational decision making.

Risk analysis is a technique that has been used by a number of firms to improve the decision-making process in a risk environment. When applied properly, risk analysis can enhance significantly the manager's understanding of the risks associated with an investment alternative. The major premise underlying risk analysis is the belief that a manager can make better decisions when he or she has a fuller understanding of the implications of the decision.

7.5 SUMMARY

Descriptive approaches were presented in this chapter for coping with risk and uncertainty conditions in performing economic analyses. In particular, break-even, sensitivity, and risk analyses were discussed. In Chapter Eight, we employ normative or prescriptive approaches for dealing with risk and uncertainty.

The extent to which supplementary analyses are justified depends largely on the magnitude of the initial investment. In particular, if an analysis has been performed using the techniques described in Chapters Four, Five, and Six and one of the alternatives is not clearly the superior alternative, a risk analysis will probably yield additional insight concerning the decision to be made. If either small amounts of money are involved or if the choice is not close using the techniques described in Chapters Four, Five, and Six under assumed certainty, then supplementary analyses are not justified. In those cases where supplementary analysis is justified and a sensitivity analysis is not sufficient to allow a decision, a risk analysis is performed.

BIBLIOGRAPHY

1. Canada, J. R., *Intermediate Economic Analysis for Management and Engineering*, Prentice-Hall, 1971.

2. Fleischer, G. A., *Risk and Uncertainty: Non-Deterministic Decision Making in Engineering Economy*, AIIE Monograph Series, American Institute of Industrial Engineers, Inc., 1975.

3. Giffin, W. C., *Introduction to Operations Engineering*, R. D. Irwin, 1971.

4. Hertz, D. B., "Risk Analysis in Capital Investments," *Harvard Business Review*, *42* (1), January–February 1964.

5. Hillier, F. S., "The Derivation of Probabilistic Information for the Evaluation of Risky Investments," *Management Science*, *9* (3), April 1963.

6. Hillier, F. S., "Supplement to 'The Derivation of Probabilistic Information Evaluation of Risky Investments,'" *Management Science*, *11* (3), January 1965.

7. Hillier, F. S., *The Evaluation of Risky Interrelated Investments*, North-Holland Publishing, 1969.

8. Hillier, F. S., "A Basic Approach to the Evaluation of Risky Interrelated Investments," Chapter 1 in *Studies in Budgeting*, R. F. Byrne, W. W. Cooper, A. Charnes, O. A. Davis, and D. A. Gilford, Editors, North-Holland Publishing, 1971.

9. Hillier, F. S., "A Basic Model for Capital Budgeting of Risky Interrelated Projects," *The Engineering Economist*, *17* (1), 1971.

10. Kaplan, S., and Barish, N., "Decision Making Allowing for Uncertainty of Future Investment Opportunities," *Management Science*, *13* (10), June 1967.

11. Klausner, R. F., "The Evaluation of Risk in Marine Capital Investments," *The Engineering Economist*, *14* (4), Summer 1969.

12. Lifson, M., "Evaluation Modeling in the Context of the Systems Decision Process," *Decision and Risk Analysis: Powerful New Tools for Management*, Proceedings of the Sixth Triennial Symposium, published by *The Engineering Economist*, 1971.

13. Miller, E. C., *Advanced Techniques for Strategic Planning*, AMA Research Study 104, American Management Association, Inc., 1971.

14. Schmidt, J. W., and Taylor, R. E., *Simulation and Analysis of Industrial Systems*, R. D. Irwin, 1970.

15. Shannon, R. E., *Systems Simulation: the Art and Science*, Prentice-Hall, 1975.

PROBLEMS

1. A plastic extrusion plant manufactures a particular product at a variable cost of $0.10 per unit, including material cost. The fixed costs associated with manufacturing the product equal $30,000/year. Determine the break-even value for annual sales if the selling price per unit is (a) $0.60, (b) $0.40, and (c) $0.20. (7.2)

2. A consulting engineer is considering two pumps to meet a demand of 15,000 gallons/minute at 12 feet total dynamic head. The specific gravity of the liquid being pumped is 1.50. Pump A operates at 70% efficiency and costs $8000; pump B operates at 75% efficiency and costs $12,000. Power costs $0.01/kilowatt-hour. Continuous pumping for 365 days/year is required (i.e., 24 hours/day). Using a *MARR* of 10% and assuming equal salvage values for both pumps, how many years of service are required for pump B to be justified economically? (*Note.* Dynamic head times gallon/minute times specific gravity divided by 3960 equals horsepower required. Horsepower times 0.746 equals kilowatts required.) (7.2)

3. Two 100-horsepower motors are under consideration by the Mighty Machinery Company. Motor Q costs $2000 and operates at 90% efficiency. Motor R costs $1500 and is 88% efficient. Annual operating and maintenance costs are estimated to be 15% of the initial purchase price. Power costs 2.4¢/kilowatt-hour. How many hours of full-load operation are necessary each year in order to justify the purchase of motor Q! Use a 15-year planning horizon; assume that salvage values will equal 20% of the initial purchase price; and let the *MARR* be 10%. (*Note.* 0.746 kilowatts = 1 horsepower.) (7.2)

4. Owners of a nationwide motel chain are considering locating a new motel in Portland, Arkansas. The complete cost of building a 150-unit motel (excluding furnishings) is $2 million; the firm estimates that the furnishings in the motel must be replaced at a cost of $750,000 every five years. Annual operating and maintenance cost for the facility is estimated to be $50,000. The average rate for a unit is anticipated to be $18/day. A 15-year planning horizon is used by the firm in evaluating new ventures of this type; a terminal salvage value of 20% of the original building cost is anticipated; furnishings are estimated to have no salvage value at the end of each 5 year replacement interval; land cost is not to be included. Determine the break-even value for the daily occupancy percentage based on a *MARR* of (a) 0%, (b) 10%, (c) 20%, and (d) 30%. (Assume that the motel will operate 365 days/year.) (7.2)

5. A manufacturing plant in Michigan has been contracting snow removal at a cost of $200/day. The past three years have produced heavy snow-falls, resulting in the

cost of snow removal being of concern to the plant manager. The plant engineer has found that a snow-removal machine can be purchased for $35,000; it is estimated to have a useful life of 10 years, and a zero salvage value at that time. Annual costs for operating and maintaining the equipment are estimated to be $8000. Determine the break-even value for the number of days per year that snow removal is required in order to justify the equipment, based on a *MARR* of (a) 0%, (b) 10%, and (c) 30%. (7.2)

6. A machine can be purchased at $t = 0$ for $10,000. The estimated life is five years, with an estimated salvage value of zero at that time. The average annual operating and maintenance expenses are expected to be $4000. If *MARR* = 20%, what must the average annual revenues be in order to be indifferent between (a) purchasing the machine, or (b) "do nothing"? (7.2)

7. A business firm is contemplating the installation of an improved material handling system between the packaging department and the finished goods warehouse. Two designs are being considered. The first consists of a driverless tractor system involving three tractors on the loop, with four trailers pulled by each tractor. The second design consists of a pallet conveyor installed between packaging and the warehouse. The driverless tractor system will have an initial equipment cost of $185,000 and annual operating and maintenance costs of $7800. The pallet conveyor has an initial cost of $220,000 and annual operating and maintenance costs of $500. The firm is not sure what planning horizon to use in the analysis; however, the salvage value estimates given in the following table have been developed for various planning horizons. Using a *MARR* of 20%, determine the break-even value for *n*, the planning horizon. (7.2)

Salvage Value Estimates

n	Driverless Tractor	Pallet Conveyor
4	$108,000	$140,000
5	90,000	125,000
6	74,000	110,000
7	60,000	95,000
8	48,000	80,000
9	38,000	70,000
10	30,000	60,000
11	23,000	50,000
12	17,000	40,000
13	12,000	30,000
14	8,000	20,000
15	5,000	20,000

8. The motor on a gas-fired furnace in a small foundry is to be replaced. Three different 15-horsepower electric motors are being considered. Motor X sells for $500 and has an efficiency rating of 90%; motor Y sells for $400 and has a rating of

85%; motor Z sells for $300 and is rated to be 80% efficient. The cost of electricity is $0.03/kilowatt-hour. An eight-year planning horizon is used, and zero salvage values are assumed for all three motors. A *MARR* of 15% is to be used. Assume that the motor selected will be loaded to capacity. Determine the range of values for annual usage of the motor (in hours) that will lead to the preference of each motor. (*Note.* 0.746 kilowatts = 1 horsepower.) (7.2)

9. An investment of $10,000 is to be made into a savings account. The interest rate to be paid each year is uncertain; however, it is estimated that it is twice as likely to be 7% as it is to be 6% and it is equally likely to be either 6% or 8%. Determine the probability distribution for the amount in the fund after 3 years, assuming the interest rate is not autocorrelated. (7.4)

10. Two condensers are being considered by the Ajax Company. A copper condenser can be purchased for $1000; annual operating and maintenance costs are estimated to be $100. Alternatively, a ferrous condenser can be purchased for $900; since the Ajax Company has not had previous experience with ferrous condensers, they are not sure what annual operating and maintenance cost estimate is appropriate. A five-year planning horizon is to be used, salvage values are estimated to be 10% of the original purchase price, and a *MARR* of 20% is to be used. Determine the break-even value for the annual operating and maintenance cost for the ferrous condenser. (7.2)

11. A firm has decided to manufacture widgets. There are two production processes available for consideration. Process A involves the purchase of a $22,000 machine that will last for 10 years and have a $2000 salvage value at that time. Annual operating and maintenance costs amount to $3000/year for the machine. In addition, using Process A requires additional costs amounting to $0.15/widget produced.
The second process, Process B, requires an investment of $12,000 in a machine that will last for 10 years and have a $2000 salvage value at that time. Annual operating and maintenance costs amount to $2000/year for the machine. Additional costs of $0.20/widget produced result when Process B is used.
With an 8% interest rate, for what annual production volume is Process A preferred? (7.2)

12. Two manufacturing methods are being considered. Method A has a fixed cost of $500 and a variable cost of $10. Method B has a fixed cost of $200 and a variable cost of $50. For what production volume would one prefer (a) Method A, and (b) Method B? (7.2)

13. The ABC Company is faced with three proposed methods for making one of their products. Method A involves the purchase of a machine for $5000. It will have a seven-year life, with a zero salvage value at that time. Using Method A involves additional costs of $0.20/unit of product produced per year. Method B involves the purchase of a machine for $10,000. It will have a $2000 salvage value when disposed of in seven years. Using Method B involves additional costs of $0.15/unit of product produced per year. Method C involves buying a machine for $8000. It will have a $2000 salvage value when disposed of in seven years. Additional costs of $0.25/unit of product per year arise when Method C is used. An 8% interest rate is

used by the ABC Company in evaluating investment alternatives. For what range of values for annual production volume is each method preferred? (7.2)

14. In problem 4 suppose the following pessimistic, most likely, and optimistic estimates are given for building cost, furnishings cost, annual operating and maintenance costs and the average rate per occupied unit.

	Pessimistic	Most Likely	Optimistic
Building cost	$3,000,000	$2,000,000	$1,500,000
Furnishings cost	1,250,000	750,000	500,000
Annual operating and			
maintenance costs	60,000	50,000	40,000
Average rate	13/day	15/day	18/day

Determine the pessimistic and optimistic limits on the break-even value for the daily occupancy percentage based on a *MARR* of 20%. Assume the motel will operate 365 days/year. (7.3)

15. In problem 7 suppose a 10-year planning horizon is specified. Perform a sensitivity analysis comparable to that given in Figure 7.3 to determine the effect of errors in estimating the *differences* in initial investment, the *differences* in annual operating and maintenance costs, and the *differences* in salvage values for the two investment alternatives. (7.3)

16. A warehouse modernization plan requires an investment of $3 million in equipment; at the end of the 15-year planning horizon, it is anticipated the equipment will have a salvage value of $600,000. Annual savings in operating and maintenance costs due to the modernization are anticipated to total $1,400,000/year. A *MARR* of 15% is used by the firm. Perform a sensitivity analysis to determine the effects on the economic feasibility of the plan due to errors in estimating the initial investment required, and the annual savings. (7.3)

17. Given an initial investment of $10,000, annual receipts of $3000, and an uncertain life for the investment, determine the probability of the investment being profitable. Use a 15% *MARR*. Let the probability distribution for the life of the investment be given as follows: (7.4)

n	$p(n)$
1	0.10
2	0.15
3	0.20
4	0.25
5	0.15
6	0.10
7	0.05

18. In problem 17 suppose the *MARR* is not known with certainty and the following probability distribution is anticipated to hold.

i	p(i)
0.10	0.20
0.15	0.60
0.20	0.20

What is the probability of the investment being profitable? (7.4)

19. In problem 17 suppose the magnitude of the annual receipts (*R*) is subject to random variation. Assume that each annual receipt will be identical in value and the annual receipt has the following probability distribution:

R	p(R)
$2000	0.20
3000	0.50
4000	0.30

What is the probability of the investment being profitable? (7.4)

20. In problem 17 suppose the minimum attractive rate of return is distributed as given in problem 18 and suppose the annual receipts are distributed as given in problem 19. Determine the probability of the investment being profitable. (7.4)

21. In problem 20 suppose the initial investment is equally likely to be either $9,000 or $10,000. Determine the probability of the investment being profitable. (7.4)

22. Suppose $n = 4$, $i = 0\%$, a $10,000 investment is made, and the receipt in year j, $j = 1, \ldots, 4$, is statistically independent and distributed as in problem 19. Determine the probability distribution for present worth. (7.4)

23. Consider an investment alternative having a six-year planning horizon and expected values and variances for statistically independent cash flows as given below:

i	$E(C_i)$	$V(C_i)$
0	-$25,000	625×10^4
1	4,000	16×10^4
2	5,000	25×10^4
3	6,000	36×10^4
4	7,000	49×10^4
5	8,000	64×10^4
6	9,000	81×10^4

Using a discount rate of 20%, determine the expected values and variances for both present worth and annual worth. Based on the central limit theorem, compute the probability of a positive present worth; compute the probability of a positive annual worth. (7.4)

24. Solve problem 23 using a discount rate of (a) 0%, (b) 30%. (7.4)

25. Solve problem 23 when the C_j, $j = 1, \ldots, 6$, are perfectly correlated. (7.4)

26. Two investment alternatives are being considered. Alternative A requires an initial investment of $13,000 in equipment; annual operating and maintenance costs are anticipated to be normally distributed, with a mean of $5000 and a standard deviation of $500; the terminal salvage value at the end of the eight-year planning horizon is anticipated to be normally distributed with a mean of $2000 and a standard deviation of $800. Alternative B requires end of year annual expenditures over the planning horizon, with the annual expenditure being normally distributed with a mean of $7500 and a standard deviation of $750. Using a *MARR* of 15%, what is the probability that Alternative A is the most economic alternative? (7.4)

27. In problem 26, suppose the *MARR* were 10% with probability 0.25, 15% with probability 0.50, and 20% with probability 0.25, what is the probability that Alternative A is the most economic alternative? (7.4)

28. Recall Example 7.10. Continue the simulation for 10 additional trials and compute the cumulative average present worth. (7.4)

29. In Example 7.10, suppose the individual had used a minimum attractive rate of return of 10% in the calculations. Would the first 10 simulation trials have yielded a positive average present worth? (7.4)

30. In Example 7.10, determine the ERR value for each trial, given a *MARR* of 10%. (7.4)

31. Two investment alternatives are under consideration. The data for the alternatives are given below. It is assumed the cash flow are not autocorrelated.

EOY	E[CF(A)]	SD[CF(A)]	E[CF(B)]	SD[CF(B)]
0	−10,000	1,000	−15,000	1,500
1	4,000	400	8,000	800
2	5,000	500	8,000	800
3	6,000	600	8,000	800
4	7,000	700	8,000	800
5	8,000	800	8,000	800
6	9,000	900	8,000	800

(a) For each alternative, determine the mean and standard deviation for present worth and annual worth using a *MARR* of 10%.
(b) For each alternative, based on the central limit theorem, compute the probability of a positive present worth using a *MARR* of 10%.

(c) Develop the mean and standard deviation for the *incremental* present worth using a *MARR* of 10%.

(d) Based on the central limit theorem, compute the probability of a positive incremental present worth using a *MARR* of 10%. (7.4)

32. Company W is considering investing $10,000 in a machine. The machine will last n years at which time it will be sold for L. Maintenance costs for this machine are estimated to increase by $200/year over its life. The maintenance cost for the first year is estimated to be $1000. The company has a 10% *MARR*. Based on the probability distributions given below for n and L, what is the expected equivalent uniform annual cost for the machine? (7.4)

n	L	$p(n)$
6	$3000	0.2
8	2000	0.4
10	1000	0.4

CHAPTER EIGHT
DECISION MODELS

8.1 INTRODUCTION

This chapter presents basic models for displaying decision situations involving risk and uncertainty and suggests criteria for choosing among alternatives. As such, these models will be *prescriptive* or *normative* in that they prescribe a decision action to be taken. The risk-analysis discussion and models presented in Chapter Seven were descriptive; they described the behavior of the measure of effectiveness under conditions of uncertainty and risk. This chapter is concerned with specifying the preferred choice from a set of feasible alternatives and represents the final step in the eight-step procedure for comparing alternatives presented in Chapter Four.

The first section of this chapter discusses a matrix model for decisions under risk and uncertainty and suggests criteria for choosing among alternatives. The second section presents a network or tree model for decision situations involving a sequence of decisions over a time horizon where the outcomes of alternatives are uncertain. The final section of the chapter offers an introductory discussion on methodology for choosing among alternatives when multiple measures of effectiveness are involved. The matrix model for decisions under risk and uncertainty is now introduced by an example concerning material handling.

Example 8.1 ————————————————————————————

For purposes of introducing a matrix model for decision making, consider the following manufacturing situation. A new product group is to be added by the ABC Company with initial production scheduled two years hence. The new product group, which will consist of a single product in varying sizes and finishes, requires that a new and separate facility be added to the existing plant. The company is presently completing plans for the new facility, and plant construction will begin in the near future. Excess floor space is planned in order to increase production if demand increases; this increase is expected, with demand reaching stability after five years of production.

The manufacturing system consists, essentially, of several machining centers that operate on a job-shop basis instead of on a production-line basis. The manufacturing sequence is fixed except for slight variations, depending on product size and finish required. Thus, a subproblem of the total manufacturing system is to determine the material handling requirements. After considerable study and discussion, it has been decided that industrial lift trucks provide the flexibility of handling required in the job-shop environment. Now, questions arise as to the type of lift truck, the number required, the attachments needed, whether they should be leased or purchased and, if leased, which of several leasing plans should be chosen, and so forth. The answer to these and other relevant questions depends primarily on the demand for the new product during the first five-year production period. This demand is uncertain, and ABC Company management can influence demand through advertising only to a limited degree.

Management could make a decision on material handling requirements based on the demand forecast for the first year of production, then evaluate the situation at the end of the first year, make a new decision for the next year, reevaluate, and so on, in sequential fashion. However, let us assume that it has been decided to choose a material handling system now for the first three-year production period and then reevaluate at the end of that period.

Given the above managerial decision, the analysts investigating the material handling requirements make the judgment that demand for the product group during the three-year planning horizon will be such that four, five, or six lift trucks are required. They further reason that the feasible alternatives are

A_1, A_2, A_3—lease four, five, or six trucks, respectively
A_4, A_5, A_6—purchase four, five, or six trucks, respectively

These lease plans do not include a maintenance contract, since the ABC Company presently employs skilled maintenance personnel. Furthermore, the analysts have not considered any lease plan with an "option to purchase" clause in order to avoid "contractual sales" income tax complexities.

The analysts select before-tax, equivalent uniform annual cost as the single measure of effectiveness for a given "A_j versus number of trucks required" combination. It is necessary to make gross estimates of these annual costs, but essentially, for leasing, there is a fixed monthly charge per truck plus a judgmental penalty cost associated with production delays when the number of trucks required exceeds the number of trucks available. On the other hand, the company can obtain a more favorable leasing charge as the number of trucks leased increases. For the purchase alternatives, similar fixed (actually fixed and variable) annual operating and maintenance costs, penalty costs, and quantity discounts are involved. After much study, somewhat gross estimations, and a liberal application of judgment, the analysts present the model of Table 8.1 to management for a final decision.

On the basis of the single, equivalent uniform annual cost measure of effectiveness, any *purchase* alternative is preferable to any *lease* alternative. For example, the purchase alternative A_4 has a lower equivalent annual cost than any lease alternative if four, five, or six trucks are required. Thus, Alternative A_4 dominates the A_1, A_2, and A_3 alternatives, and they may therefore be deleted from the decision matrix. (It is noted that this might not have been the case if after-tax annual cost values had been calculated. Furthermore, a consideration of intangible factors by management could result in a lease alternative being chosen over a purchase alternative even though annual costs are higher.) The purchase alternatives A_5 and A_6 also dominate the three lease alternatives, but no one purchase alternative dominates any other purchase alternative. That is, no single purchase alternative has lower annual costs than any other purchase alternative when viewed across all "number of trucks required" conditions, anyone of which it is assumed can occur. Dominance will be discussed in greater detail subsequently.

TABLE 8.1 Decision Model for the Lift Truck Example

Number of Trucks Required / Alternatives	4	5	6
	Equivalent Uniform Annual Costs		
A_1—Lease four	$18,000	$20,000	$24,000
A_2—Lease five	20,000	20,000	22,000
A_3—Lease six	21,000	21,000	21,000
A_4—Purchase four	12,500	14,500	18,500
A_5—Purchase five	14,000	14,000	16,000
A_6—Purchase six	15,000	15,000	15,000

A solution to the lift truck problem will not be presented now, the example has been cited to illustrate the matrix format of Table 8.1 and to suggest the model as an effective portrayal of many decision situations.

The elements in the decision model of Table 8.1 are now summarized as:

States = the number of lift trucks required in the three-year planning horizon.

Feasible alternatives = the number of lift trucks available, whether leased or purchased.

Outcomes = the results of particular "state versus alternative" combinations. In this case, the general outcomes are that the number of trucks required is less than, equal to, or greater than the number of trucks available.

Value of the outcomes = the measure of effectiveness for a particular "state versus alternative" combination. In this case, equivalent uniform annual costs.

A matrix decision model with general symbolism, adapted from Morris [15], is given in Figure 8.1 and further discussed in the next section.

A_j ╲ S_k	p_1 / S_1	p_2 / S_2	...	...	p_k / S_k	...	...	p_m / S_m
A_1	$V(\theta_{11})$	$V(\theta_{12})$	...	...	$V(\theta_{1k})$	...	...	$V(\theta_{1m})$
A_2	$V(\theta_{21})$	$V(\theta_{22})$	...	...	$V(\theta_{2k})$	...	...	$V(\theta_{2m})$
⋮	⋮	⋮			⋮			⋮
⋮	⋮	⋮			⋮			⋮
A_j	$V(\theta_{j1})$	$V(\theta_{j2})$	...	...	$V(\theta_{jk})$	...	...	$V(\theta_{jm})$
⋮	⋮	⋮			⋮			⋮
⋮	⋮	⋮			⋮			⋮
A_n	$V(\theta_{n1})$	$V(\theta_{n2})$	...	...	$V(\theta_{nk})$	...	...	$V(\theta_{nm})$

FIGURE 8.1. A matrix decision model with general symbolism.

8.2 THE MATRIX DECISION MODEL

The symbolism employed is defined as follows:

A_j = an *alternative* or *strategy* under the decision maker's control, where $j = 1, 2, \ldots, n$.

S_k = a *state* or *possible future* that can occur given that alternative A_j is chosen, where $k = 1, 2, \ldots, m$.

θ_{jk} = the *outcome* of choosing alternative A_j and having state S_k occur.

$V(\theta_{jk})$ = the *value* of outcome θ_{jk}, which may be in terms of dollars, time, distance, or utility.

p_k = the *probability* that state S_k will occur.[1]

The term p_k was not previously introduced in the lift truck decision problem but will subsequently serve as a means by which decisions may be categorized into decisions under assumed certainty, decisions under risk, and decisions under uncertainty. This categorization has essentially been made in Chapter Seven but will be repeated in order to present decision rules for the normative models of this chapter.

The matrix model, in general, describes a set of alternatives available where a single alternative is to be selected at the present time. It is implied that the outcomes possible for a given alternative do not result in the necessity for subsequent decisions at future times. If a sequence of decisions were involved, the decision tree model of a later section would probably be more appropriate.

It is also assumed in the matrix model that the alternatives are mutually exclusive, feasible, and represent the total set of alternatives the decision maker may wish to consider. Recall from the discussion in Chapter Four that mutually exclusive alternatives cannot occur together, and thus a single alternative will be chosen from the feasible set. The question now arises, "What constitutes feasibility?" and the answer to the question lies in the original statement of objectives that the alternative is to accomplish. For example, if one were considering the replacement of a production machine because of increasing maintenance costs and inadequate capacity, an objective might be that the new machine must be capable of producing 1000 units of Product A per day. Those machines incapable of satisfying this constraint of 1000 units/day are infeasible and rejected from further consideration. Those machines capable of meeting or exceeding the requirement of 1000 units/day are feasible, and the "best" replacement machine is then selected from the feasible set according to a more discriminating criterion, such as minimum estimated equivalent annual cost or maximum estimated annual profit.

[1] It is assumed that the probability of a particular state occurring does not depend on the alternative chosen by the decision maker.

Determining the "total set" of feasible alternatives is also a difficulty deserving additional comment. Rarely does the analyst have the time or even the insight to define all the possible alternatives available in a given decision situation. How to execute a search procedure and when to terminate the search are interesting and important questions, but will not be pursued here. Suffice it to say that experience plays a major role in the search procedure, and when to terminate the search is primarily a matter of judgment in the practical world of business decisions. The cost of finding an additional alternative versus the perceived benefits of an additional alternative is no doubt a subjective consideration, at least in the exercise of judgment to terminate the search.

Experience and judgment also play an important role in defining the relevant states for a given decision situation and the set of feasible alternatives involved. To convey further the meaning of a "state," other terms in the literature are *possible futures*, *states of nature*, *external conditions*, and *future events*.

In the discussion on risk analysis in Chapter Seven, states were values for the random variables of purchase price, annual operating expenses, annual revenues, service lives, and salvage values. However, in other decision situations, the states may be defined more grossly; for instance, the market demand for product A will be low, medium, or high. Furthermore, the number of states may be finite or infinite, but for the examples of this text, the number of states will be assumed finite and few in number. This seems a realistic assumption for most business and engineering decisions because, although the states may actually be a very large finite number or even infinite, the decision maker wishes to limit the scope of the problem and uses a sample of state values over the range of possible values. Using the industrial lift truck example presented earlier, knowledge of the machining processes involved and a recorded history of machining times can provide the basis for estimating values for the states in terms of the number of trucks required to service different levels of production output. This same experience can also provide the range of state values to consider. In any case, it will be assumed here that once the states are defined, they are collectively exhaustive (i.e., the total set of possible states). Furthermore,

1. The occurrence of one state precludes the occurrence of the others (i.e., they are mutually exclusive),
2. The occurrence of a state is not influenced by the alternative selected by the decision maker, and
3. The occurrence of a state is not known with certainty by the decision maker

For each alternative-versus-state combination, there is an outcome and a value associated with the outcome. In the previous industrial lift truck example, a given "number of trucks available versus number of trucks required" combination resulted in the trucks available being short, over, or equal to the trucks required. For each outcome, it was assumed that an annual equivalent dollar cost could be calculated and was the value of interest. Other value measurements may, of course, be appropriate in other decision situations. The matrix values may be one of two general kinds—objective or subjective. Objective values represent physical quantities such as dollars, hours, liters, kilograms, and so forth. With the exception of the section on multiple goals in this chapter, the subsequent examples are concerned with dollars or objective values.[2] Subjective values, on the other hand, are numbers that represent the decision maker's relative preferences; for example, a score of 90 on an examination is preferred to a score of 60. The values of 90 and 60 are subjective values assigned on an arbitrary scale of, say, 0 to 100. An example of a subjective value matrix will be given in the multiple-goals section of this chapter. In any case, as a general statement, the problems of measurement and value determination are often very substantial in real-world situations.

Let the matrix model of Figure 8.1 now be recalled in order to classify decisions on the basis of the amount of information available to the decision maker in regard to the probability of occurrence for a given state.

8.3 DECISIONS UNDER ASSUMED CERTAINTY

It is reasonable to assume in many decision situations that only one state is relevant and then treat the decision as if the state were certain to occur. This kind of case is termed a *decision under assumed certainty*. For example, it may be a matter of company policy to depreciate new equipment purchases fully in five years. If, in an economic analysis of several new equipment candidates, the measure of effectiveness is equivalent annual cost, the determination of these annual costs would be based on a five-year life for each candidate. The outcome of a five-year life and the associated annual operating costs for each candidate are assumed certain (i.e., will occur with probabilty of 1.0). The actual functional life of an equipment candidate might be 4, 5, 6, 7, 8, ..., years. However, a judgment is made by the decision maker to suppress uncertainty because of convenience or practicality and assume only the single state of a five-year life. All of the analysis in Chapters Three to Six was based on such judgments, and single-value estimates for operating costs, service life, and the like were examples of assumed certainty.

---

[2] A linear utility function for the decision maker is also assumed. That is, minimizing costs or maximizing profits are both the same as maximizing the decision maker's utility.

In terms of the matrix decision model, a decision under assumed certainty would appear as follows:

	S
A_1	$V(\theta_1)$
A_2	$V(\theta_2)$
$\vdots$	$\vdots$
A_j	$V(\theta_j)$
$\vdots$	$\vdots$
A_{n-1}	$V(\theta_{n-1})$
A_n	$V(\theta_n)$

If, in the above model, the values for each alternative were profits (or gains), the principle for choice might be "choose the alternative that maximizes profit." Conversely, if the values were costs (or losses), one might "choose the alternative that minimizes costs."

Example 8.2

Suppose that an investor is considering a $10,000 purchase of United States government securities. Different government agencies offer such securities, and they are available on the open market. Assuming that the various securities have common maturity dates, the investor judges that each agency is financially secure and payoffs are certain. A logical criterion for choice among the securities would be the effective annual yield, or internal rate of return, on the $10,000 purchase. (It is assumed there is no reason for allocating the $10,000 to different securities.) Suppose the investor considers five government securities and, for a maturity date of five years hence, has calculated the effective annual yield on each security to be

$$
\begin{array}{ll}
A_1 & 8.0\% \\
A_2 & 7.3\% \\
A_3 & 8.7\% \\
A_4 & 6.0\% \\
A_5 & 6.5\%
\end{array}
$$

Since the investor would logically desire to maximize the yield on the investment, A_3 would be chosen. Recall that equal risk and maturity dates were assumed for each security.

Example 8.3

The cash flow profiles for the three mutually exclusive investment alternatives given in Table 4.8 are repeated in Table 8.2. Considering the minimum attractive rate of return of 10% as the single state that will occur with certainty and the present worth of each alternative as the value of interest, the matrix model of this decision appears in Table 8.3. If the principle for choice is "select the alternative that maximizes the present worth," A_3 would be chosen as before in Chapter Four.

TABLE 8.2 Cash Flow Profiles for Three Mutually Exclusive Investment Alternatives ($MARR = 10\%$)

t	A_1	A_2	A_3
0	$0	−$10,000	−$15,000
1	4,000	7,000	7,000
2	4,000	7,000	8,000
3	4,000	7,000	9,000
4	4,000	7,000	10,000
5	4,000	7,000	11,000
Present worth	$15,163.20	$16,535.60	$18,397.33

TABLE 8.3 Present Worths of Three Mutually Exclusive Investment Alternatives ($MARR = 10\%$)

A_j	S_1 $i = 10\%$
A_1	$15,163.20
A_2	16,535.60
A_3	18,397.33

8.4 DECISIONS UNDER RISK

A decision situation is called a *decision under risk* when the decision maker elects to consider several states and, the probabilities of their occurrence are explicitly stated. In some decision problems, the probability values may be objectively known from historical records or objectively determined from analytical calculations. For example, if an unbiased die is rolled once, the six possible states are that one, two, three, four, five, or six dots will occur on the top face exposed. The objective probability that each state will occur prior to the roll is 1/6. The sum of the individual probabilities over all states equals the value 1.0. This property of $\Sigma p_j = 1.0$ agrees with an earlier statement that, once the set of possible states is decided on, the states are treated as if they constituted a set of mutually exclusive and collectively exhaustive events. Indeed, the property of $\Sigma p_j = 1.0$ is a necessary restriction if probability theory is to be used as a guide to decision making.

The decision maker may not have past records available to arrive at objective probability values. This was the case in the industrial truck example earlier. However, if the decision maker feels that experience and judgment are sufficient to assign probabilty values subjectively to the occurrence of each state, the decision is still treated as one under risk. The restriction of $\Sigma p_k = 1.0$ still applies in this case. Principles of choice for selecting an alternative in a decision under risk, which commonly appear in the literature on decision theory, are illustrated in Example 8.4.

Example 8.4

The three mutually exclusive investment alternatives presented in Table 8.2 will again be used to illustrate guidelines for a decision under risk. The single state previously defined was a minimum attractive rate of return equal to 10%. For the purpose of this example, the two additional states of $MARR = 20\%$ and $MARR = 30\%$ will be assumed. The realism of considering these states as probabilistic in nature may be questionable, but an argument supporting this claim can be posed. In periods of an unstable economy, the analyst may perceive the $MARR$ as uncertain and be willing to define different values with associated probabilities. The matrix model of Table 8.4 assumes that this is the case, and cell values are present worths (rounded) for a five-year planning horizon.

The states of $S_1 = 10\%$, $S_2 = 20\%$, and $S_3 = 30\%$ may have been defined by judging these values to be optimistic, most likely, and pessimistic estimates, as discussed in Chapter Seven. An $MARR$ that lowers the present worth of the investment alternative is considered a pessimistic $MARR$. The chance of occurrence for the states may have been reasoned as follows: S_3 is twice as likely to occur as S_2, which, in turn is three times as likely to occur as S_1. This reasoning can be stated as the equation $p_1 + 3p_1 + 6p_1 = 1.0$, and $p_1 = 0.10$, $p_2 = 0.30$, and $p_3 = 0.60$ determined.

8.4.1 Dominance

The *dominance* principle is described as follows. Given two alternatives, if one would always be preferred no matter which state occurs, this preferred alternative is said to dominate the other, and the dominated alternative can be deleted from further consideration. Applying this principle will not necessarily solve the decision problem by selecting a unique alternative, but it may reduce the number of alternatives for further consideration.

The dominance principle is a way of deciding which alternatives not to select. If the values are in terms of *costs*, a more formal statement of the dominance relation is: *If there exists a pair of alternatives A_j and A_l such that $V(\theta_{jk}) \leq V(\theta_{lk})$ for all k, A_j is said to dominate A_l. Alternative A_l may then be discarded from the decision problem*. If the values of a decision matrix are in terms of *gains*, the condition for A_j to dominate A_l is that $V(\theta_{jk}) \geq V(\theta_{lk})$ for all k.

In Table 8.4, no alternative has a present worth value greater than the present worth value of another alternative for all states. Thus, there is no dominant alternative for this example, but recall that there were dominated alternatives in Table 8.1.

TABLE 8.4 Matrix Model with States of *MARR* = 10%, 20%, and 30%

A_i \ MARR	$p_1 = 0.10$ S_1 10%	$p_2 = 0.30$ S_2 20%	$p_3 = 0.60$ S_3 30%
A_1	$15,163	$11,962	$9,742
A_2	$16,536	$10,934	$7,049
A_3	$18,397	$10,840	$5,679

8.4.2 Expectation-Variance Principle

If X is a discrete random variable that is defined for a finite number of values and $p(x)$ denotes the probability of a particular value occurring, then the expected or mean value of the random variable is defined as

$$E(X) = \sum_{\text{all } x} x p(x) \tag{8.1}$$

The variance of the random variable is defined as Var (X), where

$$\text{Var } (X) = \sum_{\text{all } x} [x - E(X)]^2 p(x) \qquad (8.2)$$

For the matrix model of Figure 8.1, the random variable is $V(\theta_{jk})$, and the expected value of an alternative A_j is

$$E(A_j) = \sum_{\text{all } k} V(\theta_{jk}) p_k \qquad (8.3)$$

where p_k denotes the probability of state k occurring.

The expected values for each of the alternatives of Table 8.4 are calculated as

$$E(A_1) = \$15,163(0.10) + \$11,962(0.30) + \$9742(0.60)$$
$$= \$10,950.10$$

$$E(A_2) = \$16,536(0.10) + \$10,934(0.30) + \$7049(0.60)$$
$$= \$9163.20$$

$$E(A_3) = \$18,397(0.10) + \$10,840(0.30) + \$5679(0.60)$$
$$= \$8499.10$$

If the principle of maximizing (minimizing) the expected gain (loss) is followed, alternative A_1 would be chosen, since the expected values are positive present worths. A corollary of the expected-value criterion is that, in the event of a tie in expected value for two or more alternatives, the alternative having minimum variance should be chosen. The argument for this secondary criterion is, essentially, that a larger variance is indicative of greater uncertainty, and since most decision makers have an aversion to uncertainty, the alternative with smallest variance should be chosen. Conrath [4], in an article on the implementation of statistical decision theory in practical decisions under risk, states that "... as yet we have little guidance as to what the trade-offs between mean and variance might be or ought to be." That the expected value and the variance are two independent decision criteria instead of dependent criteria is an interesting speculation (or perhaps observation) that deserves further research.

8.4.3 Most Probable Future Principle

If, in a decision under risk, one state has a probability of occurrence considerably greater than any other, the *most probable future principle* is to consider this state as certain and all other states as having a zero chance of occurrence. The decision is thereby reduced to a decision under assumed certainty. Then an alternative is chosen that maximizes (or minimizes) the measure of effectiveness being used. For Table 8.4, S_3 has a probability of 0.60 assigned and is therefore the most probable state. Assuming certainty for S_3 or $p_3 = 1.0$, the reduced matrix is

$$
\begin{array}{cc}
 & S_3 \\
A_1 & \$9742 \\
A_2 & \$7049 \\
A_3 & \$5679
\end{array}
$$

To obtain the maximum present worth, A_1 is selected as the preferred alternative.

Another example of the most probable future principle is the case of a person commuting to work. Each morning that the person leaves for work there is a nonzero probability that an accident could occur and serious injury result. However, the most probable future is that the commuter will reach work safely. The commuter considers this future certain, or else he (she) would probably not go to work!

As a final remark on this principle of choice, it seems appealing only if one state has a significantly higher probability of occurring than any other and the values in the total matrix do not differ significantly. One would not wish to suppress the consideration of a state that, albeit with low probability, would inflict a severe loss on the decision maker. An example of this is the matrix below where the positive values are profits.

	$p_1 = 0.95$	$p_2 = 0.05$
A_1	$10	$40
A_2	$50	-$5000

Following the most probable future principle, A_2 would be chosen. However, the suppression of S_2 does not alter the fact that S_2 could occur, by virtue of $p_2 = 0.05$; in that event the decision maker would suffer a loss of $5000.

8.4.4 Aspiration-Level Principle

In most real-world decisions, the complexity of the decision prevents the discovery and selection of an alternative that will yield the single "best" result, and decision makers set their goals in terms of outcomes that are good enough. That is, decision makers set aspiration levels and then evaluate alternatives against them. An interpretation of this philosophy in terms of a decision under risk is to select an alternative that maximizes the probability of achieving the desired aspiration level. Typical aspiration-level objectives might be to choose the alternative that maximizes the probability of either (1) a 25%

internal rate of return, (2) annual costs less than $8000, or (3) annual profits greater than $9500. In terms of the example of Table 8.4, let the objective be to choose the alternative that maximizes the probability that the present worth value will be equal to or greater than $8000. This objective can be symbolized as $P(PW \geq \$8000)$. From Table 8.4 it can be seen that, for A_1, the present worth will be greater than $8000 when S_1, S_2, or S_3 occurs. Thus, a present worth greater than $8000 is certain if A_1 is chosen. Because the states are mutually exclusive, the additive-probability law permits the sum of the individual state probabilities to be the probability of the joint event, "$PW \geq \$8000$" for A_1. The calculations for each alternative are

$$\text{For } A_1, \ P(PW \geq \$8000) = P(S_1) + P(S_2) + P(S_3)$$
$$= 0.10 + 0.30 + 0.60 = 1.0$$

$$\text{For } A_2, \ P(PW \geq \$8000) = P(S_1) + P(S_2)$$
$$= 0.10 + 0.30 = 0.40$$

$$\text{For } A_3, \ P(PW \geq \$8000) = P(S_1) + P(S_2)$$
$$= 0.10 + 0.30 = 0.40$$

and A_1 maximizes $P(PW \geq \$8000)$.

A variety of aspiration level objectives are, of course, possible for even a *given* decision under risk. The relevant aspiration-level objective is a decision maker's preference.

Do not interpret any of the principles for choice above or those to follow in the next section as the way an alternative *should* be selected. Instead, the principles for choice presented result from observing how people *seem* to decide on alternatives; therefore, they are offered only as possible guidelines to the decision maker.

8.5 DECISIONS UNDER UNCERTAINTY

A decision situation where several states are possible and sufficient information is not available to assign probability values to their occurrence is termed a *decision under uncertainty*. It can be argued that any decision maker, if able to define the states, should have sufficient knowledge to make at least gross subjective probability estimates for these states. On the other hand, in situations such as (1) installing new safety devices on machinery where the states are the number of injuries expected per year, or (2) the introduction of a new product where the states are units demanded per year, one may feel quite inadequate to assign probabilities to these states. This may be the case even though the analyst is not totally lacking in knowledge concerning the number of injuries expected per year or the units of new product demanded per year. The principles for choice subsequently presented are based on the premise that probabilities cannot be assigned.

8.5.1 The Laplace Principle

The philosophy of the *Laplace Principle*, named after an early nineteenth-century mathematician [6], is simply that if one cannot assign probabilities to the states, the states should be considered as equally probable. Then, consider the decision as one under risk. Applying this principle to the previous example of Table 8.4, the probability value assigned to each of the three states is 1/3. Then,

$$E(A_1) = \$15{,}163(1/3) + \$11{,}962(1/3) + \$9742(1/3)$$
$$= \$12{,}289$$

$$E(A_2) = \$16{,}536(1/3) + \$10{,}934(1/3) + \$7049(1/3)$$
$$= \$11{,}506.33$$

$$E(A_3) = \$18{,}397(1/3) + \$10{,}840(1/3) + \$5679(1/3)$$
$$= \$11{,}638.67$$

and A_1 would be chosen to maximize expected present worth.

8.5.2 Maximin and Minimax Principles

The *maximin* and *minimax* principles for choice represent a single philosophy, depending on whether the matrix values are *gains* or *losses*, respectively. These principles hold considerable appeal for the conservative or pessimistic decision maker. In applying the principles, if the matrix values are *gains*, the minimum gain associated with each alternative (when viewed over all states) is determined, and the maximum value in the set of minimum values designates the alternative to be chosen. In other words, determine the worst value possible for *each* alternative and then choose the best value from the set of worst values. Applying the maximin principle to the example problem of Table 8.4 gives

	Minimum Present Worth Value	Maximum of These
A_1	$9742	$9742
A_2	7049	
A_3	5679	

Thus, alternative A_1 is selected as the alternative that will maximize the minimum present worth value that could occur. More formally stated, the maximin principle is to

Select the alternative, j, associated with the max min $V(\theta_{jk})$.
$$\underset{j}{} \quad \underset{k}{}$$

In the case of a decision dealing with *losses* or *costs*, the minimax philosophy is also one of extreme pessimism. In applying the minimax principle to a matrix of costs, the maximum cost associated with each alternative (when viewed over all states) is determined, and the minimum value in the set of maximum values designates the alternative to be chosen. Again, this is choosing the best (or minimum) cost from the set of worst (or maximum) costs. An example to illustrate the application of the minimax principle is given in Table 8.5, where the cell values are equivalent uniform annual costs for the alternatives. Thus, from Table 8.5, alternative A_1 would be chosen. Formally stated, the minimax principle is to

Select the alternative, j, associated with the min max $V(\theta_{jk})$.
$$\underset{j}{} \quad \underset{k}{}$$

8.5.3 Maximax and Minimin Principles

As illustrated in the previous section, the maximin principle for gains and the minimax principle for losses represent a philosophy of extreme pessimism on behalf of the decision maker. For the person who is an extreme optimist in a given decision situation, an *optimistic* rule for choice among alternatives that involve *gains* is the *maximax principle*. That is, the decision maker desires to select the alternative that affords the opportunity to obtain the largest value given in the matrix. A formal statement for the maximax principle is to

Select the alternative, j, associated with the max max $V(\theta_{jk})$.
$$\underset{j}{} \quad \underset{k}{}$$

Then, if the matrix cell values were losses, an optimistic philosophy of

TABLE 8.5 Matrix of Equivalent Uniform Annual Costs

Alternatives \ States	S_1	S_2	S_3	S_4	Maximum Cost	Minimum of These
A_1	$7,000	$10,000	$ 6,500	$ 5,000	$10,000	$10,000
A_2	9,000	4,000	9,000	15,000	15,000	
A_3	4,000	6,000	12,000	8,500	12,000	
A_4	2,000	8,000	3,000	11,000	11,000	

choice is to select the alternative that affords the opportunity to obtain the minimum loss value given in the matrix. A formal statement for the *minimin principle* involving *loss* values is

Select the alternative, j, associated with the min min $V(\theta_{jk})$.
$$\quad\quad\quad\quad\quad\quad\quad\quad\quad\quad\quad\quad\quad\quad\quad\quad j \quad\quad k$$

Applying the maximax principle to the example of Table 8.4 selects alternative A_3, and applying the minimin principle to the example of Table 8.5 selects alternative A_4.

8.5.4 Hurwicz Principle

The *Hurwicz principle* considers that a decision maker's view may, in the case of gains, fall between the extreme pessimism of the maximin principle and the extreme optimism of the maximax principle and offers a method by which various levels of optimism-pessimism may be incorporated into the decision. The Hurwicz principle defines an index of optimism, α, on a scale from 0 to 1. A value of $\alpha = 0$ thus indicates zero optimism or extreme pessimism and, conversely, a value of $\alpha = 1$ indicates extreme optimism.

Assuming that a decision maker is able to reflect a degree of optimism by assigning a specific value to α, and again emphasizing that the decision is in terms of *gains*, the maximum gain value for each alternative is multiplied by α and the minimum gain value for each alternative is multiplied by $(1-\alpha)$. The linear sum of these two products for each alternative j is called the Hurwicz value, H_j, and the alternative that maximizes this value is selected. That is,

Select an index of optimism, α, such that $0 \le \alpha \le 1$. For each alternative j, compute $H_j = \alpha [max\ V(\theta_{jk})] + (1-\alpha) [min\ V(\theta_{jk})]$ and select the alternative
$$\quad\quad\quad\quad\quad\quad\quad\quad\quad\quad k \quad\quad\quad\quad\quad\quad\quad\quad\quad\quad\quad k$$
that maximizes this quantity.

Note that if $\alpha = 0$ (extreme pessimism), only the minimum gain for each alternative will be examined, and the maximum of these values would then be selected—the maximin principle of choice. If $\alpha = 1$ (extreme optimism), the maximax principle would be executed.

Suppose the decision maker concerned with the example problem of Table 8.4 was a middle-of-the-road type of person and assigns $\alpha = 0.5$. Then

$$H_1 \text{ for } A_1 = (0.5)(\$15,163) + (0.5)(\$9742) = \$12,452.50$$
$$H_2 \text{ for } A_2 = (0.5)(\$16,536) + (0.5)(\$7049) = \$11,792.50$$
$$H_3 \text{ for } A_3 = (0.5)(\$18,397) + (0.5)(\$5679) = \$12,038.00$$

Choosing the maximum of these values is to select alternative A_1.

Shortcomings of the Hurwicz principle include (1) ignoring intermediate values for each alternative, (2) the inability to select a particular alternative when two or more alternatives have the same Hurwicz value, and (3) the practical difficulty of assigning a specific value to α. This latter objection can be circumvented by the graphic solution discussed below.

It will be noted from the expression for calculating the Hurwicz value that the equation is linear in α, which ranges in value from 0 to 1.0. Any alternative thus yields a linear function that can be plotted and a preferred alternative determined for various ranges of α. An example to illustrate the procedure is given in Table 8.6, where the cell values are gains.

TABLE 8.6 Data for the Hurwicz
Example

	S_1	S_2	S_3
A_1	$ 2	$10	$30
A_2	18	18	18
A_3	24	12	10

Applying the principle for the data of Table 8.6, the Hurwicz values are

$$H_1 \text{ for } A_1 = \alpha(\$30) + (1 - \alpha)(\$2) = 28\alpha + 2$$
$$H_2 \text{ for } A_2 = \alpha(\$18) + (1 - \alpha)(\$18) = 18$$
$$H_3 \text{ for } A_3 = \alpha(\$24) + (1 - \alpha)(\$10) = 14\alpha + 10$$

These equations are plotted in Figure 8.2.

From Figure 8.2 it can be seen that if a decision maker's index of optimism is less than α_1, alternative A_2 would be preferred; if it is greater than α_1, alternative A_1 would be preferred. A specific value for α_1 can be determined graphically or analytically by, in this case, equating the Hurwicz value functions for any pair of alternatives (since there is a single intersection for the three linear functions). For example,

$$H_1 = H_2$$
$$28\alpha_1 + 2 = 18 \qquad \text{and} \qquad \alpha_1 = 4/7$$

If the decision problem involves costs or losses, a reinterpretation of the Hurwicz principle is necessary:

Select an index of optimism, α, such that $0 \le \alpha \le 1$. For each alternative, A_j, compute $H_j = \alpha[\min_k V(\theta_{jk})] + (1 - \alpha) [\max_k V(\theta_{jk})]$ and select the alternative that minimizes this quantity.

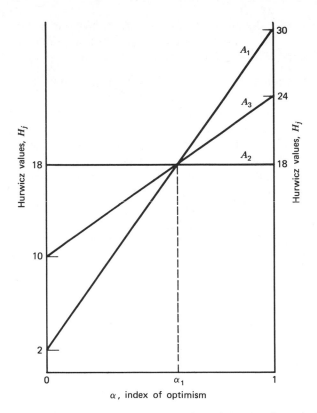

FIGURE 8.2. Graphical interpretation of the Hurwicz principle.

Note that if $\alpha = 0$, only the maximum costs would be examined for each alternative and the minimum of these would be selected—the pessimistic minimax criterion. If $\alpha = 1$, the optimistic minimin principle would be executed.

8.5.5 Savage Principle (Minimax Regret)

This principle, proposed by L. J. Savage, introduces and defines a quantity termed *regret*. A matrix consisting of regret values is first developed. Then the maximum regret value for each alternative A_j is determined, and the alternative associated with the minimum regret value is chosen from the set of maximum regret values.

If the original decision matrix values are *gains*, the procedure for determining the regret matrix is as follows.

1. For a given state S_l, search the matrix column values over all alternatives and determine the largest gain. Assign this gain a zero regret value.

2. For all other gain values *under* the given state S_l, subtract these from the largest gain value from step 1. The difference is interpreted as units of regret for a particular alternative A_j, given that state S_l occurs.

3. Repeat steps 1 and 2 for each state S_k, $k \neq l$, until the regret matrix is completed.

Once the regret matrix is developed, examine each alternative over all states and select the maximum regret value. Then *choose the alternative that minimizes the maximum regret.* This principle of choice is conservative and is very similar to the maximin principle. It is emphasized, however, that the Savage principle deals in units of regret, not gain values in the original matrix; therefore, if both the maximin principle and the minimax regret principle are applied to a common decision problem, different alternatives may be selected.

Applying the Savage principle to the present-worth example of Table 8.4, the regret matrix given in Table 8.7 is obtained.

Table 8.7 Regret Matrix for the Minimax Regret Example

	S_1	S_2	S_3
A_1	\$3234	\$0	\$0
A_2	1861	1028	2693
A_3	0	1122	4063

The maximum regret values are \$3234, \$2693, and \$4063 for alternatives A_1, A_2, and A_3, respectively. Thus, the minimum of these is \$2693, and alternative A_2 would be preferred.

In a decision problem involving costs, the "logic" for creating the regret matrix is the same as the problem involving gains.

1. For a given state S_l, search the matrix column values over all alternatives and determine the smallest cost. Assign this cost a zero regret value.

2. For all other cost values *under* the given state S_l, subtract the smallest cost value determined in step 1 from these to determine regret values.

3. Repeat steps 1 and 2 for each state S_k, $k \neq l$, until the regret matrix is completed.

After the regret matrix is obtained, the procedural steps for selecting an alternative are the same as before.

8.6 SEQUENTIAL DECISIONS

In the previous sections, the decision situations happened once. We now wish to consider situations that may require multiple decisions in sequential fashion. The reader can no doubt recall from personal experience or imagine real-world decisions that are sequential. For example, medical decisions are typically sequential. Based on the results of a routine blood analysis, a physician may require additional medical tests. The physician may then perform exploratory surgery and, based on these findings, recommend additional surgery for the patient. Following surgery, the recuperative procedure may also be conditional and involve further sequential decisions on behalf of the physician.

Sequential decisions in the business and industrial world are also common. For example, the number and type of material-handling units to purchase may depend on the forecast of product demand in each of the next five years. In the field of quality control, sampling is concerned with sequential decisions. For example, if a sample of five units is taken from a production machine, subsequent actions taken in regard to the production machine might very well depend on whether 0, 1, 2, 3, 4, or 5 defective units were found in the sample of five units.

In this section, sequential decisions under risk will be discussed, the technique of *decision-tree* representation will be used, and the logic of Bayes's theorem will be applied in the solution of the decision-tree problem.

8.6.1 Decision Trees

Magee [11, 12] is generally credited with representing sequential decisions in decision-trees and advocating the expectation criterion for solving the problem. Hespos and Strassman [8] developed stochastic decision trees incorporating risk-analysis methods and simulation techniques into the basic decision-tree methodology originated by Magee. Before presenting the graphic technique and a solution procedure, it is necessary to define certain symbolism.

Δ = a decision point, which will be labeled D_i for the ith decision point in the tree

Δ—— = a branch emanating from a decision point; represents an alternative that can be chosen at this point

0 = a fork (or node) in the tree where chance events influence the outcomes of an alternative choice

0—— = a branch representing a probabilistic outcome for a given alternative. The notation $(\theta; p)$ represents an outcome θ, having an associated value $V(\theta; p)$, which occurs with probability p. It is assumed that branches emanating from a fork in the tree represent mutually exclusive and collectively exhaustive outcomes such that their probabilities sum to 1.0

V = a value associated with a particular outcome (branch)

In Figure 8.3, a symbolic decision tree for a simple sequential decision problem is presented. The sequential decision problem depicted involves two decision points, D_1 and D_2. If alternative A_1 is chosen, it can have two outcomes, θ_1 and θ_2. If θ_1 occurs, then a decision, D_2, is required. If alternative A_{11} is chosen, there can be the two outcomes θ_{111} and θ_{112}. If alternative A_{12} is chosen at D_2, there can be the two outcomes θ_{121} and θ_{122}. For each branch in the tree there is an associated value. For example, if alternative A_1 is chosen *and* outcome θ_1 occurs *and* alternative A_{11} is chosen *and* outcome θ_{111} occurs, than value V_{111} results with the conditional probability p_{111}. The first decision to be made is at decision point D_1, and if alternative A_1 is chosen, then a second decision is required at decision point D_2. The ultimate question to answer is which alternative to choose at D_1?

The principle of *maximizing expected gain* or *minimizing expected loss* is

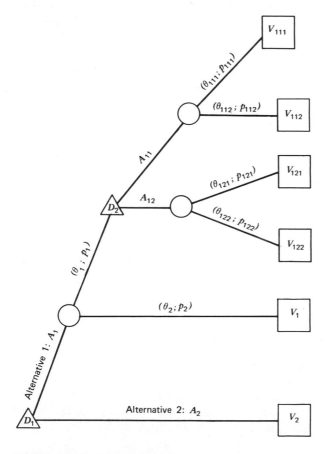

FIGURE 8.3. A symbolic decision tree.

adopted as the principle of choice at each decision point. The solution procedure advocated is to reach the best decision at the decision point (or points) most distant from the base (the first decision) of the tree. Then, replace this most distant decision point with the best expected value and work backward through decision points until the best decision is made at the initial decision point, D_1. To illustrate the solution procedure, consider the decision tree of Figure 8.3 and assume all the values are *gains*. The procedure follows.

1. At decision point D_2, calculate the expected gains for A_{11} and A_{12}. That is,

$$E(A_{11}) = p_{111}V_{111} + p_{112}V_{112}$$

and

$$E(A_{12}) = p_{121}V_{121} + p_{122}V_{122}$$

If $E(A_{11}) > E(A_{12})$, then alternative A_{11} is chosen as best at D_2. Assume this is the case.

2. Replace D_2 with $E(A_{11})$, and the new reduced decision tree becomes as shown in Figure 8.4.

3. Calculate the expected gains for alternatives A_1 and A_2. That is,

$$E(A_1) = p_1 E(A_{11}) + p_2 V_1$$

and

$$E(A_2) = V_2 \text{(a certain event with } p = 1.0\text{).}$$

If $E(A_1) > E(A_2)$, alternative A_1 is chosen as the best at decision point D_1. If $E(A_2) > E(A_1)$, then A_2 would, of course, be chosen, and if $E(A_2) = E(A_1)$, then one would be indifferent between the choice of A_1 and A_2. Given that $E(A_1) > E(A_2)$ from step 3, the optimal sequence of decisions is to choose A_1 at D_1 and then choose A_{11} later on if outcome θ_1 does occur.

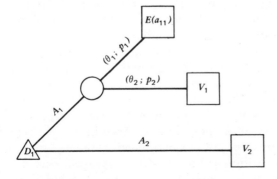

FIGURE 8.4. The reduced decision tree for Figure 8.3.

Example 8.5 ————————————————————————————

The Jax Tool and Engineering Company is a medium-size, job-order machine shop and has been in business for 10 years. After surviving two national economic recessions with shrewd and industrious management, the company has built a solid reputation for dependability and high-quality work. During the past two years, sales have sharply increased, and the company is operating near capacity. A market test has recently been made for a new product—electroplating tanks—and the response has been very favorable. However, production floor space to establish the new product line is limited in the present building. The people in management have three alternatives: (1) do nothing, (2) rearrange an area of the existing floor space, or (3) build an addition to the present plant. Also, there is the consideration of competition from one other job-order machine shop in the local area. If the electroplating tank venture is successful and the market demand cannot be fully met by the Jax Company, there is a good chance that competition will occur.

Management assigns representatives from production engineering and marketing to make a detailed study of the alternatives. Because management desires a report as soon as possible, it is necessary to make rather gross estimates on many items in the analysis. The analysts decide to use a five-year planning horizon, a minimum attractive rate of return of 30%, and an incremental present worth value for each alternative relative to the do-nothing alternative. The results of the analysis now follow.

Rearrangement does not require any new construction, but machinery must be moved, some new machinery purchased, and new storage areas created. Production would be limited to 10 tanks/week on a two-shift operation. If there is competition in the first year from the other local firm, the Jax Company would probably take no further action, and a present worth value of $50,000 is estimated for this outcome. On the other hand, if there is no competition in the first year, the Jax Company will not be able to meet the expected demand. Subjective probabilities of 0.4 and 0.6 are assigned to the outcomes of no competition of the first year and competition in the first year, respectively. If no competition, Jax management would then face the decision of either expanding the product line by building an addition or not expanding.

Expansion at this point would be expensive; a building addition plus a layout of the existing plant would be involved. However, no competition is expected if the expansion is made now and demand can be met with increased capacity. A present value of $30,000 is estimated for this outcome. If there is no expansion, two outcomes are possible. One possibility is that the other local machine shop can still enter the competition. However, a one-year advantage in the market will have been gained, and the present worth of this outcome is estimated to be $10,000. It is very unlikely that competition will occur after a two-year period, and the probability of such an event is assumed to be zero. A subjective probability value of 0.7 is assigned to the outcome of competition in the second year. The other possible outcome is, of course, no competition in the second year, with a probability of 0.3 and estimated present worth of $50,000.

The "build addition initially" alternative has higher first and recurring costs than the "new layout" alternative, but production capacity is estimated at 25 tanks/week, which should satisfy average annual demand over the next five years. With adequate production capacity initially, the chance of competition is low and about 20%. The present worth estimate in the event of no competition is $90,000. If there is competition, then the reduced sales and high annual fixed costs for the new building would result in a present worth of −$100,000. A decision-tree representation of this example is given in Figure 8.5 (values are in thousands of dollars).

Using the principle of choice of maximizing expected present worth, a best decision is sought at D_2, the decision point most distant from the base of the tree, D_1. The expected values at D_2 are

$$E(\text{expansion}) = \$30,000$$

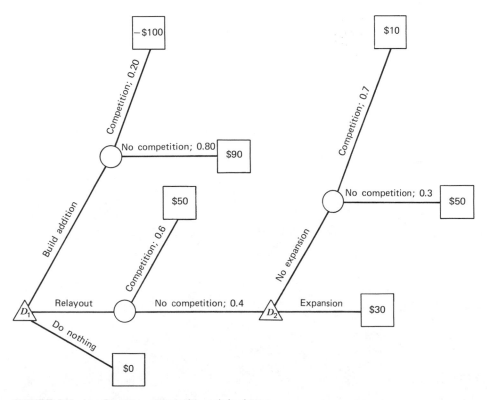

FIGURE 8.5. Jax Company example—original tree.

and

$$E(\text{no expansion}) = \$10,000(0.7) + \$50,000(0.3)$$
$$= \$22,000$$

Thus, the alternative of expansion is chosen to maximize expected present worth, and this *best* value replaces D_2 in the tree, as shown in Figure 8.6. From Figure 8.6, the expected values at D_1 are calculated as:

$$E(\text{build addition}) = -\$100,000(0.20) + \$90,000(0.80)$$
$$= \$52,000$$

$$E(\text{rearrange}) = \$50,000(0.6) + \$30,000(0.4)$$
$$= \$42,000$$

$$E(\text{do nothing}) = \$0$$

Thus, the alternative of building an addition should be chosen in order to maximize expected present worth. It is interesting to note from Figure 8.6 that if

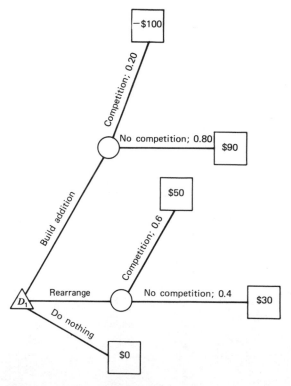

FIGURE 8.6. Jax Company example—reduced tree.

this alternative is indeed chosen, the Jax Company is exposed to the possibility of a −$100,000 present worth. This event is "smoothed" by the expectation principle, on which most decision-tree analysis is based. The −$100,000 figure is not necessarily a dollar loss, since a $MARR = 30\%$ was assumed in the calculation of the values. However, management may wish to choose the "rearrange" alternative instead because of the large negative figure; if so, they would be exercising a maximin philosophy.

Example 8.6

For further illustration of the decision-tree approach to a problem and exposure to fundamental laws of probability, consider an urn containing five white (W) balls and seven black (B) balls. Three consecutive, independent, random (without bias in the selection) draws are made from the urn without replacing any ball drawn. The probability of selecting a white or black ball on the first draw is unconditional; that is, it does not depend on a previous draw (event). The probability of selecting a white or black ball on the second draw is, however, conditional on the color of ball selected on the first draw, and the third draw is conditional on both the first and second draws. In this example, we are interested in all the possible sequences of colors for the three draws and their probabilities of occurrence. A decision-tree representation of this problem is helpful and given in Figure 8.7.

All the eight possible sequences of colors are indicated in Figure 8.7. The objective probabilities of occurrence for each sequence will be determined by a relative-frequency argument. That is, for a general event A, the probability of A, $P(A)$, is calculated as the ratio

$$P(A) = \frac{\text{number of ways favorable for the event}}{\text{total number of ways possible}}$$

In the urn example, the probability of the unconditional event "a white ball on the first draw," W_1, is given by

$$P(W_1) = \frac{5 \text{ white balls (favorable number)}}{12 \text{ balls (total number)}}$$

Similarly, the probability of the unconditional event "a black ball on the first draw," B_1, is

$$P(B_1) = \frac{7}{12}$$

For the second draw, the probability of the event "a white ball on the second draw," W_2, is conditional on whether the first draw results in a white or black ball [i.e., $P(W_2|W_1)$ or $P(W_2|B_1)$]. The probability that a white ball is

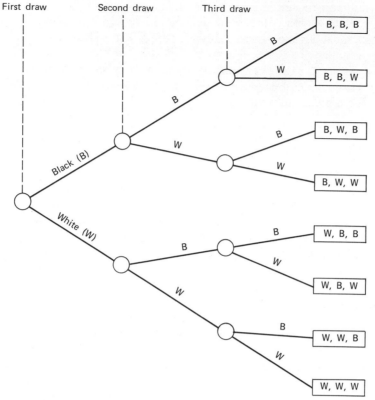

First draw Second draw Third draw

B, B, B
B, B, W
B, W, B
B, W, W
W, B, B
W, B, W
W, W, B
W, W, W

Black (B)
White (W)

FIGURE 8.7. Tree representation of urn example.

selected on the second draw given that a white ball is selected on the first draw is reasoned to be

$$P(W_2|W_1) = \frac{4 \text{ (white balls remaining after first draw)}}{11 \text{ (total balls remaining after second draw)}}$$

Similarly,

$$P(W_2|B_1) = \frac{5}{11}$$

The probability of the event "a black ball on the second draw," B_2, is also conditional on whether the first draw results in a white or black ball, and

$$P(B_2|W_1) = \frac{7}{11}$$

$$P(B_2|B_1) = \frac{6}{11}$$

From the above results, note that

$$P(W_2|W_1) + P(B_2|W_1) = 1.0$$

and

$$P(W_2|B_1) + P(B_2|B_1) = 1.0$$

If a white (or black) ball is in fact drawn first, then the second draw can only yield two possible outcomes (collectively exhaustive): either a white or black ball. Since these two outcomes are mutually exclusive (cannot occur together), the individual probabilities of occurrence are additive and sum to 1.0 or certainty. In general, for two mutually exclusive events A and B, the additive law of probability states

$$P(\text{either } A \text{ or } B) = P(A \cup B) = P(A) + P(B) \qquad (8.4)$$

where the symbol ($\cup$) stands for the union of the events. Thus, in this example,

$$P(W_2|W_1 \cup B_2|W_1) = P(W_2|W_1) + P(B_2|W_1)$$

and

$$P(W_2|B_1 \cup B_2|B_1) = P(W_2|B_1) + P(B_2|B_1)$$

The additive law is easily expanded for multiple, mutually exclusive events.

For the third draw, if the ball selected on each of the first and second draws is white, then the probability of selecting a white ball on the third draw is denoted $P(W_3|W_1, W_2)$ and reasoned, by the relative-frequency logic, to be

$$P(W_3|W_1, W_2) = \frac{3 \text{ (white balls remaining after first and second draw)}}{10 \text{ (total balls remaining after first and second draw)}}$$

The symbolism $(W_3|W_1, W_2)$ is interpreted to mean the event of drawing a white ball on the third draw given that a white ball was selected on the first draw *and* a white ball was selected on the second draw.
Similarly,

$$P(B_3|W_1, W_2) = \frac{7}{10}$$

$$P(W_3|W_1, B_2) = \frac{4}{10}$$

$$P(B_3|W_1, B_2) = \frac{6}{10}$$

$$P(W_3|B_1, W_2) = \frac{4}{10}$$

$$P(B_3|B_1, W_2) = \frac{6}{10}$$

$$P(W_3|B_1, B_2) = \frac{5}{10}$$

$$P(B_3|B_1, B_2) = \frac{5}{10}$$

A summary of all these events and their probabilities is shown in Figure 8.8.

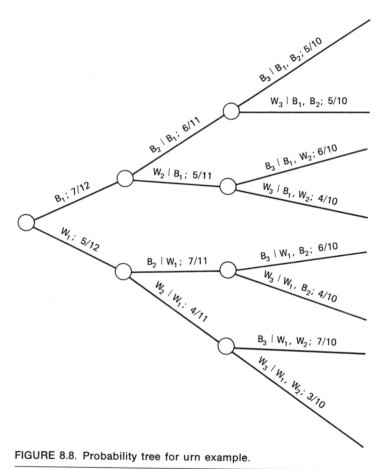

FIGURE 8.8. Probability tree for urn example.

Up to this point the probabilities of each individual outcome (event) of the example have been determined. However, the probability of each sequence (branch of the tree) occurring is yet to be determined (see Figure 8.7 to recall the possible sequences of colors). For instance, it is desired to know the probability of the joint event of a white ball being selected with each of the three draws or $P(W_1, W_2, W_3)$, which may also be written with set notation as $P(W_1 \cap W_2 \cap W_3)$ to mean the joint occurrence of these three events (conditional events in this particular case). This probability may be calculated as

$$P(W_1, W_2, W_3) = P(W_1) \cdot P(W_2|W_1) \cdot P(W_3|W_1, W_2)$$

$$= \left(\frac{5}{12}\right)\left(\frac{4}{11}\right)\left(\frac{3}{10}\right) = \frac{60}{1320}$$

A full justification of the above multiplicative calculation requires a probability theory development that is outside the scope of this textbook. However, the calculation of the joint occurrence of three conditional events is a direct extension of the conditional probability theorem presented later in this chapter. For the other possible sequences of this example,

$$P(W_1, W_2, B_3) = P(W_1) \cdot P(W_2|W_1) \cdot P(B_3|W_1, W_2)$$

$$= \left(\frac{5}{12}\right)\left(\frac{4}{11}\right)\left(\frac{7}{10}\right) = \frac{140}{1320}$$

$$P(W_1, B_2, W_3) = P(W_1) \cdot P(B_2|W_1) \cdot P(W_3|W_1, B_2)$$

$$= \left(\frac{5}{12}\right)\left(\frac{7}{11}\right)\left(\frac{4}{10}\right) = \frac{140}{1320}$$

$$P(W_1, B_2, B_3) = P(W_1) \cdot P(B_2|W_1) \cdot P(B_3|W_1, B_2)$$

$$= \left(\frac{5}{12}\right)\left(\frac{7}{11}\right)\left(\frac{6}{10}\right) = \frac{210}{1320}$$

$$P(B_1, B_2, B_3) = P(B_1) \cdot P(B_2|B_1) \cdot P(B_3|B_1, B_2)$$

$$= \left(\frac{7}{12}\right)\left(\frac{6}{11}\right)\left(\frac{5}{10}\right) = \frac{210}{1320}$$

$$P(B_1, B_2, W_3) = P(B_1) \cdot P(B_2|B_1) \cdot P(W_3|B_1, B_2)$$

$$= \left(\frac{7}{12}\right)\left(\frac{6}{11}\right)\left(\frac{5}{10}\right) = \frac{210}{1320}$$

$$P(B_1, W_2, B_3) = P(B_1) \cdot P(W_2|B_1) \cdot P(B_3|B_1, W_2)$$

$$= \left(\frac{7}{12}\right)\left(\frac{5}{11}\right)\left(\frac{6}{10}\right) = \frac{210}{1320}$$

$$P(B_1, W_2, W_3) = P(B_1) \cdot P(W_2|B_1) \cdot P(W_3|B_1, W_2)$$

$$= \left(\frac{7}{12}\right)\left(\frac{5}{11}\right)\left(\frac{4}{10}\right) = \frac{140}{1320}$$

The sum of the probabilities for each of these eight possible sequences is $1320/1320 = 1.0$, and the complete probability tree is shown in Figure 8.9.

8.6.2 The Conditional Probability Theorem

The probability of the joint occurrence of two dependent events, A and B, is given by

$$P(A \text{ and } B) = P(A \cap B) = P(A) \cdot P(B|A) \tag{8.6}$$

where $P(A)$ is an unconditional probability and $P(B|A)$ is the conditional probability of event B given that event A has occurred or will occur. If event B is not conditional on event A, then events A and B are *independent* events and Equation 8.6 may be written as

$$P(A \cap B) = P(A) \cdot P(B) \tag{8.7}$$

An extension of Equation 8.6 for three dependent events is given by

$$P(A \cap B \cap C) = P(A) \cdot P(B|A) \cdot P(C|A, B) \tag{8.8}$$

which was the relationship used in the previous urn example.

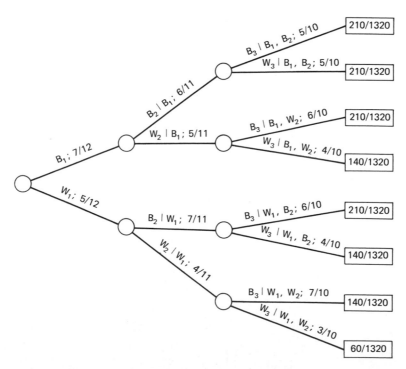

FIGURE 8.9. Complete probability tree for urn example.

358 • DECISION MODELS

For the purpose of the development to follow, Equation 8.6 is now written in its complete form as

$$P(A \cap B) = P(A) \cdot P(B|A) = P(A|B) \cdot P(B) \qquad (8.9)$$

from which it follows that

$$P(B|A) = \frac{P(A|B)P(B)}{P(A)} \qquad (8.10)$$

or

$$P(A|B) = \frac{P(B|A)P(A)}{P(B)} \qquad (8.11)$$

Example 8.7 ───

Either Equation 8.10 or 8.11 is a form of Bayes's theorem, which may provide a logical guide to sequential decision making under risk. To illustrate this claim, let us consider another urn example where one urn, U_1, contains two white and eight black balls; the other urn, U_2, contains five white and five black balls. Suppose the urn contents are known to you, but the urns themselves are hidden from your view. Now, another person selects a single ball from one of the urns. Without knowing the color of the ball drawn, which urn would you guess was selected? Then, if the person reported a black ball had been drawn, which urn would you guess had been selected?

Without benefit of the sample information concerning the color of the ball drawn, most persons would feel that it was equally likely that U_1 or U_2 had been chosen. That is, since $P(U_1) = P(U_2) = 1/2$, guess either urn. However, after the sample information of a black ball is received, intuition suggests that U_1 was the urn selected because the probability of a black ball in U_1 is greater than the probability of a black ball in U_2. We now wish to formalize this example and show that Bayes's theorem can be used to support the intuitive guess of U_1 after receiving the report.

Before receiving the report of a black ball, the prior belief that U_1 or U_2 was selected is subjectively stated as

$$P(U_1) = \text{the } prior \text{ probability of selecting } U_1$$
$$= 1/2$$

and

$$P(U_2) = \text{the } prior \text{ probability of selecting } U_2$$
$$= 1/2$$

Note that

$$W = \text{the event of a white ball being drawn}$$

and

$$B = \text{the event of a black ball being drawn}$$

Also, before receiving the sample information, the likelihood of a particular report (W or B), given that U_1 or U_2 was selected, can be determined, based on knowledge of the contents of each urn. That is,

$P(W|U_1) =$ the conditional probability of a white ball being drawn given that U_1 was selected for the draw

$$= \frac{2}{10}$$

Similarly,

$$P(B|U_1) = \frac{8}{10}$$

$$P(W|U_2) = \frac{5}{10}$$

$$P(B|U_2) = \frac{5}{10}$$

In the literature on sequential decision making, these conditional probabilities, either objectively known as above or subjectively assigned, are termed *likelihood statements*.

Ultimately, it is desired to assess the probability of whether U_1 or U_2 was chosen based on the sample information received (the report of a black ball in this case). Thus, the *posterior probabilities* of $P(U_1|B)$, $P(U_2|B)$, $P(U_1|W)$, and $P(U_2|W)$ need to be determined.

In the calculation of these posterior probabilities, it is convenient to calculate the probability of event W and event B before the ball is drawn. It is reasoned that event W could occur if either U_1 or U_2 were chosen. The same is true for the event B. Thus, B could occur if (1) U_1 were selected and B occurred, or (2) U_2 were selected and B occurred. That is,

$$P(B) = P(U_1 \text{ and } B \text{ or } U_2 \text{ and } B)$$
$$= P(U_1 \cap B \cup U_2 \cap B)$$
$$= P(U_1 \cap B) + P(U_2 \cap B)$$

Then, using Equation 8.9,

$$P(B) = P(U_1)P(B|U_1) + P(U_2)P(B|U_2)$$

and, from the previous data,

$$P(B) = \left(\frac{1}{2}\right)\left(\frac{8}{10}\right) + \left(\frac{1}{2}\right)\left(\frac{5}{10}\right) = \frac{13}{20}$$

Similarly,

$$P(W) = P(U_1)P(W|U_1) + P(U_2)P(W|U_2)$$

$$= \left(\frac{1}{2}\right)\left(\frac{2}{10}\right) + \left(\frac{1}{2}\right)\left(\frac{5}{10}\right) = \frac{7}{20}$$

With these results, the posterior probabilities may be calculated from Bayes's theorem as

$$P(U_1|B) = \frac{P(B|U_1)P(U_1)}{P(B)} = \frac{(8/10)(1/2)}{13/20} = \frac{8}{13}$$

$$P(U_2|B) = \frac{P(B|U_2)P(U_2)}{P(B)} = \frac{(5/10)(1/2)}{13/20} = \frac{5}{13}$$

$$P(U_1|W) = \frac{P(W|U_1)P(U_1)}{P(W)} = \frac{(2/10)(1/2)}{7/20} = \frac{2}{7}$$

$$P(U_2|W) = \frac{P(W|U_2)P(U_2)}{P(W)} = \frac{(5/10)(1/2)}{7/20} = \frac{5}{7}$$

It is noted from the results of $P(U_1|B) = 8/13$ and $P(U_2|B) = 5/13$ that the event of B should therefore increase the belief that U_1 was selected for taking the sample if one is to be consistent in a probabilistic sense. That is, the prior probability of U_1 was 1/2 and the posterior probability, given the event B, of U_1 is 8/13. If, after receiving the information of B, the guess had been U_1, then the above calculations would support the intuitive guess. If a second sample of a ball is taken from an urn, then the posterior probabilities $P(U_1|B)$ and $P(U_2|B)$ become the prior probabilities $P(U_1)$ and $P(U_2)$ for the second trial.

8.6.3 The Value of Perfect Information

In order to use the logic of Bayes's theorem in a more practical way to aid decision making, it is of interest to evaluate the value of the sample information in a monetary sense. Recalling the previous example, suppose the person who drew the ball from the urn had offered to pay a $10 prize if the correct urn were guessed. However, the person will charge for the sample information of whether a white or black ball is drawn. The question then is how much should be paid for the *sample information*. Before answering this particular question, a rationale for determining the *value of perfect information* is considered (i.e., the knowledge of exactly which urn was selected for the draw).

The *prior* probabilities of $P(U_1) = P(U_2) = 1/2$ are now recalled and written with the more suggestive notation of $PR(U_1) = PR(U_2) = 1/2$ [16]. Without any sample information, one should be indifferent as to which urn is guessed and the *prior expected profit* (*PEP*) would be $5, as determined from the matrix of Table 8.8.

Table 8.8 Matrix for the Urn Example

		$PR(U_1) = 1/2$	$PR(U_2) = 1/2$
Actual Urn Selected Urn Guessed		U_1	U_2
	U_1	$10	$ 0
	U_2	$ 0	$10

The prior expected profit is $5, as determined from either of the two calculations below.

$$E(\text{profit}|\text{guess } U_1) = \left(\frac{1}{2}\right)(\$10) + \left(\frac{1}{2}\right)(\$0) = \$5$$

$$E(\text{profit}|\text{guess } U_2) = \left(\frac{1}{2}\right)(\$0) + \left(\frac{1}{2}\right)(\$10) = \$5$$

If these expected profits were different, the alternative with the larger expected value would be chosen, and the prior expected profit is the larger value.

Now, given that the person drawing the ball states that U_1 was selected and that the person is perfectly reliable, U_1 would obviously be guessed and a $10 payoff received. The same is true, of course, for the report of U_2. The *expected profit given perfect information* (*EP|PI*) is thus $10 in this example. It is now reasoned that the maximum amount one should pay for perfect information is $10 − $5 (expected profit without any information) = $5. We can formalize the *expected value of perfect information* (*EVPI*) as the difference between the *expected profit given perfect information* (*EP|PI*) and the *prior expected profit* (*PEP*), or

$$EVPI = EP|PI - PEP \tag{8.12}$$

8.6.4 The Value of Imperfect Information

In most real-world decision situations, information is incomplete and not perfectly reliable, as in the urn example just discussed. The report that a black ball had been drawn gave additional information to the decision maker, but still did not provide the absolute answer as to which urn was selected. In order to define the *expected value of sample information (EVSI)*, the posterior probability (with new notation) results are repeated below.

For the event of B, the posterior probabilities were

$$PO(U_1|B) = \frac{8}{13}$$

and

$$PO(U_2|B) = \frac{5}{13}$$

Thus, since $8/13 > 5/13$, the best guess is that U_1 was selected. The actual urn chosen could have been U_1 or U_2, and the expected profit given a guess of U_1 is

$$E(\text{profit}|\text{guess } U_1) = (\text{expected profit}|\text{guess } U_1 \text{ and is } U_1)PO(U_1)$$
$$+ (\text{expected profit}|\text{guess } U_1 \text{ and is } U_2)PO(U_2)$$

$$= (\$10) \left(\frac{8}{13}\right) + (\$0) \left(\frac{5}{13}\right)$$

$$= \frac{\$80}{13}$$

For the event of W, the posterior probabilities were

$$PO(U_1|W) = \frac{2}{7}$$

$$PO(U_2|W) = \frac{5}{7}$$

Thus, if the report had been W, and since $5/7 > 2/7$, the best guess would be U_2 with

$$E(\text{profit}|\text{guess } U_2) = (\text{expected profit}|\text{guess } U_2 \text{ and is } U_1)PO(U_1)$$
$$+ (\text{expected profit}|\text{guess } U_2 \text{ and is } U_2)PO(U_2)$$

$$= (\$0) \left(\frac{2}{7}\right) + (\$10) \left(\frac{5}{7}\right)$$

$$= \frac{\$50}{7}$$

However, *before* the sample is taken and the report is given, both events B and W are possible with $P(B) = 13/20$ and $P(W) = 7/20$. The *expected profit* given sample information $(EP|SI)$ is

$$EP|SI = (\text{expected profit}|B)P(B) + (\text{expected profit}|W)P(W)$$

$$= \left(\frac{\$80}{13}\right)\left(\frac{13}{20}\right) + \left(\frac{\$50}{7}\right)\left(\frac{7}{20}\right)$$

$$= \$6.50$$

In the above calculations, both expected profit values (expected profit|B) and (expected profit|W) assume that the best guess would, in fact, be made for the reports of B and W, respectively.

Finally, it is reasoned that the *expected value of sample information* is the difference between the *expected value given sample information* and the *prior expected profit*, or

$$EVSI = EP|SI - PEP$$
$$= \$6.50 - \$5 \qquad\qquad (8.13)$$
$$= \$1.50$$

which is the maximum amount one should pay for the sample information of this example. This value is, of course, less than the $5 value for perfect information.

A decision-tree representation of this example is given in Figure 8.10 for a sample information charge of $1.50.

From Figure 8.10, the appropriate calculations at D_2 are

$$E(\text{guess } U_1) = (\$8.50)\left(\frac{8}{13}\right) + (-\$1.50)\left(\frac{5}{13}\right)$$

$$= \frac{\$60.5}{13}$$

$$E(\text{guess } U_2) = (-\$1.50)\left(\frac{8}{13}\right) + (\$8.50)\left(\frac{5}{13}\right)$$

$$= \frac{\$30.5}{13}$$

Since $60.5/13 > $30.5/13$, the best guess given B would be U_1.

The appropriate calculations at D_3 are

$$E(\text{guess } U_1) = (\$8.50)\left(\frac{2}{7}\right) + (-\$1.50)\left(\frac{5}{7}\right)$$

$$= \frac{\$6.5}{7}$$

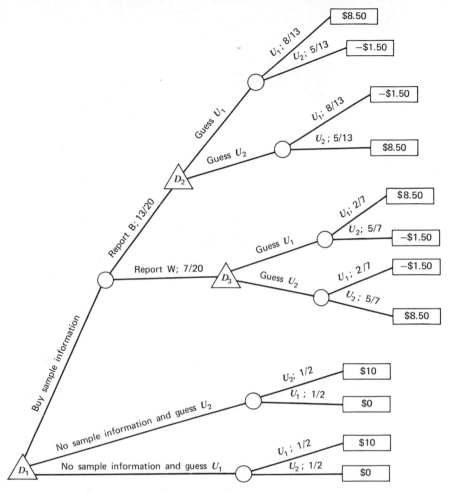

FIGURE 8.10. Decision tree for urn example.

$$E(\text{guess } U_2) = (-\$1.50) \left(\frac{2}{7}\right) + (\$8.50) \left(\frac{5}{7}\right)$$

$$= \frac{\$39.5}{7}$$

Since $\$39.5/7 > \$6.5/7$, the best guess given W would be U_2.

Replacing D_2 and D_3 with these best expected values yields the reduced decision tree of Figure 8.11. From Figure 8.11, the appropriate calculations at D_1 are

$$E(\text{buy sample information}) = \left(\frac{\$60.5}{13}\right)\left(\frac{13}{20}\right) + \left(\frac{\$39.5}{7}\right)\left(\frac{7}{20}\right)$$

$$= \$5$$

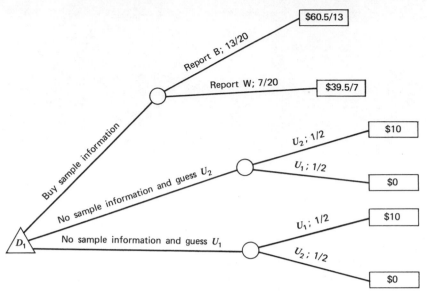

FIGURE 8.11. Reduced decision tree for urn example.

E(no sample information and guess U_2)

$$= E(\text{no sample information and guess } U_1)$$
$$= \$5$$

Thus, one is indifferent between the three alternatives at D_1 given that the charge for sample information is $1.50. If the charge is less than this amount, the decision maker should choose the "buy sample information" alternative in order to maximize expected profits.

For a similar sequential decision problem involving only costs, the basic relationships to determine expected value of information become

$EVPI$ = prior expected cost − expected cost given perfect information

$$= PEC - EC|PI \qquad (8.14)$$

$EVSI$ = prior expected cost − expected cost given sample information

$$= PEC - EC|SI \qquad (8.15)$$

Example 8.3[3]

The editor of the American Book Company is considering a manuscript for a historical novel. The editor estimates the book can be marketed for $5, with the

[3] This example was presented at the Sixth Triennial Symposium, Engineering Economy Division, ASEE, June 19–20, 1971, in a paper entitled "Introduction to Decision Theory" by Barnard E. Smith. The article appears in the publication, *Decision and Risk Analysis: Powerful New Tools For Management*, The Engineering Economist, Stevens Institute of Technology, Hoboken, N.J., and is presented here by the permission of the publisher.

company receiving $4 and the author $1. Since other publishers are interested in the manuscript, the author desires a commitment from the American Book Company in the very near future. American has already invested $300 in a preliminary review of the manuscript and a quick market survey. The editor considers this manuscript competitive with other recently published historical novels and the author, although relatively new, has other published works that have received good reviews.

The editor identifies the immediate alternatives of (1) reject the manuscript, (2) accept the manuscript, or (3) obtain an additional review of the manuscript by an expert. An additional review will cost $400, and the expert's opinion will not perfectly predict the historical novel's success in the market. The editor reasons that at best the expert can only rate the manuscript good, fair, or poor.

If the manuscript is accepted for publication, the editor speculates on the possible market outcomes; there could be (1) a low market demand, or (2) the novel could be a best-seller. Additionally, there is the question of movie rights. There may be no interest in filming the novel, but, on the other hand, a best-seller could yield $30,000 for the movie rights. Even if there is low market demand, a film company might pay $4,000 for the rights to make a low-budget, grade B film.

The American Book Company has available historical data on 60 recent manuscripts of similar books, given in Table 8.9.

Table 8.9 Sixty Manuscript Decisions/Outcomes

Decision/Outcome When Marketed	No Outside Expert Review	Outside Reviewer's Evaluation			Total
		Good	Fair	Poor	
Not published	5	0	14	12	31
Low market demand	2	0	4	3	9
Best-seller	1	2	4	0	7
Best-seller + movie	1	4	3	0	8
Low market + B movie	1	0	3	1	5
Total	10	6	28	16	60

From available sales figures, the editor can estimate reasonably well the revenues given a particular outcome for a published book. He considers the present worth of after-tax revenues over the effective life of a book using a 10% expected rate of return. For a best-seller, a present worth of $70,000 is estimated, and a low-demand publication yields an estimated $5000 present worth. Offers for movie rights, if any, are usually made one year after the manuscript is under contract, and these after-tax revenues should be added to the book sales (assume that $4000 for a grade B film and $30,000 for a "best-seller film" are appropriate figures). It costs $15,000 to publish a book.

If it is assumed the editor has a linear utility function for dollars and uses a "maximize present worth" principle for choice among alternatives, what decision should be made at the present time?

The editor faces the immediate alternatives of

A_1—reject the manuscript without any additional review by an expert
A_2—accept the manuscript without obtaining an expert's review
A_3—obtain an expert's review

The value of alternative A_1 is the $300 cost, which has already been spent on a preliminary review [i.e., $V(A_1) = \$300$]. Indeed, the $300 for a preliminary review is a sunk cost and applies whether alternative A_1, A_2, or A_3 is chosen at the present time. This $300 cost can therefore be dropped from consideration, and $V(A_1) = 0$ with this adjustment.

For alternative A_2 there could be four possible market outcomes, defined as

θ_1—low market demand

θ_2—a best-seller

θ_3—best-seller plus movie rights

θ_4—low market demand plus grade B movie

Thus, the value associated with each of these outcomes is determined as

$$V(\theta_1|A_2) = \text{revenues} - \text{publication costs}$$
$$= \$5000 - \$15,000 \qquad\qquad = -\$10,000$$
$$V(\theta_2|A_2) = \$70,000 - \$15,000 \qquad\qquad = \$55,000$$
$$V(\theta_3|A_2) = \$70,000 + \$30,000 - \$15,000 = \$85,000$$
$$V(\theta_4|A_2) = \$5000 + \$4000 - \$15,000 \qquad = -\$6000$$

For alternative A_3, the expert's evaluations could be either

Z_1—Good

Z_2—Fair

Z_3—Poor

For *each* of these evaluations, the editor must make one of two possible decisions, defined as

X_1—reject the manuscript

X_2—accept the manuscript

Given the decision of X_1 for *each* of the three possible evaluations (Z_j), the cost to the American Book Company would be the same—the additional cost of an expert's review. That is,

$$V(X_1|A_3, Z_j) = -\$400, \qquad \text{for } j = 1, 2, 3$$

Then, for the decision of X_2 associated with each of the reviewer's evaluations, there are the four possible market outcomes defined previously. The value for any *given* market outcome *and* decision X_2 is the same for all Z_j evaluations. This value is equal to the values for alternative A_2 minus \$400 for the expert's review. That is,

$$V(\theta_1|A_3, Z, X_2) = -\$10,000 - \$400 = -\$10,400$$
$$V(\theta_2|A_3, Z, X_2) = \$55,000 - \$400 \quad = \$54,600$$
$$V(\theta_3|A_3, Z, X_2) = \$85,000 - \$400 \quad = \$84,600$$
$$V(\theta_4|A_3, Z, X_2) = -\$6000 - \$400 \quad = -\$6400$$

A decision-tree representation of this example, including probability values for the various outcomes, is given by Figure 8.12, where the values are in thousands of dollars.

The probability values for the various outcomes of the decision tree given in Figure 8.12 are computed using the data of Table 8.9. First, consider alternative A_2 (accept for publication without further review). It is noted from Table 8.9 that out of the ten manuscripts that were not reviewed by an outside expert, five were published. Two of the five had low market demand, one was a best-seller, one was a best-seller plus movie rights, and another had low market demand plus the rights sold for a grade B movie. Thus, by a relative frequency logic, the following probabilities can be determined:

$$P(\theta_1|A_2) = \frac{2}{5}$$

$$P(\theta_2|A_2) = \frac{1}{5}$$

$$P(\theta_3|A_2) = \frac{1}{5}$$

$$P(\theta_4|A_2) = \frac{1}{5}$$

For the 50 manuscripts that were reviewed by an expert, 6 received a good evaluation, 28 received a fair evaluation, and 16 received a poor evaluation. Thus,

$$P(Z_1) = \frac{6}{50}$$

$$P(Z_2) = \frac{28}{50}$$

$$P(Z_3) = \frac{16}{50}$$

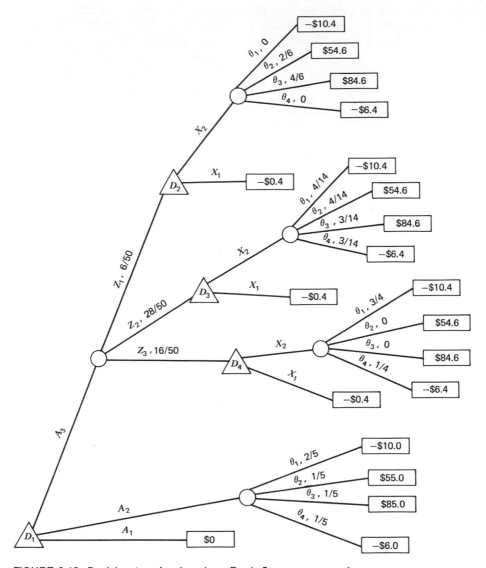

FIGURE 8.12. Decision tree for American Book Company example.

Similarly, the following conditional probabilities may be determined from the data of Table 8.9:

$$P(\theta_1|Z_1) = 0, \qquad P(\theta_1|Z_2) = \frac{4}{14}, \qquad P(\theta_1|Z_3) = \frac{3}{4}$$

$$P(\theta_2|Z_1) = \frac{2}{6}, \qquad P(\theta_2|Z_2) = \frac{4}{14}, \qquad P(\theta_2|Z_3) = 0$$

$$P(\theta_3|Z_1) = \frac{4}{6}, \qquad P(\theta_3|Z_2) = \frac{3}{14}, \qquad P(\theta_3|Z_3) = 0$$

$$P(\theta_4|Z_1) = 0, \qquad P(\theta_4|Z_2) = \frac{3}{14}, \qquad P(\theta_4|Z_3) = \frac{1}{4}$$

It is recognized that these probability values are estimates but would seem to be less gross than purely subjective evaluations made without any historical data as a basis.

The solution to the decision tree of Figure 8.12 follows. The appropriate calculations for D_2 are

$$E(X_1) = -\$0.4 = -\$400$$

$$E(X_2) = (-\$10.4)(0) + (\$54.6)\left(\frac{2}{6}\right) + (\$84.6)\left(\frac{4}{6}\right) + (-\$6.4)(0)$$

$$= \$74.6 = \$74,600$$

and the best decision at D_2 is therefore X_2 (accept the manuscript). The calculations for D_3 are

$$E(X_1) = -\$400$$

$$E(X_2) = (-\$10.4)\left(\frac{4}{14}\right) + (\$54.6)\left(\frac{4}{14}\right) + (\$84.6)\left(\frac{3}{14}\right) + (-\$6.4)\left(\frac{3}{14}\right)$$

$$= \$29.386 = \$29,386$$

and the best decision at D_3 is therefore X_2 (accept the manuscript). The calculations for D_4 are

$$E(X_1) = -\$400$$

$$E(X_2) = (-\$10.4)\left(\frac{3}{4}\right) + (\$54.6)(0) + (\$84.6)(0) + (-\$6.4)\left(\frac{1}{4}\right)$$

$$= -\$9.4 = -\$9,400$$

and the best decision at D_4 is therefore X_1 (reject the manuscript).

The reduced decision tree is given by Figure 8.13, and the appropriate calculations for D_1 are

$$E(A_3) = (\$74,600)\left(\frac{6}{50}\right) + (\$29,386)\left(\frac{28}{50}\right) + (-\$400)\left(\frac{16}{50}\right)$$

$$= \$25,280$$

$$E(A_2) = (-\$10,000)\left(\frac{2}{5}\right) + (\$55,000)\left(\frac{1}{5}\right) + (\$85,000)\left(\frac{1}{5}\right) + (-\$6000)\left(\frac{1}{5}\right)$$

$$= \$22,800$$

$$E(A_1) = \$0$$

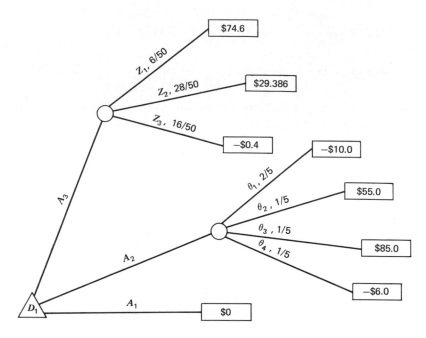

FIGURE 8.13. Reduced decision tree for American Book Company example (value in thousands of dollars).

and the best decision at D_1 in order to maximize the expected present worth is alternative A_3 (obtain an expert's opinion). Then, if the evaluation is either good or fair, accept the manuscript for publication; but if the evaluation is poor, reject the manuscript. The expected value of sample information in this example is the difference between $E(A_3)$ and $E(A_2)$ plus the $400 paid the expert, or $2880.

$$EVSI = EV|SI - PEP + \$400$$
$$= \$25,280 - \$22,800 + \$400$$
$$= \$2880$$

Clearly the expert's opinion is well worth the $400 fee in this example.

8.6.5 Sequential Decisions—Summary Comments

The decision-tree representation of a sequential-decision situation is, in our opinion, an effective device for visualization. By virtue of developing the tree, serious attention is given to explicitly identifying and describing alternatives, to assessing the outcomes that might occur as a result of the alternative choice, and to making any necessary future decisions. Even if the probabilities for outcomes cannot be adequately assessed or values for the "branches" accurately estimated, considerable insight into the decision situation may be achieved and decision making aided. On the other hand, whether the expectation principle of choice among alternatives is an appropriate criterion or not is a matter of personal preference, and any real-world sequential-decision situation may readily become too large for effective description or convenient solution. As the number of alternatives, the number of outcomes per alternative, and the number of decision points increase, so do the uncertainty of data and the burden of calculation. Furthermore, the expense in time and money to obtain the information required may become prohibitive when compared with the benefits expected from the analysis. An evaluation of such a trade-off is largely, if not primarily, a matter of judgment.

8.7 MULTIPLE OBJECTIVES

In order to initiate a discussion of multiple objectives and the confounding effect they have on the decision-making process, quotations from Miller and Starr [13], given in Chapter Two, are now paraphrased: "Being unable to describe goals satisfactorily in terms of one objective, people customarily maintain various objectives, and multiple objectives are frequently in conflict with each other." Because organizations are made up of people, this statement would seem to apply to an organizational unit as well as an individual. Accepting this premise, then, managers who are responsible for the organization's behavior and performance usually expect a decision to achieve multiple objectives and evaluate recommendations made to them on this basis.

Establishing the relevant objectives is an important problem for a given decision maker, and the problem is further complicated by conflicting objectives and the difficulty (if not impossibility) of obtaining a measurement of these objectives.

At the outset, we acknowledge that the discussion to follow is somewhat superficial, and we have drawn heavily from Churchman, Ackoff, and Arnoff [3, Chapters 5 and 6], Miller and Starr [13, Chapters 1–3 and pp. 263–241], Kepner and Tregoe [9], and Morris [15, Chapter 7]. The literature on the subject of multiple objectives and the theory of measurement is very extensive, since these factors are involved in all areas of human endeavor. The primary literature reference sources are the general areas of psychology, statistics, mathematics, and management. The previous references have been cited because they are concerned with managerial decision making instead of the psychological motivation stimulating the multiple objectives or the mathematical theory of measurement. For the reader particularly interested in the subject of multiple objectives from a managerial decision-making point of view, the cited references are recommended readings.

Kepner and Tregoe [9] state that "making the best decision will involve a sequence of procedures based on the following seven concepts:

1. The objectives of a decision must be established first.
2. The objectives are classified as to importance.
3. Alternative actions are developed.
4. The alternatives are evaluated against the established objectives.
5. The choice of the alternative best able to achieve all the objectives represents the tentative decision.
6. The tentative decision is explored for future possible adverse consequences.
7. The effects of the final decision are controlled by taking other actions to prevent possible adverse consequences from becoming problems, and by making sure the actions decided on are carried out."

For a single objective, Chapters Three to Eight of this text have been concerned primarily with steps 3 to 6 of the above procedure. This particular section of Chapter Eight will be concerned mainly with steps 2 and 4 of the above procedure. Churchman, Ackoff, and Arnoff [3] present an interesting discussion on establishing the objectives of a decision. Morse [17] states that objectives are derived from two general areas: the *results* expected to come from a decision, and the *resources* available for expenditure in carrying out a decision. The results expected may be to achieve certain things, but it may also be desired to maintain certain things. For example, over the next five years, management may wish to increase sales by 40% relative to the past year and render better customer service, retaining good employee morale and the company's good financial credit rating. In any case, for the subsequent discussion, it will be assumed that objectives can be precisely stated and are relevant to the decision.

8.7.1 Classifying Objectives According to Importance

Assessing the relative importance of multiple objectives, measuring the outcomes of alternatives in terms of these objectives, obtaining a common measure for multiple outcomes having different measures, and combining all these into a single measure of merit form the heart of the multiple-objective problem and the source of much difficulty and controversy. We cannot offer a solution to the problem, but can only discuss approaches to it, citing cautions at appropriate points.

When objectives can be sorted by classes such that each objective in the class is equivalent, then this is nothing more than a classification system or *nominal* system of measurement. However, when objectives can be sorted into classes that can then be ordered, this is termed an *ordinal* system of measurement.

An illustration of a nominal classification of objectives is the concept advanced by Kepner and Tregoe where objectives first are categorized on a "must" or "desirable" basis. For example, a university library must serve the faculty and student population, but perhaps it is only desirable to serve residents of the local community. Once objectives are classed on a "must" versus "desirable" basis, the desirable objectives may be considered only subjectively thereafter, whereas the must objectives are deserving of greater attention and, if possible, quantification. Some of the approaches taken to quantify objectives will now be treated; in the discussion, the term *goal* is used synonymously with *objective* and the symbolism adopted to denote the kth goal or objective of interest is G_k.

8.7.2 Ranking

If a decision maker is capable of stating preferences among objectives, objectives may be ranked in terms of relative importance. A technique to assist the decision maker in making consistent preference statements is the method of paired comparisons. In order to illustrate the method, assume that four goals (objectives) are relevant: G_1, G_2, G_3, and G_4. The method of paired comparisons enumerates all possible pairs $[n(n-1)/2]$, makes preference statements, and deduces a ranking for each goal. In this example, the possible pairs are:

$$G_1 \text{ versus } G_2 \qquad G_2 \text{ versus } G_3 \qquad G_3 \text{ versus } G_4$$
$$G_1 \text{ versus } G_3 \qquad G_2 \text{ versus } G_4$$
$$G_1 \text{ versus } G_4$$

If the symbol $>$ represents "preferred to" and the symbol $<$ represents "not preferred to," suppose the results of the decision about the pairs were:

$$G_1 > G_2 \qquad G_2 < G_3 \qquad G_3 > G_4$$
$$G_1 < G_3 \qquad G_2 > G_4$$
$$G_1 > G_4$$

Rewriting so that all comparisons are of the "preferred to" variety yields:

$$G_3 > G_1 \qquad G_1 > G_2 \qquad G_2 > G_4$$
$$G_3 > G_2 \qquad G_1 > G_4$$
$$G_3 > G_4$$

It can then be concluded that G_3 is preferred to all others, G_1 is preferred to two others, G_2 is preferred to one other, and G_4 is preferred to zero others and the following ranks should be assigned: $G_3 = \text{I}$, $G_1 = \text{II}$, $G_2 = \text{III}$, and $G_4 = \text{IV}$.

If, in the above procedure, the preference statements had been $G_1 > G_2$, $G_2 > G_4$ and $G_1 < G_4$, the decision maker should reconsider the inconsistent judgments. These results are "intransitive" in a mathematical sense.

The ranking method of classifying objectives according to relative importance is certainly useful; indeed, ranking was implicitly used in the comparison of alternative investments under conditions of assumed certainty in the earlier chapters of this textbook. There are, however, obvious deficiencies in the ranking method. For instance, the ranks do not indicate the *extent* to which one objective is preferred over another. Furthermore, because rankings are an *ordinal scale of measurement*, the mathematical operations of addition, subtraction, multiplication, division, and so forth, cannot be performed. Therefore a higher-order scale of measurement (interval or ratio scale) is generally required to determine choice among alternatives under conditions of risk and uncertainty [15]. Unfortunately, accepted methods for measuring objectives on a ratio scale do not currently exist.

8.7.3 Weighting Objectives

Before proceeding with a discussion of certain methods to assign relative weights to multiple objectives, a distinction between objectives and outcomes should be made. This is perhaps most easily done by means of an example where a machine-replacement situation is of interest, and the following two objectives are defined:

G_1: minimize the equivalent annual cost over a five-year planning horizon
G_2: minimize the number of production personnel transferred to other jobs within the firm

Let

θ_{jk} = the outcome resulting from taking the *j*th alternative course of action in terms of the *k*th objective or goal

$V(\theta_{jk})$ = the value of the outcome resulting from selecting the *j*th alternative in terms of the *k*th objective

Suppose, in this example, that

A_1 = the alternative of keeping the present machining center
A_2 = the alternative of replacing the present center with a new machining center

The outcomes for each of these alternatives in terms of the two objectives, G_1 and G_2, are:

$$\theta_{11} = \text{an annual cost of } \$80,000$$
$$\theta_{12} = \text{no persons transferred}$$
$$\theta_{21} = \text{an annual cost of } \$60,000$$
$$\theta_{22} = \text{two persons transferred}$$

In order to compare the two alternatives in a quantitative fashion, values must be assigned to each outcome; the values must be weighted in some way to reflect the relative importance of the two objectives, and some type of aggregate number must be determined for each alternative to enable a comparison and final choice. For example, the end result (without regard for validity at this point) might appear as

$$V(A_1) = w_1 V(\theta_{11}) + w_2 V(\theta_{12})$$
$$V(A_2) = w_1 V(\theta_{21}) + w_2 V(\theta_{22})$$

where $V(A_2) > V(A_1)$.

The w_k (or w_1, w_2) values are weighting factors identifying the relative importance of the objectives G_k (or G_1, G_2) and hence are applied to the appropriate outcomes (θ_{jk}) of the jth alternative. Determining the weighting factors is thus one issue in quantifying multiple objectives. The values $V(\theta_{jk})$ may be in different dimensions and, if so, may have to be transformed to another common dimension. Determining the transformation function is another issue in quantifying multiple objectives. Finally, the values for each alternative, $V(A_j)$, were determined by a linear function. Determining the mathematical functional form of the model is yet another issue in quantifying multiple objectives. Again, all these issues will not be resolved in this chapter, but some approaches that have been taken will be presented.

One method for assigning relative weights to multiple objectives in order to indicate the extent by which one objective is preferred to another is simply to assign weights by judgment. That is, in the case of four objectives (G_1, G_2, G_3, G_4), the decision maker may reason that G_3 is twice as important as G_1, three times as important as G_2 and five times as important as G_4. Assigning a weighting value of 1.00 to G_3, it can be reasoned that the set of weighting values is

$$\text{For } G_3: \quad w_3 = 1.000$$
$$\text{For } G_1: \quad w_1 = 0.500$$
$$\text{For } G_2: \quad w_2 = 0.333$$
$$\text{For } G_4: \quad w_4 = 0.200$$

These weights may be transformed to a scale from 0 to 1.0, or "normalized," by dividing each weight by the sum of all weights; that is,

$$w_3' = \frac{1.000}{2.033} = 0.4918$$

$$w_1' = \frac{0.500}{2.033} = 0.2459$$

$$w_2' = \frac{0.333}{2.033} = 0.1639$$

$$w_4' = \frac{0.200}{2.033} = 0.0984$$

Another method of assigning weights to objectives by judgment, but still keeping consistency, is adapted from Churchman, Ackoff, and Arnoff [3], and the commentary on this method given by Morris [15]. The references state that the method is applicable to weight objectives or outcomes. For a decision involving multiple objectives $G_1, \ldots, G_m$ to be evaluated, this method requires two major assumptions. Paraphrasing Morris, these are:

1. It must be possible for the decision maker to think about and judge the value of any combination of objectives. That is, it must be possible to consider not only the importance or weight value for, say, G_1 but also the weighting value for sums of objectives.

2. Values are assumed to be additive. Given the individual weighting values for, say, G_1 and G_2, it is assumed that the weighting value for both objectives is the sum of their individual weighting values.

The general procedure for the method now follows.

1. Rank the objectives in order of importance, where G_1 indicates the most important, G_2 the next most important, and so forth, and G_m is the least important.

2. Assign the weighting value of 1.00 to G_1 (i.e., $w_1 = 1.00$) and weighting values to the other objectives to reflect their importance relative to G_1. These two steps actually complete the judgmental process, and the following steps serve to refine the initial judgments and aid consistency.

3. Compare G_1 to the linear combination of all other objectives. That is, compare G_1 versus $(G_2 + G_3 + \cdots + G_m)$.
   (a) If $G_1 > (G_2 + G_3 + \cdots + G_m)$, adjust (if necessary) the value of w_1 such that $w_1 > (w_2 + w_3 + \cdots + w_m)$. Attempt, in all adjustments of the procedural steps, to keep the relative values within the objective *group* invariant. Proceed to step 4.
   (b) If $G_1 = (G_2 + G_3 + \cdots + G_m)$, adjust w_1 so that $w_1 = (w_2 + \cdots + w_m)$ and proceed to step 4.
   (c) If $G_1 < (G_2 + G_3 + \cdots + G_m)$, adjust (if necessary) the value of w_1 so that $w_1 < (w_2 + \cdots + w_m)$.
      (1) Now compare G_1 versus $(G_2 + G_3 + \cdots + G_{m-1})$.
         [a] If $G_1 > (G_2 + G_3 + \cdots + G_{m-1})$, adjust (if necessary) the values so that $w_1 > (G_2 + G_3 + \cdots + G_{m-1})$ and proceed to step 4.
         [b] If $G_1 = (G_2 + G_3 + \cdots + G_{m-1})$, adjust (if necessary) the values such that $w_1 = (w_2 + w_3 + \cdots + w_{m-1})$ and proceed to step 4.
         [c] If $G_1 < (G_2 + G_3 + \cdots + G_{m-1})$, adjust (if necessary) the values such that $w_1 < (w_2 + w_3 + \cdots + w_{m-1})$. Then, compare G_1 versus $(G_2 + G_3 + \cdots + G_{m-2})$, and so forth, either until G_1 is preferred or equal to the rest, then proceed to step 4, or until the

comparison of G_1 versus $(G_2 + G_3)$ is completed, then proceed
to step 4.

4. Compare G_2 versus $(G_3 + G_4 + \cdots + G_m)$ and proceed as in step 3.

5. Continue until the comparison of G_{m-2} versus $(G_{m-1} + G_m)$ is completed.

6. If desired, convert each w_k into a normalized value, dividing w_k by the
sum $\Sigma_{k=1}^{m} w_k$.

Example 8.9

Assume that it is desired to determine weighting values for four objectives
where these have been ranked in order of importance such that G_1 is most
important and G_4 is the least important. Step 1 of the procedure is thus
accomplished.

Step 2. The objectives are tentatively assigned the weights $w_1 = 1.00$, $w_2 = 0.80$, $w_3 = 0.50$, and $w_4 = 0.20$.

Step 3. Assume that G_1 is preferred to the linear combination of the other
objectives. That is, $G_1 > (G_2 + G_3 + G_4)$. Since $w_1 < (w_2 + w_2 + w_3)$, it
is necessary to adjust w_1 to be greater than the sum of 1.50—say,
$w_1 = 1.75$. Then proceed to step 4.

Step 4. Comparing G_2 versus $(G_3 + G_4)$, assume that $G_2 < (G_3 + G_4)$, and the
step 3 procedure is repeated. Since the weighting values do not
reflect this [i.e., $w_2 = 0.80$ is not less than $(0.50 + 0.20)$], w_2 is
adjusted to be 0.65. [The adjustment to decrease w_2 does not violate
the previous preference of $G_1 > (G_2 + G_3 + G_4)$ with the new value
of $w_1 = 1.75$ to reflect this preference.] Since, in this example of
only four objectives, the comparison of G_{m-2} versus $(G_{m-1} + G_m)$ or
G_2 versus $(G_3 + G_4)$ has been completed at this point, step 5 of the
procedure has been done.

If it is desired to normalize the results, new weighting values are calculated
as

$$w_1' = 1.75/(1.75 + 0.65 + 0.50 + 0.20) = 0.5645$$
$$w_2' = 0.90/3.10 \qquad\qquad\qquad = 0.2097$$
$$w_3' = 0.50/3.10 \qquad\qquad\qquad = 0.1613$$
$$w_4' = 0.20/3.10 \qquad\qquad\qquad = \underline{0.0645}$$
$$\text{Total} \qquad\qquad 1.0000$$

For a small number of objectives, this method is relatively easy to use, but
becomes cumbersome as the number of objectives increases. However, it is
believed that many real-world problems reduce to a few primary objectives. It
should also be again emphasized that the method still depends on the decision
maker's judgment in assigning the relative weights.

8.7.4 Determining the Value of Multiple Objectives

In the previous section on weighting objectives, an example of a machine replacement situation was hypothesized to introduce the discussion. This example is now recalled, where the objectives were:

G_1: minimize the equivalent annual cost over a five-year planning horizon
G_2: minimize the number of production personnel transferred to other jobs within the firm

For the two alternatives of A_1 (keep present machining center) and A_2 (replace with new machining center), the outcomes were

A_1: \$80,000 annual cost (θ_{11}); no persons transferred (θ_{12})
A_2: \$60,000 annual cost (θ_{21}); two persons transferred (θ_{22})

Let it be assumed that, by some method, the objectives have been weighted as $w_1 = 0.60$ and $w_2 = 0.40$.

One method to assign values to the outcomes of the two alternatives is to make a second series of judgments as to the degree to which each outcome succeeds in meeting the objective. Suppose these judgments are made in the form of numbers on an arbitrary scale from zero to one where the higher the scale value, the closer to meeting the desired objective. Assume the results of such judgments are as given below.

Outcome	Value
θ_{11}	0.50
θ_{12}	1.00
θ_{21}	0.85
θ_{22}	0.60

Then, assuming that a linear model is appropriate, the values for each alternative are calculated as

$$V(A_1) = w_1 V(\theta_{11}) + w_2 V(\theta_{12})$$
$$= (0.60)(0.50) + (0.40)(1.00) = 0.70$$

$$V(A_2) = w_1 V(\theta_{21}) + w_2 V(\theta_{22})$$
$$= (0.60)(0.85) + (0.40)(0.60) = 0.75$$

Thus, since $V(A_2) > V(A_1)$, alternative A_2 would be chosen, and the present machining center would be replaced.

Clearly, these results are sensitive to the weighting values assigned by judgment. Furthermore, the magnitude of the difference in values between

$V(A_1)$ and $V(A_2)$ is a function of the arbitrary scale (from zero to one) chosen. The linear model assumed is also questionable [13]. All these points are criticisms of this method of quantifying multiple objectives. In the literature on the theory of utility functions, it is argued that if a decision maker's utility function can be established for each objective, then values of outcomes (e.g., $80,000 annual cost) can be converted to a utility value (e.g., $80,000 annual cost equals 0.75) from the decision maker's utility curve. Such a utility value could be determined for each objective, and our model for the alternatives above would become

$$U(A_1) = w_1 U(\theta_{11}) + w_2 U(\theta_{12})$$

$$U(A_2) = w_1 U(\theta_{21}) + w_2 U(\theta_{22})$$

Then, the decision maker should choose to maximize the weighted utility value. Whether such utility functions for each objective can be readily obtained for a single decision maker is still questionable in our opinion, and even if this could be accomplished, there is yet another question of whether interaction between multiple objectives would be totally missed. As stated in Chapter Two, the study of value measurement, and utility theory in particular, is interesting, but questions remain on the practical application and implementation of utility theory to business and engineering problems. In any event, it is judged that a treatment of utility theory is outside the scope of this textbook.

8.8 SUMMARY

In this chapter, some prescriptive or normative models for decisions under risk and uncertainty have been presented. For decisions under risk and uncertainty where the matrix model is appropriate, criteria that are commonly cited in the literature for choosing among mutually exclusive alternatives were discussed. None of these criteria were cited as the best or optimal basis for selecting an alternative; they merely represent observations on how decision makers seem to decide.

The decision-tree model for a sequence of decisions where the outcomes of alternatives are chance events was also presented, and the expectation-variance criterion was suggested as a guideline for choice among alternatives. Some fundamental laws of probability were reviewed, and Bayes's theorem of conditional probability was illustrated as a guideline for revising one's opinion in sequential decisions under risk when only imperfect information is available to the decision maker.

The last section of the chapter was an introductory presentation of some approaches in the literature for quantifying multiple objectives. The discussion offered and examples used were restricted to decisions under assumed certainty. However, the methods illustrated are easily extended to decisions under risk and uncertainty and the matrix model used for analysis.

BIBLIOGRAPHY

1. Agee, Marvin H., Taylor, Robert E., and Torgersen, Paul E., *Quantitative Analysis For Management Decision*, Prentice-Hall, 1976.

2. Chernoff, Herman, and Moses, Lincoln E., *Elementary Decision Theory*, Wiley, 1959.

3. Churchman, C. West, Ackoff, Russell L., and Arnoff, E. Leonard, *Introduction to Operations Research*, Wiley, 1957.

4. Conrath, David W., "From Statistical Decision Theory to Practice: Some Problems with the Transition," *Management Science*, **19** (8), April 1973. Reprinted in EE Monograph Series No. 2, entitled *Risk and Uncertainty: Non-Deterministic Decision Making in Engineering Economy*, American Institute of Industrial Engineers, 1975, pp. 32–37.

5. de Neufville, Richard, and Stafford, Joseph H., *Systems Analysis for Engineers and Managers*, McGraw-Hill, 1958.

6. Fishburn, Peter C., "Decision Under Uncertainty: An Introductory Exposition," EE Monograph Series No. 2, entitled *Risk and Uncertainty: Non-Deterministic Decision Making in Engineering Economy*, American Institute of Industrial Engineers, Atlanta, Georgia, 1975, pp. 19–31.

7. Goetz, Billy, E., "Perplexing Problems in Decision Theory," *The Engineering Economist*, **14** (3), ASEE, Spring 1969.

8. Hespos, Richard F., and Strassman, Paul A., "Stochastic Decision Trees for the Analysis of Investment Decisions," *Management Science*, **11** (10), August 1965. Reprinted in EE Monograph Series No. 2, entitled *Risk and Uncertainty: Non-Deterministic Decision Making in Engineering Economy*, American Institute of Industrial Engineers, 1975, pp. 48–55.

9. Kepner, Charles H., and Tregoe, Benjamin B., *The Rational Manager*, McGraw-Hill, 1965.

10. Luce, R. Duncan, and Raiffa, Howard, *Games and Decisions: Introduction and Critical Survey*, Wiley, 1958.

11. Magee, John F., "Decision Trees for Decision Making," *Harvard Business Review*, **42** (4), July–August 1964.

12. Magee, John F., "How to Use Decision Trees in Capital Investment," *Harvard Business Review*, **42** (5), September–October 1964.

13. Miller, David W., and Starr, Martin K., *Executive Decisions and Operations Research*, Second Edition, Prentice-Hall, 1969.

14. Miller, Ernest C., *Advanced Techniques for Strategic Planning*, AMA Research Study 104, American Management Association, 1971.

15. Morris, William T., *The Analysis of Management Decisions*, Revised Edition, Richard D. Irwin, 1964.

16. Morris, William T., *Management Science: A Bayesian Introduction*, Prentice-Hall, 1968.

17. Morse, Phillip M., *Library Effectiveness: A Systems Approach*, M.I.T. Press, 1968.

18. Spiegel, Murray R., *Schaum's Outline of Theory and Problems of Statistics*, Schaum Publishing, 1961.

PROBLEMS

1. Assume the following decision matrix where the cell values are units of gain.

	S_1	S_2	S_3	S_4
A_1	$4	$ 2	$0	$1
A_2	− 4	10	3	7
A_3	4	8	2	3
A_4	2	4	9	5

If a person's index of optimism is estimated as $\alpha = 0.25$, predict the person's choice of alternative by applying the various principles of choice for a decision under uncertainty. (8.4.1, 8.5)

2. Analyze the cost matrix below by the use of the Hurwicz principle. (8.5.4)

	S_1	S_2	S_3
A_1	$5	$5	$5
A_2	9	8	0
A_3	4	6	3

3. Apply the various decision under uncertainty principles of choice to the matrix below. The values in the matrix are cost units. (8.4.1, 8.5)

	S_1	S_2	S_3	S_4
A_1	$13	$13	$5	$9
A_2	8	8	8	8
A_3	0	21	5	5
A_4	9	17	5	5
A_5	5	7	7	5

4. A particular production department supervisor is known to be very conservative in business matters. For the decision model below, where values are equivalent annual profits (in thousands of dollars) for a $MARR = 15\%$ and a five-year planning horizon, predict the supervisor's choice of an alternative by each of the principles for a decision under uncertainty. (8.4.1, 8.5)

	S_1	S_2	S_3	S_4
A_1	$16	$16	$ 8	$12
A_2	9	20	10	6
A_3	15	10	8	9
A_4	3	14	6	11
A_5	8	16	9	12

5. Assume the following decision matrix.

	p_1	p_2	p_3
	S_1	S_2	S_3
A_1	$ 20	$100	$1200
A_2	190	190	190
A_3	500	120	100

(a) Treat the decision as one under uncertainty and the values in the matrix as costs.

   (1) If α = an index of optimism, for what value of α is one indifferent between A_2 and A_3? (8.5.4)

   (2) Which alternative would be chosen by the Savage principle? (8.5.5)

(b) Treat the decision as one under risk where $p_1 = 0.20$, $p_2 = 0.70$, and $p_3 = 0.10$ and the values in the matrix are profits. Which alternative would be chosen by the expectation-variance principle? (8.4.2)

6. Given the decision matrix shown below (with *cost* elements), determine the preferred alternative using the following principles of choice. (8.5)

   (a) Minimax.

   (b) Minimin.

   (c) Minimax regret.

   (d) Expected cost, if each future state is expected to occur with equal probabiiity.

   (e) Hurwicz (with $\alpha = 0.40$).

	S_1	S_2	S_3
a_1	$120	$50	$ 10
a_2	60	60	60
a_3	70	50	60
a_4	20	20	150

7. Given the decision matrix shown below, determine the recommended alternative under the following principles of choice: expectation, most probable future, maximax, maximin, and minimax regret. Entries in the matrix are profits. (8.4, 8.5)

	$p_1 = 0.1$ S_1	$p_2 = 0.3$ S_2	$p_3 = 0.4$ S_3	$p_4 = 0.2$ S_4
A_1	−$ 50	$100	$200	$400
A_2	− 20	50	500	100
A_3	100	− 100	50	100
A_4	200	− 50	50	200

8. Shown below is a matrix of costs for three investment alternatives under three future states. Determine the preferred alternative using the following decision rules. (8.5.2, 8.5.3, 8.5.5)

   (a) Minimax.

(b) Minimin.

(c) Minimax regret.

	S_1	S_2	S_3
a_1	$300	$200	$100
a_2	150	180	200
a_3	210	110	175

9. An analysis yields a decision under risk given in the matrix below, where the matrix values are annual profits in thousands of dollars.

$p_k =$ $S_k =$	0.1 S_1	0.1 S_2	0.4 S_3	0.2 S_4	0.1 S_5	0.1 S_6
A_1	$12	$ 5	-$ 8	-$3	$ 6	$ 9
A_2	7	0	1	5	20	7
A_3	3	3	7	9	- 5	5
A_4	0	12	15	2	8	- 200
A_5	- 10	22	9	0	4	12

(a) Which alternative should be chosen in order to minimize the probability of a loss? (8.4.4)

(b) Which alternative should be chosen in order to maximize the probability of an annual profit of at least $9000? (8.4.4)

(c) Which alternative should be chosen in order to maximize the probability that annual profits will be between $3000 and $10,000? (8.4.4)

(d) Would the most probable future principle be a reasonable one to follow in selecting an alternative? (8.4.3)

10. An electronics firm has recently received a United States government contract to produce a certain quantity of expensive electronic guidance systems. The contracted number of units can be produced in two years, and the contract terminates then. The firm won the contract with cost estimates based on using present manufacturing equipment. If the firm had more specialized equipment (with almost zero salvage value immediately after purchase), the unit cost of production would be reduced, and profit per unit would thus be increased. However, this would be true only if the contract were for at least four years instead of two years. The firm feels there is a 50% chance for an additional two-year contract, a 30% chance for an additional four-year contract, and a 20% chance of no contract renewal.

Using only present equipment, total profit on this job for a two-year, four-year, and six-year contract is estimated as $40,000, $80,000 and $120,000, respectively. If the specialized equipment is purchased, total profit on this job for two-year, four-year, and six-year contract is estimated as −$100,000, $90,000, and $180,000. If an expectation principle is used, should the specialized equipment be purchased? Does the most probable future criterion seem reasonable in this decision? (8.4.1, 8.4.2, 8.4.3)

11. A highway construction firm is considering the purchase of a used mobile crane. Two such cranes are available, A_1 and A_2. If either is purchased, the firm will use a six-year depreciation schedule. The cranes differ slightly in capacity, age, and mechanical condition, but both are presently in operating condition and have the capacity to do the job expected. The firm expects that a major overhaul of each will eventually be required, but when this will be necessary is uncertain. Estimates of the operating expenses (excluding labor) for each crane are given below:

	A_1		A_2	
Year	Expense Schedule 1	Expense Schedule 2	Expense Schedule 1	Expense Schedule 2
1	$ 5,000	$ 5,000	$14,000	$ 6,000
2	5,000	11,000	6,000	6,000
3	5,000	5,000	6,000	14,000
4	11,000	5,000	6,000	6,000
5	5,000	5,000	6,000	6,000
6	5,000	5,000	6,000	6,000

Other data for the two used cranes are:

	A_1	A_2
First cost	$30,000	$20,000
Salvage value at the end of six years	12,000	10,000

If the firm uses a before-tax $MARR = 20\%$ and a present worth method of comparing alternatives, which crane would be purchased if the Laplace principle were used? (8.4.1, 8.5.1)

12. Suppose a coin is biased such that, when the coin is tossed, the probability of a head (H) showing is 0.60. Now assume that the coin is tossed three consecutive times.
   (a) Sketch a probability tree for the three tosses and determine the probability values for each of the eight possible outcomes: HHH, HHT, and so on. For the

three tosses, the following states are now defined:

S_1 = the event of three heads (3H)

S_2 = the event of two heads, one tail—occurring in any order (2H, T)

S_3 = the event of two tails, one head—occurring in any order (2T, H)

S_4 = the event of three tails (3T) (8.6.1, 8.6.2)

(b) Create a decision matrix where four alternatives are to guess the events defined above. Now assume that before tossing the coin three times, a person offers a $4 prize if the (3H) event is guessed correctly, a $3 prize if the (2H, T) event is guessed correctly, a $4 prize if the (2T, H) event is guessed correctly, and a $5 prize if the (3T) event is guessed correctly. However, there is a $1 charge to play the game. If the game is played, would the expectation principle and the most probable future principle select the same alternative? (8.2, 8.4.2, 8.4.3)

13.* A vehicle manufacturer plans to develop and operate a public bus system for a community. The manufacturer's purpose is to demonstrate the profitability of the bus system and then sell the system to the community. The manufacturer's most serious concern is whether the Public Utilities Commission will authorize a competing bus system. It is rationalized that there is a small chance that such an event will happen; therefore, the manufacturer subjectively assigns a probability of 0.2 to the event of competition.

A detailed analysis by the manufacturer results in an estimate of 50 vehicles needed in the event of no competition and 25 vehicles needed if competition results. Furthermore, it is reasoned that operating either a 50-vehicle or a 25-vehicle system will not affect the probability of the Public Utilities Commission authorizing or not authorizing competition.

If the manufacturer initially develops a 50-vehicle system and there is no competition, it is estimated the system will yield a $250,000 profit. In the event of competition, a $120,000 loss is projected.

If a 25-vehicle system is initially developed and competition occurs, the manufacturer plans to sell the system as quickly as possible and estimates a $25,000 profit in this instance. On the other hand, if there is no competition, the 25-vehicle system will be inadequate. Poor service would therefore result, and the value of the demonstration would be reduced, thereby adversely affecting the sales price. Thus, in the event of no competition, the manufacturer reasons that another decision must be faced: whether to expand from 25 vehicles to 50 vehicles or not to expand. If there is no expansion, the manufacturer would sell soon to avoid the bad publicity for expected poor service, estimating a $35,000 profit in this case. If, on the other hand, the system is expanded and the total system is sold as soon as possible, two outcomes are predicted. One outcome is that poor service could result during the expansion and a net $10,000 loss would result (a probability of 0.10 is subjectively assigned to this event). The second outcome is that poor service would not result, and a net $140,000 profit is expected in this instance. Create a decision-tree model of this situation and solve by use of the expectation principle. (8.6.1, 8.6.2, 8.4.2)

---

* Problem 13 taken from deNeufville and Stafford [5] by permission of the publisher.

14. Company Able purchases two models, Model R and Model G, of Product Tau from three suppliers, A, B, and C. These suppliers have produced, and are expected to continue to produce, according to the following table:

Supplier	Proportion of Total Product Tau	Proportion of Product Tau by Model	
		R	G
A	0.60	0.20	0.80
B	0.10	0.55	0.45
C	0.30	0.60	0.40

The Quality Control Department of Company Able has recently determined that many type R models have been defective and intends to send a representative to each of the suppliers to observe their manufacturing procedures.

Using the logic of Bayes's theorem, which supplier is the most likely, next most likely, and least likely defect producer? (8.6.1, 8.6.2)

15. Let it be supposed that, out of your sight, an experimenter rolls a green, red, and white die. The experimenter covers two of these. Data concerning the dice are that (1) the green die has five surfaces marked H, one surface marked T, (2) the red die has two surfaces marked H, four surfaces marked T, and (3) the white die has three surfaces marked H, three surfaces marked T. Also

A = the event that the green die is uncovered
B = the event that the red die is uncovered
C = the event that the white die is uncovered
h = the event that a surface marked H is showing on the uppermost surface
    of the uncovered die
t = the event that a surface marked T is showing

Now, the experimenter reports that the letter H is showing on the uppermost surface of the uncovered die.

(a) Using the logic of Bayes's theorem, what is the best guess concerning the color of the uncovered die? (8.6.1, 8.6.2)

(b) Assume that the experimenter, after the report but before the guess, offers to pay $6 if the green die is guessed correctly, $15 if the red die is guessed correctly, and $15 if the white die is guessed correctly. However, there is a $3 charge to play the game. Given that the game is played, model this situation as a decision under risk matrix and determine the best guess by the expectation principle. (8.2, 8.4.2)

16.* After receiving somewhat unreliable information concerning enemy troop movement along a supply route, a military commander of an artillery battalion

* Problem 16 taken from Agee, Taylor, and Torgersen [1] by permission of the publisher.

faces the following combat decision (assume the matrix values are loss units, arbitrarily chosen).

	$p_1 = 0.7$ $S_1 =$ Troop Movement	$p_2 = 0.3$ $S_2 =$ No Troop Movement
$A_1 =$ bombard supply route	0	4
$A_2 =$ not bombard supply route	12	0

The commander can choose between A_1 and A_2 without additional information, or he can send out a reconnaissance plane for additional information. If the plane is sent out, he reasons the mutually exclusive and collectively exhaustive outcomes of the flight are:

Outcome O_1: the plane is shot down, with a loss of 3 units.

Outcome O_2: the plane returns safely with negative information about troop movement, with a loss of 0.2 units.

Outcome O_3: the plane returns safely with a report of suspicious activity, with a loss of 0.2 units.

The commander assigns the following subjective conditional probabilities to these outcomes:

$$P(O_1|S_1) = 0.4 \qquad P(O_1|S_2) = 0.2$$
$$P(O_2|S_1) = 0.1 \qquad P(O_2|S_2) = 0.8$$
$$P(O_3|S_1) = 0.5 \qquad P(O_3|S_2) = 0.0$$

(a) Determine the posterior probabilities. (8.6.2)
(b) Create a decision tree for this problem. (8.6.1)
(c) Using the minimizing expected loss principle of choice, what action should the commander take? (8.4.2)
(d) What is the expected value of sample information in this problem? (8.6.3, 8.6.4)

17. A manufacturer is considering the possibility of introducing a new product and the advisability of a test marketing prior to making the final decision. The alternatives are

$$a_1 = \text{market the product}$$

$$a_2 = \text{do not market the product}$$

For simplicity, only three possible futures are considered; they are shown below together with the prior probabilities associated with them.

	Profit	Prior Probability
S_1: the product captures 10% of the market	$10,000,000	0.70
S_2: the product captures 3% of the market	1,000,000	0.10
S_3: the product captures less than 1% of the market	− 5,000,000	0.20

If the test marketing is made, three possible results are considered.

Z_1: test sales of more than 10% of the market

Z_2: test sales of 5 to 10% of the market

Z_3: test sales of less than 5% of the market

The conditional probabilities of the test results are

	Z_1	Z_2	Z_3
$P(Z\|S_1)$	0.6	0.3	0.1
$P(Z\|S_2)$	0.3	0.6	0.1
$P(Z\|S_3)$	0.1	0.1	0.8

(a) Determine the prior expected profit. (8.6.1, 8.6.2, 8.6.3)
(b) Determine *EVPI*. (8.6.3)
(c) Determine *EVSI*. (8.6.4)
(d) If the test marketing costs $250,000, what would be the expected net gain from the sample information? (8.6.4)

18. A textile firm in a rural town owns two trucks that transport raw materials from and finished goods to a nearby city. One of the trucks (tractor and trailer) is eight years old, and annual maintenance expenses are increasing significantly. Management is considering the purchase of a new truck but, because of a mild national economic recession, textile sales have declined during the past two years, and management is reluctant to make a large capital expenditure at present. Although in the midst of a recession, inflation is also occurring; if the decision to purchase is delayed for a year, the cost of a new truck will probably increase 10% or more.

Adopting a two-year planning horizon, the following estimates and judgments are made.

Present Truck

The present market value is approximately $9000. If kept one more year, operating (excluding labor and depreciation) and maintenance expenses are estimated as either $6000, $10,000, or $12,000 with subjective probabilities of 0.25, 0.50, and

0.25, respectively. The salvage value at the end of the year is about $8000. If kept one more year, a new decision to keep an additional year or buy a new truck will be made.

If the truck is kept two additional years, operating and maintenance expenses (excluding depreciation) for the second year are estimated to be either $7000, $10,000, or $15,000 with subjective probabilities of 0.10, 0.70, and 0.20, respectively. The salvage value at the end of the second year is about $7500.

New Truck

A new truck can be purchased now for $30,000. Average annual operating, maintenance, and depreciation expenses for each of the first two years are estimated to be $15,000 with reasonable certainty.

If the purchase of the truck is delayed for one year, an increase in purchase price is virtually certain. It is judged there is a 50–50 chance that the purchase price will be $32,000 or $34,000. If it is $32,000, total operating, maintenance, and depreciation expenses will be $16,000; if it is $34,000, total expenses will be $17,500.

For simplicity, assume $MARR = 0\%$ and (a) create a decision-tree model for this situation, and then (b) determine the best present decision by the expectation principle. (8.6.1, 8.4.2)

19. The Jax Tool and Engineering Company is experiencing considerable problems with in-process material handling and storage of finished goods. A new layout of machinery is possible, but this alternative is subjectively judged by management to be prohibitively expensive, and the alternative is discarded. They wish, therefore, to consider a new material-handling system and identify the following objectives.

G_1: minimize annual costs over a 10-year planning horizon
G_2: minimize the disruption of production during the installation of the new system
G_3: install material-handling equipment that has flexibility to meet different handling requirements
G_4: the material-handling equipment should have a high index of repairability
G_5: material-handling personnel should require little training in order to operate the equipment

Management ranks these goals in the following order of importance: $G_1 = I$, $G_3 = II$, $G_2 = III$, $G_5 = IV$, and $G_4 = V$. Using the Churchman, Ackoff, and Arnoff method of weighting objectives, assign weights to these five objectives. (8.7.4)

20. Suppose a decision may be modeled as

	$p_1 = 0.6$ S_1 Contract Renewed	$p_2 = 0.4$ S_2 Contract Not Renewed
A_1: use present equipment	$V(\theta_{11})$	$V(\theta_{12})$
A_2: purchase new equipment	$V(\theta_{21})$	$V(\theta_{22})$

Assume that analysis produces the following descriptions of the outcomes:

θ_{11} = no new capital required, 20 people required for at least four years, average annual profit units are 85

θ_{12} = no new capital required, 20 people required for two years and then laid off, average annual profit units are 67

θ_{21} = two units of new capital required, 8 people required for at least four years, average annual profit units are 150

θ_{22} = two units of new capital required, 8 people required for two years and then laid off, average annual profit units are 400

Presume the relevant values for objectives are

V(profit units) $\qquad = +0.003$ (profit units) $+ 1.0$

V(capital units required) $= e^{-u}$, where u = the number of capital units required

V(work-years required) $= -0.0145$ (work-years required) $+ 1.0$

Furthermore, the goals of "capital units required" and "annual profit units" are considered equally important, but the labor consideration is only half as important as either of the other two.

On the basis of this information, which alternative would you recommend? (8.2, 8.7, 8.7.2)

APPENDIX A
DISCRETE COMPOUNDING

SECTION I—DISCRETE COMPOUND INTEREST FACTORS

SECTION II—GEOMETRIC SERIES FACTORS

SECTION I—DISCRETE COMPOUND INTEREST FACTORS

TABLE A.1 Discrete Compounding: i = 1/4%

	Single payment		Uniform series				Gradient series	
	Compound amount factor	Present worth factor	Compound amount factor	Sinking fund factor	Present worth factor	Capital recovery factor	Uniform series factor	Present worth factor
n	To find F given P $F/P\ i,n$	To find P given F $P/F\ i,n$	To find F given A $F/A\ i,n$	To find A given F $A/F\ i,n$	To find P given A $P/A\ i,n$	To find A given P $A/P\ i,n$	To find A given G $A/G\ i,n$	To find P given G $P/G\ i,n$
1	1.0025	0.9975	1.0000	1.0000	0.9975	1.0025	0.0000	0.0000
2	1.0050	0.9950	2.0025	0.4994	1.9925	0.5019	0.4994	0.9950
3	1.0075	0.9925	3.0075	0.3325	2.9851	0.3350	0.9983	2.9801
4	1.0100	0.9901	4.0150	0.2491	3.9751	0.2516	1.4969	5.9503
5	1.0125	0.9876	5.0251	0.1990	4.9627	0.2015	1.9950	9.9007
6	1.0151	0.9851	6.0376	0.1656	5.9478	0.1681	2.4927	14.8263
7	1.0176	0.9827	7.0527	0.1418	6.9305	0.1443	2.9900	20.7223
8	1.0202	0.9802	8.0704	0.1239	7.9107	0.1264	3.4869	27.5839
9	1.0227	0.9778	9.0905	0.1100	8.8885	0.1125	3.9834	35.4061
10	1.0253	0.9753	10.1133	0.0989	9.8639	0.1014	4.4794	44.1842
11	1.0278	0.9729	11.1385	0.0898	10.8368	0.0923	4.9750	53.9133
12	1.0304	0.9705	12.1664	0.0822	11.8073	0.0847	5.4702	64.5886
13	1.0330	0.9681	13.1968	0.0758	12.7753	0.0783	5.9650	76.2053
14	1.0356	0.9656	14.2298	0.0703	13.7410	0.0728	6.4594	88.7587
15	1.0382	0.9632	15.2654	0.0655	14.7042	0.0680	6.9534	102.2441
16	1.0408	0.9608	16.3035	0.0613	15.6650	0.0638	7.4469	116.6566
17	1.0434	0.9584	17.3443	0.0577	16.6235	0.0602	7.9401	131.9917
18	1.0460	0.9561	18.3876	0.0544	17.5795	0.0569	8.4328	148.2445
19	1.0486	0.9537	19.4336	0.0515	18.5332	0.0540	8.9251	165.4105
20	1.0512	0.9513	20.4822	0.0488	19.4845	0.0513	9.4170	183.4850
21	1.0538	0.9489	21.5334	0.0464	20.4334	0.0489	9.9085	202.4634
22	1.0565	0.9466	22.5872	0.0443	21.3800	0.0468	10.3995	222.3409
23	1.0591	0.9442	23.6437	0.0423	22.3241	0.0448	10.8901	243.1131
24	1.0618	0.9418	24.7028	0.0405	23.2660	0.0430	11.3804	264.7752
25	1.0644	0.9395	25.7646	0.0388	24.2055	0.0413	11.8702	287.3230
26	1.0671	0.9371	26.8290	0.0373	25.1426	0.0398	12.3596	310.7515
27	1.0697	0.9348	27.8961	0.0358	26.0774	0.0383	12.8485	335.0564
28	1.0724	0.9325	28.9658	0.0345	27.0099	0.0370	13.3371	360.2332
29	1.0751	0.9301	30.0382	0.0333	27.9400	0.0358	13.8252	386.2774
30	1.0778	0.9278	31.1133	0.0321	28.8679	0.0346	14.3130	413.1846
31	1.0805	0.9255	32.1911	0.0311	29.7934	0.0336	14.8003	440.9500
32	1.0832	0.9232	33.2715	0.0301	30.7166	0.0326	15.2872	469.5696
33	1.0859	0.9209	34.3547	0.0291	31.6375	0.0316	15.7736	499.0386
34	1.0886	0.9186	35.4406	0.0282	32.5561	0.0307	16.2597	529.3525
35	1.0913	0.9163	36.5292	0.0274	33.4724	0.0299	16.7454	560.5073
40	1.1050	0.9050	42.0132	0.0238	38.0199	0.0263	19.1673	728.7398
45	1.1189	0.8937	47.5661	0.0210	42.5109	0.0235	21.5789	917.3399
50	1.1330	0.8826	53.1887	0.0188	46.9462	0.0213	23.9802	1125.7764
55	1.1472	0.8717	58.8819	0.0170	51.3264	0.0195	26.3710	1353.5286
60	1.1616	0.8609	64.6467	0.0155	55.6524	0.0180	28.7514	1600.0845
65	1.1762	0.8502	70.4839	0.0142	59.9246	0.0167	31.1215	1864.9424
70	1.1910	0.8396	76.3944	0.0131	64.1439	0.0156	33.4812	2147.6094
75	1.2059	0.8292	82.3792	0.0121	68.3108	0.0146	35.8305	2447.6062
80	1.2211	0.8189	88.4392	0.0113	72.4260	0.0138	38.1694	2764.4558
85	1.2364	0.8088	94.5753	0.0106	76.4901	0.0131	40.4980	3097.6958
90	1.2520	0.7987	100.7885	0.0099	80.5038	0.0124	42.8162	3446.8687
95	1.2677	0.7888	107.0796	0.0093	84.4677	0.0118	45.1241	3811.5296
100	1.2836	0.7790	113.4499	0.0088	88.3825	0.0113	47.4216	4191.2383

TABLE A.2 Discrete Compounding: i = 1/3%

	Single payment		Uniform series				Gradient series	
	Compound amount factor	Present worth factor	Compound amount factor	Sinking fund factor	Present worth factor	Capital recovery factor	Uniform series factor	Present worth factor
n	To find F given P $F\|P\ i,n$	To find P given F $P\|F\ i,n$	To find F given A $F\|A\ i,n$	To find A given F $A\|F\ i,n$	To find P given A $P\|A\ i,n$	To find A given P $A\|P\ i,n$	To find A given G $A\|G\ i,n$	To find P given G $P\|G\ i,n$
1	1.0033	0.9967	1.0000	1.0000	0.9967	1.0033	0.0000	0.0000
2	1.0067	0.9934	2.0033	0.4992	1.9900	0.5025	0.4992	0.9934
3	1.0100	0.9901	3.0100	0.3322	2.9801	0.3356	0.9978	2.9735
4	1.0134	0.9868	4.0200	0.2488	3.9669	0.2521	1.4958	5.9338
5	1.0168	0.9835	5.0334	0.1987	4.9504	0.2020	1.9933	9.8678
6	1.0202	0.9802	6.0502	0.1653	5.9306	0.1686	2.4903	14.7690
7	1.0236	0.9770	7.0704	0.1414	6.9076	0.1448	2.9867	20.6308
8	1.0270	0.9737	8.0940	0.1235	7.8813	0.1269	3.4825	27.4469
9	1.0304	0.9705	9.1209	0.1096	8.8518	0.1130	3.9778	35.2109
10	1.0338	0.9673	10.1513	0.0985	9.8191	0.1018	4.4725	43.9163
11	1.0373	0.9641	11.1852	0.0894	10.7831	0.0927	4.9667	53.5569
12	1.0407	0.9609	12.2225	0.0818	11.7440	0.0851	5.4603	64.1263
13	1.0442	0.9577	13.2632	0.0754	12.7017	0.0787	5.9534	75.6182
14	1.0477	0.9545	14.3074	0.0699	13.6561	0.0732	6.4459	88.0264
15	1.0512	0.9513	15.3551	0.0651	14.6074	0.0685	6.9379	101.3447
16	1.0547	0.9481	16.4063	0.0610	15.5556	0.0643	7.4293	115.5669
17	1.0582	0.9450	17.4610	0.0573	16.5006	0.0606	7.9201	130.6869
18	1.0617	0.9419	18.5192	0.0540	17.4424	0.0573	8.4104	146.6985
19	1.0653	0.9387	19.5809	0.0511	18.3812	0.0544	8.9002	163.5956
20	1.0688	0.9355	20.6462	0.0484	19.3168	0.0518	9.3894	181.3722
21	1.0724	0.9325	21.7150	0.0461	20.2493	0.0494	9.8780	200.0223
22	1.0760	0.9294	22.7874	0.0439	21.1787	0.0472	10.3661	219.5398
23	1.0795	0.9263	23.8633	0.0419	22.1050	0.0452	10.8536	239.9187
24	1.0831	0.9232	24.9429	0.0401	23.0283	0.0434	11.3406	261.1531
25	1.0868	0.9202	26.0260	0.0384	23.9484	0.0418	11.8270	283.2371
26	1.0904	0.9171	27.1128	0.0369	24.8655	0.0402	12.3128	306.1650
27	1.0940	0.9141	28.2032	0.0355	25.7796	0.0388	12.7981	329.9309
28	1.0977	0.9110	29.2972	0.0341	26.6906	0.0375	13.2829	354.5286
29	1.1013	0.9080	30.3948	0.0329	27.5986	0.0362	13.7671	379.9529
30	1.1050	0.9050	31.4961	0.0317	28.5035	0.0351	14.2507	406.1975
31	1.1087	0.9020	32.6011	0.0307	29.4056	0.0340	14.7338	433.2568
32	1.1124	0.8990	33.7098	0.0297	30.3046	0.0330	15.2164	461.1255
33	1.1161	0.8960	34.8222	0.0287	31.2006	0.0321	15.6983	489.7974
34	1.1198	0.8930	35.9382	0.0278	32.0936	0.0312	16.1798	519.2671
35	1.1235	0.8901	37.0580	0.0270	32.9837	0.0303	16.6606	549.5291
40	1.1424	0.8754	42.7132	0.0234	37.3898	0.0267	19.0567	712.5259
45	1.1615	0.8609	48.4633	0.0206	41.7232	0.0240	21.4389	894.5000
50	1.1810	0.8467	54.3099	0.0184	45.9851	0.0217	23.8073	1094.7810
55	1.2008	0.8327	60.2546	0.0166	50.1767	0.0199	26.1619	1312.7149
60	1.2210	0.8190	66.2990	0.0151	54.2991	0.0184	28.5026	1547.6648
65	1.2415	0.8055	72.4448	0.0138	58.3535	0.0171	30.8295	1799.0078
70	1.2623	0.7922	78.6937	0.0127	62.3409	0.0160	33.1427	2066.1438
75	1.2835	0.7791	85.0475	0.0118	66.2626	0.0151	35.4420	2348.4783
80	1.3050	0.7663	91.5079	0.0109	70.1196	0.0143	37.7275	2645.4382
85	1.3269	0.7536	98.0767	0.0102	73.9129	0.0135	39.9993	2956.4656
90	1.3492	0.7412	104.7557	0.0095	77.6436	0.0129	42.2574	3281.0142
95	1.3718	0.7290	111.5467	0.0090	81.3128	0.0123	44.5017	3618.5535
100	1.3948	0.7169	118.4517	0.0084	84.9214	0.0118	46.7322	3968.5662

TABLE A.3 Discrete Compounding: i = 5/12%

	Single payment		Uniform series				Gradient series	
	Compound amount factor	Present worth factor	Compound amount factor	Sinking fund factor	Present worth factor	Capital recovery factor	Uniform series factor	Present worth factor
n	To find F given P $F/P\ i,n$	To find P given F $P/F\ i,n$	To find F given A $F/A\ i,n$	To find A given F $A/F\ i,n$	To find P given A $P/A\ i,n$	To find A given P $A/P\ i,n$	To find A given G $A/G\ i,n$	To find P given G $P/G\ i,n$
1	1.0042	0.9959	1.0000	1.0000	0.9959	1.0042	0.0000	0.0000
2	1.0084	0.9917	2.0042	0.4990	1.9876	0.5031	0.4990	0.9917
3	1.0126	0.9876	3.0125	0.3319	2.9752	0.3361	0.9972	2.9669
4	1.0168	0.9835	4.0251	0.2484	3.9587	0.2526	1.4948	5.9174
5	1.0210	0.9794	5.0418	0.1983	4.9381	0.2025	1.9917	9.8351
6	1.0253	0.9754	6.0628	0.1649	5.9135	0.1691	2.4879	14.7119
7	1.0295	0.9713	7.0881	0.1411	6.8848	0.1452	2.9834	20.5398
8	1.0338	0.9673	8.1176	0.1232	7.8521	0.1274	3.4782	27.3108
9	1.0381	0.9633	9.1515	0.1093	8.8153	0.1134	3.9723	35.0170
10	1.0425	0.9593	10.1896	0.0981	9.7746	0.1023	4.4657	43.6504
11	1.0468	0.9553	11.2321	0.0890	10.7299	0.0932	4.9584	53.2033
12	1.0512	0.9513	12.2789	0.0814	11.6812	0.0856	5.4505	63.6679
13	1.0555	0.9474	13.3300	0.0750	12.6286	0.0792	5.9418	75.0365
14	1.0599	0.9434	14.3856	0.0695	13.5721	0.0737	6.4324	87.3014
15	1.0644	0.9395	15.4455	0.0647	14.5116	0.0689	6.9224	100.4549
16	1.0688	0.9356	16.5099	0.0606	15.4472	0.0647	7.4116	114.4894
17	1.0732	0.9318	17.5786	0.0569	16.3790	0.0611	7.9002	129.3974
18	1.0777	0.9279	18.6519	0.0536	17.3069	0.0578	8.3881	145.1715
19	1.0822	0.9240	19.7296	0.0507	18.2309	0.0549	8.8753	161.8042
20	1.0867	0.9202	20.8118	0.0480	19.1511	0.0522	9.3618	179.2881
21	1.0912	0.9164	21.8985	0.0457	20.0675	0.0498	9.8476	197.6158
22	1.0958	0.9126	22.9898	0.0435	20.9801	0.0477	10.3327	216.7800
23	1.1004	0.9088	24.0856	0.0415	21.8889	0.0457	10.8171	236.7735
24	1.1049	0.9050	25.1859	0.0397	22.7939	0.0439	11.3008	257.5889
25	1.1095	0.9013	26.2909	0.0380	23.6952	0.0422	11.7838	279.2195
26	1.1142	0.8975	27.4004	0.0365	24.5927	0.0407	12.2662	301.6577
27	1.1188	0.8938	28.5146	0.0351	25.4865	0.0392	12.7478	324.8967
28	1.1235	0.8901	29.6334	0.0337	26.3766	0.0379	13.2288	348.9295
29	1.1282	0.8864	30.7569	0.0325	27.2630	0.0367	13.7090	373.7488
30	1.1329	0.8827	31.8850	0.0314	28.1457	0.0355	14.1886	399.3477
31	1.1376	0.8791	33.0179	0.0303	29.0248	0.0345	14.6675	425.7195
32	1.1423	0.8754	34.1554	0.0293	29.9002	0.0334	15.1456	452.8574
33	1.1471	0.8718	35.2978	0.0283	30.7720	0.0325	15.6231	480.7546
34	1.1519	0.8682	36.4448	0.0274	31.6402	0.0316	16.0999	509.4041
35	1.1567	0.8647	37.5967	0.0266	32.5047	0.0308	16.5760	538.7991
40	1.1810	0.8468	43.4283	0.0230	36.7740	0.0272	18.9462	696.7280
45	1.2058	0.8294	49.3825	0.0203	40.9555	0.0244	21.2991	872.3147
50	1.2311	0.8123	55.4618	0.0180	45.0509	0.0222	23.6347	1064.7654
55	1.2570	0.7956	61.6637	0.0162	49.0621	0.0204	25.9531	1273.3125
60	1.2834	0.7792	68.0061	0.0147	52.9907	0.0189	28.2542	1497.2117
65	1.3103	0.7632	74.4766	0.0134	56.8385	0.0176	30.5382	1735.7422
70	1.3378	0.7475	81.0830	0.0123	60.6071	0.0165	32.8049	1988.2095
75	1.3660	0.7321	87.8282	0.0114	64.2982	0.0156	35.0544	2253.9358
80	1.3946	0.7170	94.7151	0.0106	67.9134	0.0147	37.2868	2532.2720
85	1.4239	0.7023	101.7467	0.0098	71.4541	0.0140	39.5021	2822.5840
90	1.4539	0.6878	108.9259	0.0092	74.9220	0.0133	41.7002	3124.2639
95	1.4844	0.6737	116.2561	0.0086	78.3186	0.0128	43.8813	3436.7183
100	1.5156	0.6598	123.7402	0.0081	81.6452	0.0122	46.0453	3759.3760

TABLE A.4 Discrete Compounding: i = 1/2%

	Single payment		Uniform series				Gradient series	
	Compound amount factor	Present worth factor	Compound amount factor	Sinking fund factor	Present worth factor	Capital recovery factor	Uniform series factor	Present worth factor
n	To find F given P F|P i,n	To find P given F P|F i,n	To find F given A F|A i,n	To find A given F A|F i,n	To find P given A P|A i,n	To find A given P A|P i,n	To find A given G A|G i,n	To find P given G P|G i,n
1	1.0050	0.9950	1.0000	1.0000	0.9950	1.0050	0.0000	0.0000
2	1.0100	0.9901	2.0050	0.4988	1.9851	0.5038	0.4988	0.9901
3	1.0151	0.9851	3.0150	0.3317	2.9702	0.3367	0.9967	2.9604
4	1.0202	0.9802	4.0301	0.2481	3.9505	0.2531	1.4938	5.9011
5	1.0253	0.9754	5.0503	0.1980	4.9259	0.2030	1.9900	9.8026
6	1.0304	0.9705	6.0755	0.1646	5.8954	0.1696	2.4855	14.6552
7	1.0355	0.9657	7.1059	0.1407	6.8621	0.1457	2.9801	20.4493
8	1.0407	0.9609	8.1414	0.1228	7.8230	0.1278	3.4738	27.1755
9	1.0459	0.9561	9.1821	0.1089	8.7791	0.1139	3.9668	34.8244
10	1.0511	0.9513	10.2280	0.0978	9.7304	0.1028	4.4589	43.3865
11	1.0564	0.9466	11.2792	0.0887	10.6770	0.0937	4.9501	52.8526
12	1.0617	0.9419	12.3356	0.0811	11.6189	0.0861	5.4406	63.2136
13	1.0670	0.9372	13.3972	0.0746	12.5562	0.0796	5.9302	74.4602
14	1.0723	0.9326	14.4642	0.0691	13.4887	0.0741	6.4190	86.5835
15	1.0777	0.9279	15.5365	0.0644	14.4166	0.0694	6.9069	99.5743
16	1.0831	0.9233	16.6142	0.0602	15.3399	0.0652	7.3940	113.4238
17	1.0885	0.9187	17.6973	0.0565	16.2586	0.0615	7.8803	128.1231
18	1.0939	0.9141	18.7858	0.0532	17.1728	0.0582	8.3658	143.6634
19	1.0994	0.9096	19.8797	0.0503	18.0824	0.0553	8.8504	160.0360
20	1.1049	0.9051	20.9791	0.0477	18.9874	0.0527	9.3342	177.2322
21	1.1104	0.9006	22.0840	0.0453	19.8880	0.0503	9.8172	195.2434
22	1.1160	0.8961	23.1944	0.0431	20.7841	0.0481	10.2993	214.0610
23	1.1216	0.8916	24.3104	0.0411	21.6757	0.0461	10.7806	233.6767
24	1.1272	0.8872	25.4320	0.0393	22.5629	0.0443	11.2611	254.0820
25	1.1328	0.8828	26.5591	0.0377	23.4456	0.0427	11.7407	275.2583
26	1.1385	0.8784	27.6919	0.0361	24.3240	0.0411	12.2195	297.2278
27	1.1442	0.8740	28.8304	0.0347	25.1980	0.0397	12.6975	319.9522
28	1.1499	0.8697	29.9745	0.0334	26.0677	0.0384	13.1747	343.4329
29	1.1556	0.8653	31.1244	0.0321	26.9330	0.0371	13.6510	367.6624
30	1.1614	0.8610	32.2800	0.0310	27.7941	0.0360	14.1265	392.6323
31	1.1672	0.8567	33.4414	0.0299	28.6508	0.0349	14.6012	418.3347
32	1.1730	0.8525	34.6086	0.0289	29.5033	0.0339	15.0750	444.7615
33	1.1789	0.8482	35.7817	0.0279	30.3515	0.0329	15.5480	471.9053
34	1.1848	0.8440	36.9606	0.0271	31.1955	0.0321	16.0202	499.7581
35	1.1907	0.8398	38.1454	0.0262	32.0354	0.0312	16.4915	528.3120
40	1.2208	0.8191	44.1588	0.0226	36.1722	0.0276	18.8359	681.3345
45	1.2516	0.7990	50.3242	0.0199	40.2072	0.0249	21.1595	850.7629
50	1.2832	0.7793	56.6452	0.0177	44.1428	0.0227	23.4624	1035.6965
55	1.3156	0.7601	63.1258	0.0158	47.9814	0.0208	25.7447	1235.2686
60	1.3489	0.7414	69.7700	0.0143	51.7256	0.0193	28.0064	1448.6458
65	1.3829	0.7231	76.5821	0.0131	55.3775	0.0181	30.2475	1675.0271
70	1.4178	0.7053	83.5661	0.0120	58.9394	0.0170	32.4680	1913.6414
75	1.4536	0.6879	90.7265	0.0110	62.4136	0.0160	34.6679	2163.7520
80	1.4903	0.6710	98.0677	0.0102	65.8023	0.0152	36.8474	2424.6448
85	1.5280	0.6545	105.5942	0.0095	69.1075	0.0145	39.0065	2695.6382
90	1.5666	0.6383	113.3109	0.0088	72.3313	0.0138	41.1451	2976.0767
95	1.6061	0.6226	121.2224	0.0082	75.4757	0.0132	43.2633	3265.3296
100	1.6467	0.6073	129.3336	0.0077	78.5426	0.0127	45.3613	3562.7935

TABLE A.5 Discrete Compounding: i = 7/12%

	Single payment		Uniform series				Gradient series	
	Compound amount factor	Present worth factor	Compound amount factor	Sinking fund factor	Present worth factor	Capital recovery factor	Uniform series factor	Present worth factor
n	To find F given P $F\|P\ i,n$	To find P given F $P\|F\ i,n$	To find F given A $F\|A\ i,n$	To find A given F $A\|F\ i,n$	To find P given A $P\|A\ i,n$	To find A given P $A\|P\ i,n$	To find A given G $A\|G\ i,n$	To find P given G $P\|G\ i,n$
1	1.0058	0.9942	1.0000	1.0000	0.9942	1.0058	0.0000	0.0000
2	1.0117	0.9834	2.0058	0.4985	1.9826	0.5044	0.4985	0.9884
3	1.0176	0.9827	3.0175	0.3314	2.9653	0.3372	0.9961	2.9538
4	1.0235	0.9770	4.0351	0.2478	3.9423	0.2537	1.4927	5.8848
5	1.0295	0.9713	5.0587	0.1977	4.9137	0.2035	1.9884	9.7702
6	1.0355	0.9657	6.0882	0.1643	5.8794	0.1701	2.4830	14.5987
7	1.0416	0.9601	7.1237	0.1404	6.8395	0.1462	2.9767	20.3593
8	1.0476	0.9545	8.1653	0.1225	7.7940	0.1283	3.4695	27.0411
9	1.0537	0.9490	9.2129	0.1085	8.7430	0.1144	3.9612	34.6331
10	1.0599	0.9435	10.2666	0.0974	9.6865	0.1032	4.4520	43.1245
11	1.0661	0.9380	11.3265	0.0883	10.6245	0.0941	4.9418	52.5048
12	1.0723	0.9326	12.3926	0.0807	11.5571	0.0865	5.4307	62.7632
13	1.0785	0.9272	13.4649	0.0743	12.4843	0.0801	5.9186	73.8893
14	1.0848	0.9218	14.5434	0.0688	13.4061	0.0746	6.4055	85.8727
15	1.0912	0.9165	15.6283	0.0640	14.3225	0.0698	6.8914	98.7030
16	1.0975	0.9111	16.7194	0.0598	15.2337	0.0656	7.3764	112.3700
17	1.1039	0.9059	17.8170	0.0561	16.1395	0.0620	7.8604	126.8636
18	1.1104	0.9005	18.9209	0.0529	17.0401	0.0587	8.3435	142.1738
19	1.1168	0.8954	20.0313	0.0499	17.9355	0.0558	8.8255	158.2906
20	1.1234	0.8902	21.1481	0.0473	18.8257	0.0531	9.3066	175.2041
21	1.1299	0.8850	22.2715	0.0449	19.7107	0.0507	9.7868	192.9045
22	1.1365	0.8799	23.4014	0.0427	20.5906	0.0486	10.2660	211.3821
23	1.1431	0.8748	24.5379	0.0408	21.4654	0.0465	10.7442	230.6274
24	1.1498	0.8697	25.6810	0.0389	22.3351	0.0448	11.2214	250.6308
25	1.1565	0.8647	26.8308	0.0373	23.1998	0.0431	11.6977	271.3826
26	1.1633	0.8597	27.9874	0.0357	24.0594	0.0416	12.1730	292.8740
27	1.1700	0.8547	29.1506	0.0343	24.9141	0.0401	12.6473	315.0955
28	1.1769	0.8497	30.3207	0.0330	25.7638	0.0388	13.1206	338.0376
29	1.1837	0.8448	31.4975	0.0317	26.6085	0.0376	13.5930	361.6917
30	1.1905	0.8399	32.6813	0.0306	27.4485	0.0364	14.0645	386.0481
31	1.1976	0.8350	33.8719	0.0295	28.2835	0.0354	14.5349	411.0986
32	1.2046	0.8302	35.0695	0.0285	29.1137	0.0343	15.0044	436.8340
33	1.2116	0.8254	36.2741	0.0276	29.9390	0.0334	15.4730	463.2456
34	1.2187	0.8206	37.4857	0.0267	30.7595	0.0325	15.9405	490.3240
35	1.2258	0.8158	38.7043	0.0258	31.5754	0.0317	16.4071	518.0615
40	1.2619	0.7924	44.9051	0.0223	35.5840	0.0281	18.7257	666.3342
45	1.2992	0.7697	51.2889	0.0195	39.4777	0.0253	21.0201	829.8254
50	1.3375	0.7477	57.8611	0.0173	43.2599	0.0231	23.2904	1007.5413
55	1.3770	0.7262	64.6271	0.0155	46.9336	0.0213	25.5368	1198.5317
60	1.4176	0.7054	71.5929	0.0140	50.5020	0.0198	27.7591	1401.8899
65	1.4595	0.6852	78.7642	0.0127	53.9691	0.0185	29.9575	1616.7498
70	1.5025	0.6655	86.1471	0.0116	57.3349	0.0174	32.1320	1842.2847
75	1.5469	0.6465	93.7479	0.0107	60.6052	0.0165	34.2827	2077.7070
80	1.5925	0.6279	101.5730	0.0098	63.7817	0.0157	36.4095	2322.2639
85	1.6395	0.6099	109.6289	0.0091	66.8672	0.0150	38.5127	2575.2383
90	1.6879	0.5925	117.9226	0.0085	69.8643	0.0143	40.5922	2835.9470
95	1.7377	0.5755	126.4611	0.0079	72.7754	0.0137	42.6482	3133.7390
100	1.7893	0.5590	135.2515	0.0074	75.6031	0.0132	44.6806	3377.9934

TABLE A.6 Discrete Compounding: i = 2/3%

	Single payment		Uniform series				Gradient series	
	Compound amount factor	Present worth factor	Compound amount factor	Sinking fund factor	Present worth factor	Capital recovery factor	Uniform series factor	Present worth factor
n	To find F given P $F\|P\ i,n$	To find P given F $P\|F\ i,n$	To find F given A $F\|A\ i,n$	To find A given F $A\|F\ i,n$	To find P given A $P\|A\ i,n$	To find A given P $A\|P\ i,n$	To find A given G $A\|G\ i,n$	To find P given G $P\|G\ i,n$
1	1.0067	0.9934	1.0000	1.0000	0.9934	1.0067	0.0000	0.0000
2	1.0134	0.9868	2.0067	0.4983	1.9802	0.5050	0.4983	0.9868
3	1.0201	0.9803	3.0200	0.3311	2.9604	0.3378	0.9956	2.9473
4	1.0269	0.9738	4.0402	0.2475	3.9342	0.2542	1.4917	5.8686
5	1.0338	0.9673	5.0671	0.1974	4.9015	0.2040	1.9867	9.7379
6	1.0407	0.9609	6.1009	0.1639	5.8625	0.1706	2.4806	14.5425
7	1.0476	0.9545	7.1416	0.1400	6.8170	0.1467	2.9734	20.2698
8	1.0546	0.9482	8.1892	0.1221	7.7652	0.1288	3.4651	26.9075
9	1.0616	0.9420	9.2438	0.1082	8.7072	0.1148	3.9557	34.4431
10	1.0687	0.9357	10.3054	0.0970	9.6429	0.1037	4.4452	42.8645
11	1.0758	0.9295	11.3741	0.0879	10.5724	0.0945	4.9336	52.1597
12	1.0830	0.9234	12.4499	0.0803	11.4958	0.0870	5.4208	62.3167
13	1.0902	0.9172	13.5329	0.0739	12.4130	0.0806	5.9070	73.3236
14	1.0975	0.9112	14.6231	0.0684	13.3242	0.0751	6.3920	85.1688
15	1.1048	0.9051	15.7206	0.0636	14.2293	0.0703	6.8760	97.8408
16	1.1122	0.8991	16.8254	0.0594	15.1285	0.0661	7.3588	111.3279
17	1.1196	0.8932	17.9376	0.0557	16.0217	0.0624	7.8406	125.6189
18	1.1270	0.8873	19.0572	0.0525	16.9089	0.0591	8.3212	140.7026
19	1.1346	0.8814	20.1842	0.0495	17.7903	0.0562	8.8007	156.5677
20	1.1421	0.8755	21.3188	0.0469	18.6659	0.0536	9.2791	173.2034
21	1.1497	0.8698	22.4609	0.0445	19.5357	0.0512	9.7564	190.5986
22	1.1574	0.8640	23.6107	0.0424	20.3997	0.0490	10.2327	208.7427
23	1.1651	0.8583	24.7681	0.0404	21.2579	0.0470	10.7078	227.6248
24	1.1729	0.8526	25.9332	0.0386	22.1105	0.0452	11.1818	247.2345
25	1.1807	0.8470	27.1061	0.0369	22.9575	0.0436	11.6546	267.5613
26	1.1885	0.8413	28.2868	0.0354	23.7988	0.0420	12.1264	288.5947
27	1.1965	0.8358	29.4754	0.0339	24.6346	0.0406	12.5971	310.3247
28	1.2045	0.8302	30.6719	0.0326	25.4648	0.0393	13.0667	332.7410
29	1.2125	0.8247	31.8763	0.0314	26.2896	0.0380	13.5352	355.8338
30	1.2206	0.8193	33.0888	0.0302	27.1089	0.0369	14.0025	379.5928
31	1.2287	0.8138	34.3094	0.0291	27.9227	0.0358	14.4688	404.0081
32	1.2369	0.8085	35.5382	0.0281	28.7312	0.0348	14.9340	429.0703
33	1.2452	0.8031	36.7751	0.0272	29.5343	0.0339	15.3980	454.7698
34	1.2535	0.7978	38.0203	0.0263	30.3320	0.0330	15.8610	481.0967
35	1.2618	0.7925	39.2737	0.0255	31.1246	0.0321	16.3229	508.0418
40	1.3045	0.7666	45.6675	0.0219	35.0090	0.0286	18.6157	651.7161
45	1.3495	0.7416	52.2773	0.0191	38.7666	0.0258	20.8810	809.4832
50	1.3941	0.7173	59.1104	0.0169	42.4013	0.0236	23.1188	980.2688
55	1.4412	0.6939	66.1743	0.0151	45.9173	0.0218	25.3293	1163.0535
60	1.4898	0.6712	73.4769	0.0136	49.3184	0.0203	27.5125	1356.8711
65	1.5402	0.6493	81.0261	0.0123	52.6084	0.0190	29.6684	1550.8049
70	1.5922	0.6281	88.8303	0.0113	55.7909	0.0179	31.7971	1773.9871
75	1.6460	0.6075	96.8982	0.0103	58.8693	0.0170	33.8987	1995.5950
80	1.7016	0.5877	105.2385	0.0095	61.8472	0.0162	35.9734	2224.8511
85	1.7591	0.5685	113.8607	0.0088	64.7277	0.0154	38.0211	2461.0191
90	1.8185	0.5499	122.7741	0.0081	67.5142	0.0148	40.0420	2703.4014
95	1.8799	0.5319	131.9986	0.0076	70.2096	0.0142	42.0362	2951.3406
100	1.9434	0.5146	141.5144	0.0071	72.8169	0.0137	44.0038	3204.2144

TABLE A.7 Discrete Compounding: i = 3/4%

	Single payment		Uniform series				Gradient series	
	Compound amount factor	Present worth factor	Compound amount factor	Sinking fund factor	Present worth factor	Capital recovery factor	Uniform series factor	Present worth factor
n	To find F given P $F\|P\ i,n$	To find P given F $P\|F\ i,n$	To find F given A $F\|A\ i,n$	To find A given F $A\|F\ i,n$	To find P given A $P\|A\ i,n$	To find A given P $A\|P\ i,n$	To find A given G $A\|G\ i,n$	To find P given G $P\|G\ i,n$
1	1.0075	0.9926	1.0000	1.0000	0.9926	1.0075	0.0000	0.0000
2	1.0151	0.9852	2.0075	0.4981	1.9777	0.5056	0.4981	0.9852
3	1.0227	0.9778	3.0226	0.3308	2.9556	0.3383	0.9950	2.9408
4	1.0303	0.9706	4.0452	0.2472	3.9261	0.2547	1.4907	5.8525
5	1.0381	0.9633	5.0756	0.1970	4.8894	0.2045	1.9851	9.7058
6	1.0459	0.9562	6.1136	0.1636	5.8456	0.1711	2.4782	14.4866
7	1.0537	0.9490	7.1595	0.1397	6.7946	0.1472	2.9701	20.1808
8	1.0616	0.9420	8.2132	0.1218	7.7366	0.1293	3.4608	26.7747
9	1.0696	0.9350	9.2748	0.1078	8.6716	0.1153	3.9502	34.2544
10	1.0776	0.9280	10.3443	0.0967	9.5996	0.1042	4.4384	42.6064
11	1.0857	0.9211	11.4219	0.0876	10.5207	0.0951	4.9253	51.8174
12	1.0938	0.9142	12.5076	0.0800	11.4349	0.0875	5.4110	61.8740
13	1.1020	0.9074	13.6014	0.0735	12.3423	0.0810	5.8954	72.7632
14	1.1103	0.9007	14.7034	0.0680	13.2430	0.0755	6.3786	84.4720
15	1.1186	0.8940	15.8137	0.0632	14.1370	0.0707	6.8606	96.9876
16	1.1270	0.8873	16.9323	0.0591	15.0243	0.0666	7.3413	110.2973
17	1.1354	0.8807	18.0593	0.0554	15.9050	0.0629	7.8207	124.3887
18	1.1440	0.8742	19.1947	0.0521	16.7792	0.0596	8.2989	139.2494
19	1.1525	0.8676	20.3387	0.0492	17.6468	0.0567	8.7759	154.8670
20	1.1612	0.8612	21.4912	0.0465	18.5080	0.0540	9.2516	171.2296
21	1.1699	0.8548	22.6524	0.0441	19.3628	0.0516	9.7261	188.3252
22	1.1787	0.8484	23.8223	0.0420	20.2112	0.0495	10.1994	206.1420
23	1.1875	0.8421	25.0010	0.0400	21.0533	0.0475	10.6714	224.6682
24	1.1964	0.8358	26.1885	0.0382	21.8891	0.0457	11.1422	243.8923
25	1.2054	0.8296	27.3849	0.0365	22.7188	0.0440	11.6117	263.8027
26	1.2144	0.8234	28.5903	0.0350	23.5422	0.0425	12.0800	284.3887
27	1.2235	0.8173	29.8047	0.0336	24.3595	0.0411	12.5470	305.6387
28	1.2327	0.8112	31.0282	0.0322	25.1707	0.0397	13.0128	327.5415
29	1.2420	0.8052	32.2609	0.0310	25.9759	0.0385	13.4774	350.0867
30	1.2513	0.7992	33.5029	0.0298	26.7751	0.0373	13.9407	373.2629
31	1.2607	0.7932	34.7542	0.0288	27.5683	0.0363	14.4028	397.0601
32	1.2701	0.7873	36.0148	0.0278	28.3557	0.0353	14.8636	421.4673
33	1.2795	0.7815	37.2349	0.0268	29.1371	0.0343	15.3232	446.4744
34	1.2892	0.7757	38.5646	0.0259	29.9128	0.0334	15.7816	472.0711
35	1.2989	0.7699	39.8538	0.0251	30.6827	0.0326	16.2387	498.2471
40	1.3483	0.7415	46.4465	0.0215	34.4469	0.0290	18.5058	637.4692
45	1.3997	0.7145	53.2901	0.0188	38.0732	0.0263	20.7421	789.7173
50	1.4530	0.6883	60.3943	0.0166	41.5665	0.0241	22.9476	953.8486
55	1.5083	0.6630	67.7688	0.0148	44.9316	0.0223	25.1223	128.7869
60	1.5657	0.6387	75.4241	0.0133	48.1734	0.0208	27.2665	313.5188
65	1.6253	0.6153	83.3708	0.0120	51.2963	0.0195	29.3801	507.0911
70	1.6872	0.5927	91.6201	0.0109	54.3045	0.0184	31.4634	708.6062
75	1.7514	0.5710	100.1833	0.0100	57.2027	0.0175	33.5163	917.2222
80	1.8180	0.5500	109.0725	0.0092	59.9944	0.0167	35.5391	132.1470
85	1.8873	0.5299	118.3001	0.0085	62.6838	0.0160	37.5318	352.6367
90	1.9591	0.5104	127.8789	0.0078	65.2746	0.0153	39.4946	577.9951
95	2.0337	0.4917	137.8224	0.0073	67.7704	0.0148	41.4277	807.5696
100	2.1111	0.4737	148.1444	0.0068	70.1746	0.0143	43.3311	040.7454

TABLE A.8 Discrete Compounding: i = 1%

	Single payment		Uniform series				Gradient series	
	Compound amount factor	Present worth factor	Compound amount factor	Sinking fund factor	Present worth factor	Capital recovery factor	Uniform series factor	Present worth factor
n	To find F given P $F/P\ i,n$	To find P given F $P/F\ i,n$	To find F given A $F/A\ i,n$	To find A given F $A/F\ i,n$	To find P given A $P/A\ i,n$	To find A given P $A/P\ i,n$	To find A given G $A/G\ i,n$	To find P given G $P/G\ i,n$
1	1.0100	0.9901	1.0000	1.0000	0.9901	1.0100	0.0000	0.0000
2	1.0201	0.9803	2.0100	0.4975	1.9704	0.5075	0.4975	0.9803
3	1.0303	0.9706	3.0301	0.3300	2.9410	0.3400	0.9934	2.9215
4	1.0406	0.9610	4.0604	0.2463	3.9020	0.2563	1.4876	5.8044
5	1.0510	0.9515	5.1010	0.1960	4.8534	0.2060	1.9801	9.6103
6	1.0615	0.9420	6.1520	0.1625	5.7955	0.1725	2.4710	14.3205
7	1.0721	0.9327	7.2135	0.1386	6.7282	0.1486	2.9602	19.9168
8	1.0829	0.9235	8.2857	0.1207	7.6517	0.1307	3.4478	26.3812
9	1.0937	0.9143	9.3685	0.1067	8.5660	0.1167	3.9337	33.6959
10	1.1046	0.9053	10.4622	0.0956	9.4713	0.1056	4.4179	41.8435
11	1.1157	0.8963	11.5668	0.0865	10.3676	0.0965	4.9005	50.8067
12	1.1268	0.8874	12.6825	0.0788	11.2551	0.0888	5.3815	60.5687
13	1.1381	0.8787	13.8093	0.0724	12.1337	0.0824	5.8607	71.1126
14	1.1495	0.8700	14.9474	0.0669	13.0037	0.0769	6.3384	82.4221
15	1.1610	0.8613	16.0969	0.0621	13.8651	0.0721	6.8143	94.4810
16	1.1726	0.8528	17.2579	0.0579	14.7179	0.0679	7.2886	107.2733
17	1.1843	0.8444	18.4304	0.0543	15.5623	0.0643	7.7613	120.7834
18	1.1961	0.8360	19.6147	0.0510	16.3983	0.0610	8.2323	134.9956
19	1.2081	0.8277	20.8109	0.0481	17.2260	0.0581	8.7017	149.8950
20	1.2202	0.8195	22.0190	0.0454	18.0455	0.0554	9.1694	165.4663
21	1.2324	0.8114	23.2392	0.0430	18.8570	0.0530	9.6354	181.6949
22	1.2447	0.8034	24.4716	0.0409	19.6604	0.0509	10.0998	198.5662
23	1.2572	0.7954	25.7163	0.0389	20.4558	0.0489	10.5626	216.0660
24	1.2697	0.7876	26.9735	0.0371	21.2434	0.0471	11.0237	234.1800
25	1.2824	0.7798	28.2432	0.0354	22.0232	0.0454	11.4831	252.8944
26	1.2953	0.7720	29.5256	0.0339	22.7952	0.0439	11.9409	272.1956
27	1.3082	0.7644	30.8209	0.0324	23.5596	0.0424	12.3971	292.0701
28	1.3213	0.7568	32.1291	0.0311	24.3164	0.0411	12.8516	312.5045
29	1.3345	0.7493	33.4504	0.0299	25.0658	0.0399	13.3044	333.4861
30	1.3478	0.7419	34.7849	0.0287	25.8077	0.0387	13.7557	355.0020
31	1.3613	0.7346	36.1327	0.0277	26.5423	0.0377	14.2052	377.0391
32	1.3749	0.7273	37.4941	0.0267	27.2696	0.0367	14.6532	399.5857
33	1.3887	0.7201	38.8690	0.0257	27.9897	0.0357	15.0995	422.6289
34	1.4026	0.7130	40.2577	0.0248	28.7027	0.0348	15.5441	446.1570
35	1.4166	0.7059	41.6603	0.0240	29.4085	0.0340	15.9871	470.1582
40	1.4889	0.6717	48.8864	0.0205	32.8347	0.0305	18.1776	596.8560
45	1.5648	0.6391	56.4811	0.0177	36.0945	0.0277	20.3273	733.7036
50	1.6446	0.6080	64.4632	0.0155	39.1961	0.0255	22.4363	879.4175
55	1.7285	0.5785	72.8525	0.0137	42.1472	0.0237	24.5049	1032.8147
60	1.8167	0.5504	81.6697	0.0122	44.9550	0.0222	26.5333	1192.8062
65	1.9094	0.5237	90.9366	0.0110	47.6266	0.0210	28.5217	1358.3901
70	2.0068	0.4983	100.6763	0.0099	50.1685	0.0199	30.4703	1528.6475
75	2.1091	0.4741	110.9128	0.0090	52.5871	0.0190	32.3793	1702.7327
80	2.2167	0.4511	121.6715	0.0082	54.8882	0.0182	34.2492	1879.8767
85	2.3298	0.4292	132.9789	0.0075	57.0777	0.0175	36.0801	2059.3694
90	2.4486	0.4084	144.8632	0.0069	59.1609	0.0169	37.8724	2240.5662
95	2.5735	0.3886	157.3537	0.0064	61.1430	0.0164	39.6265	2422.8799
100	2.7048	0.3697	170.4813	0.0059	63.0289	0.0159	41.3426	2605.7759

TABLE A.9 Discrete Compounding: $i = 3/2\%$

	Single payment		Uniform series				Gradient series	
	Compound amount factor	Present worth factor	Compound amount factor	Sinking fund factor	Present worth factor	Capital recovery factor	Uniform series factor	Present worth factor
n	To find F given P $F/P\ i,n$	To find P given F $P/F\ i,n$	To find F given A $F/A\ i,n$	To find A given F $A/F\ i,n$	To find P given A $P/A\ i,n$	To find A given P $A/P\ i,n$	To find A given G $A/G\ i,n$	To find P given G $P/G\ i,n$
1	1.0150	0.9852	1.0000	1.0000	0.9852	1.0150	0.0000	0.0000
2	1.0302	0.9707	2.0150	0.4963	1.9559	0.5113	0.4963	0.9707
3	1.0457	0.9563	3.0452	0.3284	2.9122	0.3434	0.9901	2.8833
4	1.0614	0.9422	4.0909	0.2444	3.8544	0.2594	1.4814	5.7098
5	1.0773	0.9283	5.1523	0.1941	4.7826	0.2091	1.9702	9.4229
6	1.0934	0.9145	6.2296	0.1605	5.6972	0.1755	2.4566	13.9956
7	1.1098	0.9010	7.3230	0.1366	6.5982	0.1516	2.9405	19.4018
8	1.1265	0.8877	8.4328	0.1186	7.4859	0.1336	3.4219	25.6157
9	1.1434	0.8746	9.5593	0.1046	8.3605	0.1196	3.9008	32.6125
10	1.1605	0.8617	10.7027	0.0934	9.2222	0.1084	4.3772	40.3675
11	1.1779	0.8489	11.8633	0.0843	10.0711	0.0993	4.8512	48.8568
12	1.1956	0.8364	13.0412	0.0767	10.9075	0.0917	5.3227	58.0571
13	1.2136	0.8240	14.2368	0.0702	11.7315	0.0852	5.7917	67.9454
14	1.2318	0.8118	15.4504	0.0647	12.5434	0.0797	6.2582	78.4994
15	1.2502	0.7999	16.6821	0.0599	13.3432	0.0749	6.7223	89.6974
16	1.2690	0.7880	17.9324	0.0558	14.1313	0.0708	7.1839	101.5178
17	1.2880	0.7764	19.2014	0.0521	14.9076	0.0671	7.6431	113.9399
18	1.3073	0.7649	20.4894	0.0488	15.6726	0.0638	8.0997	126.9434
19	1.3270	0.7536	21.7967	0.0459	16.4262	0.0609	8.5539	140.5084
20	1.3469	0.7425	23.1237	0.0432	17.1686	0.0582	9.0057	154.6153
21	1.3671	0.7315	24.4705	0.0409	17.9001	0.0559	9.4550	169.2453
22	1.3876	0.7207	25.8376	0.0387	18.6208	0.0537	9.9018	184.3797
23	1.4084	0.7100	27.2251	0.0367	19.3309	0.0517	10.3462	200.0005
24	1.4295	0.6995	28.6335	0.0349	20.0304	0.0499	10.7881	216.0900
25	1.4509	0.6892	30.0630	0.0333	20.7196	0.0483	11.2276	232.6310
26	1.4727	0.6790	31.5140	0.0317	21.3986	0.0467	11.6646	249.6065
27	1.4948	0.6690	32.9867	0.0303	22.0676	0.0453	12.0992	267.0000
28	1.5172	0.6591	34.4815	0.0290	22.7267	0.0440	12.5313	284.7957
29	1.5400	0.6494	35.9987	0.0278	23.3761	0.0428	12.9610	302.9778
30	1.5631	0.6398	37.5387	0.0266	24.0158	0.0416	13.3883	321.5308
31	1.5865	0.6303	39.1018	0.0256	24.6461	0.0406	13.8131	340.4402
32	1.6103	0.6210	40.6883	0.0246	25.2671	0.0395	14.2355	359.6909
33	1.6345	0.6118	42.2986	0.0236	25.8790	0.0386	14.6555	379.2691
34	1.6590	0.6028	43.9331	0.0228	26.4817	0.0378	15.0731	399.1604
35	1.6839	0.5939	45.5921	0.0219	27.0756	0.0369	15.4882	419.3521
40	1.8140	0.5513	54.2679	0.0184	29.9158	0.0334	17.5277	524.3567
45	1.9542	0.5117	63.6142	0.0157	32.5523	0.0307	19.5074	635.0110
50	2.1052	0.4750	73.6828	0.0136	34.9997	0.0286	21.4277	749.9634
55	2.2679	0.4409	84.5296	0.0118	37.2715	0.0268	23.2894	868.0283
60	2.4432	0.4093	96.2146	0.0104	39.3803	0.0254	25.0930	988.1675
65	2.6320	0.3799	108.8027	0.0092	41.3378	0.0242	26.8393	1139.4751
70	2.8355	0.3527	122.3637	0.0082	43.1549	0.0232	28.5290	1231.1658
75	3.0546	0.3274	136.9727	0.0073	44.8416	0.0223	30.1631	1352.5601
80	3.2907	0.3039	152.7108	0.0065	46.4073	0.0215	31.7423	1473.0742
85	3.5450	0.2821	169.6651	0.0059	47.8607	0.0209	33.2676	1592.2095
90	3.8189	0.2619	187.9298	0.0053	49.2099	0.0203	34.7399	1709.5440
95	4.1141	0.2431	207.6061	0.0048	50.4622	0.0198	36.1602	1824.7214
100	4.4320	0.2256	228.8029	0.0044	51.6247	0.0194	37.5295	1937.4495

TABLE A.10 Discrete Compounding: i = 2%

	Single payment		Uniform series				Gradient series	
	Compound amount factor	Present worth factor	Compound amount factor	Sinking fund factor	Present worth factor	Capital recovery factor	Uniform series factor	Present worth factor
n	To find F given P $F\|P\ i,n$	To find P given F $P\|F\ i,n$	To find F given A $F\|A\ i,n$	To find A given F $A\|F\ i,n$	To find P given A $P\|A\ i,n$	To find A given P $A\|P\ i,n$	To find A given G $A\|G\ i,n$	To find P given G $P\|G\ i,n$
1	1.0200	0.9804	1.0000	1.0000	0.9804	1.0200	0.0000	0.0000
2	1.0404	0.9612	2.0200	0.4950	1.9416	0.5150	0.4950	0.9612
3	1.0612	0.9423	3.0604	0.3268	2.8839	0.3468	0.9868	2.8458
4	1.0824	0.9238	4.1216	0.2426	3.8077	0.2626	1.4752	5.6173
5	1.1041	0.9057	5.2040	0.1922	4.7135	0.2122	1.9604	9.2403
6	1.1262	0.8880	6.3081	0.1585	5.6014	0.1785	2.4423	13.6801
7	1.1487	0.8706	7.4343	0.1345	6.4720	0.1545	2.9208	18.9035
8	1.1717	0.8535	8.5830	0.1165	7.3255	0.1365	3.3961	24.8779
9	1.1951	0.8368	9.7546	0.1025	8.1622	0.1225	3.8681	31.5720
10	1.2190	0.8203	10.9497	0.0913	8.9826	0.1113	4.3367	38.9551
11	1.2434	0.8043	12.1687	0.0822	9.7868	0.1022	4.8021	46.9977
12	1.2682	0.7885	13.4121	0.0746	10.5753	0.0946	5.2642	55.6712
13	1.2936	0.7730	14.6803	0.0681	11.3484	0.0881	5.7231	64.9475
14	1.3195	0.7579	15.9739	0.0626	12.1062	0.0826	6.1786	74.7999
15	1.3459	0.7430	17.2934	0.0578	12.8493	0.0778	6.6309	85.2021
16	1.3728	0.7284	18.6393	0.0537	13.5777	0.0737	7.0799	96.1288
17	1.4002	0.7142	20.0121	0.0500	14.2919	0.0700	7.5256	107.5554
18	1.4282	0.7002	21.4123	0.0467	14.9920	0.0667	7.9681	119.4581
19	1.4568	0.6864	22.8406	0.0438	15.6785	0.0638	8.4073	131.8138
20	1.4859	0.6730	24.2974	0.0412	16.3514	0.0612	8.8433	144.6003
21	1.5157	0.6598	25.7833	0.0388	17.0112	0.0588	9.2760	157.7958
22	1.5460	0.6468	27.2990	0.0366	17.6580	0.0566	9.7055	171.3794
23	1.5769	0.6342	28.8450	0.0347	18.2922	0.0547	10.1317	185.3309
24	1.6084	0.6217	30.4219	0.0329	18.9139	0.0529	10.5547	199.6304
25	1.6406	0.6095	32.0303	0.0312	19.5235	0.0512	10.9745	214.2592
26	1.6734	0.5976	33.6709	0.0297	20.1210	0.0497	11.3910	229.1987
27	1.7069	0.5859	35.3443	0.0283	20.7069	0.0483	11.8043	244.4311
28	1.7410	0.5744	37.0512	0.0270	21.2813	0.0470	12.2145	259.9390
29	1.7758	0.5631	38.7922	0.0258	21.8444	0.0458	12.6214	275.7063
30	1.8114	0.5521	40.5681	0.0246	22.3965	0.0446	13.0251	291.7163
31	1.8476	0.5412	42.3794	0.0236	22.9377	0.0436	13.4257	307.9536
32	1.8845	0.5306	44.2270	0.0226	23.4683	0.0426	13.8230	324.4033
33	1.9222	0.5202	46.1116	0.0217	23.9886	0.0417	14.2172	341.0505
34	1.9607	0.5100	48.0338	0.0208	24.4986	0.0408	14.6083	357.9816
35	1.9999	0.5000	49.9945	0.0200	24.9986	0.0400	14.9961	374.8826
40	2.2080	0.4529	60.4020	0.0166	27.3555	0.0366	16.8885	461.9929
45	2.4379	0.4102	71.8927	0.0139	29.4902	0.0339	18.7034	551.5650
50	2.6916	0.3715	84.5794	0.0118	31.4236	0.0318	20.4420	642.3604
55	2.9717	0.3365	98.5865	0.0101	33.1748	0.0301	22.1057	733.3525
60	3.2810	0.3048	114.0515	0.0088	34.7609	0.0288	23.6961	823.6973
65	3.6225	0.2761	131.1261	0.0076	36.1975	0.0276	25.2147	912.7085
70	3.9996	0.2500	149.9779	0.0067	37.4986	0.0267	26.6632	999.8342
75	4.4158	0.2265	170.7917	0.0059	38.6771	0.0259	28.0434	1084.6392
80	4.8754	0.2051	193.7719	0.0052	39.7445	0.0252	29.3572	1166.7866
85	5.3829	0.1858	219.1439	0.0046	40.7113	0.0246	30.6064	1246.0239
90	5.9431	0.1683	247.1566	0.0040	41.5869	0.0240	31.7929	1322.1699
95	6.5617	0.1524	278.0847	0.0036	42.3800	0.0236	32.9189	1395.1033
100	7.2446	0.1380	312.2322	0.0032	43.0984	0.0232	33.9863	1464.7527

TABLE A.11 Discrete Compounding: i = 5/2%

	Single payment		Uniform series				Gradient series	
	Compound amount factor	Present worth factor	Compound amount factor	Sinking fund factor	Present worth factor	Capital recovery factor	Uniform series factor	Present worth factor
n	To find F given P $F/P\ i,n$	To find P given F $P/F\ i,n$	To find F given A $F/A\ i,n$	To find A given F $A/F\ i,n$	To find P given A $P/A\ i,n$	To find A given P $A/P\ i,n$	To find A given G $A/G\ i,n$	To find P given G $P/G\ i,n$
1	1.0250	0.9756	1.0000	1.0000	0.9756	1.0250	0.0000	0.0000
2	1.0506	0.9518	2.0250	0.4938	1.9274	0.5188	0.4938	0.9518
3	1.0769	0.9286	3.0756	0.3251	2.8560	0.3501	0.9835	2.8090
4	1.1038	0.9060	4.1525	0.2408	3.7620	0.2558	1.4691	5.5269
5	1.1314	0.8839	5.2563	0.1902	4.6458	0.2152	1.9506	9.0623
6	1.1597	0.8623	6.3877	0.1565	5.5081	0.1815	2.4280	13.3738
7	1.1887	0.8413	7.5474	0.1325	6.3494	0.1575	2.9013	18.4214
8	1.2184	0.8207	8.7361	0.1145	7.1701	0.1395	3.3704	24.1666
9	1.2489	0.8007	9.9545	0.1005	7.9709	0.1255	3.8355	30.5724
10	1.2801	0.7812	11.2034	0.0893	8.7521	0.1143	4.2965	37.6032
11	1.3121	0.7621	12.4835	0.0801	9.5142	0.1051	4.7534	45.2246
12	1.3449	0.7436	13.7956	0.0725	10.2578	0.0975	5.2062	53.4038
13	1.3785	0.7254	15.1404	0.0660	10.9832	0.0910	5.6549	62.1086
14	1.4130	0.7077	16.5190	0.0605	11.6909	0.0855	6.0995	71.3093
15	1.4483	0.6905	17.9319	0.0558	12.3814	0.0808	6.5401	80.9758
16	1.4845	0.6736	19.3802	0.0516	13.0550	0.0766	6.9766	91.0802
17	1.5216	0.6572	20.8647	0.0479	13.7122	0.0729	7.4091	101.5953
18	1.5597	0.6412	22.3863	0.0447	14.3534	0.0597	7.8375	112.4950
19	1.5987	0.6255	23.9460	0.0418	14.9789	0.0668	8.2619	123.7545
20	1.6386	0.6103	25.5447	0.0391	15.5892	0.0641	8.6823	135.3497
21	1.6796	0.5954	27.1833	0.0368	16.1845	0.0618	9.0986	147.2574
22	1.7216	0.5809	28.8629	0.0346	16.7654	0.0596	9.5110	159.4556
23	1.7646	0.5667	30.5844	0.0327	17.3321	0.0577	9.9193	171.9229
24	1.8087	0.5529	32.3490	0.0309	17.8850	0.0559	10.3237	184.6390
25	1.8539	0.5394	34.1578	0.0293	18.4244	0.0543	10.7241	197.5844
26	1.9003	0.5262	36.0117	0.0278	18.9506	0.0528	11.1205	210.7403
27	1.9478	0.5134	37.9120	0.0264	19.4640	0.0514	11.5130	224.0887
28	1.9965	0.5009	39.8598	0.0251	19.9649	0.0501	11.9015	237.6124
29	2.0464	0.4887	41.8563	0.0239	20.4536	0.0489	12.2861	251.2949
30	2.0976	0.4767	43.9027	0.0228	20.9303	0.0478	12.6668	265.1204
31	2.1500	0.4651	46.0003	0.0217	21.3954	0.0467	13.0436	279.0737
32	2.2038	0.4538	48.1503	0.0208	21.8492	0.0458	13.4166	293.1406
33	2.2589	0.4427	50.3540	0.0199	22.2919	0.0449	13.7856	307.3071
34	2.3153	0.4319	52.6129	0.0190	22.7238	0.0440	14.1508	321.5601
35	2.3732	0.4214	54.9282	0.0182	23.1452	0.0432	14.5122	335.8867
40	2.6851	0.3724	67.4026	0.0148	25.1028	0.0398	16.2620	408.2219
45	3.0379	0.3292	81.5161	0.0123	26.8330	0.0373	17.9185	480.8069
50	3.4371	0.2909	97.4843	0.0103	28.3623	0.0353	19.4839	552.6079
55	3.8888	0.2572	115.5509	0.0087	29.7140	0.0337	20.9608	622.8279
60	4.3998	0.2273	135.9915	0.0074	30.9087	0.0324	22.3518	690.8655
65	4.9780	0.2009	159.1183	0.0063	31.9645	0.0313	23.6600	756.2805
70	5.6321	0.1776	185.2841	0.0054	32.8979	0.0304	24.8881	818.7642
75	6.3722	0.1569	214.8882	0.0047	33.7227	0.0297	26.0393	878.1150
80	7.2096	0.1387	248.3826	0.0040	34.4518	0.0290	27.1167	934.2180
85	8.1570	0.1226	286.2783	0.0035	35.0962	0.0285	28.1235	987.0269
90	9.2289	0.1084	329.1541	0.0030	35.6658	0.0280	29.0629	1036.5498
95	10.4416	0.0958	377.6641	0.0026	36.1692	0.0276	29.9382	1082.8379
100	11.8137	0.0846	432.5486	0.0023	36.6141	0.0273	30.7525	1125.9746

TABLE A.12 Discrete Compounding: i = 3%

	Single payment		Uniform series				Gradient series	
	Compound amount factor	Present worth factor	Compound amount factor	Sinking fund factor	Present worth factor	Capital recovery factor	Uniform series factor	Present worth factor
n	To find F given P F/P i,n	To find P given F P/F i,n	To find F given A F/A i,n	To find A given F A/F i,n	To find P given A P/A i,n	To find A given P A/P i,n	To find A given G A/G i,n	To find P given G P/G i,n
1	1.0300	0.9709	1.0000	1.0000	0.9709	1.0300	0.0000	0.0000
2	1.0609	0.9426	2.0300	0.4926	1.9135	0.5226	0.4926	0.9426
3	1.0927	0.9151	3.0909	0.3235	2.8286	0.3535	0.9803	2.7729
4	1.1255	0.8885	4.1836	0.2390	3.7171	0.2690	1.4631	5.4383
5	1.1593	0.8626	5.3091	0.1884	4.5797	0.2184	1.9409	8.8888
6	1.1941	0.8375	6.4684	0.1546	5.4172	0.1846	2.4138	13.0762
7	1.2299	0.8131	7.6625	0.1305	6.2303	0.1605	2.8819	17.9547
8	1.2668	0.7894	8.8923	0.1125	7.0197	0.1425	3.3450	23.4806
9	1.3048	0.7664	10.1591	0.0984	7.7861	0.1284	3.8032	29.6119
10	1.3439	0.7441	11.4639	0.0872	8.5302	0.1172	4.2565	36.3088
11	1.3842	0.7224	12.8078	0.0781	9.2526	0.1081	4.7049	43.5330
12	1.4258	0.7014	14.1920	0.0705	9.9540	0.1005	5.1485	51.2482
13	1.4685	0.6810	15.6178	0.0640	10.6350	0.0940	5.5872	59.4196
14	1.5126	0.6611	17.0863	0.0585	11.2961	0.0885	6.0210	68.0141
15	1.5580	0.6419	18.5989	0.0538	11.9379	0.0838	6.4500	77.0002
16	1.6047	0.6232	20.1569	0.0496	12.5611	0.0796	6.8742	86.3477
17	1.6528	0.6050	21.7616	0.0460	13.1661	0.0760	7.2936	96.0280
18	1.7024	0.5874	23.4144	0.0427	13.7535	0.0727	7.7081	106.0136
19	1.7535	0.5703	25.1169	0.0398	14.3238	0.0698	8.1179	116.2788
20	1.8061	0.5537	26.8704	0.0372	14.8775	0.0672	8.5229	126.7986
21	1.8603	0.5375	28.6765	0.0349	15.4150	0.0649	8.9231	137.5496
22	1.9161	0.5219	30.5368	0.0327	15.9369	0.0627	9.3186	148.5093
23	1.9736	0.5067	32.4529	0.0308	16.4436	0.0608	9.7093	159.6566
24	2.0328	0.4919	34.4265	0.0290	16.9355	0.0590	10.0954	170.9710
25	2.0938	0.4775	36.4593	0.0274	17.4131	0.0574	10.4768	182.4336
26	2.1566	0.4637	38.5530	0.0259	17.8768	0.0559	10.8535	194.0259
27	2.2213	0.4502	40.7096	0.0246	18.3270	0.0546	11.2255	205.7309
28	2.2879	0.4371	42.9309	0.0233	18.7641	0.0533	11.5930	217.5319
29	2.3566	0.4243	45.2188	0.0221	19.1885	0.0521	11.9558	229.4136
30	2.4273	0.4120	47.5754	0.0210	19.6004	0.0510	12.3141	241.3612
31	2.5001	0.4000	50.0027	0.0200	20.0004	0.0500	12.6678	253.3609
32	2.5751	0.3883	52.5028	0.0190	20.3888	0.0490	13.0169	265.3992
33	2.6523	0.3770	55.0778	0.0182	20.7658	0.0482	13.3616	277.4641
34	2.7319	0.3660	57.7302	0.0173	21.1318	0.0473	13.7018	289.5435
35	2.8139	0.3554	60.4621	0.0165	21.4872	0.0465	14.0375	301.6265
40	3.2620	0.3066	75.4013	0.0133	23.1148	0.0433	15.6502	361.7498
45	3.7816	0.2644	92.7199	0.0108	24.5187	0.0408	17.1556	420.6323
50	4.3839	0.2281	112.7968	0.0089	25.7298	0.0389	18.5575	477.4802
55	5.0821	0.1968	136.0716	0.0073	26.7744	0.0373	19.8600	531.7410
60	5.8916	0.1697	163.0534	0.0061	27.6756	0.0361	21.0674	583.0525
65	6.8300	0.1464	194.3327	0.0051	28.4529	0.0351	22.1841	631.2009
70	7.9178	0.1263	230.5940	0.0043	29.1234	0.0343	23.2145	676.0867
75	9.1789	0.1089	272.6306	0.0037	29.7018	0.0337	24.1634	717.6978
80	10.6409	0.0940	321.3628	0.0031	30.2008	0.0331	25.0353	756.0864
85	12.3357	0.0811	377.8567	0.0026	30.6312	0.0326	25.8349	791.3528
90	14.3005	0.0699	443.3486	0.0023	31.0024	0.0323	26.5667	823.6301
95	16.5782	0.0603	519.2717	0.0019	31.3227	0.0319	27.2351	853.0740
100	19.2186	0.0520	607.2876	0.0016	31.5989	0.0316	27.8444	879.8540

TABLE A.13 Discrete Compounding: i = 4%

	Single payment		Uniform series				Gradient series	
	Compound amount factor	Present worth factor	Compound amount factor	Sinking fund factor	Present worth factor	Capital recovery factor	Uniform series factor	Present worth factor
n	To find F given P $F\|P\ i,n$	To find P given F $P\|F\ i,n$	To find F given A $F\|A\ i,n$	To find A given F $A\|F\ i,n$	To find P given A $P\|A\ i,n$	To find A given P $A\|P\ i,n$	To find A given G $A\|G\ i,n$	To find P given G $P\|G\ i,n$
1	1.0400	0.9615	1.0000	1.0000	0.9615	1.0400	0.0000	0.0000
2	1.0816	0.9245	2.0400	0.4902	1.8861	0.5302	0.4902	0.9246
3	1.1249	0.8890	3.1216	0.3203	2.7751	0.3603	0.9739	2.7025
4	1.1699	0.8548	4.2465	0.2355	3.6299	0.2755	1.4510	5.2670
5	1.2167	0.8219	5.4163	0.1846	4.4518	0.2246	1.9216	8.5547
6	1.2653	0.7903	6.6330	0.1508	5.2421	0.1908	2.3857	12.5062
7	1.3159	0.7599	7.8983	0.1266	6.0021	0.1666	2.8433	17.0657
8	1.3686	0.7307	9.2142	0.1085	6.7327	0.1485	3.2944	22.1806
9	1.4233	0.7026	10.5828	0.0945	7.4353	0.1345	3.7391	27.8013
10	1.4802	0.6756	12.0061	0.0833	8.1109	0.1233	4.1773	33.8814
11	1.5395	0.6496	13.4864	0.0741	8.7605	0.1141	4.6090	40.3772
12	1.6010	0.6246	15.3258	0.0666	9.3851	0.1066	5.0343	47.2477
13	1.6651	0.6006	16.6268	0.0601	9.9856	0.1001	5.4533	54.4546
14	1.7317	0.5775	18.2919	0.0547	10.5631	0.0947	5.8659	61.9618
15	1.8009	0.5553	20.0236	0.0499	11.1184	0.0899	6.2721	69.7355
16	1.8730	0.5339	21.8245	0.0458	11.6523	0.0858	6.6720	77.7441
17	1.9479	0.5134	23.6975	0.0422	12.1657	0.0822	7.0656	85.9581
18	2.0258	0.4936	25.6454	0.0390	12.6593	0.0790	7.4530	94.3498
19	2.1068	0.4746	27.6712	0.0361	13.1339	0.0761	7.8342	102.8933
20	2.1911	0.4564	29.7781	0.0336	13.5903	0.0736	8.2091	111.5646
21	2.2788	0.4388	31.9692	0.0313	14.0292	0.0713	8.5779	120.3413
22	2.3699	0.4220	34.2480	0.0292	14.4511	0.0692	8.9407	129.2024
23	2.4647	0.4057	36.6179	0.0273	14.8568	0.0673	9.2973	138.1284
24	2.5633	0.3901	39.0826	0.0256	15.2470	0.0656	9.6479	147.1011
25	2.6658	0.3751	41.6459	0.0240	15.6221	0.0640	9.9925	156.1039
26	2.7725	0.3607	44.3117	0.0226	15.9828	0.0626	10.3312	165.1212
27	2.8834	0.3468	47.0842	0.0212	16.3296	0.0612	10.6640	174.1384
28	2.9987	0.3335	49.9676	0.0200	16.6631	0.0600	10.9909	183.1423
29	3.1187	0.3207	52.9663	0.0189	16.9837	0.0589	11.3120	192.1205
30	3.2434	0.3083	56.0849	0.0178	17.2920	0.0578	11.6274	201.0618
31	3.3731	0.2965	59.3283	0.0169	17.5885	0.0559	11.9371	209.9556
32	3.5081	0.2851	62.7015	0.0159	17.8736	0.0559	12.2411	218.7924
33	3.6484	0.2741	66.2095	0.0151	18.1476	0.0551	12.5396	227.5634
34	3.7943	0.2636	69.8579	0.0143	18.4112	0.0543	12.8324	236.2606
35	3.9461	0.2534	73.6522	0.0136	18.6646	0.0536	13.1198	244.8767
40	4.8010	0.2083	95.0255	0.0105	19.7928	0.0505	14.4765	286.5303
45	5.8412	0.1712	121.0293	0.0083	20.7200	0.0483	15.7047	325.4026
50	7.1067	0.1407	152.6670	0.0066	21.4822	0.0466	16.8122	361.1636
55	8.6464	0.1157	191.1591	0.0052	22.1086	0.0452	17.8070	393.6887
60	10.5196	0.0951	237.9906	0.0042	22.6235	0.0442	18.6972	422.9966
65	12.7987	0.0781	294.9683	0.0034	23.0467	0.0434	19.4909	449.2012
70	15.5716	0.0642	364.2903	0.0027	23.3945	0.0427	20.1961	472.4788
75	18.9453	0.0528	448.6311	0.0022	23.6804	0.0422	20.8206	493.0408
80	23.0498	0.0434	551.2449	0.0018	23.9154	0.0418	21.3718	511.1160
85	28.0436	0.0357	676.0899	0.0015	24.1085	0.0415	21.8569	526.9382
90	34.1193	0.0293	827.9832	0.0012	24.2673	0.0412	22.2826	540.7368
95	41.5114	0.0241	1012.7844	0.0010	24.3978	0.0410	22.6550	552.7305
100	50.5049	0.0198	1237.6235	0.0008	24.5050	0.0408	22.9800	563.1248

TABLE A.14 Discrete Compounding: i = 5%

	Single payment		Uniform series				Gradient series	
	Compound amount factor	Present worth factor	Compound amount factor	Sinking fund factor	Present worth factor	Capital recovery factor	Uniform series factor	Present worth factor
n	To find F given P F/P i,n	To find P given F P/F i,n	To find F given A F/A i,n	To find A given F A/F i,n	To find P given A P/A i,n	To find A given P A/P i,n	To find A given G A/G i,n	To find P given G P/G i,n
1	1.0500	0.9524	1.0000	1.0000	0.9524	1.0500	0.0000	0.0000
2	1.1025	0.9070	2.0500	0.4878	1.8594	0.5378	0.4878	0.9070
3	1.1576	0.8638	3.1525	0.3172	2.7232	0.3672	0.9675	2.6347
4	1.2155	0.8227	4.3101	0.2320	3.5460	0.2820	1.4391	5.1028
5	1.2763	0.7835	5.5256	0.1810	4.3295	0.2310	1.9025	8.2369
6	1.3401	0.7462	6.8019	0.1470	5.0757	0.1970	2.3579	11.9680
7	1.4071	0.7107	8.1420	0.1228	5.7864	0.1728	2.8052	16.2321
8	1.4775	0.6768	9.5491	0.1047	6.4632	0.1547	3.2445	20.9700
9	1.5513	0.6446	11.0266	0.0907	7.1078	0.1407	3.6758	26.1268
10	1.6289	0.6139	12.5779	0.0795	7.7217	0.1295	4.0991	31.6520
11	1.7103	0.5847	14.2068	0.0704	8.3064	0.1204	4.5144	37.4988
12	1.7959	0.5568	15.9171	0.0628	8.8633	0.1128	4.9219	43.6241
13	1.8856	0.5303	17.7130	0.0565	9.3936	0.1065	5.3215	49.9879
14	1.9799	0.5051	19.5986	0.0510	9.8986	0.1010	5.7133	56.5538
15	2.0789	0.4810	21.5786	0.0463	10.3797	0.0963	6.0973	63.2880
16	2.1829	0.4581	23.6575	0.0423	10.8378	0.0923	6.4736	70.1597
17	2.2920	0.4363	25.8404	0.0387	11.2741	0.0887	6.8423	77.1405
18	2.4066	0.4155	28.1324	0.0355	11.6896	0.0855	7.2034	84.2043
19	2.5270	0.3957	30.5390	0.0327	12.0853	0.0827	7.5569	91.3275
20	2.6533	0.3769	33.0660	0.0302	12.4622	0.0802	7.9030	98.4884
21	2.7860	0.3589	35.7193	0.0280	12.8212	0.0780	8.2416	105.6672
22	2.9253	0.3418	38.5052	0.0260	13.1630	0.0760	8.5730	112.8461
23	3.0715	0.3256	41.4305	0.0241	13.4886	0.0741	8.8971	120.0086
24	3.2251	0.3101	44.5020	0.0225	13.7986	0.0725	9.2140	127.1402
25	3.3864	0.2953	47.7271	0.0210	14.0939	0.0710	9.5238	134.2275
26	3.5557	0.2812	51.1135	0.0196	14.3752	0.0696	9.8266	141.2585
27	3.7335	0.2678	54.6691	0.0183	14.6430	0.0683	10.1224	148.2225
28	3.9201	0.2551	58.4026	0.0171	14.8981	0.0671	10.4114	155.1101
29	4.1161	0.2429	62.3227	0.0160	15.1411	0.0660	10.6936	161.9126
30	4.3219	0.2314	66.4388	0.0151	15.3725	0.0651	10.9691	168.6225
31	4.5380	0.2204	70.7608	0.0141	15.5928	0.0641	11.2381	175.2333
32	4.7649	0.2099	75.2988	0.0133	15.8027	0.0633	11.5005	181.7391
33	5.0032	0.1999	80.0638	0.0125	16.0025	0.0625	11.7566	188.1351
34	5.2533	0.1904	85.0670	0.0118	16.1929	0.0618	12.0063	194.4168
35	5.5160	0.1813	90.3203	0.0111	16.3742	0.0611	12.2498	200.5806
40	7.0400	0.1420	120.7997	0.0083	17.1591	0.0583	13.3775	229.5451
45	8.9850	0.1113	159.7001	0.0063	17.7741	0.0563	14.3644	255.3145
50	11.4674	0.0872	209.3479	0.0048	18.2559	0.0548	15.2233	277.9148
55	14.6356	0.0683	272.7124	0.0037	18.6335	0.0537	15.9664	297.5103
60	18.6792	0.0535	353.5835	0.0028	18.9293	0.0528	16.6062	314.3430
65	23.8399	0.0419	456.7979	0.0022	19.1611	0.0522	17.1541	328.6909
70	30.4264	0.0329	588.5283	0.0017	19.3427	0.0517	17.6212	340.8408
75	38.8327	0.0258	756.6533	0.0013	19.4850	0.0513	18.0176	351.0720
80	49.5614	0.0202	971.2285	0.0010	19.5965	0.0510	18.3526	359.6460
85	63.2543	0.0158	1245.0867	0.0008	19.6838	0.0508	18.6346	366.8005
90	80.7303	0.0124	1594.6067	0.0006	19.7523	0.0506	18.8712	372.7488
95	103.0346	0.0097	2040.6926	0.0005	19.8059	0.0505	19.0689	377.6773
100	131.5012	0.0076	2610.0239	0.0004	19.8479	0.0504	19.2337	381.7490

TABLE A.15 Discrete Compounding: i = 6%

	Single payment		Uniform series				Gradient series	
	Compound amount factor	Present worth factor	Compound amount factor	Sinking fund factor	Present worth factor	Capital recovery factor	Uniform series factor	Present worth factor
n	To find F given P $F/P\ i,n$	To find P given F $P/F\ i,n$	To find F given A $F/A\ i,n$	To find A given F $A/F\ i,n$	To find P given A $P/A\ i,n$	To find A given P $A/P\ i,n$	To find A given G $A/G\ i,n$	To find P given G $P/G\ i,n$
1	1.0600	0.9434	1.0000	1.0000	0.9434	1.0600	0.0000	0.0000
2	1.1236	0.8900	2.0600	0.4854	1.8334	0.5454	0.4854	0.8900
3	1.1910	0.8396	3.1836	0.3141	2.6730	0.3741	0.9612	2.5692
4	1.2625	0.7921	4.3746	0.2286	3.4651	0.2886	1.4272	4.9455
5	1.3382	0.7473	5.6371	0.1774	4.2124	0.2374	1.8836	7.9345
6	1.4185	0.7050	6.9753	0.1434	4.9173	0.2034	2.3304	11.4594
7	1.5036	0.6651	8.3938	0.1191	5.5824	0.1791	2.7676	15.4497
8	1.5938	0.6274	9.8975	0.1010	6.2098	0.1610	3.1952	19.8416
9	1.6895	0.5919	11.4913	0.0870	6.8017	0.1470	3.6133	24.5768
10	1.7908	0.5584	13.1808	0.0759	7.3601	0.1359	4.0220	29.6023
11	1.8983	0.5268	14.9716	0.0668	7.8869	0.1268	4.4213	34.8702
12	2.0122	0.4970	16.8699	0.0593	8.3838	0.1193	4.8113	40.3369
13	2.1329	0.4688	18.8821	0.0530	8.8527	0.1130	5.1920	45.9629
14	2.2609	0.4423	21.0151	0.0476	9.2950	0.1076	5.5635	51.7128
15	2.3966	0.4173	23.2760	0.0430	9.7122	0.1030	5.9260	57.5546
16	2.5404	0.3936	25.6725	0.0390	10.1059	0.0990	6.2794	63.4592
17	2.6928	0.3714	28.2129	0.0354	10.4773	0.0954	6.6240	69.4011
18	2.8543	0.3503	30.9057	0.0324	10.8276	0.0924	6.9597	75.3569
19	3.0256	0.3305	33.7600	0.0296	11.1581	0.0896	7.2867	81.3062
20	3.2071	0.3118	36.7856	0.0272	11.4699	0.0872	7.6051	87.2304
21	3.3996	0.2942	39.9927	0.0250	11.7641	0.0850	7.9151	93.1136
22	3.6035	0.2775	43.3923	0.0230	12.0416	0.0830	8.2166	98.9412
23	3.8197	0.2618	46.9958	0.0213	12.3034	0.0813	8.5099	104.7007
24	4.0489	0.2470	50.8156	0.0197	12.5504	0.0797	8.7951	110.3812
25	4.2919	0.2330	54.8645	0.0182	12.7834	0.0782	9.0722	115.9731
26	4.5494	0.2198	59.1564	0.0169	13.0032	0.0769	9.3414	121.4684
27	4.8223	0.2074	63.7058	0.0157	13.2105	0.0757	9.6029	126.8599
28	5.1117	0.1956	68.5281	0.0146	13.4062	0.0746	9.8568	132.1420
29	5.4184	0.1846	73.6398	0.0136	13.5907	0.0736	10.1032	137.3095
30	5.7435	0.1741	79.0582	0.0126	13.7648	0.0726	10.3422	142.3587
31	6.0881	0.1643	84.8017	0.0118	13.9291	0.0718	10.5740	147.2864
32	6.4534	0.1550	90.8898	0.0110	14.0840	0.0710	10.7988	152.0901
33	6.8406	0.1462	97.3432	0.0103	14.2302	0.0703	11.0166	156.7680
34	7.2510	0.1379	104.1838	0.0096	14.3681	0.0696	11.2276	161.3191
35	7.6861	0.1301	111.4347	0.0090	14.4982	0.0690	11.4319	165.7427
40	10.2857	0.0972	154.7619	0.0065	15.0463	0.0665	12.3590	185.9568
45	13.7646	0.0727	212.7435	0.0047	15.4558	0.0647	13.1413	203.1096
50	18.4202	0.0543	290.3357	0.0034	15.7619	0.0634	13.7964	217.4573
55	24.6503	0.0406	394.1719	0.0025	15.9905	0.0625	14.3411	229.3222
60	32.9877	0.0303	533.1279	0.0019	16.1614	0.0619	14.7909	239.0427
65	44.1450	0.0227	719.0828	0.0014	16.2891	0.0614	15.1601	246.9450
70	59.0759	0.0169	967.9319	0.0010	16.3845	0.0610	15.4613	253.3271
75	79.0569	0.0126	1300.9485	0.0008	16.4558	0.0608	15.7058	258.4526
80	105.7959	0.0095	1746.5984	0.0006	16.5091	0.0606	15.9033	262.5491
85	141.5788	0.0071	2342.9807	0.0004	16.5489	0.0604	16.0620	265.8093
90	189.4044	0.0053	3141.0735	0.0003	16.5787	0.0603	16.1891	268.3945
95	253.5462	0.0039	4209.1015	0.0002	16.6009	0.0602	16.2905	270.4373
100	339.3018	0.0029	5638.3683	0.0002	16.6175	0.0602	16.3711	272.0469

TABLE A.16 Discrete Compounding: i = 7%

	Single payment		Uniform series				Gradient series	
	Compound amount factor	Present worth factor	Compound amount factor	Sinking fund factor	Present worth factor	Capital recovery factor	Uniform series factor	Present worth factor
n	To find F given P $F\|P\ i,n$	To find P given F $P\|F\ i,n$	To find F given A $F\|A\ i,n$	To find A given F $A\|F\ i,n$	To find P given A $P\|A\ i,n$	To find A given P $A\|P\ i,n$	To find A given G $A\|G\ i,n$	To find P given G $P\|G\ i,n$
1	1.0700	0.9346	1.0000	1.0000	0.9346	1.0700	0.0000	0.0000
2	1.1449	0.8734	2.0700	0.4831	1.8080	0.5531	0.4831	0.8734
3	1.2250	0.8163	3.2149	0.3111	2.6243	0.3811	0.9549	2.5060
4	1.3108	0.7629	4.4399	0.2252	3.3872	0.2952	1.4155	4.7947
5	1.4026	0.7130	5.7507	0.1739	4.1002	0.2439	1.8650	7.6467
6	1.5007	0.6663	7.1533	0.1398	4.7665	0.2098	2.3032	10.9784
7	1.6058	0.6227	8.6540	0.1156	5.3893	0.1856	2.7304	14.7149
8	1.7182	0.5820	10.2598	0.0975	5.9713	0.1675	3.1465	18.7889
9	1.8385	0.5439	11.9780	0.0835	6.5152	0.1535	3.5517	23.1404
10	1.9672	0.5083	13.8164	0.0724	7.0236	0.1424	3.9461	27.7156
11	2.1049	0.4751	15.7836	0.0634	7.4987	0.1334	4.3296	32.4665
12	2.2522	0.4440	17.8885	0.0559	7.9427	0.1259	4.7025	37.3506
13	2.4098	0.4150	20.1406	0.0497	8.3577	0.1197	5.0648	42.3302
14	2.5785	0.3878	22.5505	0.0443	8.7455	0.1143	5.4167	47.3718
15	2.7590	0.3624	25.1290	0.0398	9.1079	0.1098	5.7583	52.4461
16	2.9522	0.3387	27.8881	0.0359	9.4466	0.1059	6.0897	57.5271
17	3.1588	0.3166	30.8402	0.0324	9.7632	0.1024	6.4110	62.5923
18	3.3799	0.2959	33.9990	0.0294	10.0591	0.0994	6.7225	67.6220
19	3.6165	0.2765	37.3790	0.0268	10.3356	0.0968	7.0242	72.5991
20	3.8697	0.2584	40.9955	0.0244	10.5940	0.0944	7.3163	77.5091
21	4.1406	0.2415	44.8652	0.0223	10.8355	0.0923	7.5990	82.3393
22	4.4304	0.2257	49.0057	0.0204	11.0612	0.0904	7.8725	87.0793
23	4.7405	0.2109	53.4361	0.0187	11.2722	0.0887	8.1369	91.7201
24	5.0724	0.1971	58.1767	0.0172	11.4693	0.0872	8.3923	96.2545
25	5.4274	0.1842	63.2490	0.0158	11.6536	0.0858	8.6391	100.6765
26	5.8074	0.1722	68.6765	0.0146	11.8258	0.0846	8.8773	104.9813
27	6.2139	0.1609	74.4838	0.0134	11.9867	0.0834	9.1072	109.1655
28	6.6488	0.1504	80.6977	0.0124	12.1371	0.0824	9.3289	113.2264
29	7.1143	0.1406	87.3465	0.0114	12.2777	0.0814	9.5427	117.1621
30	7.6123	0.1314	94.4608	0.0106	12.4090	0.0806	9.7487	120.9718
31	8.1451	0.1228	102.0730	0.0098	12.5318	0.0798	9.9471	124.6550
32	8.7153	0.1147	110.2181	0.0091	12.6466	0.0791	10.1381	128.2120
33	9.3253	0.1072	118.9334	0.0084	12.7538	0.0784	10.3219	131.6435
34	9.9781	0.1002	128.2587	0.0078	12.8540	0.0778	10.4987	134.9507
35	10.6766	0.0937	138.2368	0.0072	12.9477	0.0772	10.6687	138.1352
40	14.9745	0.0668	199.6350	0.0050	13.3317	0.0750	11.4233	152.2927
45	21.0024	0.0476	285.7490	0.0035	13.6055	0.0735	12.0360	163.7559
50	29.4570	0.0339	406.5236	0.0025	13.8007	0.0725	12.5287	172.9051
55	41.3150	0.0242	575.9282	0.0017	13.9399	0.0717	12.9215	180.1243
60	57.9464	0.0173	813.5200	0.0012	14.0392	0.0712	13.2321	185.7677
65	81.2728	0.0123	1146.7546	0.0009	14.1099	0.0709	13.4760	190.1452
70	113.9893	0.0088	1614.1333	0.0006	14.1604	0.0706	13.6662	193.5185
75	159.8759	0.0063	2269.6558	0.0004	14.1964	0.0704	13.8136	196.1035
80	224.2342	0.0045	3189.0608	0.0003	14.2220	0.0703	13.9273	198.0748
85	314.5000	0.0032	4478.5703	0.0002	14.2403	0.0702	14.0146	199.5717
90	441.1025	0.0023	6287.1797	0.0002	14.2533	0.0702	14.0812	200.7042
95	618.6692	0.0016	8823.8477	0.0001	14.2626	0.0701	14.1319	201.5581
100	867.7156	0.0012	12381.6524	0.0001	14.2693	0.0701	14.1703	202.2001

TABLE A.17 Discrete Compounding: i = 8%

	Single payment		Uniform series				Gradient series	
	Compound amount factor	Present worth factor	Compound amount factor	Sinking fund factor	Present worth factor	Capital recovery factor	Uniform series factor	Present worth factor
n	To find F given P $F\|P\ i,n$	To find P given F $P\|F\ i,n$	To find F given A $F\|A\ i,n$	To find A given F $A\|F\ i,n$	To find P given A $P\|A\ i,n$	To find A given P $A\|P\ i,n$	To find A given G $A\|G\ i,n$	To find P given G $P\|G\ i,n$
1	1.0800	0.9259	1.0000	1.0000	0.9259	1.0800	0.0000	0.0000
2	1.1664	0.8573	2.0800	0.4808	1.7833	0.5608	0.4808	0.8573
3	1.2597	0.7938	3.2464	0.3080	2.5771	0.3880	0.9487	2.4450
4	1.3605	0.7350	4.5061	0.2219	3.3121	0.3019	1.4040	4.6501
5	1.4693	0.6806	5.8666	0.1705	3.9927	0.2505	1.8465	7.3724
6	1.5869	0.6302	7.3359	0.1363	4.6229	0.2163	2.2763	10.5233
7	1.7138	0.5835	8.9228	0.1121	5.2064	0.1921	2.6937	14.0242
8	1.8509	0.5403	10.6366	0.0940	5.7466	0.1740	3.0985	17.8061
9	1.9990	0.5002	12.4876	0.0801	6.2469	0.1601	3.4910	21.8081
10	2.1589	0.4632	14.4866	0.0690	6.7101	0.1490	3.8713	25.9768
11	2.3316	0.4289	16.6455	0.0601	7.1390	0.1401	4.2395	30.2657
12	2.5182	0.3971	18.9771	0.0527	7.5361	0.1327	4.5957	34.6339
13	2.7196	0.3677	21.4953	0.0465	7.9038	0.1265	4.9402	39.0463
14	2.9372	0.3405	24.2149	0.0413	8.2442	0.1213	5.2731	43.4723
15	3.1722	0.3152	27.1521	0.0368	8.5595	0.1168	5.5945	47.8857
16	3.4259	0.2919	30.3243	0.0330	8.8514	0.1130	5.9046	52.2640
17	3.7000	0.2703	33.7502	0.0296	9.1216	0.1096	6.2037	56.5883
18	3.9960	0.2502	37.4502	0.0267	9.3719	0.1067	6.4920	60.8426
19	4.3157	0.2317	41.4463	0.0241	9.6036	0.1041	6.7697	65.0134
20	4.6610	0.2145	45.7620	0.0219	9.8181	0.1019	7.0369	69.0898
21	5.0338	0.1987	50.4229	0.0198	10.0168	0.0998	7.2940	73.0629
22	5.4365	0.1839	55.4567	0.0180	10.2007	0.0980	7.5412	76.9257
23	5.8715	0.1703	60.8933	0.0164	10.3711	0.0964	7.7786	80.6726
24	6.3412	0.1577	66.7647	0.0150	10.5288	0.0950	8.0066	84.2997
25	6.8485	0.1460	73.1059	0.0137	10.6748	0.0937	8.2254	87.8041
26	7.3964	0.1352	79.9544	0.0125	10.8100	0.0925	8.4352	91.1842
27	7.9881	0.1252	87.3507	0.0114	10.9352	0.0914	8.6363	94.4390
28	8.6271	0.1159	95.3388	0.0105	11.0511	0.0905	8.8289	97.5687
29	9.3173	0.1073	103.9659	0.0096	11.1584	0.0896	9.0133	100.5739
30	10.0627	0.0994	113.2831	0.0088	11.2578	0.0888	9.1897	103.4558
31	10.8677	0.0920	123.3458	0.0081	11.3498	0.0881	9.3584	106.2163
32	11.7371	0.0852	134.2134	0.0075	11.4350	0.0875	9.5197	108.8575
33	12.6760	0.0789	145.9505	0.0069	11.5139	0.0869	9.6737	111.3819
34	13.6901	0.0730	158.6266	0.0063	11.5869	0.0863	9.8208	113.7924
35	14.7853	0.0676	172.3167	0.0058	11.6546	0.0858	9.9611	116.0920
40	21.7245	0.0460	259.0564	0.0039	11.9246	0.0839	10.5699	126.0422
45	31.9204	0.0313	386.5054	0.0026	12.1084	0.0826	11.0447	133.7331
50	46.9016	0.0213	573.7698	0.0017	12.2335	0.0817	11.4107	139.5928
55	68.9139	0.0145	848.9224	0.0012	12.3186	0.0812	11.6902	144.0065
60	101.2570	0.0099	1253.2122	0.0008	12.3766	0.0808	11.9015	147.3000
65	148.7796	0.0067	1847.2463	0.0005	12.4160	0.0805	12.0602	149.7387
70	218.6061	0.0046	2720.0767	0.0004	12.4428	0.0804	12.1783	151.5326
75	321.2039	0.0031	4002.5518	0.0002	12.4611	0.0802	12.2658	152.8449
80	471.9541	0.0021	5886.9258	0.0002	12.4735	0.0802	12.3301	153.8001
85	693.4555	0.0014	8655.6953	0.0001	12.4820	0.0801	12.3772	154.4925
90	1018.9136	0.0010	12723.9219	0.0001	12.4877	0.0801	12.4116	154.9925
95	1497.1182	0.0007	18701.4805	0.0001	12.4917	0.0801	12.4365	155.3525
100	2199.7656	0.0005	27484.4649	0.0000	12.4943	0.0800	12.4545	155.6107

TABLE A.18 Discrete Compounding: i = 9%

	Single payment		Uniform series				Gradient series	
	Compound amount factor	Present worth factor	Compound amount factor	Sinking fund factor	Present worth factor	Capital recovery factor	Uniform series factor	Present worth factor
n	To find F given P $F/P\ i,n$	To find P given F $P/F\ i,n$	To find F given A $F/A\ i,n$	To find A given F $A/F\ i,n$	To find P given A $P/A\ i,n$	To find A given P $A/P\ i,n$	To find A given G $A/G\ i,n$	To find P given G $P/G\ i,n$
1	1.0900	0.9174	1.0000	1.0000	0.9174	1.0900	0.0000	0.0000
2	1.1881	0.8417	2.0900	0.4785	1.7591	0.5685	0.4785	0.8417
3	1.2950	0.7722	3.2781	0.3051	2.5313	0.3951	0.9426	2.3860
4	1.4116	0.7084	4.5731	0.2187	3.2397	0.3087	1.3925	4.5113
5	1.5386	0.6499	5.9847	0.1671	3.8897	0.2571	1.8282	7.1110
6	1.6771	0.5963	7.5233	0.1329	4.4859	0.2229	2.2498	10.0924
7	1.8280	0.5470	9.2004	0.1087	5.0330	0.1987	2.6574	13.3746
8	1.9926	0.5019	11.0285	0.0907	5.5348	0.1807	3.0512	16.8877
9	2.1719	0.4604	13.0210	0.0768	5.9952	0.1668	3.4312	20.5711
10	2.3674	0.4224	15.1929	0.0658	6.4177	0.1558	3.7978	24.3728
11	2.5804	0.3875	17.5603	0.0569	6.8052	0.1469	4.1510	28.2481
12	2.8127	0.3555	20.1407	0.0497	7.1607	0.1397	4.4910	32.1590
13	3.0658	0.3262	22.9534	0.0436	7.4869	0.1336	4.8182	36.0731
14	3.3417	0.2992	26.0192	0.0384	7.7862	0.1284	5.1326	39.9633
15	3.6425	0.2745	29.3609	0.0341	8.0607	0.1241	5.4346	43.8069
16	3.9703	0.2519	33.0034	0.0303	8.3126	0.1203	5.7245	47.5849
17	4.3276	0.2311	36.9737	0.0270	8.5436	0.1170	6.0024	51.2821
18	4.7171	0.2120	41.3013	0.0242	8.7556	0.1142	6.2687	54.8860
19	5.1417	0.1945	46.0184	0.0217	8.9501	0.1117	6.5236	58.3868
20	5.6044	0.1784	51.1601	0.0195	9.1285	0.1095	6.7674	61.7770
21	6.1088	0.1637	56.7645	0.0176	9.2922	0.1076	7.0006	65.0510
22	6.6586	0.1502	62.8733	0.0159	9.4424	0.1059	7.2232	68.2048
23	7.2579	0.1378	69.5319	0.0144	9.5802	0.1044	7.4357	71.2360
24	7.9111	0.1264	76.7898	0.0130	9.7066	0.1030	7.6384	74.1433
25	8.6231	0.1160	84.7009	0.0118	9.8226	0.1018	7.8316	76.9265
26	9.3992	0.1064	93.3239	0.0107	9.9290	0.1007	8.0156	79.5863
27	10.2451	0.0976	102.7231	0.0097	10.0266	0.0997	8.1906	82.1241
28	11.1671	0.0895	112.9681	0.0089	10.1161	0.0989	8.3571	84.5419
29	12.1722	0.0822	124.1352	0.0081	10.1983	0.0981	8.5154	86.8423
30	13.2677	0.0754	136.3074	0.0073	10.2737	0.0973	8.6657	89.0280
31	14.4618	0.0691	149.5751	0.0067	10.3428	0.0967	8.8083	91.1025
32	15.7633	0.0634	164.0368	0.0061	10.4062	0.0961	8.9436	93.0691
33	17.1820	0.0582	179.8002	0.0056	10.4644	0.0956	9.0718	94.9315
34	18.7284	0.0534	196.9822	0.0051	10.5178	0.0951	9.1933	96.6935
35	20.4140	0.0490	215.7106	0.0046	10.5668	0.0946	9.3083	98.3590
40	31.4094	0.0318	337.8821	0.0030	10.7574	0.0930	9.7957	105.3762
45	48.3272	0.0207	525.8582	0.0019	10.8812	0.0919	10.1603	110.5561
50	74.3574	0.0134	815.0825	0.0012	10.9617	0.0912	10.4295	114.3251
55	114.4081	0.0087	1260.0903	0.0008	11.0140	0.0908	10.6261	117.0362
60	176.0310	0.0057	1944.7836	0.0005	11.0480	0.0905	10.7683	118.9683
65	270.8455	0.0037	2998.2830	0.0003	11.0701	0.0903	10.8702	120.3344
70	416.7293	0.0024	4619.2149	0.0002	11.0845	0.0902	10.9427	121.2942
75	641.1897	0.0016	7113.2188	0.0001	11.0938	0.0901	10.9940	121.9646
80	986.5496	0.0010	10950.5547	0.0001	11.0999	0.0901	11.0299	122.4307
85	1517.9287	0.0007	16854.7695	0.0001	11.1038	0.0901	11.0551	122.7533
90	2335.5198	0.0004	25939.1328	0.0000	11.1064	0.0900	11.0726	122.9758
95	3593.4878	0.0003	39916.5430	0.0000	11.1080	0.0900	11.0847	123.1287
100	5529.0234	0.0002	61422.5391	0.0000	11.1091	0.0900	11.0930	123.2335

TABLE A.19 Discrete Compounding: i = 10%

	Single payment		Uniform series				Gradient series	
	Compound amount factor	Present worth factor	Compound amount factor	Sinking fund factor	Present worth factor	Capital recovery factor	Uniform series factor	Present worth factor
n	To find F given P $F\|P\ i,n$	To find P given F $P\|F\ i,n$	To find F given A $F\|A\ i,n$	To find A given F $A\|F\ i,n$	To find P given A $P\|A\ i,n$	To find A given P $A\|P\ i,n$	To find A given G $A\|G\ i,n$	To find P given G $P\|G\ i,n$
1	1.1000	0.9091	1.0000	1.0000	0.9091	1.1000	0.0000	0.0000
2	1.2100	0.8264	2.1000	0.4762	1.7355	0.5762	0.4762	0.8264
3	1.3310	0.7513	3.3100	0.3021	2.4869	0.4021	0.9366	2.3291
4	1.4641	0.6830	4.6410	0.2155	3.1699	0.3155	1.3812	4.3781
5	1.6105	0.6209	6.1051	0.1638	3.7908	0.2638	1.8101	6.8618
6	1.7716	0.5645	7.7156	0.1296	4.3553	0.2296	2.2236	9.6842
7	1.9487	0.5132	9.4872	0.1054	4.8684	0.2054	2.6216	12.7631
8	2.1436	0.4665	11.4359	0.0874	5.3349	0.1874	3.0045	16.0287
9	2.3579	0.4241	13.5795	0.0736	5.7590	0.1736	3.3724	19.4215
10	2.5937	0.3855	15.9374	0.0627	6.1446	0.1627	3.7255	22.8913
11	2.8531	0.3505	18.5312	0.0540	6.4951	0.1540	4.0641	26.3963
12	3.1384	0.3186	21.3843	0.0468	6.8137	0.1468	4.3884	29.9012
13	3.4523	0.2897	24.5227	0.0408	7.1034	0.1408	4.6988	33.3772
14	3.7975	0.2633	27.9750	0.0357	7.3667	0.1357	4.9955	36.8005
15	4.1772	0.2394	31.7725	0.0315	7.6061	0.1315	5.2789	40.1520
16	4.5950	0.2176	35.9497	0.0278	7.8237	0.1278	5.5493	43.4164
17	5.0545	0.1978	40.5447	0.0247	8.0216	0.1247	5.8071	46.5820
18	5.5599	0.1799	45.5992	0.0219	8.2014	0.1219	6.0526	49.6396
19	6.1159	0.1635	51.1591	0.0195	8.3649	0.1195	6.2861	52.5827
20	6.7275	0.1486	57.2750	0.0175	8.5136	0.1175	6.5081	55.4069
21	7.4002	0.1351	64.0025	0.0156	8.6487	0.1156	6.7189	58.1095
22	8.1403	0.1228	71.4027	0.0140	8.7715	0.1140	6.9189	60.6893
23	8.9543	0.1117	79.5430	0.0126	8.8832	0.1126	7.1085	63.1462
24	9.8497	0.1015	88.4973	0.0113	8.9847	0.1113	7.2881	65.4813
25	10.8347	0.0923	98.3470	0.0102	9.0770	0.1102	7.4580	67.6964
26	11.9182	0.0839	109.1817	0.0092	9.1609	0.1092	7.6186	69.7941
27	13.1100	0.0763	121.0998	0.0083	9.2372	0.1083	7.7704	71.7773
28	14.4210	0.0693	134.2098	0.0075	9.3066	0.1075	7.9137	73.6496
29	15.8631	0.0630	148.6308	0.0067	9.3696	0.1067	8.0489	75.4147
30	17.4494	0.0573	164.4939	0.0061	9.4269	0.1061	8.1762	77.0766
31	19.1943	0.0521	181.9432	0.0055	9.4790	0.1055	8.2962	78.6396
32	21.1138	0.0474	201.1376	0.0050	9.5264	0.1050	8.4091	80.1078
33	23.2251	0.0431	222.2513	0.0045	9.5694	0.1045	8.5152	81.4856
34	25.5476	0.0391	245.4765	0.0041	9.6086	0.1041	8.6149	82.7773
35	28.1024	0.0356	271.0239	0.0037	9.6442	0.1037	8.7086	83.9872
40	45.2592	0.0221	442.5921	0.0023	9.7791	0.1023	9.0962	88.9526
45	72.8904	0.0137	718.9058	0.0014	9.8628	0.1014	9.3740	92.4545
50	117.3906	0.0085	1163.9070	0.0009	9.9148	0.1009	9.5704	94.8889
55	189.0588	0.0053	1880.5872	0.0005	9.9471	0.1005	9.7075	96.5620
60	304.4810	0.0033	3034.8110	0.0003	9.9672	0.1003	9.8023	97.7011
65	490.3694	0.0020	4893.6953	0.0002	9.9796	0.1002	9.8672	98.4706
70	789.7449	0.0013	7887.4531	0.0001	9.9873	0.1001	9.9113	98.9871
75	1271.8921	0.0008	12708.9258	0.0001	9.9921	0.1001	9.9410	99.3318
80	2048.3936	0.0005	20473.9531	0.0000	9.9951	0.1000	9.9609	99.5607
85	3298.9583	0.0003	32979.6094	0.0000	9.9970	0.1000	9.9742	99.7121
90	5313.0039	0.0002	53120.0742	0.0000	9.9981	0.1000	9.9831	99.8119
95	8556.6484	0.0001	85556.5000	0.0000	9.9988	0.1000	9.9889	99.8774
100	13780.5625	0.0001	137795.6875	0.0000	9.9993	0.1000	9.9927	99.9202

	Single payment		Uniform series				Gradient series	
	Compound amount factor	Present worth factor	Compound amount factor	Sinking fund factor	Present worth factor	Capital recovery factor	Uniform series factor	Present worth factor
n	To find F given P $F/P\ i,n$	To find P given F $P/F\ i,n$	To find F given A $F/A\ i,n$	To find A given F $A/F\ i,n$	To find P given A $P/A\ i,n$	To find A given P $A/P\ i,n$	To find A given G $A/G\ i,n$	To find P given G $P/G\ i,n$
1	1.1200	0.8929	1.0000	1.0000	0.8929	1.1200	0.0000	0.0000
2	1.2544	0.7972	2.1200	0.4717	1.6901	0.5917	0.4717	0.7972
3	1.4049	0.7118	3.3744	0.2963	2.4018	0.4163	0.9246	2.2208
4	1.5735	0.6355	4.7793	0.2092	3.0373	0.3292	1.3589	4.1273
5	1.7623	0.5674	6.3528	0.1574	3.6048	0.2774	1.7746	6.3970
6	1.9738	0.5066	8.1152	0.1232	4.1114	0.2432	2.1720	8.9302
7	2.2107	0.4523	10.0890	0.0991	4.5638	0.2191	2.5515	11.6443
8	2.4760	0.4039	12.2997	0.0813	4.9676	0.2013	2.9131	14.4715
9	2.7731	0.3606	14.7757	0.0677	5.3283	0.1877	3.2574	17.3563
10	3.1058	0.3220	17.5487	0.0570	5.6502	0.1770	3.5847	20.2541
11	3.4785	0.2875	20.6546	0.0484	5.9377	0.1684	3.8953	23.1289
12	3.8960	0.2567	24.1331	0.0414	6.1944	0.1614	4.1897	25.9523
13	4.3635	0.2292	28.0291	0.0357	6.4235	0.1557	4.4683	28.7024
14	4.8871	0.2046	32.3926	0.0309	6.6282	0.1509	4.7317	31.3624
15	5.4736	0.1827	37.2797	0.0268	6.8109	0.1468	4.9803	33.9202
16	6.1304	0.1631	42.7533	0.0234	6.9740	0.1434	5.2147	36.3670
17	6.8660	0.1456	48.8836	0.0205	7.1196	0.1405	5.4353	38.6973
18	7.6900	0.1300	55.7497	0.0179	7.2497	0.1379	5.6427	40.9080
19	8.6128	0.1161	63.4396	0.0158	7.3658	0.1358	5.8375	42.9979
20	9.6463	0.1037	72.0524	0.0139	7.4694	0.1339	6.0202	44.9676
21	10.8038	0.0925	81.6987	0.0122	7.5620	0.1322	6.1913	46.8188
22	12.1003	0.0826	92.5025	0.0108	7.6446	0.1308	6.3514	48.5543
23	13.5523	0.0738	104.6028	0.0096	7.7184	0.1296	6.5010	50.1776
24	15.1786	0.0659	118.1551	0.0085	7.7843	0.1285	6.6406	51.6929
25	17.0000	0.0588	133.3337	0.0075	7.8431	0.1275	6.7708	53.1047
26	19.0400	0.0525	150.3338	0.0067	7.8957	0.1267	6.8921	54.4177
27	21.3249	0.0469	169.3738	0.0059	7.9426	0.1259	7.0049	55.6369
28	23.8838	0.0419	190.6987	0.0052	7.9844	0.1252	7.1098	56.7674
29	26.7499	0.0374	214.5825	0.0047	8.0218	0.1247	7.2071	57.8141
30	29.9599	0.0334	241.3324	0.0041	8.0552	0.1241	7.2974	58.7821
31	33.5551	0.0298	271.2922	0.0037	8.0850	0.1237	7.3811	59.6761
32	37.5817	0.0266	304.8472	0.0033	8.1116	0.1233	7.4586	60.5010
33	42.0915	0.0238	342.4290	0.0029	8.1354	0.1229	7.5302	61.2613
34	47.1424	0.0212	384.5203	0.0026	8.1566	0.1226	7.5965	61.9613
35	52.7995	0.0189	431.6629	0.0023	8.1755	0.1223	7.6577	62.6052
40	93.0508	0.0107	767.0901	0.0013	8.2438	0.1213	7.8988	65.1159
45	163.9872	0.0061	1358.2275	0.0007	8.2825	0.1207	8.0572	66.7343
50	289.0012	0.0035	2400.0127	0.0004	8.3045	0.1204	8.1597	67.7625

TABLE A.21 Discrete Compounding: i = 15%

	Single payment		Uniform series				Gradient series	
	Compound amount factor	Present worth factor	Compound amount factor	Sinking fund factor	Present worth factor	Capital recovery factor	Uniform series factor	Present worth factor
n	To find F given P $F\backslash P\ i,n$	To find P given F $P\backslash F\ i,n$	To find F given A $F\backslash A\ i,n$	To find A given F $A\backslash F\ i,n$	To find P given A $P\backslash A\ i,n$	To find A given P $A\backslash P\ i,n$	To find A given G $A\backslash G\ r,n$	To find P given G $P\backslash G\ r,n$
1	1.1500	0.8696	1.0000	1.0000	0.8696	1.1500	0.0000	0.0000
2	1.3225	0.7561	2.1500	0.4651	1.6257	0.6151	0.4651	0.7561
3	1.5209	0.6575	3.4725	0.2880	2.2832	0.4380	0.9071	2.0712
4	1.7490	0.5718	4.9934	0.2003	2.8550	0.3503	1.3263	3.7864
5	2.0114	0.4972	6.7424	0.1483	3.3522	0.2983	1.7228	5.7751
6	2.3131	0.4323	8.7537	0.1142	3.7845	0.2642	2.0972	7.9368
7	2.6600	0.3759	11.0668	0.0904	4.1604	0.2404	2.4498	10.1924
8	3.0590	0.3269	13.7268	0.0729	4.4873	0.2229	2.7813	12.4807
9	3.5179	0.2843	16.7858	0.0596	4.7716	0.2096	3.0922	14.7548
10	4.0456	0.2472	20.3037	0.0493	5.0188	0.1993	3.3832	16.9795
11	4.6524	0.2149	24.3493	0.0411	5.2337	0.1911	3.6549	19.1289
12	5.3502	0.1869	29.0017	0.0345	5.4206	0.1845	3.9082	21.1849
13	6.1528	0.1625	34.3519	0.0291	5.5831	0.1791	4.1438	23.1352
14	7.0757	0.1413	40.5047	0.0247	5.7245	0.1747	4.3624	24.9725
15	8.1371	0.1229	47.5804	0.0210	5.8474	0.1710	4.5650	26.6930
16	9.3576	0.1069	55.7175	0.0179	5.9542	0.1679	4.7522	28.2960
17	10.7613	0.0929	65.0751	0.0154	6.0472	0.1654	4.9251	29.7828
18	12.3754	0.0808	75.8363	0.0132	6.1280	0.1632	5.0843	31.1565
19	14.2318	0.0703	88.2118	0.0113	6.1982	0.1613	5.2307	32.4213
20	16.3665	0.0611	102.4436	0.0098	6.2593	0.1598	5.3651	33.5822
21	18.8215	0.0531	118.8100	0.0084	6.3125	0.1584	5.4883	34.6448
22	21.6447	0.0462	137.6315	0.0073	6.3587	0.1573	5.6010	35.6150
23	24.8914	0.0402	159.2763	0.0063	6.3988	0.1563	5.7040	36.4988
24	28.6252	0.0349	184.1677	0.0054	6.4338	0.1554	5.7979	37.3023
25	32.9189	0.0304	212.7929	0.0047	6.4642	0.1547	5.8834	38.0314
26	37.8568	0.0264	245.7118	0.0041	6.4906	0.1541	5.9612	38.6918
27	43.5353	0.0230	283.5684	0.0035	6.5135	0.1535	6.0319	39.2890
28	50.0656	0.0200	327.1038	0.0031	6.5335	0.1531	6.0960	39.8283
29	57.5754	0.0174	377.1694	0.0027	6.5509	0.1527	6.1541	40.3146
30	66.2117	0.0151	434.7449	0.0023	6.5660	0.1523	6.2066	40.7526
31	76.1435	0.0131	500.9566	0.0020	6.5791	0.1520	6.2541	41.1465
32	87.5650	0.0114	577.1001	0.0017	6.5905	0.1517	6.2970	41.5006
33	100.6998	0.0099	664.6650	0.0015	6.6005	0.1515	6.3357	41.8184
34	115.8047	0.0086	765.3648	0.0013	6.6091	0.1513	6.3705	42.1034
35	133.1754	0.0075	881.1694	0.0011	6.6166	0.1511	6.4019	42.3537
40	267.8633	0.0037	1779.0879	0.0006	6.6418	0.1506	6.5168	43.2830
45	538.7685	0.0019	3585.1248	0.0003	6.6543	0.1503	6.5830	43.8051
50	1083.6563	0.0009	7217.7070	0.0001	6.6605	0.1501	6.6205	44.0958

TABLE A.22 Discrete Compounding: i = 20%

	Single payment		Uniform series				Gradient series	
	Compound amount factor	Present worth factor	Compound amount factor	Sinking fund factor	Present worth factor	Capital recovery factor	Uniform series factor	Present worth factor
n	To find F given P $F\|P\ i,n$	To find P given F $P\|F\ i,n$	To find F given A $F\|A\ i,n$	To find A given F $A\|F\ i,n$	To find P given A $P\|A\ i,n$	To find A given P $A\|P\ i,n$	To find A given G $A\|G\ i,n$	To find P given G $P\|G\ i,n$
1	1.2000	0.8333	1.0000	1.0000	0.8333	1.2000	0.0000	0.0000
2	1.4400	0.6944	2.2000	0.4545	1.5278	0.6545	0.4545	0.6944
3	1.7280	0.5787	3.6400	0.2747	2.1065	0.4747	0.8791	1.8519
4	2.0736	0.4823	5.3680	0.1863	2.5887	0.3863	1.2742	3.2986
5	2.4883	0.4019	7.4416	0.1344	2.9906	0.3344	1.6405	4.9061
6	2.9860	0.3349	9.9299	0.1007	3.3255	0.3007	1.9788	6.5806
7	3.5832	0.2791	12.9159	0.0774	3.6045	0.2774	2.2902	8.2551
8	4.2998	0.2326	16.4991	0.0606	3.8372	0.2606	2.5756	9.8831
9	5.1598	0.1938	20.7989	0.0481	4.0310	0.2481	2.8364	11.4335
10	6.1917	0.1615	25.9587	0.0385	4.1925	0.2385	3.0739	12.8871
11	7.4301	0.1346	32.1504	0.0311	4.3271	0.2311	3.2893	14.2330
12	8.9161	0.1122	39.5805	0.0253	4.4392	0.2253	3.4841	15.4667
13	10.6993	0.0935	48.4966	0.0206	4.5327	0.2206	3.6597	16.5883
14	12.8392	0.0779	59.1959	0.0169	4.6106	0.2169	3.8175	17.6008
15	15.4070	0.0649	72.0351	0.0139	4.6755	0.2139	3.9588	18.5095
16	18.4884	0.0541	87.4421	0.0114	4.7296	0.2114	4.0851	19.3208
17	22.1861	0.0451	105.9305	0.0094	4.7746	0.2094	4.1976	20.0419
18	26.6233	0.0375	128.1166	0.0078	4.8122	0.2078	4.2975	20.6805
19	31.9480	0.0313	154.7399	0.0065	4.8435	0.2065	4.3861	21.2439
20	38.3376	0.0261	186.6879	0.0054	4.8696	0.2054	4.4643	21.7395
21	46.0051	0.0217	225.0255	0.0044	4.8913	0.2044	4.5334	22.1742
22	55.2061	0.0181	271.0305	0.0037	4.9094	0.2037	4.5941	22.5546
23	66.2474	0.0151	326.2366	0.0031	4.9245	0.2031	4.6475	22.8867
24	79.4968	0.0126	392.4839	0.0025	4.9371	0.2025	4.6943	23.1760
25	95.3962	0.0105	471.9807	0.0021	4.9476	0.2021	4.7352	23.4276
26	114.4754	0.0087	567.3770	0.0018	4.9563	0.2018	4.7709	23.6460
27	137.3705	0.0073	681.8525	0.0015	4.9636	0.2015	4.8020	23.8353
28	164.8446	0.0061	819.2229	0.0012	4.9697	0.2012	4.8291	23.9991
29	197.8135	0.0051	984.0676	0.0010	4.9747	0.2010	4.8527	24.1406
30	237.3752	0.0042	1181.8811	0.0008	4.9789	0.2008	4.8731	24.2528
31	284.8513	0.0035	1419.2573	0.0007	4.9824	0.2007	4.8908	24.3681
32	341.8215	0.0029	1704.1087	0.0006	4.9854	0.2006	4.9061	24.4588
33	410.1860	0.0024	2045.9295	0.0005	4.9878	0.2005	4.9194	24.5368
34	492.2232	0.0020	2456.1167	0.0004	4.9898	0.2004	4.9308	24.6038
35	590.9680	0.0017	2948.3391	0.0003	4.9915	0.2003	4.9406	24.6614
40	1469.7708	0.0007	7343.8516	0.0001	4.9966	0.2001	4.9728	24.8469
45	3657.2590	0.0003	18281.3008	0.0001	4.9985	0.2001	4.9877	24.9316
50	9100.4336	0.0001	45497.1641	0.0000	4.9995	0.2000	4.9945	24.9698

TABLE A.23 Discrete Compounding: i = 25%

	Single payment		Uniform series				Gradient series	
	Compound amount factor	Present worth factor	Compound amount factor	Sinking fund factor	Present worth factor	Capital recovery factor	Uniform series factor	Present worth factor
n	To find F given P $F\|P\ i,n$	To find P given F $P\|F\ i,n$	To find F given A $F\|A\ i,n$	To find A given F $A\|F\ i,n$	To find P given A $P\|A\ i,n$	To find A given P $A\|P\ i,n$	To find A given G $A\|G\ i,n$	To find P given G $P\|G\ i,n$
1	1.2500	0.8000	1.0000	1.0000	0.8000	1.2500	0.0000	0.0000
2	1.5625	0.6400	2.2500	0.4444	1.4400	0.6944	0.4444	0.6400
3	1.9531	0.5120	3.8125	0.2623	1.9520	0.5123	0.8525	1.6640
4	2.4414	0.4096	5.7656	0.1734	2.3616	0.4234	1.2249	2.8928
5	3.0518	0.3277	8.2070	0.1218	2.6893	0.3718	1.5631	4.2035
6	3.8147	0.2621	11.2588	0.0888	2.9514	0.3388	1.8683	5.5142
7	4.7684	0.2097	15.0735	0.0663	3.1611	0.3163	2.1424	6.7725
8	5.9605	0.1678	19.8419	0.0504	3.3289	0.3004	2.3872	7.9469
9	7.4506	0.1342	25.8023	0.0388	3.4631	0.2888	2.6048	9.0207
10	9.3132	0.1074	33.2529	0.0301	3.5705	0.2801	2.7971	9.9870
11	11.6415	0.0859	42.5661	0.0235	3.6564	0.2735	2.9663	10.8460
12	14.5519	0.0687	54.2077	0.0184	3.7251	0.2684	3.1145	11.6020
13	18.1899	0.0550	68.7596	0.0145	3.7801	0.2645	3.2437	12.2617
14	22.7374	0.0440	86.9495	0.0115	3.8241	0.2615	3.3559	12.8334
15	28.4217	0.0352	109.6868	0.0091	3.8593	0.2591	3.4530	13.3260
16	35.5271	0.0281	138.1085	0.0072	3.8874	0.2572	3.5366	13.7482
17	44.4089	0.0225	173.6356	0.0058	3.9099	0.2558	3.6084	14.1085
18	55.5112	0.0180	218.0446	0.0046	3.9279	0.2546	3.6698	14.4147
19	69.3889	0.0144	273.5557	0.0037	3.9424	0.2537	3.7222	14.6741
20	86.7362	0.0115	342.9446	0.0029	3.9539	0.2529	3.7667	14.8932
21	108.4202	0.0092	429.6807	0.0023	3.9631	0.2523	3.8045	15.0777
22	135.5252	0.0074	538.1008	0.0019	3.9705	0.2519	3.8365	15.2326
23	169.4065	0.0059	673.6262	0.0015	3.9764	0.2515	3.8634	15.3625
24	211.7582	0.0047	843.0327	0.0012	3.9811	0.2512	3.8861	15.4711
25	264.6978	0.0038	1054.7910	0.0009	3.9849	0.2509	3.9052	15.5618
26	330.8721	0.0030	1319.4888	0.0008	3.9879	0.2508	3.9212	15.6373
27	413.5901	0.0024	1650.3611	0.0006	3.9903	0.2506	3.9346	15.7002
28	516.9378	0.0019	2063.9502	0.0005	3.9923	0.2505	3.9457	15.7524
29	646.2346	0.0015	2580.9390	0.0004	3.9938	0.2504	3.9551	15.7957
30	807.7935	0.0012	3227.1726	0.0003	3.9950	0.2503	3.9628	15.8316
31	1009.7417	0.0010	4034.9663	0.0002	3.9960	0.2502	3.9693	15.8614
32	1262.1773	0.0008	5044.7070	0.0002	3.9968	0.2502	3.9746	15.8859
33	1577.7217	0.0006	6306.8828	0.0002	3.9975	0.2502	3.9791	15.9062
34	1972.1519	0.0005	7884.6055	0.0001	3.9980	0.2501	3.9828	15.9229
35	2465.1887	0.0004	9856.7578	0.0001	3.9984	0.2501	3.9858	15.9367

TABLE A.24 Discrete Compounding: i = 30%

	Single payment		Uniform series				Gradient series	
	Compound amount factor	Present worth factor	Compound amount factor	Sinking fund factor	Present worth factor	Capital recovery factor	Uniform series factor	Present worth factor
n	To find F given P $F\backslash P\ i,n$	To find P given F $P\backslash F\ i,n$	To find F given A $F\backslash A\ i,n$	To find A given F $A\backslash F\ i,n$	To find P given A $P\backslash A\ i,n$	To find A given P $A\backslash P\ i,n$	To find A given G $A\backslash G\ i,n$	To find P given G $P\backslash G\ i,n$
1	1.3000	0.7692	1.0000	1.0000	0.7692	1.3000	0.0000	0.0000
2	1.6900	0.5917	2.3000	0.4348	1.3609	0.7348	0.4348	0.5917
3	2.1970	0.4552	3.9900	0.2506	1.8161	0.5506	0.8271	1.5020
4	2.8561	0.3501	6.1870	0.1616	2.1662	0.4616	1.1783	2.5524
5	3.7129	0.2693	9.0431	0.1106	2.4356	0.4105	1.4903	3.6297
6	4.8268	0.2072	12.7560	0.0784	2.6427	0.3784	1.7654	4.6656
7	6.2749	0.1594	17.5828	0.0569	2.8021	0.3569	2.0063	5.6218
8	8.1573	0.1226	23.8577	0.0419	2.9247	0.3419	2.2156	6.4800
9	10.6045	0.0943	32.0150	0.0312	3.0190	0.3312	2.3963	7.2344
10	13.7858	0.0725	42.6195	0.0235	3.0915	0.3235	2.5512	7.8872
11	17.9216	0.0558	56.4053	0.0177	3.1473	0.3177	2.6833	8.4452
12	23.2981	0.0429	74.3269	0.0135	3.1903	0.3135	2.7952	8.9173
13	30.2875	0.0330	97.6250	0.0102	3.2233	0.3102	2.8895	9.3135
14	39.3737	0.0254	127.9124	0.0078	3.2487	0.3078	2.9685	9.6437
15	51.1859	0.0195	167.2862	0.0060	3.2682	0.3060	3.0344	9.9172
16	66.5416	0.0150	218.4721	0.0046	3.2832	0.3046	3.0892	10.1426
17	86.5041	0.0115	285.0137	0.0035	3.2948	0.3035	3.1345	10.3276
18	112.4553	0.0089	371.5176	0.0027	3.3037	0.3027	3.1718	10.4788
19	146.1919	0.0068	483.9729	0.0021	3.3105	0.3021	3.2025	10.6019
20	190.0494	0.0053	630.1648	0.0016	3.3158	0.3016	3.2275	10.7019
21	247.0643	0.0040	820.2144	0.0012	3.3198	0.3012	3.2480	10.7828
22	321.1836	0.0031	1067.2788	0.0009	3.3230	0.3009	3.2646	10.8482
23	417.5386	0.0024	1388.4624	0.0007	3.3254	0.3007	3.2781	10.9009
24	542.8001	0.0018	1806.0000	0.0006	3.3272	0.3006	3.2890	10.9433
25	705.6401	0.0014	2348.7998	0.0004	3.3286	0.3004	3.2979	10.9773
26	917.3323	0.0011	3054.4414	0.0003	3.3297	0.3003	3.3050	11.0045
27	1192.5320	0.0008	3971.7727	0.0003	3.3305	0.3003	3.3107	11.0263
28	1550.2915	0.0006	5164.3047	0.0002	3.3312	0.3002	3.3153	11.0437
29	2015.3775	0.0005	6714.5938	0.0001	3.3317	0.3001	3.3189	11.0576
30	2619.9920	0.0004	8729.9727	0.0001	3.3321	0.3001	3.3219	11.0687
31	3405.9902	0.0003	11349.9688	0.0001	3.3324	0.3001	3.3242	11.0775
32	4427.7852	0.0002	14755.9570	0.0001	3.3326	0.3001	3.3261	11.0845
33	5756.1211	0.0002	19183.7461	0.0001	3.3328	0.3001	3.3276	11.0901
34	7482.9570	0.0001	24939.8672	0.0000	3.3329	0.3000	3.3288	11.0945
35	9727.8438	0.0001	32422.8086	0.0000	3.3330	0.3000	3.3297	11.0980

TABLE A.25 Discrete Compounding: i = 5%

j = increase in bonus every year

		Geometric series present worth factor, $(P|A_1 \ i,j,n)$			
n	j = 4%	j = 6%	j = 8%	j = 10%	j = 15%
1	0.9524	0.9524	0.9524	0.9524	0.9524
2	1.8957	1.9138	1.9320	1.9501	1.9955
3	2.8300	2.8844	2.9396	2.9954	3.1379
4	3.7554	3.8643	3.9759	4.0904	4.3891
5	4.6721	4.8535	5.0419	5.2375	5.7595
6	5.5799	5.8521	6.1383	6.4393	7.2604
7	6.4792	6.8602	7.2661	7.6983	8.9043
8	7.3699	7.8779	8.4261	9.0173	10.7047
9	8.2521	8.9053	9.6192	10.3991	12.6765
10	9.1258	9.9425	10.8464	11.8467	14.8362
11	9.9913	10.9896	12.1087	13.3632	17.2016
12	10.8485	12.0466	13.4070	14.9519	19.7922
13	11.6976	13.1137	14.7425	16.6163	22.6295
14	12.5386	14.1910	16.1161	18.3599	25.7371
15	13.3715	15.2785	17.5289	20.1866	29.1407
16	14.1966	16.3764	18.9821	22.1002	32.8683
17	15.0137	17.4848	20.4769	24.1050	36.9510
18	15.8231	18.6037	22.0143	26.2052	41.4226
19	16.6248	19.7332	23.5956	28.4054	46.3199
20	17.4189	20.8736	25.2222	30.7105	51.6837
21	18.2054	22.0247	26.8952	33.1253	57.5584
22	18.9844	23.1869	28.6160	35.6550	63.9925
23	19.7559	24.3601	30.3860	38.3053	71.0394
24	20.5202	25.5445	32.2066	41.0817	78.7575
25	21.2771	26.7401	34.0791	43.9904	87.2106
26	22.0269	27.9472	36.0052	47.0375	96.4687
27	22.7695	29.1657	37.9863	50.2298	106.6085
28	23.5050	30.3959	40.0240	53.5740	117.7141
29	24.2335	31.6377	42.1199	57.0776	129.8774
30	24.9551	32.8914	44.2757	60.7479	143.1990
31	25.6698	34.1571	46.4931	64.5931	157.7894
32	26.3777	35.4348	48.7739	68.6213	173.7693
33	27.0789	36.7246	51.1198	72.8414	191.2712
34	27.7734	38.0267	53.5327	77.2624	210.4399
35	28.4612	39.3413	56.0146	81.8939	231.4341
36	29.1426	40.6683	58.5674	86.7460	254.4279
37	29.8174	42.0080	61.1932	91.8292	279.6113
38	30.4858	43.3605	63.8939	97.1543	307.1934
39	31.1478	44.7258	66.6718	102.7331	337.4024
40	31.8036	46.1042	69.5291	108.5775	370.4883
41	32.4531	47.4957	72.4681	114.7002	406.7251
42	33.0964	48.9004	75.4910	121.1145	446.4133
43	33.7335	50.3185	78.6002	127.8343	489.8814
44	34.3647	51.7501	81.7983	134.8740	537.4890
45	34.9898	53.1953	85.0878	142.2490	589.6309
46	35.6089	54.6543	88.4713	149.9751	646.7385
47	36.2221	56.1272	91.9514	158.0691	709.2852
48	36.8296	57.6141	95.5309	166.5486	777.7886
49	37.4312	59.1152	99.2128	175.4319	852.8159
50	38.0271	60.6306	102.9998	184.7382	934.9838

TABLE A.26 Discrete Compounding: i = 5%

	Geometric series future worth factor $(F\|A_1\ i,j,n)$				
n	$j = 4\%$	$j = 6\%$	$j = 8\%$	$j = 10\%$	$j = 15\%$
1	1.0000	1.0000	1.0000	1.0000	1.0000
2	2.0900	2.1100	2.1300	2.1500	2.2000
3	3.2761	3.3391	3.4029	3.4675	3.6325
4	4.5648	4.6971	4.8328	4.9719	5.3350
5	5.9629	6.1944	6.4349	6.6846	7.3508
6	7.4777	7.8423	8.2260	8.6293	9.7297
7	9.1169	9.6530	10.2241	10.8323	12.5292
8	10.8886	11.6393	12.4492	13.3227	15.8157
9	12.8016	13.8151	14.9225	16.1324	19.6655
10	14.8650	16.1953	17.6677	19.2970	24.1666
11	17.0885	18.7959	20.7100	22.8555	29.4205
12	19.4824	21.6340	24.0771	26.8514	35.5439
13	22.0576	24.7279	27.7991	31.3324	42.6714
14	24.8255	28.0972	31.9087	36.3513	50.9577
15	27.7985	31.7630	36.4414	41.9664	60.5813
16	30.9893	35.7477	41.4356	48.2420	71.7474
17	34.4118	40.0754	46.9333	55.2490	84.6924
18	38.0803	44.7720	52.9800	63.0659	99.6883
19	42.0101	49.8649	59.6250	71.7792	117.0481
20	46.2175	55.3838	66.9220	81.4840	137.1323
21	50.7195	61.3601	74.9290	92.2857	160.3555
22	55.5342	67.8277	83.7093	104.3002	187.1947
23	60.6808	74.8226	93.3313	117.6555	218.1992
24	66.1796	82.3835	103.8693	132.4925	254.0036
25	72.0519	90.5516	115.4039	148.9669	295.3257
26	78.3203	99.3710	128.0226	167.2499	343.0110
27	85.0088	108.8889	141.8201	187.5306	398.0183
28	92.1426	119.1557	156.8991	210.0171	461.4546
29	99.7484	130.2252	173.3712	234.9390	534.5928
30	107.8544	142.1548	191.3570	262.5488	618.8980
31	116.4906	155.0061	210.9876	293.1257	716.0545
32	125.6882	168.8445	232.4046	326.9763	828.0007
33	135.4807	183.7401	255.7619	364.4390	956.9658
34	145.9031	199.7677	281.2258	405.8860	1105.5137
35	156.9926	217.0071	308.9773	451.7278	1276.5942
36	168.7883	235.5435	339.2117	502.4168	1473.5994
37	181.3316	255.4679	372.1404	558.4502	1700.4302
38	194.6663	276.8772	407.9929	620.3767	1961.5759
39	208.8385	299.8755	447.0178	688.7998	2262.1983
40	223.8967	324.5728	489.4841	764.3845	2608.2334
41	239.8926	351.0872	535.6829	847.8630	3006.5088
42	256.8801	379.5442	585.9295	940.0413	3464.8767
43	274.9170	410.0786	643.5654	1041.8069	3992.3694
44	294.0632	442.8330	699.9602	1154.1372	4599.3711
45	314.3831	477.9600	764.5142	1278.1082	5297.8359
46	335.9434	515.6226	834.6604	1414.9038	6101.4961
47	358.8154	555.9944	910.8674	1565.8284	7026.1563
48	383.0740	599.2600	993.6428	1732.3167	8089.9844
49	408.7981	645.6170	1083.5354	1915.9487	9313.8828
50	436.0715	695.2752	1181.1397	2118.4656	10721.8906

TABLE A.27 Discrete Compounding: i = 8%

	Geometric series present worth factor, $(P|A_1\ i,j,n)$				
n	$j = 4\%$	$j = 6\%$	$j = 8\%$	$j = 10\%$	$j = 15\%$
1	0.9259	0.9259	0.9259	0.9259	0.9259
2	1.8176	1.8347	1.8519	1.8690	1.9119
3	2.6762	2.7267	2.7778	2.8295	2.9617
4	3.5030	3.6021	3.7037	3.8079	4.0796
5	4.2992	4.4613	4.6296	4.8043	5.2699
6	5.0659	5.3046	5.5556	5.8192	6.5374
7	5.8042	6.1323	6.4815	6.8529	7.8871
8	6.5151	6.9447	7.4074	7.9057	9.3242
9	7.1997	7.7420	8.3333	8.9780	10.8545
10	7.8590	8.5246	9.2593	10.0702	12.4839
11	8.4939	9.2926	10.1852	11.1826	14.2190
12	9.1052	10.0465	11.1111	12.3157	16.0665
13	9.6939	10.7863	12.0370	13.4696	18.0338
14	10.2608	11.5125	12.9630	14.6450	20.1286
15	10.8067	12.2252	13.8889	15.8421	22.3592
16	11.3324	12.9248	14.8148	17.0614	24.7343
17	11.8386	13.6114	15.7407	18.3033	27.2634
18	12.3260	14.2852	16.6667	19.5682	29.9564
19	12.7954	14.9466	17.5926	20.8565	32.8239
20	13.2475	15.5957	18.5185	22.1686	35.8773
21	13.6827	16.2329	19.4444	23.5051	39.1286
22	14.1019	16.8582	20.3704	24.8663	42.5906
23	14.5055	17.4719	21.2963	26.2527	46.2771
24	14.8942	18.0743	22.2222	27.6648	50.2024
25	15.2685	18.6655	23.1481	29.1030	54.3822
26	15.6289	19.2458	24.0741	30.5679	58.8329
27	15.9760	19.8153	25.0000	32.0599	63.5721
28	16.3102	20.3743	25.9259	33.5795	68.6184
29	16.6321	20.9229	26.8519	35.1273	73.9919
30	16.9420	21.4614	27.7778	36.7037	79.7136
31	17.2404	21.9899	28.7037	38.3094	85.8061
32	17.5278	22.5086	29.6296	39.9447	92.2935
33	17.8046	23.0177	30.5556	41.6104	99.2014
34	18.0711	23.5173	31.4815	43.3069	106.5570
35	18.3277	24.0078	32.4074	45.0348	114.3894
36	18.5748	24.4891	33.3333	46.7947	122.7295
37	18.8128	24.9615	34.2593	48.5872	131.6101
38	19.0420	25.4252	35.1852	50.4128	141.0663
39	19.2626	25.8803	36.1111	52.2723	151.1355
40	19.4751	26.3269	37.0370	54.1663	161.8572
41	19.6797	26.7653	37.9630	56.0953	173.2739
42	19.8768	27.1956	38.8889	58.0600	185.4305
43	20.0665	27.6179	39.8148	60.0611	198.3751
44	20.2493	28.0324	40.7407	62.0993	212.1587
45	20.4252	28.4392	41.6667	64.1752	226.8356
46	20.5947	28.8385	42.5926	66.2895	242.4639
47	20.7578	29.2304	43.5185	68.4431	259.1050
48	20.9149	29.6150	44.4444	70.6364	276.8247
49	21.0662	29.9925	45.3704	72.8704	295.6929
50	21.2119	30.3630	46.2963	75.1458	315.7842

TABLE A.28 Discrete Compounding: i = 8%

	Geometric series future worth factor $(F|A_1\ i,j,n)$				
n	$j = 4\%$	$j = 6\%$	$j = 8\%$	$j = 10\%$	$j = 15\%$
1	1.0000	1.0000	1.0000	1.0000	1.0000
2	2.1200	2.1400	2.1600	2.1800	2.2300
3	3.3712	3.4348	3.4992	3.5644	3.7309
4	4.7658	4.9006	5.0388	5.1806	5.5502
5	6.3169	6.5551	6.8024	7.0591	7.7433
6	8.0389	8.4178	8.8160	9.2343	10.3741
7	9.9473	10.5097	11.1081	11.7446	13.5171
8	12.0590	12.8541	13.7106	14.6329	17.2585
9	14.3923	15.4763	16.6584	17.9471	21.6982
10	16.9670	18.4039	19.9900	21.7409	26.9519
11	19.8046	21.6670	23.7482	26.0739	33.1536
12	22.9284	25.2987	27.9797	31.0129	40.4583
13	26.3638	29.3348	32.7362	36.6324	49.0452
14	30.1379	33.8145	38.0747	43.0152	59.1216
15	34.2806	38.7805	44.0579	50.2539	70.9270
16	38.8240	44.2795	50.7547	58.4515	84.7382
17	43.8029	50.3623	58.2410	67.7226	100.8749
18	49.2551	57.0840	66.6003	78.1949	119.7061
19	55.2213	64.5051	75.9243	90.0104	141.6581
20	61.7458	72.6911	86.3140	103.3271	167.2225
21	68.8766	81.7135	97.8801	118.3207	196.9668
22	76.6655	91.6501	110.7443	135.1866	231.5456
23	85.1687	102.5857	125.0403	154.1418	271.7139
24	94.4469	114.6122	140.9150	175.4274	318.3425
25	104.5659	127.8301	158.5294	199.3114	372.4351
26	115.5970	142.3484	178.0602	226.0910	435.1487
27	127.6172	158.2857	199.7014	256.0962	507.8174
28	140.7100	175.7709	223.6656	289.6941	591.9783
29	154.9655	194.9443	250.1859	327.2905	689.4021
30	170.4814	215.9582	279.5181	369.3369	802.1296
31	187.3633	238.9783	311.9421	416.3333	932.5117
32	205.7255	264.1846	347.7651	468.8342	1083.2561
33	225.6916	291.7727	387.3235	527.4546	1257.4817
34	247.3953	321.9551	430.9854	592.8762	1458.7798
35	270.9810	354.9627	479.1543	665.8538	1691.2864
36	296.6057	391.0457	532.2720	747.2246	1959.7647
37	324.4380	430.4766	590.8218	837.9153	2269.6975
38	354.6614	473.5508	655.3333	938.9524	2627.3982
39	387.4729	520.5891	726.3853	1051.4729	3040.1326
40	423.0872	571.9397	804.6113	1176.7354	3516.2688
41	461.7351	627.9807	890.7046	1316.1333	4065.4336
42	503.6670	689.1221	985.4236	1471.2092	4698.7070
43	549.1531	755.8088	1089.5969	1643.6697	5428.8555
44	598.4861	828.5239	1204.1311	1835.4031	6270.5508
45	651.9815	907.7913	1330.0176	2048.4990	7240.6875
46	709.9310	994.1792	1468.3396	2285.2688	8358.7149
47	772.8545	1088.3042	1620.2808	2548.2703	9646.9922
48	841.0005	1190.8345	1787.1343	2840.3296	11131.2734
49	914.8511	1302.4949	1970.3152	3164.5728	12841.1758
50	994.8726	1424.0720	2171.3679	3524.4575	14810.7813

TABLE A.29 Discrete Compounding: i = 10%

	Geometric series present worth factor, $(P|A_1, i,j,n)$				
n	$j = 4\%$	$j = 6\%$	$j = 8\%$	$j = 10\%$	$j = 15\%$
1	0.9091	0.9091	0.9091	0.9091	0.9091
2	1.7686	1.7851	1.8017	1.8182	1.8595
3	2.5812	2.6293	2.6780	2.7273	2.8531
4	3.3495	3.4428	3.5384	3.6364	3.8919
5	4.0759	4.2267	4.3831	4.5455	4.9779
6	4.7627	4.9321	5.2125	5.4545	6.1133
7	5.4120	5.7100	6.0269	6.3636	7.3002
8	6.0259	6.4115	6.8264	7.2727	8.5411
9	6.6063	7.0874	7.6113	8.1818	9.8385
10	7.1550	7.7388	8.3820	9.0909	11.1948
11	7.6738	8.3664	9.1387	10.0000	12.6127
12	8.1644	8.9713	9.8817	10.9091	14.0951
13	8.6281	9.5542	10.6111	11.8182	15.6449
14	9.0666	10.1158	11.3273	12.7273	17.2651
15	9.4811	10.6571	12.0304	13.6364	18.9590
16	9.8731	11.1786	12.7208	14.5455	20.7298
17	10.2436	11.6812	13.3986	15.4545	22.5812
18	10.5940	12.1656	14.0640	16.3636	24.5167
19	10.9252	12.6323	14.7174	17.2727	26.5402
20	11.2384	13.0820	15.3589	18.1818	28.6556
21	11.5345	13.5154	15.9888	19.0909	30.8673
22	11.8144	13.9330	16.6071	20.0000	33.1794
23	12.0791	14.3354	17.2143	20.9091	35.5966
24	12.3293	14.7232	17.8104	21.8182	38.1238
25	12.5659	15.0969	18.3957	22.7273	40.7658
26	12.7896	15.4571	18.9703	23.6364	43.5278
27	13.0011	15.8041	19.5345	24.5455	46.4155
28	13.2010	16.1385	20.0884	25.4545	49.4344
29	13.3900	16.4607	20.6322	26.3636	52.5905
30	13.5688	16.7712	21.1662	27.2727	55.8900
31	13.7377	17.0705	21.6904	28.1818	59.3396
32	13.8975	17.3588	22.2052	29.0909	62.9459
33	14.0485	17.6367	22.7105	30.0000	66.7162
34	14.1914	17.9044	23.2067	30.9091	70.6578
35	14.3264	18.1624	23.6938	31.8182	74.7787
36	14.4540	18.4111	24.1721	32.7273	79.0868
37	14.5747	18.6507	24.6417	33.6364	83.5907
38	14.6888	18.8816	25.1028	34.5455	88.2994
39	14.7967	19.1041	25.5555	35.4545	93.2221
40	14.8987	19.3185	25.9999	36.3636	98.3685
41	14.9951	19.5251	26.4363	37.2727	103.7489
42	15.0863	19.7241	26.8647	38.1818	109.3733
43	15.1725	19.9160	27.2854	39.0909	115.2545
44	15.2540	20.1009	27.6984	40.0000	121.4024
45	15.3311	20.2790	28.1038	40.9091	127.8298
46	15.4039	20.4507	28.5020	41.8182	134.5493
47	15.4728	20.6161	28.8928	42.7273	141.5743
48	15.5379	20.7755	29.2766	43.6364	148.9186
49	15.5995	20.9292	29.6534	44.5455	156.5967
50	15.6577	21.0772	30.0233	45.4545	164.6238

TABLE A.30 Discrete Compounding: i = 10%

	Geometric series future worth factor $(F|A_1\ i,j,n)$				
n	$j = 4\%$	$j = 6\%$	$j = 8\%$	$j = 10\%$	$j = 15\%$
1	1.0000	1.0000	1.0000	1.0000	1.0000
2	2.1400	2.1600	2.1800	2.2000	2.2500
3	3.4356	3.4996	3.5644	3.6300	3.7975
4	4.9040	5.0406	5.1806	5.3240	5.6981
5	6.5643	6.8071	7.0591	7.3205	8.0169
6	8.4374	8.8260	9.2343	9.6631	10.8300
7	10.5464	11.1272	11.7446	12.4009	14.2261
8	12.9170	13.7435	14.6329	15.5897	18.3087
9	15.5773	16.7117	17.9471	19.2923	23.1986
10	18.5583	20.0724	21.7409	23.5795	29.0363
11	21.8944	23.8704	26.0739	28.5312	35.9855
12	25.6233	28.1558	31.0129	34.2374	44.2364
13	29.7366	32.9836	36.6324	40.7996	54.0103
14	34.4304	38.4149	43.0152	48.3318	65.5641
15	39.6051	44.5172	50.2539	56.9624	79.1962
16	45.3665	51.3655	58.4515	66.8359	95.2529
17	51.7761	59.0424	67.7226	78.1145	114.1358
18	58.9317	67.6394	78.1949	90.9804	136.3106
19	66.8176	77.2577	90.0104	105.6383	162.3171
20	75.6063	88.0091	103.3271	122.3181	192.7806
21	85.3580	100.0171	113.3207	141.2774	228.4252
22	96.1726	113.4183	135.1866	162.8053	270.0891
23	108.1597	128.3637	154.1418	187.2261	318.7427
24	121.4404	145.0198	175.4274	214.9030	375.5086
25	136.1477	163.5706	199.3114	246.2431	441.6846
26	152.4283	184.2197	226.0910	281.7019	518.7720
27	170.4436	207.1911	256.0962	321.7903	608.5059
28	190.3713	232.7325	289.6941	367.0794	712.8916
29	212.4072	261.1172	327.2905	418.2083	834.2465
30	236.7666	292.6475	369.3369	475.8921	975.2466
31	263.6865	327.6558	416.3333	540.9307	1138.9829
32	293.4282	366.5093	468.8342	614.2183	1329.0247
33	326.2793	409.6138	527.4546	696.7537	1549.4922
34	362.5554	457.4158	592.8762	789.6543	1805.1406
35	402.6055	510.4082	665.8538	894.1672	2101.4590
36	446.8120	569.1353	747.2246	1011.6863	2444.7803
37	495.5972	634.1958	837.9153	1143.7676	2842.4094
38	549.4248	706.2515	938.9524	1292.1482	3302.7759
39	608.8062	786.0310	1051.4729	1458.7673	3835.5967
40	674.3032	874.3377	1176.7354	1645.7888	4452.0781
41	746.5344	972.0569	1316.1333	1855.6255	5165.1484
42	826.1809	1080.1655	1471.2092	2090.9744	5989.7070
43	913.9920	1199.7393	1643.6697	2354.8350	6942.9258
44	1010.7915	1331.9634	1835.4031	2650.5584	8044.6094
45	1117.4871	1478.1453	2048.4990	2981.8782	9317.5625
46	1235.0772	1639.7244	2285.2688	3352.9568	10788.0859
47	1364.6597	1818.2864	2548.2703	3768.4319	12486.4805
48	1507.4434	2015.5806	2840.3296	4233.4688	14447.6524
49	1664.7581	2233.5327	3164.6728	4753.8359	16711.8125
50	1838.0671	2474.2639	3524.4575	5335.9375	19325.3047

(handwritten margin notes)

Sum of yrs' digits

life = 6 yr $21 = \dfrac{n(n+1)}{2}$

Value = $\dfrac{6}{21}$

$\dfrac{5}{21}$

$D_p = \dfrac{5}{21} \times P - F$

$\dfrac{5}{21} \times P - F$

$B_t = B - D_t$

Declining balance

$p = \dfrac{1}{n}\ \text{or}\ \dfrac{2}{n}$

$D_t = p\ B_{t-1}$

$= p$

Sinking Fund

$A = (P-F)(A/F\ i, n)$

$B_t = P - A(F/A\ i, t)$

$D_t = (P-F)(A/F\ i, n)(F/p\ i,\ t-1)$

(bottom handwritten notes)

amt. by which payment k reduces unpaid principle

$$E_k = [A](P/F\ i, n-k+1)$$

$$= [P(A/P\ i, n)](PF\ i, n-k+1)$$

interest payment $I_k = A - E_k$

$A =$ amt. borrowed $(A/p\ i, n)$

TABLE A.31 Discrete Compounding: i = 15%

	Geometric series future worth factor $(F|A_1\ i,j,n)$				
n	$j = 4\%$	$j = 6\%$	$j = 8\%$	$j = 10\%$	$j = 15\%$
1	0.8696	0.8696	0.8696	0.8696	0.8696
2	1.6560	1.6711	1.6862	1.7013	1.7391
3	2.3671	2.4099	2.4531	2.4969	2.6087
4	3.0103	3.0908	3.1734	3.2579	3.4783
5	3.5919	3.7185	3.8498	3.9858	4.3478
6	4.1179	4.2971	4.4850	4.6821	5.2174
7	4.5936	4.8303	5.0816	5.3481	6.0870
8	5.0237	5.3219	5.6418	5.9851	6.9565
9	5.4128	5.7749	6.1680	6.5945	7.8261
10	5.7646	6.1926	6.6621	7.1773	8.6957
11	6.0828	6.5775	7.1261	7.7348	9.5652
12	6.3705	6.9323	7.5619	8.2681	10.4348
13	6.6307	7.2593	7.9712	8.7782	11.3043
14	6.8660	7.5608	8.3556	9.2661	12.1739
15	7.0789	7.8386	8.7165	9.7328	13.0435
16	7.2713	8.0947	9.0555	10.1792	13.9130
17	7.4454	8.3308	9.3739	10.6062	14.7826
18	7.6023	8.5484	9.6729	11.0146	15.6522
19	7.7451	8.7489	9.9537	11.4053	16.5217
20	7.8738	8.9338	10.2173	11.7790	17.3913
21	7.9903	9.1042	10.4650	12.1364	18.2609
22	8.0955	9.2613	10.6976	12.4783	19.1304
23	8.1907	9.4060	10.9160	12.8053	20.0000
24	8.2768	9.5395	11.1211	13.1181	20.8696
25	8.3547	9.6625	11.3137	13.4173	21.7391
26	8.4251	9.7759	11.4946	13.7035	22.6087
27	8.4888	9.8804	11.6645	13.9773	23.4783
28	8.5464	9.9767	11.8241	14.2392	24.3478
29	8.5985	10.0655	11.9739	14.4896	25.2174
30	8.6456	10.1473	12.1146	14.7292	26.0870
31	8.6882	10.2227	12.2468	14.9584	26.9565
32	8.7267	10.2922	12.3709	15.1776	27.8261
33	8.7615	10.3563	12.4874	15.3872	28.6957
34	8.7930	10.4154	12.5969	15.5878	29.5652
35	8.8215	10.4698	12.6997	15.7796	30.4348
36	8.8473	10.5200	12.7962	15.9631	31.3043
37	8.8706	10.5663	12.8869	16.1386	32.1739
38	8.8917	10.6089	12.9720	16.3065	33.0435
39	8.9107	10.6482	13.0520	16.4671	33.9130
40	8.9280	10.6845	13.1271	16.6207	34.7826
41	8.9436	10.7178	13.1976	16.7676	35.6522
42	8.9577	10.7486	13.2639	16.9082	36.5217
43	8.9704	10.7770	13.3261	17.0426	37.3913
44	8.9819	10.8031	13.3845	17.1712	38.2609
45	8.9923	10.8272	13.4393	17.2942	39.1304
46	9.0018	10.8495	13.4908	17.4118	40.0000
47	9.0103	10.8699	13.5392	17.5244	40.8696
48	9.0180	10.8888	13.5847	17.6320	41.7391
49	9.0250	10.9062	13.6273	17.7349	42.6087
50	9.0313	10.9222	13.6674	17.8334	43.4783

TABLE A.32 Discrete Compounding: i = 15%

	Geometric series present worth factor, $(P/A_1\ i,j,n)$				
n	$j = 4\%$	$j = 6\%$	$j = 8\%$	$j = 10\%$	$j = 15\%$
1	1.0000	1.0000	1.0000	1.0000	1.0000
2	2.1900	2.2100	2.2300	2.2500	2.3000
3	3.6001	3.6551	3.7309	3.7975	3.9675
4	5.2650	5.4059	5.5502	5.6981	6.0835
5	7.2246	7.4792	7.7433	8.0169	8.7450
6	9.5249	9.9394	10.3741	10.8300	12.0681
7	12.2190	12.8488	13.5171	14.2261	16.1914
8	15.3678	16.2797	17.2585	18.3087	21.2802
9	19.0415	20.3155	21.6982	23.1986	27.5312
10	23.3210	25.0523	26.9519	29.0363	35.1788
11	28.2994	30.6010	33.1536	35.9855	44.5011
12	34.0838	37.0895	40.4583	44.2364	55.8287
13	40.7974	44.6651	49.0452	54.0103	69.5532
14	48.5821	53.4978	59.1216	65.5641	86.1390
15	57.6011	63.7834	70.9270	79.1962	106.1355
16	68.0422	75.7474	84.7382	95.2529	130.1929
17	80.1215	89.6499	100.8749	114.1358	159.0795
18	94.0876	105.7901	119.7061	136.3106	193.7026
19	110.2265	124.5129	141.6581	162.3171	235.1335
20	128.8673	146.2155	167.2225	192.7806	284.6353
21	150.3885	171.3550	196.9668	228.4252	343.6970
22	175.2256	200.4578	231.5456	270.0891	414.0730
23	203.8794	234.1300	271.7139	318.7427	497.8239
24	236.9260	273.0691	318.3425	375.5086	597.3945
25	275.0281	318.0784	372.4351	441.6846	715.6289
26	318.9480	370.0820	435.1487	518.7720	855.8921
27	369.5630	430.1438	507.8174	608.5059	1022.1328
28	427.8806	499.4678	591.9783	712.8916	1218.9880
29	495.0615	579.5225	689.4021	834.2466	1451.9019
30	572.4395	671.8694	802.1296	975.2466	1727.2622
31	661.5486	778.3931	932.5117	1138.9829	2052.5630
32	764.1541	901.2402	1083.2561	1329.0247	2436.5903
33	882.2852	1042.8799	1257.4817	1549.4922	2889.6448
34	1013.2764	1206.1524	1458.7798	1805.1436	3423.7903
35	1174.8123	1394.3262	1691.2864	2101.4590	4053.1648
36	1354.9802	1611.1611	1959.7647	2444.7808	4794.3125
37	1562.3311	1860.9822	2269.6975	2842.4094	5666.6094
38	1800.9487	2148.7647	2627.3982	3302.7759	6692.7266
39	2075.5296	2480.2351	3040.1326	3835.5967	7899.1797
40	2391.4751	2861.9727	3516.2688	4452.0781	9316.9805
41	2754.9966	3301.5552	4065.4336	5165.1484	10982.3945
42	3173.2400	3807.6912	4698.7070	5989.7070	12937.7930
43	3654.4192	4390.3984	5428.8555	6942.9258	15232.7149
44	4207.9805	5061.2109	6270.5508	8044.6094	17925.0078
45	4844.7930	5833.3789	7240.6875	9317.5625	21082.2539
46	5577.3555	6722.1484	8358.7149	10788.0859	24733.3594
47	6420.0352	7745.0625	9646.9922	12486.4805	29120.4336
48	7389.3555	8922.2891	11131.2734	14447.6524	34201.0352
49	8504.3320	10277.0234	12841.1758	16711.8125	40150.5781
50	9786.8125	11835.9570	14810.7813	19325.3047	47115.4688

APPENDIX B
CONTINUOUS COMPOUNDING

SECTION I—CONTINUOUS COMPOUNDING INTEREST FACTORS

SECTION II—CONTINUOUS COMPOUNDING, CONTINUOUS FLOW INTEREST FACTORS

SECTION III—GEOMETRIC SERIES FACTORS

SECTION I—CONTINUOUS COMPOUNDING INTEREST FACTORS

TABLE B.1 Continuous Compounding: r = 1/2%

	Single payment		Uniform series				Gradient series	
	Compound amount factor	Present worth factor	Compound amount factor	Sinking fund factor	Present worth factor	Capital recovery factor	Uniform series factor	Present worth factor
n	To find F given P $F/P\ r,n$	To find P given F $P/F\ r,n$	To find F given A $F/A\ r,n$	To find A given F $A/F\ r,n$	To find P given A $P/A\ r,n$	To find A given P $A/P\ r,n$	To find A given G $A/G\ r,n$	To find P given G $P/G\ r,n$
1	1.0050	0.9950	1.0000	1.0000	0.9950	1.0050	0.0000	0.0000
2	1.0100	0.9901	2.0049	0.4988	1.9850	0.5038	0.4988	0.9771
3	1.0151	0.9851	3.0150	0.3317	2.9702	0.3367	0.9967	2.9539
4	1.0202	0.9802	4.0301	0.2481	3.9503	0.2531	1.4938	5.8784
5	1.0253	0.9753	5.0502	0.1980	4.9255	0.2030	1.9900	9.7732
6	1.0305	0.9704	6.0755	0.1646	5.8960	0.1596	2.4854	14.6235
7	1.0356	0.9655	7.1060	0.1407	6.8615	0.1457	2.9800	20.4148
8	1.0408	0.9608	8.1417	0.1228	7.8225	0.1278	3.4738	27.1690
9	1.0460	0.9560	9.1825	0.1089	8.7784	0.1139	3.9667	34.7988
10	1.0513	0.9512	10.2285	0.0978	9.7297	0.1028	4.4588	43.3628
11	1.0565	0.9465	11.2799	0.0887	10.6762	0.0937	4.9500	52.8465
12	1.0618	0.9418	12.3364	0.0811	11.6180	0.0861	5.4404	63.1395
13	1.0672	0.9371	13.3982	0.0746	12.5550	0.0796	5.9300	74.4440
14	1.0725	0.9324	14.4654	0.0691	13.4874	0.0741	6.4188	86.5656
15	1.0779	0.9277	15.5379	0.0644	14.4152	0.0594	6.9067	99.5504
16	1.0833	0.9231	16.6157	0.0602	15.3382	0.0652	7.3938	113.3847
17	1.0887	0.9185	17.6990	0.0565	16.2568	0.0615	7.8800	128.0895
18	1.0942	0.9139	18.7877	0.0532	17.1707	0.0582	8.3654	143.6162
19	1.0997	0.9094	19.8818	0.0503	18.0800	0.0553	8.8500	159.9857
20	1.1052	0.9048	20.9815	0.0477	18.9849	0.0527	9.3338	177.1844
21	1.1107	0.9003	22.0867	0.0453	19.8853	0.0503	9.8167	195.1986
22	1.1163	0.8958	23.1975	0.0431	20.7811	0.0481	10.2988	214.0148
23	1.1219	0.8914	24.3137	0.0411	21.6725	0.0461	10.7800	233.6195
24	1.1275	0.8869	25.4357	0.0393	22.5594	0.0443	11.2605	254.0333
25	1.1331	0.8825	26.5632	0.0376	23.4419	0.0427	11.7401	275.2088
26	1.1388	0.8781	27.6963	0.0361	24.3200	0.0411	12.2188	297.1670
27	1.1445	0.8737	28.8350	0.0347	25.1936	0.0397	12.6968	319.8611
28	1.1503	0.8694	29.9796	0.0334	26.0631	0.0384	13.1739	343.3442
29	1.1560	0.8650	31.1299	0.0321	26.9281	0.0371	13.6501	367.5703
30	1.1618	0.8607	32.2859	0.0310	27.7888	0.0360	14.1256	392.5261
31	1.1677	0.8564	33.4479	0.0299	28.6453	0.0349	14.6002	418.2317
32	1.1735	0.8521	34.6155	0.0289	29.4974	0.0339	15.0739	444.6414
33	1.1794	0.8479	35.7890	0.0279	30.3453	0.0330	15.5469	471.7749
34	1.1853	0.8437	36.9684	0.0271	31.1889	0.0321	16.0190	499.6194
35	1.1912	0.8395	38.1537	0.0262	32.0284	0.0312	16.4903	528.1621
40	1.2214	0.8187	44.1699	0.0226	36.1633	0.0277	18.8342	681.1308
45	1.2523	0.7985	50.3385	0.0199	40.1961	0.0249	21.1574	850.4431
50	1.2840	0.7788	56.6632	0.0176	44.1294	0.0227	23.4598	1035.2771
55	1.3165	0.7595	63.1480	0.0158	47.9655	0.0208	25.7416	1234.7127
60	1.3499	0.7408	69.7970	0.0143	51.7069	0.0193	28.0027	1447.9358
65	1.3840	0.7225	76.6143	0.0131	55.3560	0.0181	30.2431	1574.1472
70	1.4191	0.7047	83.6042	0.0120	58.9149	0.0170	32.4629	1912.5662
75	1.4550	0.6873	90.7709	0.0110	62.3859	0.0160	34.6621	2162.4270
80	1.4918	0.6703	98.1193	0.0102	65.7713	0.0152	36.8408	2423.0816
85	1.5296	0.6538	105.6535	0.0095	69.0731	0.0145	38.9990	2693.7920
90	1.5683	0.6376	113.3786	0.0088	72.2934	0.0138	41.1368	2973.9312
95	1.6080	0.6219	121.2990	0.0082	75.4341	0.0133	43.2541	3262.8384
100	1.6487	0.6065	129.4202	0.0077	78.4974	0.0127	45.3510	3559.9534

	Single payment		Uniform series				Gradient series	
	Compound amount factor	Present worth factor	Compound amount factor	Sinking fund factor	Present worth factor	Capital recovery factor	Uniform series factor	Present worth factor
n	To find F given P $F\|P\ r,n$	To find P given F $P\|F\ r,n$	To find F given A $F\|A\ r,n$	To find A given F $A\|F\ r,n$	To find P given A $P\|A\ r,n$	To find A given P $A\|P\ r,n$	To find A given G $A\|G\ r,n$	To find P given G $P\|G\ r,n$
1	1.0075	0.9925	1.0000	1.0000	0.9925	1.0075	0.0000	0.0000
2	1.0151	0.9851	2.0077	0.4981	1.9778	0.5355	0.4981	1.0114
3	1.0228	0.9778	3.0229	0.3308	2.9557	0.3383	0.9950	2.9787
4	1.0305	0.9704	4.0457	0.2472	3.9262	0.2547	1.4906	5.8965
5	1.0382	0.9632	5.0764	0.1970	4.8895	0.2045	1.9850	9.7757
6	1.0460	0.9560	6.1147	0.1635	5.8456	0.1711	2.4781	14.5620
7	1.0539	0.9489	7.1608	0.1396	6.7945	0.1472	2.9700	20.2564
8	1.0618	0.9418	8.2149	0.1217	7.7365	0.1293	3.4606	26.8834
9	1.0698	0.9347	9.2768	0.1078	8.6713	0.1153	3.9500	34.3758
10	1.0779	0.9277	10.3468	0.0966	9.5991	0.1042	4.4381	42.7384
11	1.0860	0.9208	11.4248	0.0875	10.5201	0.0951	4.9250	51.9661
12	1.0942	0.9139	12.5108	0.0799	11.4341	0.0875	5.4106	62.0226
13	1.1024	0.9071	13.6052	0.0735	12.3413	0.0810	5.8950	72.9335
14	1.1107	0.9003	14.7077	0.0680	13.2417	0.0755	6.3781	84.6479
15	1.1191	0.8936	15.8186	0.0632	14.1354	0.0707	6.8600	97.1757
16	1.1275	0.8869	16.9378	0.0590	15.0225	0.0666	7.3407	110.4967
17	1.1360	0.8803	18.0654	0.0554	15.9028	0.0629	7.8200	124.5913
18	1.1445	0.8737	19.2014	0.0521	16.7765	0.0596	8.2982	139.4543
19	1.1532	0.8672	20.3461	0.0491	17.6439	0.0567	8.7751	155.0807
20	1.1618	0.8607	21.4994	0.0465	18.5047	0.0540	9.2507	171.4510
21	1.1706	0.8543	22.6613	0.0441	19.3591	0.0517	9.7251	188.5457
22	1.1794	0.8479	23.8321	0.0420	20.2071	0.0495	10.1983	206.3743
23	1.1883	0.8416	25.0116	0.0400	21.0488	0.0475	10.6702	224.9033
24	1.1972	0.8353	26.2000	0.0382	21.8841	0.0457	11.1408	244.1280
25	1.2062	0.8290	27.3973	0.0365	22.7132	0.0440	11.6102	264.0296
26	1.2153	0.8228	28.6037	0.0350	23.5361	0.0425	12.0784	284.6172
27	1.2245	0.8167	29.8192	0.0335	24.3530	0.0411	12.5453	305.8721
28	1.2337	0.8106	31.0438	0.0322	25.1636	0.0397	13.0110	327.7759
29	1.2430	0.8045	32.2776	0.0310	25.9682	0.0385	13.4754	350.3103
30	1.2523	0.7985	33.5207	0.0298	26.7668	0.0374	13.9386	373.4841
31	1.2618	0.7925	34.7732	0.0288	27.5595	0.0363	14.4005	397.2793
32	1.2712	0.7866	36.0351	0.0278	28.3462	0.0353	14.8612	421.6777
33	1.2808	0.7808	37.3065	0.0268	29.1270	0.0343	15.3207	446.6751
34	1.2905	0.7749	38.5873	0.0259	29.9020	0.0334	15.7789	472.2537
35	1.3002	0.7691	39.8780	0.0251	30.6712	0.0326	16.2359	498.4214
40	1.3499	0.7408	46.4783	0.0215	34.4320	0.0290	18.5021	637.5711
45	1.4014	0.7135	53.3307	0.0188	38.0542	0.0263	20.7374	789.7029
50	1.4550	0.6873	60.4449	0.0165	41.5432	0.0241	22.9418	953.6831
55	1.5106	0.6620	67.8311	0.0147	44.9038	0.0223	25.1153	1128.4336
60	1.5683	0.6376	75.4995	0.0132	48.1407	0.0208	27.2582	1312.9395
65	1.6282	0.6142	83.4609	0.0120	51.2584	0.0195	29.3704	1506.2349
70	1.6905	0.5916	91.7266	0.0109	54.2614	0.0184	31.4521	1707.4382
75	1.7551	0.5698	100.3080	0.0100	57.1538	0.0175	33.5034	1915.6895
80	1.8221	0.5488	109.2174	0.0092	59.9398	0.0167	35.5244	2130.2127
85	1.8917	0.5286	118.4672	0.0084	62.6233	0.0160	37.5153	2350.2559
90	1.9640	0.5092	128.0705	0.0078	65.2080	0.0153	39.4762	2575.1262
95	2.0391	0.4904	138.0407	0.0072	67.6975	0.0148	41.4072	2834.1648
100	2.1170	0.4724	148.3919	0.0067	70.0955	0.0143	43.3084	3036.7615

TABLE B.3 Continuous Compounding: r = 1%

	Single payment		Uniform series				Gradient series	
	Compound amount factor	Present worth factor	Compound amount factor	Sinking fund factor	Present worth factor	Capital recovery factor	Uniform series factor	Present worth factor
n	To find F given P $F\|P\ r,n$	To find P given F $P\|F\ r,n$	To find F given A $F\|A\ r,n$	To find A given F $A\|F\ r,n$	To find P given A $P\|A\ r,n$	To find A given P $A\|P\ r,n$	To find A given G $A\|G\ r,n$	To find P given G $P\|G\ r,n$
1	1.0100	0.9901	1.0000	1.0000	0.9901	1.0100	0.0000	0.0000
2	1.0202	0.9802	2.0100	0.4975	1.9703	0.5075	0.4975	0.9811
3	1.0305	0.9704	3.0303	0.3300	2.9407	0.3401	0.9933	2.9231
4	1.0408	0.9608	4.0608	0.2463	3.9015	0.2563	1.4875	5.8153
5	1.0513	0.9512	5.1016	0.1960	4.8528	0.2061	1.9800	9.6196
6	1.0618	0.9418	6.1530	0.1625	5.7947	0.1726	2.4708	14.3348
7	1.0725	0.9324	7.2148	0.1386	6.7271	0.1487	2.9600	19.9324
8	1.0833	0.9231	8.2873	0.1207	7.6502	0.1307	3.4475	26.3934
9	1.0942	0.9139	9.3707	0.1067	8.5641	0.1168	3.9333	33.7377
10	1.1052	0.9048	10.4649	0.0956	9.4690	0.1056	4.4175	41.8564
11	1.1163	0.8958	11.5701	0.0864	10.3649	0.0965	4.9000	50.8208
12	1.1275	0.8869	12.6865	0.0788	11.2519	0.0889	5.3809	60.5824
13	1.1388	0.8781	13.8140	0.0724	12.1300	0.0824	5.8600	71.1232
14	1.1503	0.8694	14.9528	0.0669	12.9994	0.0759	6.3376	82.4251
15	1.1618	0.8607	16.1031	0.0621	13.8601	0.0721	6.8134	94.4784
16	1.1735	0.8521	17.2650	0.0579	14.7123	0.0680	7.2876	107.2653
17	1.1853	0.8437	18.4386	0.0542	15.5560	0.0643	7.7601	120.7682
18	1.1972	0.8353	19.6239	0.0510	16.3913	0.0610	8.2310	134.9698
19	1.2092	0.8270	20.8212	0.0480	17.2183	0.0581	8.7002	149.8607
20	1.2214	0.8187	22.0305	0.0454	18.0370	0.0554	9.1677	165.4160
21	1.2337	0.8106	23.2519	0.0430	18.8477	0.0531	9.6336	181.6342
22	1.2461	0.8025	24.4857	0.0408	19.6502	0.0509	10.0978	198.4909
23	1.2586	0.7945	25.7317	0.0389	20.4447	0.0489	10.5604	215.9695
24	1.2712	0.7866	26.9904	0.0371	21.2314	0.0471	11.0213	234.0683
25	1.2840	0.7788	28.2617	0.0354	22.0103	0.0454	11.4805	252.7537
26	1.2969	0.7711	29.5457	0.0338	22.7813	0.0439	11.9381	272.0393
27	1.3100	0.7634	30.8428	0.0324	23.5447	0.0425	12.3941	291.8938
28	1.3231	0.7558	32.1528	0.0311	24.3006	0.0412	12.8484	312.3042
29	1.3364	0.7483	33.4759	0.0299	25.0488	0.0399	13.3010	333.2546
30	1.3499	0.7408	34.8124	0.0287	25.7897	0.0388	13.7520	354.7434
31	1.3634	0.7334	36.1623	0.0277	26.5231	0.0377	14.2013	376.7481
32	1.3771	0.7261	37.5258	0.0266	27.2493	0.0367	14.6490	399.2669
33	1.3910	0.7189	38.9029	0.0257	27.9682	0.0358	15.0950	422.2713
34	1.4049	0.7118	40.2940	0.0248	28.6801	0.0349	15.5394	445.7661
35	1.4191	0.7047	41.6990	0.0240	29.3848	0.0340	15.9821	469.7297
40	1.4918	0.6703	48.9386	0.0204	32.8045	0.0305	18.1710	596.2322
45	1.5683	0.6376	56.5494	0.0177	36.0575	0.0277	20.3190	732.7742
50	1.6487	0.6065	64.5505	0.0155	39.1519	0.0255	22.4261	878.1565
55	1.7333	0.5769	72.9617	0.0137	42.0953	0.0238	24.4926	1031.1660
60	1.8221	0.5488	81.8042	0.0122	44.8951	0.0223	26.5187	1190.7100
65	1.9155	0.5220	91.1001	0.0110	47.5585	0.0210	28.5045	1355.7915
70	2.0138	0.4966	100.8727	0.0099	50.0919	0.0200	30.4505	1525.4907
75	2.1170	0.4724	111.1462	0.0090	52.5018	0.0190	32.3567	1698.9614
80	2.2255	0.4493	121.9464	0.0082	54.7941	0.0183	34.2235	1875.4336
85	2.3396	0.4274	133.3005	0.0075	56.9745	0.0176	36.0513	2054.2031
90	2.4596	0.4066	145.2366	0.0069	59.0488	0.0169	37.8402	2234.6223
95	2.5857	0.3867	157.7847	0.0063	61.0219	0.0164	39.5907	2416.1087
100	2.7183	0.3679	170.9762	0.0058	62.8987	0.0159	41.3032	2598.1294

TABLE B.4 Continuous Compounding: r = 3/2%

	Single payment		Uniform series				Gradient series	
	Compound amount factor	Present worth factor	Compound amount factor	Sinking fund factor	Present worth factor	Capital recovery factor	Uniform series factor	Present worth factor
n	To find F given P $F\|P\ r,n$	To find P given F $P\|F\ r,n$	To find F given A $F\|A\ r,n$	To find A given F $A\|F\ r,n$	To find P given A $P\|A\ r,n$	To find A given P $A\|P\ r,n$	To find A given G $A\|G\ r,n$	To find P given G $P\|G\ r,n$
1	1.0151	0.9851	1.0000	1.0000	0.9851	1.0151	0.0000	0.0000
2	1.0305	0.9704	2.0151	0.4963	1.9555	0.5114	0.4963	0.9684
3	1.0460	0.9560	3.0456	0.3283	2.9116	0.3435	0.9900	2.8820
4	1.0618	0.9418	4.0916	0.2444	3.8534	0.2595	1.4813	5.7097
5	1.0779	0.9277	5.1535	0.1940	4.7811	0.2092	1.9700	9.4210
6	1.0942	0.9139	6.2313	0.1605	5.6950	0.1756	2.4563	13.9898
7	1.1107	0.9003	7.3256	0.1365	6.5954	0.1516	2.9400	19.3941
8	1.1275	0.8869	8.4363	0.1185	7.4823	0.1336	3.4213	25.6047
9	1.1445	0.8737	9.5638	0.1046	8.3560	0.1197	3.9000	32.5928
10	1.1618	0.8607	10.7083	0.0934	9.2168	0.1085	4.3763	40.3412
11	1.1794	0.8479	11.8702	0.0842	10.0647	0.0994	4.8501	48.8216
12	1.1972	0.8353	13.0496	0.0766	10.8999	0.0917	5.3213	58.0101
13	1.2153	0.8228	14.2468	0.0702	11.7228	0.0853	5.7901	67.8831
14	1.2337	0.8105	15.4622	0.0647	12.5334	0.0798	6.2564	78.4240
15	1.2523	0.7985	16.6958	0.0599	13.3319	0.0750	6.7202	89.6029
16	1.2712	0.7866	17.9482	0.0557	14.1185	0.0708	7.1816	101.4038
17	1.2905	0.7749	19.2195	0.0520	14.8935	0.0671	7.6404	113.8041
18	1.3100	0.7634	20.5099	0.0488	15.6569	0.0639	8.0967	126.7817
19	1.3298	0.7520	21.8199	0.0458	16.4089	0.0609	8.5506	140.3179
20	1.3499	0.7408	23.1497	0.0432	17.1497	0.0583	9.0020	154.3943
21	1.3703	0.7298	24.4995	0.0408	17.8795	0.0559	9.4509	168.9895
22	1.3910	0.7189	25.8698	0.0387	18.5984	0.0538	9.8973	184.0884
23	1.4120	0.7082	27.2608	0.0367	19.3067	0.0518	10.3413	199.6702
24	1.4333	0.6977	28.6728	0.0349	20.0043	0.0500	10.7828	215.7173
25	1.4550	0.6873	30.1061	0.0332	20.6916	0.0483	11.2218	232.2125
26	1.4770	0.6771	31.5612	0.0317	21.3687	0.0468	11.6584	249.1408
27	1.4993	0.6670	33.0382	0.0303	22.0357	0.0454	12.0925	266.4829
28	1.5220	0.6570	34.5375	0.0290	22.6928	0.0441	12.5241	284.2246
29	1.5450	0.6473	36.0595	0.0277	23.3400	0.0428	12.9533	302.3491
30	1.5683	0.6376	37.6045	0.0266	23.9777	0.0417	13.3800	320.8403
31	1.5920	0.6281	39.1728	0.0255	24.6058	0.0406	13.8043	339.6843
32	1.6161	0.6188	40.7648	0.0245	25.2246	0.0396	14.2261	358.8677
33	1.6405	0.6096	42.3809	0.0236	25.8342	0.0387	14.6455	378.3740
34	1.6653	0.6005	44.0215	0.0227	26.4347	0.0378	15.0625	398.1926
35	1.6905	0.5915	45.6868	0.0219	27.0262	0.0370	15.4770	418.3059
40	1.8221	0.5488	54.3985	0.0184	29.8545	0.0335	17.5131	522.8696
45	1.9640	0.5092	63.7888	0.0157	32.4785	0.0308	19.4890	632.9990
50	2.1170	0.4724	73.9104	0.0135	34.9128	0.0286	21.4052	747.3428
55	2.2819	0.4382	84.8203	0.0118	37.1713	0.0269	23.2622	864.7144
60	2.4596	0.4065	96.5800	0.0104	39.2665	0.0255	25.0609	984.0840
65	2.6512	0.3772	109.2555	0.0092	41.2104	0.0243	26.8018	1104.5462
70	2.8576	0.3499	122.9183	0.0081	43.0138	0.0232	28.4859	1225.3215
75	3.0802	0.3247	137.6452	0.0073	44.6869	0.0224	30.1140	1345.7358
80	3.3201	0.3012	153.5190	0.0065	46.2391	0.0216	31.6869	1465.2097
85	3.5787	0.2794	170.6293	0.0059	47.6791	0.0210	33.2056	1583.2522
90	3.8574	0.2592	189.0721	0.0053	49.0151	0.0204	34.6710	1699.4431
95	4.1579	0.2405	208.9513	0.0048	50.2546	0.0199	36.0842	1813.4382
100	4.4817	0.2231	230.3787	0.0043	51.4045	0.0195	37.4462	1924.9439

TABLE B.5 Continuous Compounding: r = 2%

	Single payment		Uniform series				Gradient series	
	Compound amount factor	Present worth factor	Compound amount factor	Sinking fund factor	Present worth factor	Capital recovery factor	Uniform series factor	Present worth factor
n	To find F given P $F/P\ r,n$	To find P given F $P/F\ r,n$	To find F given A $F/A\ r,n$	To find A given F $A/F\ r,n$	To find P given A $P/A\ r,n$	To find A given P $A/P\ r,n$	To find A given G $A/G\ r,n$	To find P given G $P/G\ r,n$
1	1.0202	0.9802	1.0000	1.0000	0.9802	1.0202	0.0000	0.0000
2	1.0408	0.9608	2.0202	0.4950	1.9410	0.5152	0.4950	0.9633
3	1.0618	0.9418	3.0611	0.3267	2.8828	0.3469	0.9867	2.8480
4	1.0833	0.9231	4.1229	0.2425	3.8060	0.2527	1.4750	5.6178
5	1.1052	0.9048	5.2063	0.1921	4.7108	0.2123	1.9600	9.2389
6	1.1275	0.8869	6.3115	0.1584	5.5978	0.1786	2.4417	13.6761
7	1.1503	0.8694	7.4390	0.1344	6.4672	0.1546	2.9200	18.8930
8	1.1735	0.8521	8.5893	0.1164	7.3193	0.1366	3.3950	24.8598
9	1.1972	0.8353	9.7629	0.1024	8.1546	0.1226	3.8667	31.5433
10	1.2214	0.8187	10.9601	0.0912	8.9734	0.1114	4.3351	38.9129
11	1.2461	0.8025	12.1816	0.0821	9.7759	0.1023	4.8002	46.9405
12	1.2712	0.7866	13.4277	0.0745	10.5626	0.0947	5.2619	55.5945
13	1.2969	0.7711	14.6989	0.0680	11.3336	0.0882	5.7203	64.8478
14	1.3231	0.7558	15.9959	0.0625	12.0895	0.0827	6.1754	74.6755
15	1.3499	0.7408	17.3191	0.0577	12.8303	0.0779	6.6272	85.0477
16	1.3771	0.7261	18.6690	0.0536	13.5565	0.0738	7.0757	95.9421
17	1.4049	0.7118	20.0462	0.0499	14.2683	0.0701	7.5209	107.3315
18	1.4333	0.6977	21.4511	0.0466	14.9660	0.0668	7.9628	119.1926
19	1.4623	0.6839	22.8845	0.0437	15.6498	0.0639	8.4014	131.5040
20	1.4918	0.6703	24.3468	0.0411	16.3202	0.0613	8.8368	144.2414
21	1.5220	0.6570	25.8387	0.0387	16.9772	0.0589	9.2688	157.3841
22	1.5527	0.6440	27.3607	0.0365	17.6213	0.0567	9.6976	170.9100
23	1.5841	0.6313	28.9135	0.0346	18.2526	0.0548	10.1231	184.7990
24	1.6161	0.6188	30.4976	0.0328	18.8714	0.0530	10.5453	199.0326
25	1.6487	0.6065	32.1137	0.0311	19.4780	0.0513	10.9643	213.5910
26	1.6820	0.5945	33.7625	0.0296	20.0725	0.0498	11.3800	228.4552
27	1.7160	0.5827	35.4446	0.0282	20.6553	0.0484	11.7925	243.6082
28	1.7507	0.5712	37.1606	0.0269	21.2265	0.0471	12.2018	259.0325
29	1.7860	0.5599	38.9113	0.0257	21.7864	0.0459	12.6078	274.7100
30	1.8221	0.5488	40.6974	0.0246	22.3352	0.0448	13.0106	290.6267
31	1.8589	0.5379	42.5196	0.0235	22.8732	0.0437	13.4102	306.7671
32	1.8965	0.5273	44.3786	0.0225	23.4005	0.0427	13.8065	323.1145
33	1.9348	0.5169	46.2751	0.0216	23.9174	0.0418	14.1997	339.6546
34	1.9739	0.5066	48.2100	0.0207	24.4240	0.0409	14.5897	356.3743
35	2.0138	0.4966	50.1839	0.0199	24.9206	0.0401	14.9765	373.2595
40	2.2255	0.4493	60.6681	0.0165	27.2599	0.0367	16.8630	459.7251
45	2.4596	0.4066	72.2549	0.0138	29.3767	0.0340	18.6714	548.5459
50	2.7183	0.3679	85.0603	0.0118	31.2919	0.0320	20.4028	638.4905
55	3.0042	0.3329	99.2125	0.0101	33.0250	0.0303	22.0586	728.5418
60	3.3201	0.3012	114.8530	0.0087	34.5931	0.0289	23.6409	817.8630
65	3.6693	0.2725	132.1335	0.0076	36.0120	0.0278	25.1507	905.7793
70	4.0552	0.2466	151.2420	0.0066	37.2958	0.0268	26.5399	991.7483
75	4.4817	0.2231	172.3545	0.0058	38.4575	0.0260	27.9604	1075.3445
80	4.9530	0.2019	195.6875	0.0051	39.5085	0.0253	29.2640	1156.2412
85	5.4739	0.1827	221.4745	0.0045	40.4597	0.0247	30.5528	1234.1951
90	6.0496	0.1653	249.9734	0.0040	41.3233	0.0242	31.6766	1339.0337
95	6.6859	0.1496	281.4692	0.0036	42.0990	0.0238	32.7937	1380.6438
100	7.3891	0.1353	316.2781	0.0032	42.8036	0.0234	33.8499	1448.9627

TABLE B.6 Continuous Compounding: r = 3%

	Single payment		Uniform series				Gradient series	
	Compound amount factor	Present worth factor	Compound amount factor	Sinking fund factor	Present worth factor	Capital recovery factor	Uniform series factor	Present worth factor
n	To find F given P $F\|P\ r,n$	To find P given F $P\|F\ r,n$	To find F given A $F\|A\ r,n$	To find A given F $A\|F\ r,n$	To find P given A $P\|A\ r,n$	To find A given P $A\|P\ r,n$	To find A given G $A\|G\ r,n$	To find P given G $P\|G\ r,n$
1	1.0305	0.9704	1.0000	1.0000	0.9704	1.0305	0.0000	0.0000
2	1.0618	0.9418	2.0305	0.4925	1.9123	0.5229	0.4925	0.9432
3	1.0942	0.9139	3.0923	0.3234	2.8262	0.3538	0.9800	2.7715
4	1.1275	0.8869	4.1866	0.2389	3.7132	0.2693	1.4625	5.4338
5	1.1618	0.8607	5.3141	0.1882	4.5739	0.2185	1.9400	8.8772
6	1.1972	0.8353	6.4760	0.1544	5.4092	0.1849	2.4125	13.0546
7	1.2337	0.8106	7.6732	0.1303	6.2198	0.1508	2.8801	17.9191
8	1.2712	0.7866	8.9069	0.1123	7.0064	0.1427	3.3427	23.4263
9	1.3100	0.7634	10.1782	0.0982	7.7698	0.1287	3.8002	29.5343
10	1.3499	0.7408	11.4882	0.0870	8.5107	0.1175	4.2529	36.2023
11	1.3910	0.7189	12.8381	0.0779	9.2295	0.1083	4.7005	43.3922
12	1.4333	0.6977	14.2291	0.0703	9.9273	0.1007	5.1433	51.0675
13	1.4770	0.6771	15.6625	0.0638	10.6044	0.0943	5.5811	59.1932
14	1.5220	0.6570	17.1395	0.0583	11.2615	0.0888	6.0139	67.7359
15	1.5683	0.6376	18.6615	0.0536	11.8991	0.0840	6.4419	76.6634
16	1.6161	0.6188	20.2298	0.0494	12.5179	0.0799	6.8649	85.9458
17	1.6653	0.6005	21.8460	0.0458	13.1184	0.0762	7.2831	95.5548
18	1.7160	0.5827	23.5113	0.0425	13.7012	0.0730	7.6964	105.4622
19	1.7683	0.5655	25.2274	0.0396	14.2667	0.0701	8.1048	115.6426
20	1.8221	0.5488	26.9957	0.0370	14.8156	0.0675	8.5084	126.0705
21	1.8776	0.5326	28.8179	0.0347	15.3482	0.0652	8.9072	136.7235
22	1.9348	0.5169	30.6955	0.0326	15.8650	0.0630	9.3012	147.5779
23	1.9937	0.5015	32.6304	0.0306	16.3666	0.0611	9.6904	158.6136
24	2.0544	0.4868	34.6241	0.0289	16.8534	0.0593	10.0748	169.8097
25	2.1170	0.4724	36.6786	0.0273	17.3258	0.0577	10.4545	181.1472
26	2.1815	0.4584	38.7957	0.0258	17.7842	0.0562	10.8294	192.6081
27	2.2479	0.4449	40.9772	0.0244	18.2290	0.0549	11.1996	204.1752
28	2.3164	0.4317	43.2252	0.0231	18.6608	0.0536	11.5652	215.8319
29	2.3869	0.4190	45.5416	0.0220	19.0797	0.0524	11.9261	227.5534
30	2.4596	0.4066	47.9286	0.0209	19.4863	0.0513	12.2823	239.3545
31	2.5345	0.3946	50.3883	0.0198	19.8809	0.0503	12.6339	251.1920
32	2.6117	0.3829	52.9228	0.0189	20.2638	0.0493	12.9810	263.0618
33	2.6912	0.3716	55.5346	0.0180	20.6354	0.0485	13.3235	274.9531
34	2.7732	0.3606	58.2259	0.0172	20.9960	0.0475	13.6614	286.8535
35	2.8577	0.3499	60.9992	0.0164	21.3459	0.0468	13.9948	298.7520
40	3.3201	0.3012	76.1851	0.0131	22.9465	0.0436	15.5953	357.8789
45	3.8574	0.2592	93.8285	0.0107	24.3241	0.0411	17.0874	415.6582
50	4.4817	0.2231	114.3273	0.0087	25.5099	0.0392	18.4750	471.3181
55	5.2070	0.1920	138.1435	0.0072	26.5304	0.0377	19.7623	524.3274
60	6.0496	0.1653	165.8139	0.0060	27.4089	0.0365	20.9538	574.3455
65	7.0287	0.1423	197.9624	0.0051	28.1649	0.0355	22.0541	621.1765
70	8.1662	0.1225	235.3136	0.0042	28.8157	0.0347	23.0677	664.7380
75	9.4877	0.1054	278.7092	0.0036	29.3758	0.0340	23.9996	705.0325
80	11.0232	0.0907	329.1282	0.0030	29.8579	0.0335	24.8543	742.1245
85	12.8071	0.0781	387.7065	0.0026	30.2728	0.0330	25.6368	776.1245
90	14.8797	0.0672	455.7647	0.0022	30.6299	0.0326	26.3516	807.1746
95	17.2878	0.0578	534.8369	0.0019	30.9373	0.0323	27.0032	835.4363
100	20.0855	0.0498	626.7063	0.0016	31.2019	0.0320	27.5963	861.0838

TABLE B.7 Continuous Compounding: r = 4%

	Single payment		Uniform series				Gradient series	
	Compound amount factor	Present worth factor	Compound amount factor	Sinking fund factor	Present worth factor	Capital recovery factor	Uniform series factor	Present worth factor
n	To find F given P $F\|P\ r,n$	To find P given F $P\|F\ r,n$	To find F given A $F\|A\ r,n$	To find A given F $A\|F\ r,n$	To find P given A $P\|A\ r,n$	To find A given P $A\|P\ r,n$	To find A given G $A\|G\ r,n$	To find P given G $P\|G\ r,n$
1	1.0408	0.9608	1.0000	1.0000	0.9608	1.0408	0.0000	0.0000
2	1.0833	0.9231	2.0408	0.4900	1.8839	0.5308	0.4900	0.9229
3	1.1275	0.8869	3.1241	0.3201	2.7708	0.3609	0.9733	2.6972
4	1.1735	0.8521	4.2516	0.2352	3.6230	0.2760	1.4500	5.2537
5	1.2214	0.8187	5.4251	0.1843	4.4417	0.2251	1.9201	8.5286
6	1.2712	0.7866	6.6465	0.1505	5.2283	0.1913	2.3834	12.4620
7	1.3231	0.7558	7.9178	0.1263	5.9841	0.1571	2.8402	16.9968
8	1.3771	0.7261	9.2409	0.1082	6.7103	0.1490	3.2904	22.0800
9	1.4333	0.6977	10.6180	0.0942	7.4080	0.1350	3.7339	27.6513
10	1.4918	0.6703	12.0514	0.0830	8.0783	0.1238	4.1709	33.6944
11	1.5527	0.6440	13.5432	0.0738	8.7223	0.1146	4.6013	40.1348
12	1.6161	0.6188	15.0959	0.0662	9.3411	0.1071	5.0252	46.9415
13	1.6820	0.5945	16.7120	0.0598	9.9356	0.1006	5.4425	54.0758
14	1.7507	0.5712	18.3941	0.0544	10.5069	0.0952	5.8534	61.5019
15	1.8221	0.5488	20.1447	0.0496	11.0557	0.0905	6.2578	69.1851
16	1.8965	0.5273	21.9669	0.0455	11.5830	0.0863	6.6558	77.0948
17	1.9739	0.5066	23.8633	0.0419	12.0896	0.0827	7.3473	85.2006
18	2.0544	0.4868	25.8372	0.0387	12.5763	0.0795	7.4326	93.4756
19	2.1383	0.4677	27.8917	0.0359	13.0440	0.0767	7.8114	101.8938
20	2.2255	0.4493	30.0300	0.0333	13.4933	0.0741	8.1840	110.4309
21	2.3164	0.4317	32.2555	0.0310	13.9250	0.0718	8.5503	119.0652
22	2.4109	0.4148	34.5719	0.0289	14.3398	0.0597	8.9104	127.7759
23	2.5093	0.3985	36.9828	0.0270	14.7384	0.0579	9.2644	136.5434
24	2.6117	0.3829	39.4921	0.0253	15.1213	0.0661	9.6122	145.3501
25	2.7183	0.3679	42.1038	0.0238	15.4891	0.0645	9.9539	154.1791
26	2.8292	0.3535	44.8221	0.0223	15.8426	0.0631	10.2895	163.0156
27	2.9447	0.3396	47.6513	0.0210	16.1822	0.0618	10.6191	171.8453
28	3.0649	0.3263	50.5960	0.0198	16.5085	0.0606	10.9431	180.6549
29	3.1899	0.3135	53.6609	0.0186	16.8219	0.0594	11.2609	189.4326
30	3.3201	0.3012	56.8508	0.0176	17.1231	0.0584	11.5730	198.1672
31	3.4556	0.2894	60.1710	0.0166	17.4125	0.0574	11.8792	206.8489
32	3.5966	0.2780	63.6266	0.0157	17.6906	0.0565	12.1797	215.4680
33	3.7434	0.2671	67.2233	0.0149	17.9577	0.0557	12.4746	224.0165
34	3.8962	0.2567	70.9667	0.0141	18.2144	0.0549	12.7638	232.4864
35	4.0552	0.2466	74.8629	0.0134	18.4610	0.0542	13.0475	240.8708
40	4.9530	0.2019	96.3629	0.0103	19.5563	0.0511	14.3845	281.3096
45	6.0496	0.1653	123.7337	0.0081	20.4530	0.0489	15.5918	318.9024
50	7.3891	0.1353	156.5538	0.0064	21.1873	0.0472	16.6775	353.3518
55	9.0250	0.1108	196.6404	0.0051	21.7884	0.0459	17.6498	384.5620
60	11.0232	0.0907	245.6023	0.0041	22.2805	0.0449	18.5172	412.5757
65	13.4637	0.0743	305.4041	0.0033	22.6835	0.0441	19.2882	437.5259
70	16.4446	0.0608	378.4465	0.0026	23.0134	0.0435	19.9710	459.6030
75	20.0855	0.0498	467.6609	0.0021	23.2835	0.0429	20.5737	479.0286
80	24.5325	0.0408	576.6275	0.0017	23.5045	0.0425	21.1038	496.0391
85	29.9641	0.0334	709.7195	0.0014	23.6857	0.0422	21.5607	510.8709
90	36.5982	0.0273	872.2788	0.0011	23.8339	0.0420	21.9751	523.7556
95	44.7012	0.0224	1070.8289	0.0009	23.9553	0.0417	22.3295	534.9114
100	54.5981	0.0183	1313.3386	0.0008	24.0546	0.0416	22.6376	544.5418

TABLE B.8 Continuous Compounding: r = 5%

	Single payment		Uniform series				Gradient series	
	Compound—amount factor	Present worth factor	Compound—amount factor	Sinking-fund factor	Present worth factor	Capital-recovery factor	Uniform series factor	Present worth factor
n	To find F given P $F\|P\ r,n$	To find P given F $P\|F\ r,n$	To find F given A $F\|A\ r,n$	To find A given F $A\|F\ r,n$	To find P given A $P\|A\ r,n$	To find A given P $A\|P\ r,n$	To find A given G $A\|G\ r,n$	To find P given G $P\|G\ r,n$
1	1.0513	0.9512	1.0000	1.0000	0.9512	1.0513	0.0000	0.0000
2	1.1052	0.9048	2.0513	0.4875	1.8561	0.5388	0.4875	0.9051
3	1.1618	0.8607	3.1565	0.3168	2.7168	0.3681	0.9667	2.6268
4	1.2214	0.8187	4.3183	0.2316	3.5355	0.2828	1.4375	5.0831
5	1.2840	0.7788	5.5397	0.1805	4.3144	0.2318	1.9001	8.1987
6	1.3499	0.7408	6.8238	0.1465	5.0552	0.1978	2.3544	11.9929
7	1.4191	0.7047	8.1737	0.1223	5.7599	0.1736	2.8004	16.1313
8	1.4918	0.6703	9.5927	0.1042	6.4302	0.1555	3.2382	20.8237
9	1.5683	0.6376	11.0846	0.0902	7.0678	0.1415	3.6678	25.9250
10	1.6487	0.6065	12.6529	0.0790	7.6744	0.1303	4.0892	31.3840
11	1.7333	0.5769	14.3016	0.0699	8.2513	0.1212	4.5025	37.1537
12	1.8221	0.5488	16.0349	0.0624	8.8002	0.1136	4.9077	43.1907
13	1.9155	0.5220	17.8571	0.0560	9.3222	0.1073	5.3049	49.4556
14	2.0138	0.4966	19.7726	0.0506	9.8188	0.1018	5.6941	55.9113
15	2.1170	0.4724	21.7864	0.0459	10.2912	0.0972	6.0753	62.5247
16	2.2255	0.4493	23.9034	0.0418	10.7405	0.0931	6.4487	69.2648
17	2.3396	0.4274	26.1290	0.0383	11.1679	0.0895	6.8143	76.1037
18	2.4596	0.4066	28.4687	0.0351	11.5745	0.0864	7.1720	83.0155
19	2.5857	0.3867	30.9283	0.0323	11.9612	0.0836	7.5221	89.9770
20	2.7183	0.3679	33.5140	0.0298	12.3291	0.0811	7.8646	96.9669
21	2.8577	0.3499	36.2324	0.0276	12.6791	0.0789	8.1996	103.9659
22	3.0042	0.3329	39.0900	0.0256	13.0119	0.0769	8.5270	110.9562
23	3.1582	0.3166	42.0942	0.0238	13.3285	0.0750	8.8471	117.9224
24	3.3201	0.3012	45.2525	0.0221	13.6298	0.0734	9.1599	124.8500
25	3.4903	0.2865	48.5726	0.0206	13.9163	0.0719	9.4654	131.7263
26	3.6693	0.2725	52.0630	0.0192	14.1888	0.0705	9.7638	138.5397
27	3.8574	0.2592	55.7323	0.0179	14.4481	0.0692	10.0551	145.2801
28	4.0552	0.2466	59.5898	0.0168	14.6947	0.0681	10.3395	151.9384
29	4.2631	0.2346	63.6451	0.0157	14.9292	0.0670	10.6170	158.5065
30	4.4817	0.2231	67.9082	0.0147	15.1524	0.0660	10.8877	164.9774
31	4.7115	0.2122	72.3900	0.0138	15.3646	0.0651	11.1517	171.3450
32	4.9530	0.2019	77.1015	0.0130	15.5665	0.0642	11.4091	177.6038
33	5.2070	0.1920	82.0546	0.0122	15.7586	0.0635	11.6601	183.7496
34	5.4739	0.1827	87.2616	0.0115	15.9413	0.0627	11.9046	189.7782
35	5.7546	0.1738	92.7356	0.0108	16.1150	0.0621	12.1429	195.6867
40	7.3891	0.1353	124.6146	0.0080	16.8648	0.0593	13.2435	223.3517
45	9.4877	0.1054	165.5481	0.0060	17.4486	0.0573	14.2024	247.8167
50	12.1825	0.0821	218.1077	0.0046	17.9034	0.0559	15.0329	269.1433
55	15.6426	0.0639	285.5952	0.0035	18.2575	0.0548	15.7480	287.5237
60	20.0855	0.0498	372.2512	0.0027	18.5333	0.0540	16.3604	303.2171
65	25.7903	0.0388	483.5200	0.0021	18.7481	0.0533	16.8822	316.5132
70	33.1154	0.0302	626.3919	0.0016	18.9154	0.0529	17.3245	327.7048
75	42.5211	0.0235	809.8430	0.0012	19.0457	0.0525	17.6979	337.0720
80	54.5980	0.0183	1045.3989	0.0010	19.1472	0.0522	18.0116	344.8750
85	70.1054	0.0143	1347.8584	0.0007	19.2262	0.0520	18.2742	351.3464
90	90.0171	0.0111	1736.2239	0.0006	19.2877	0.0518	18.4931	356.6943
95	115.5842	0.0087	2234.8960	0.0004	19.3356	0.0517	18.6751	361.0991
100	148.4130	0.0067	2875.2031	0.0003	19.3730	0.0516	18.8253	364.7161

TABLE B.9 Continuous Compounding: r = 6%

	Single payment		Uniform series				Gradient series	
	Compound amount factor	Present worth factor	Compound amount factor	Sinking fund factor	Present worth factor	Capital recovery factor	Uniform series factor	Present worth factor
n	To find F given P $F\|P\ r,n$	To find P given F $P\|F\ r,n$	To find F given A $F\|A\ r,n$	To find A given F $A\|F\ r,n$	To find P given A $P\|A\ r,n$	To find A given P $A\|P\ r,n$	To find A given G $A\|G\ r,n$	To find P given G $P\|G\ r,n$
1	1.0618	0.9418	1.0000	1.0000	0.9418	1.0618	0.0000	0.0000
2	1.1275	0.8869	2.0618	0.4850	1.8287	0.5468	0.4850	0.8870
3	1.1972	0.8353	3.1893	0.3135	2.6640	0.3754	0.9600	2.5576
4	1.2712	0.7866	4.3866	0.2280	3.4506	0.2898	1.4251	4.9176
5	1.3499	0.7408	5.6578	0.1767	4.1914	0.2386	1.8802	7.8809
6	1.4333	0.6977	7.0077	0.1427	4.8891	0.2045	2.3254	11.3693
7	1.5220	0.6570	8.4410	0.1185	5.5461	0.1803	2.7607	15.3117
8	1.6161	0.6188	9.9630	0.1004	6.1649	0.1622	3.1862	19.6432
9	1.7160	0.5827	11.5791	0.0864	6.7477	0.1482	3.6020	24.3053
10	1.8221	0.5488	13.2951	0.0752	7.2965	0.1371	4.0080	29.2446
11	1.9348	0.5169	15.1172	0.0661	7.8134	0.1280	4.4043	34.4133
12	2.0544	0.4868	17.0520	0.0586	8.3001	0.1205	4.7911	39.7677
13	2.1815	0.4584	19.1065	0.0523	8.7585	0.1142	5.1684	45.2686
14	2.3164	0.4317	21.2879	0.0470	9.1902	0.1088	5.5363	50.8809
15	2.4596	0.4066	23.6043	0.0424	9.5968	0.1042	5.8949	56.5729
16	2.6117	0.3829	26.0639	0.0384	9.9797	0.1002	6.2442	62.3164
17	2.7732	0.3606	28.6756	0.0349	10.3403	0.0967	6.5845	68.0860
18	2.9447	0.3396	31.4489	0.0318	10.6799	0.0936	6.9156	73.8592
19	3.1268	0.3198	34.3935	0.0291	10.9997	0.0909	7.2379	79.6160
20	3.3201	0.3012	37.5203	0.0267	11.3009	0.0885	7.5514	85.3387
21	3.5254	0.2837	40.8405	0.0245	11.5846	0.0863	7.8562	91.0118
22	3.7434	0.2671	44.3659	0.0225	11.8517	0.0844	8.1525	96.6217
23	3.9749	0.2516	48.1093	0.0208	12.1033	0.0826	8.4403	102.1565
24	4.2207	0.2369	52.0843	0.0192	12.3402	0.0810	8.7199	107.6058
25	4.4817	0.2231	56.3050	0.0178	12.5633	0.0796	8.9912	112.9610
26	4.7588	0.2101	60.7867	0.0165	12.7735	0.0783	9.2546	118.2144
27	5.0531	0.1979	65.5455	0.0153	12.9714	0.0771	9.5101	123.3598
28	5.3656	0.1864	70.5985	0.0142	13.1578	0.0760	9.7578	128.3920
29	5.6973	0.1755	75.9642	0.0132	13.3333	0.0750	9.9980	133.3066
30	6.0496	0.1653	81.6616	0.0122	13.4986	0.0741	10.2307	138.1003
31	6.4237	0.1557	87.7113	0.0114	13.6542	0.0732	10.4560	142.7705
32	6.8210	0.1466	94.1350	0.0106	13.8009	0.0725	10.6743	147.3153
33	7.2427	0.1381	100.9560	0.0099	13.9389	0.0717	10.8855	151.7336
34	7.6906	0.1300	108.1988	0.0092	14.0690	0.0711	11.0899	156.0246
35	8.1662	0.1225	115.8894	0.0086	14.1914	0.0705	11.2876	160.1881
40	11.0232	0.0907	162.0922	0.0062	14.7047	0.0680	12.1839	179.1175
45	14.8797	0.0672	224.4594	0.0045	15.0849	0.0663	12.9295	195.0420
50	20.0855	0.0498	308.6460	0.0032	15.3666	0.0651	13.5519	208.2476
55	27.1126	0.0369	422.2864	0.0024	15.5753	0.0642	14.0654	219.0738
60	36.5982	0.0273	575.6851	0.0017	15.7299	0.0636	14.4862	227.8672
65	49.4024	0.0202	782.7515	0.0013	15.8444	0.0631	14.8288	234.9540
70	66.6863	0.0150	1062.2620	0.0009	15.9292	0.0628	15.1060	240.6283
75	90.0171	0.0111	1439.5615	0.0007	15.9921	0.0625	15.3291	245.1461
80	121.5103	0.0082	1949.8623	0.0005	16.0387	0.0623	15.5078	248.7259
85	164.0218	0.0061	2636.3472	0.0004	16.0732	0.0622	15.6503	251.5503
90	221.4063	0.0045	3564.3550	0.0003	16.0987	0.0621	15.7633	253.7704
95	298.8670	0.0033	4817.0274	0.0002	16.1176	0.0620	15.8527	255.5099
100	403.4288	0.0025	6507.9649	0.0002	16.1317	0.0620	15.9232	256.8679

TABLE B.10 Continuous Compounding: r = 7%

	Single payment		Uniform series				Gradient series	
	Compound amount factor	Present worth factor	Compound amount factor	Sinking fund factor	Present worth factor	Capital recovery factor	Uniform series factor	Present worth factor
n	To find F given P $F\|P\ r,n$	To find P given F $P\|F\ r,n$	To find F given A $F\|A\ r,n$	To find A given F $A\|F\ r,n$	To find P given A $P\|A\ r,n$	To find A given P $A\|P\ r,n$	To find A given G $A\|G\ r,n$	To find P given G $P\|G\ r,n$
1	1.0725	0.9324	1.0000	1.0000	0.9324	1.0725	0.0000	0.0000
2	1.1503	0.8694	2.0725	0.4825	1.8018	0.5550	0.4825	0.8694
3	1.2337	0.8106	3.2228	0.3103	2.6123	0.3828	0.9534	2.4907
4	1.3231	0.7558	4.4565	0.2244	3.3681	0.2969	1.4126	4.7581
5	1.4191	0.7047	5.7796	0.1730	4.0728	0.2455	1.8603	7.5769
6	1.5220	0.6570	7.1987	0.1389	4.7299	0.2114	2.2904	10.8622
7	1.6323	0.6126	8.7206	0.1147	5.3425	0.1872	2.7211	14.5380
8	1.7507	0.5712	10.3530	0.0966	5.9137	0.1691	3.1344	18.5365
9	1.8776	0.5326	12.1037	0.0826	6.4463	0.1551	3.5364	22.7973
10	2.0138	0.4966	13.9813	0.0715	6.9429	0.1440	3.9272	27.2666
11	2.1598	0.4630	15.9950	0.0625	7.4059	0.1350	4.3069	31.8968
12	2.3164	0.4317	18.1548	0.0551	7.8376	0.1276	4.6755	36.6456
13	2.4843	0.4025	20.4712	0.0488	8.2401	0.1214	5.0333	41.4760
14	2.6645	0.3753	22.9555	0.0436	8.6155	0.1161	5.3804	46.3551
15	2.8576	0.3499	25.6200	0.0390	8.9654	0.1115	5.7168	51.2542
16	3.0649	0.3263	28.4776	0.0351	9.2917	0.1076	6.0428	56.1485
17	3.2871	0.3042	31.5425	0.0317	9.5959	0.1042	6.3585	61.0161
18	3.5254	0.2837	34.8296	0.0287	9.8795	0.1012	6.6640	65.8382
19	3.7810	0.2645	38.3550	0.0261	10.1440	0.0986	6.9596	70.5989
20	4.0552	0.2466	42.1361	0.0237	10.3906	0.0962	7.2453	75.2843
21	4.3492	0.2299	46.1913	0.0216	10.6206	0.0942	7.5215	79.8828
22	4.6646	0.2144	50.5406	0.0198	10.8349	0.0923	7.7881	84.3849
23	5.0028	0.1999	55.2052	0.0181	11.0348	0.0906	8.0456	88.7824
24	5.3655	0.1864	60.2080	0.0166	11.2212	0.0891	8.2940	93.0690
25	5.7546	0.1738	65.5736	0.0153	11.3950	0.0878	8.5335	97.2397
26	6.1719	0.1620	71.3282	0.0140	11.5570	0.0865	8.7643	101.2903
27	6.6194	0.1511	77.5001	0.0129	11.7081	0.0854	8.9867	105.2182
28	7.0993	0.1409	84.1195	0.0119	11.8489	0.0844	9.2009	109.0214
29	7.6141	0.1313	91.2189	0.0110	11.9803	0.0835	9.4070	112.6988
30	8.1662	0.1225	98.8330	0.0101	12.1027	0.0826	9.6052	116.2501
31	8.7583	0.1142	106.9991	0.0093	12.2169	0.0819	9.7958	119.6754
32	9.3933	0.1065	115.7574	0.0086	12.3234	0.0811	9.9790	122.9757
33	10.0744	0.0993	125.1508	0.0080	12.4226	0.0805	10.1550	126.1520
34	10.8049	0.0925	135.2253	0.0074	12.5152	0.0799	10.3239	129.2062
35	11.5883	0.0863	146.0302	0.0068	12.6015	0.0794	10.4860	132.1402
40	16.4446	0.0608	213.0062	0.0047	12.9529	0.0772	11.2017	145.0952
45	23.3360	0.0429	308.0496	0.0032	13.2006	0.0758	11.7769	155.4626
50	33.1154	0.0302	442.9231	0.0023	13.3751	0.0748	12.2347	163.6412
55	46.9930	0.0213	634.3174	0.0016	13.4981	0.0741	12.5957	170.0194
60	66.6863	0.0150	905.9192	0.0011	13.5848	0.0736	12.8781	174.9474
65	94.6323	0.0106	1291.3401	0.0008	13.6459	0.0733	13.0973	178.7255
70	134.2896	0.0074	1838.2783	0.0005	13.6889	0.0731	13.2664	181.6031
75	190.5661	0.0052	2614.4207	0.0004	13.7192	0.0729	13.3959	183.7825
80	270.4258	0.0037	3715.8174	0.0003	13.7406	0.0728	13.4946	185.4251
85	383.7527	0.0026	5278.7734	0.0002	13.7557	0.0727	13.5695	186.6580
90	544.5711	0.0018	7496.7149	0.0001	13.7663	0.0726	13.6260	187.5799
95	772.7832	0.0013	10644.1289	0.0001	13.7738	0.0726	13.6685	188.2669
100	1096.6313	0.0009	15110.5234	0.0001	13.7790	0.0726	13.7003	188.7774

TABLE B.11 Continuous Compounding: r = 8%

	Single payment		Uniform series				Gradient series	
	Compound amount factor	Present worth factor	Compound amount factor	Sinking fund factor	Present worth factor	Capital recovery factor	Uniform series factor	Present worth factor
n	To find F given P $F\|P\ r,n$	To find P given F $P\|F\ r,n$	To find F given A $F\|A\ r,n$	To find A given F $A\|F\ r,n$	To find P given A $P\|A\ r,n$	To find A given P $A\|P\ r,n$	To find A given G $A\|G\ r,n$	To find P given G $P\|G\ r,n$
1	1.0833	0.9231	1.0000	1.0000	0.9231	1.0833	0.0000	0.0000
2	1.1735	0.8521	2.0833	0.4800	1.7753	0.5633	0.4800	0.8523
3	1.2712	0.7866	3.2568	0.3070	2.5619	0.3903	0.9467	2.4257
4	1.3771	0.7261	4.5281	0.2208	3.2881	0.3041	1.4002	4.6043
5	1.4918	0.6703	5.9052	0.1693	3.9584	0.2526	1.8404	7.2856
6	1.6161	0.6188	7.3971	0.1352	4.5772	0.2185	2.2676	10.3796
7	1.7507	0.5712	9.0131	0.1109	5.1484	0.1942	2.6817	13.8069
8	1.8965	0.5273	10.7638	0.0929	5.6757	0.1762	3.0829	17.4981
9	2.0544	0.4868	12.6603	0.0790	6.1624	0.1623	3.4713	21.3922
10	2.2255	0.4493	14.7148	0.0680	6.6118	0.1512	3.8470	25.4363
11	2.4109	0.4148	16.9404	0.0590	7.0265	0.1423	4.2102	29.5842
12	2.6117	0.3829	19.3513	0.0517	7.4095	0.1350	4.5611	33.7961
13	2.8292	0.3535	21.9630	0.0455	7.7629	0.1288	4.8998	38.0376
14	3.0649	0.3263	24.7922	0.0403	8.0892	0.1236	5.2265	42.2793
15	3.3201	0.3012	27.8571	0.0359	8.3904	0.1192	5.5415	46.4961
16	3.5966	0.2780	31.1772	0.0321	8.6684	0.1154	5.8449	50.6668
17	3.8962	0.2567	34.7739	0.0288	8.9251	0.1120	6.1369	54.7734
18	4.2207	0.2369	38.6702	0.0259	9.1620	0.1091	6.4178	58.8013
19	4.5722	0.2187	42.8909	0.0233	9.3808	0.1065	6.6879	62.7381
20	4.9530	0.2019	47.4632	0.0211	9.5827	0.1044	6.9473	66.5742
21	5.3656	0.1854	52.4162	0.0191	9.7690	0.1024	7.1963	70.3018
22	5.8124	0.1720	57.7818	0.0173	9.9411	0.1006	7.4352	73.9148
23	6.2965	0.1588	63.5943	0.0157	10.0999	0.0990	7.6642	77.4088
24	6.8210	0.1466	69.8909	0.0143	10.2465	0.0975	7.8836	80.7808
25	7.3891	0.1353	76.7119	0.0130	10.3818	0.0963	8.0937	84.0289
26	8.0045	0.1249	84.1010	0.0119	10.5068	0.0952	8.2947	87.1522
27	8.6711	0.1153	92.1056	0.0109	10.6221	0.0941	8.4870	90.1507
28	9.3933	0.1065	100.7768	0.0099	10.7286	0.0932	8.6707	93.0252
29	10.1757	0.0983	110.1702	0.0091	10.8268	0.0924	8.8461	95.7769
30	11.0232	0.0907	120.3459	0.0083	10.9175	0.0916	9.0136	98.4077
31	11.9413	0.0837	131.3692	0.0076	11.0013	0.0909	9.1734	100.9200
32	12.9358	0.0773	143.3106	0.0070	11.0786	0.0903	9.3257	103.3165
33	14.0132	0.0714	156.2465	0.0064	11.1500	0.0897	9.4708	105.6001
34	15.1803	0.0659	170.2598	0.0059	11.2158	0.0892	9.6090	107.7740
35	16.4446	0.0608	185.4402	0.0054	11.2766	0.0887	9.7405	109.8415
40	24.5325	0.0408	282.5491	0.0035	11.5174	0.0868	10.3069	118.7094
45	36.5982	0.0273	427.4192	0.0023	11.6787	0.0856	10.7426	125.4605
50	54.5981	0.0183	643.5400	0.0016	11.7869	0.0848	11.0738	130.5267
55	81.4508	0.0123	965.9544	0.0010	11.8594	0.0843	11.3230	134.2852
60	121.5102	0.0082	1446.9397	0.0007	11.9080	0.0840	11.5088	137.0475
65	181.2720	0.0055	2164.4846	0.0005	11.9405	0.0837	11.6461	139.0622
70	270.4258	0.0037	3234.9375	0.0003	11.9624	0.0836	11.7469	140.5217
75	403.4277	0.0025	4831.8594	0.0002	11.9770	0.0835	11.8203	141.5732
80	601.8438	0.0017	7214.1953	0.0001	11.9868	0.0834	11.8735	142.3272
85	897.8455	0.0011	10768.2227	0.0001	11.9934	0.0834	11.9119	142.8655
90	1339.4280	0.0007	16070.2070	0.0001	11.9978	0.0833	11.9394	143.2483
95	1998.1919	0.0005	23979.3399	0.0000	12.0008	0.0833	11.9591	143.5198
100	2980.9519	0.0003	35779.6094	0.0000	12.0028	0.0833	11.9731	143.7116

TABLE B.12 Continuous Compounding: r = 9%

n	Single payment		Uniform series				Gradient series	
	Compound amount factor	Present worth factor	Compound amount factor	Sinking fund factor	Present worth factor	Capital recovery factor	Uniform series factor	Present worth factor
	To find F given P $F/P\ r,n$	To find P given F $P/F\ r,n$	To find F given A $F/A\ r,n$	To find A given F $A/F\ r,n$	To find P given A $P/A\ r,n$	To find A given P $A/P\ r,n$	To find A given G $A/G\ r,n$	To find P given G $P/G\ r,n$
1	1.0942	0.9139	1.0000	1.0000	0.9139	1.0942	0.0000	0.0000
2	1.1972	0.8353	2.0942	0.4775	1.7492	0.5717	0.4775	0.8354
3	1.3100	0.7634	3.2914	0.3038	2.5126	0.3980	0.9401	2.3623
4	1.4333	0.5977	4.6014	0.2173	3.2103	0.3115	1.3878	4.4553
5	1.5683	0.6376	6.0347	0.1657	3.8479	0.2599	1.8206	7.0060
6	1.7160	0.5827	7.6031	0.1315	4.4307	0.2257	2.2338	9.9198
7	1.8776	0.5326	9.3191	0.1073	4.9633	0.2015	2.6424	13.1154
8	2.0544	0.4868	11.1967	0.0893	5.4500	0.1835	3.0316	16.5227
9	2.2479	0.4449	13.2512	0.0755	5.8949	0.1696	3.4065	20.0817
10	2.4596	0.4066	15.4991	0.0645	6.3015	0.1587	3.7674	23.7409
11	2.6912	0.3716	17.9587	0.0557	6.6730	0.1499	4.1145	27.4567
12	2.9447	0.3396	20.6500	0.0484	7.0126	0.1426	4.4479	31.1924
13	3.2220	0.3104	23.5947	0.0424	7.3230	0.1366	4.7680	34.9168
14	3.5254	0.2837	26.8167	0.0373	7.6067	0.1315	5.0750	38.6044
15	3.8574	0.2592	30.3421	0.0330	7.8659	0.1271	5.3691	42.2338
16	4.2207	0.2369	34.1996	0.0292	8.1028	0.1234	5.6537	45.7878
17	4.6132	0.2165	38.4203	0.0260	8.3194	0.1202	5.9201	49.2524
18	5.0531	0.1979	43.0385	0.0232	8.5173	0.1174	6.1776	52.6167
19	5.5290	0.1809	48.0917	0.0208	8.6981	0.1150	6.4234	55.8724
20	6.0496	0.1653	53.6207	0.0186	8.8634	0.1128	6.6579	59.0131
21	6.5194	0.1511	59.6704	0.0168	9.0145	0.1109	6.8815	62.0346
22	7.2427	0.1381	66.2898	0.0151	9.1525	0.1093	7.0945	64.9341
23	7.9248	0.1262	73.5326	0.0136	9.2788	0.1078	7.2972	67.7102
24	8.6711	0.1153	81.4575	0.0123	9.3941	0.1064	7.4900	70.3627
25	9.4877	0.1054	90.1287	0.0111	9.4995	0.1053	7.6732	72.8923
26	10.3812	0.0963	99.6165	0.0100	9.5958	0.1042	7.8471	75.3006
27	11.3589	0.0880	109.9978	0.0091	9.6839	0.1033	8.0122	77.5896
28	12.4286	0.0805	121.3567	0.0082	9.7643	0.1024	8.1686	79.7620
29	13.5990	0.0735	133.7854	0.0075	9.8379	0.1015	8.3168	81.8210
30	14.8797	0.0672	147.3846	0.0068	9.9051	0.1010	8.4572	83.7699
31	16.2810	0.0614	162.2643	0.0062	9.9665	0.1003	8.5899	85.6126
32	17.8143	0.0561	178.5456	0.0056	10.0226	0.0998	8.7155	87.3528
33	19.4919	0.0513	196.3600	0.0051	10.0739	0.0993	8.8340	88.9946
34	21.3275	0.0469	215.8520	0.0046	10.1208	0.0988	8.9460	90.5418
35	23.3360	0.0429	237.1797	0.0042	10.1637	0.0984	9.0516	91.9988
40	36.5982	0.0273	378.0064	0.0026	10.3286	0.0968	9.4950	98.0703
45	57.3974	0.0174	599.8667	0.0017	10.4337	0.0958	9.8207	102.4673
50	90.0170	0.0111	945.2449	0.0011	10.5007	0.0952	10.0569	105.6061
55	141.1747	0.0071	1438.4739	0.0007	10.5435	0.0948	10.2262	107.8213
60	221.4060	0.0045	2340.4255	0.0004	10.5707	0.0946	10.3464	109.3700
65	347.2334	0.0029	3676.5518	0.0003	10.5881	0.0944	10.4309	110.4444
70	544.5706	0.0018	5772.0117	0.0002	10.5992	0.0943	10.4898	111.1848
75	854.0567	0.0012	9058.3555	0.0001	10.6063	0.0943	10.5307	111.6923
80	1339.4273	0.0007	14212.3633	0.0001	10.6108	0.0942	10.5588	112.0385
85	2100.6399	0.0005	22295.4570	0.0000	10.6137	0.0942	10.5781	112.2735
90	3294.4590	0.0003	34972.2617	0.0000	10.6155	0.0942	10.5913	112.4326
95	5166.7344	0.0002	54853.4492	0.0000	10.6167	0.0942	10.6002	112.5398
100	8103.0847	0.0001	86033.3125	0.0000	10.6174	0.0942	10.6063	112.6119

TABLE B.13 Continuous Compounding: r = 10%

n	Single payment Compound amount factor To find F given P F|P r,n	Present worth factor To find P given F P|F r,n	Uniform series Compound amount factor To find F given A F|A r,n	Sinking fund factor To find A given F A|F r,n	Present worth factor To find P given A P|A r,n	Capital recovery factor To find A given P A|P r,n	Gradient series Uniform series factor To find A given G A|G r,n	Present worth factor To find P given G P|G r,n
1	1.1052	0.9048	1.0000	1.0000	0.9048	1.1052	0.0000	0.0000
2	1.2214	0.8187	2.1052	0.4750	1.7236	0.5802	0.4750	0.8188
3	1.3499	0.7408	3.3266	0.3006	2.4644	0.4058	0.9334	2.3005
4	1.4918	0.6703	4.6765	0.2138	3.1347	0.3190	1.3754	4.3115
5	1.6487	0.6065	6.1683	0.1621	3.7413	0.2573	1.8009	6.7377
6	1.8221	0.5488	7.8170	0.1279	4.2901	0.2331	2.2101	9.4818
7	2.0138	0.4966	9.6392	0.1037	4.7867	0.2389	2.6033	12.4614
8	2.2255	0.4493	11.6529	0.0858	5.2360	0.1910	2.9836	15.6067
9	2.4596	0.4055	13.8785	0.0721	5.6426	0.1772	3.3423	18.8593
10	2.7183	0.3679	16.3381	0.0612	6.0104	0.1664	3.6886	22.1702
11	3.0042	0.3329	19.0564	0.0525	6.3433	0.1576	4.0198	25.4990
12	3.3201	0.3012	22.0606	0.0453	6.6445	0.1505	4.3362	28.8122
13	3.6693	0.2725	25.3807	0.0394	6.9171	0.1446	4.6331	32.0826
14	4.0552	0.2456	29.0500	0.0344	7.1637	0.1396	4.9260	35.2884
15	4.4817	0.2231	33.1052	0.0302	7.3868	0.1354	5.2001	38.4123
16	4.9530	0.2019	37.5869	0.0266	7.5887	0.1318	5.4608	41.4408
17	5.4739	0.1827	42.5400	0.0235	7.7714	0.1287	5.7086	44.3637
18	6.0496	0.1653	48.0140	0.0208	7.9367	0.1260	5.9437	47.1738
19	6.6859	0.1496	54.0637	0.0185	8.0862	0.1237	6.1667	49.8661
20	7.3890	0.1353	60.7496	0.0165	8.2216	0.1216	6.3780	52.4375
21	8.1662	0.1225	68.1387	0.0147	8.3440	0.1198	6.5779	54.8866
22	9.0250	0.1108	76.3049	0.0131	8.4548	0.1183	6.7669	57.2135
23	9.9742	0.1003	85.3300	0.0117	8.5551	0.1169	6.9454	59.4193
24	11.0232	0.0907	95.3042	0.0105	8.6458	0.1157	7.1139	61.5058
25	12.1825	0.0821	106.3274	0.0094	8.7279	0.1146	7.2727	63.4759
26	13.4637	0.0743	118.5099	0.0084	8.8022	0.1136	7.4223	65.3327
27	14.8797	0.0672	131.9737	0.0076	8.8694	0.1127	7.5630	67.0801
28	16.4446	0.0608	145.8535	0.0068	8.9302	0.1120	7.6954	68.7220
29	18.1741	0.0550	163.2982	0.0061	8.9852	0.1113	7.8197	70.2626
30	20.0855	0.0498	181.4725	0.0055	9.0350	0.1107	7.9365	71.7065
31	22.1979	0.0450	201.5581	0.0050	9.0800	0.1101	8.0459	73.0579
32	24.5325	0.0408	223.7561	0.0045	9.1208	0.1096	8.1485	74.3216
33	27.1126	0.0369	248.2887	0.0040	9.1577	0.1092	8.2446	75.5018
34	29.9641	0.0334	275.4011	0.0036	9.1911	0.1088	8.3345	76.6032
35	33.1154	0.0302	305.3655	0.0033	9.2213	0.1084	8.4185	77.6299
40	54.5981	0.0183	509.6309	0.0020	9.3342	0.1071	8.7620	81.7875
45	90.0170	0.0111	846.4080	0.0012	9.4028	0.1064	9.0028	84.6519
50	148.4128	0.0067	1401.6589	0.0007	9.4443	0.1059	9.1691	86.5970
55	244.6914	0.0041	2317.1118	0.0004	9.4695	0.1056	9.2826	87.9028
60	403.4275	0.0025	3826.4400	0.0003	9.4848	0.1054	9.3592	88.7713
65	665.1397	0.0015	6314.8984	0.0002	9.4941	0.1053	9.4105	89.3444
70	1096.6299	0.0009	10417.6758	0.0001	9.4997	0.1053	9.4444	89.7202
75	1808.0366	0.0006	17182.0117	0.0001	9.5031	0.1052	9.4668	89.9551
80	2980.9470	0.0003	28334.5117	0.0000	9.5052	0.1052	9.4815	90.1241
85	4914.7461	0.0002	46721.8399	0.0000	9.5065	0.1052	9.4910	90.2267
90	8103.0508	0.0001	77037.4375	0.0000	9.5072	0.1052	9.4972	90.2927
95	13359.6719	0.0001	127019.4375	0.0000	9.5077	0.1052	9.5012	90.3352
100	22026.3750	0.0000	209425.8750	0.0000	9.5080	0.1052	9.5038	90.3623

TABLE B.14 Continuous Compounding: r = 12%

n	Single payment Compound amount factor — To find F given P — $F\|P\ r,n$	Single payment Present worth factor — To find P given F — $P\|F\ r,n$	Uniform series Compound amount factor — To find F given A — $F\|A\ r,n$	Uniform series Sinking fund factor — To find A given F — $A\|F\ r,n$	Uniform series Present worth factor — To find P given A — $P\|A\ r,n$	Uniform series Capital recovery factor — To find A given P — $A\|P\ r,n$	Gradient series Uniform series factor — To find A given G — $A\|G\ r,n$	Gradient series Present worth factor — To find P given G — $P\|G\ r,n$
1	1.1275	0.8869	1.0000	1.0000	0.8869	1.1275	0.0000	0.0000
2	1.2712	0.7866	2.1275	0.4700	1.6735	0.5975	0.4700	0.7866
3	1.4333	0.6977	3.3987	0.2942	2.3712	0.4217	0.9202	2.1820
4	1.6161	0.6188	4.8321	0.2070	2.9900	0.3344	1.3506	4.0383
5	1.8221	0.5488	6.4481	0.1551	3.5388	0.2826	1.7615	6.2336
6	2.0544	0.4868	8.2703	0.1209	4.0256	0.2484	2.1531	8.6673
7	2.3164	0.4317	10.3247	0.0969	4.4573	0.2244	2.5257	11.2576
8	2.6117	0.3829	12.6411	0.0791	4.8402	0.2066	2.8796	13.9379
9	2.9447	0.3396	15.2528	0.0656	5.1798	0.1931	3.2153	16.6546
10	3.3201	0.3012	18.1974	0.0550	5.4810	0.1824	3.5332	19.3654
11	3.7434	0.2671	21.5176	0.0465	5.7481	0.1740	3.8337	22.0368
12	4.2207	0.2369	25.2610	0.0396	5.9850	0.1671	4.1174	24.6430
13	4.7588	0.2101	29.4817	0.0339	6.1952	0.1614	4.3848	27.1646
14	5.3656	0.1864	34.2405	0.0292	6.3815	0.1567	4.6364	29.5875
15	6.0496	0.1653	39.6060	0.0252	6.5468	0.1527	4.8728	31.9017
16	6.8210	0.1465	45.6557	0.0219	6.6934	0.1494	5.0946	34.1008
17	7.6906	0.1300	52.4767	0.0191	6.8235	0.1466	5.3025	36.1812
18	8.6711	0.1153	60.1673	0.0166	6.9388	0.1441	5.4969	38.1418
19	9.7767	0.1023	68.8384	0.0145	7.0411	0.1420	5.6785	39.9829
20	11.0232	0.0907	78.6151	0.0127	7.1318	0.1402	5.8480	41.7065
21	12.4286	0.0805	89.6382	0.0112	7.2123	0.1387	6.0058	43.3157
22	14.0132	0.0714	102.0668	0.0098	7.2836	0.1373	6.1527	44.8143
23	15.7998	0.0633	116.0800	0.0086	7.3469	0.1361	6.2893	46.2067
24	17.8142	0.0561	131.8797	0.0076	7.4031	0.1351	6.4160	47.4978
25	20.0855	0.0498	149.6940	0.0067	7.4528	0.1342	6.5334	48.6927
26	22.6463	0.0442	169.7795	0.0059	7.4970	0.1334	6.6422	49.7966
27	25.5337	0.0392	192.4259	0.0052	7.5362	0.1327	6.7428	50.8149
28	28.7891	0.0347	217.9596	0.0046	7.5709	0.1321	6.8357	51.7528
29	32.4597	0.0308	246.7487	0.0041	7.6017	0.1315	6.9215	52.6154
30	36.5982	0.0273	279.2080	0.0036	7.6290	0.1311	7.0006	53.4078
31	41.2643	0.0242	315.8064	0.0032	7.6533	0.1307	7.0734	54.1348
32	46.5254	0.0215	357.0705	0.0028	7.6748	0.1303	7.1404	54.8011
33	52.4572	0.0191	403.5960	0.0025	7.6938	0.1300	7.2020	55.4111
34	59.1453	0.0169	456.0532	0.0022	7.7107	0.1297	7.2586	55.9691
35	66.6862	0.0150	515.1987	0.0019	7.7257	0.1294	7.3105	56.4789
40	121.5101	0.0082	945.2014	0.0011	7.7788	0.1286	7.5114	58.4297
45	221.4058	0.0045	1728.7168	0.0006	7.8079	0.1281	7.6392	59.6460
50	403.4273	0.0025	3156.3743	0.0003	7.8239	0.1278	7.7191	60.3934

TABLE B.15 Continuous Compounding: r = 15%

	Single payment		Uniform series				Gradient series	
	Compound amount factor	Present worth factor	Compound amount factor	Sinking fund factor	Present worth factor	Capital recovery factor	Uniform series factor	Present worth factor
n	To find F given P $F\|P\ r,n$	To find P given F $P\|F\ r,n$	To find F given A $F\|A\ r,n$	To find A given F $A\|F\ r,n$	To find P given A $P\|A\ r,n$	To find A given P $A\|P\ r,n$	To find A given G $A\|G\ r,n$	To find P given G $P\|G\ r,n$
1	1.1618	0.8607	1.0000	1.0000	0.8607	1.1618	0.0000	0.0000
2	1.3499	0.7408	2.1618	0.4626	1.6015	0.5244	0.4626	0.7408
3	1.5683	0.6376	3.5117	0.2848	2.2392	0.4466	0.9004	2.0161
4	1.8221	0.5488	5.0800	0.1968	2.7880	0.3587	1.3137	3.6626
5	2.1170	0.4724	6.9021	0.1449	3.2603	0.3067	1.7029	5.5520
6	2.4596	0.4066	9.0191	0.1109	3.6669	0.2727	2.0685	7.5849
7	2.8576	0.3499	11.4788	0.0871	4.0169	0.2490	2.4110	9.6845
8	3.3201	0.3012	14.3364	0.0698	4.3180	0.2316	2.7311	11.7929
9	3.8574	0.2592	17.6565	0.0566	4.5773	0.2185	3.0295	13.8669
10	4.4817	0.2231	21.5140	0.0465	4.8004	0.2083	3.3070	15.8750
11	5.2070	0.1920	25.9957	0.0385	4.9925	0.2003	3.5645	17.7955
12	6.0496	0.1653	31.2027	0.0320	5.1578	0.1939	3.8028	19.6138
13	7.0287	0.1423	37.2523	0.0268	5.3000	0.1887	4.0228	21.3211
14	8.1662	0.1225	44.2810	0.0226	5.4225	0.1844	4.2255	22.9131
15	9.4877	0.1054	52.4472	0.0191	5.5279	0.1809	4.4119	24.3887
16	11.0232	0.0907	61.9350	0.0161	5.6186	0.1780	4.5829	25.7494
17	12.8071	0.0781	72.9582	0.0137	5.6967	0.1755	4.7394	26.9987
18	14.8797	0.0672	85.7653	0.0117	5.7639	0.1735	4.8823	28.1413
19	17.2878	0.0578	100.6450	0.0099	5.8217	0.1718	5.0126	29.1824
20	20.0855	0.0498	117.9328	0.0085	5.8715	0.1703	5.1312	30.1284
21	23.3360	0.0429	138.0184	0.0072	5.9144	0.1691	5.2390	30.9854
22	27.1126	0.0369	161.3545	0.0062	5.9513	0.1680	5.3367	31.7600
23	31.5004	0.0317	188.4672	0.0053	5.9830	0.1571	5.4251	32.4584
24	36.5982	0.0273	219.9677	0.0045	6.0103	0.1664	5.5050	33.0869
25	42.5210	0.0235	256.5657	0.0039	6.0339	0.1657	5.5771	33.6513
26	49.4024	0.0202	299.0869	0.0033	6.0541	0.1652	5.6420	34.1573
27	57.3974	0.0174	348.4893	0.0029	6.0715	0.1647	5.7004	34.6103
28	66.6863	0.0150	405.8870	0.0025	6.0865	0.1643	5.7529	35.0152
29	77.4784	0.0129	472.5735	0.0021	6.0994	0.1639	5.8000	35.3766
30	90.0170	0.0111	550.0520	0.0018	6.1105	0.1637	5.8421	35.6988
31	104.5849	0.0096	640.0693	0.0016	6.1201	0.1634	5.8799	35.9856
32	121.5102	0.0082	744.6546	0.0013	6.1283	0.1632	5.9136	36.2407
33	141.1748	0.0071	866.1650	0.0012	6.1354	0.1630	5.9437	36.4674
34	154.0217	0.0051	1007.3403	0.0010	6.1415	0.1528	5.9706	36.6686
35	190.5660	0.0052	1171.3626	0.0009	6.1468	0.1627	5.9945	36.8470
40	403.4280	0.0025	2486.6751	0.0004	6.1639	0.1622	6.0798	37.4749
45	854.0574	0.0012	5271.1914	0.0002	6.1719	0.1620	6.1264	37.8120
50	1808.0383	0.0006	11166.0195	0.0001	6.1758	0.1619	6.1515	37.9903

	Single payment		Uniform series				Gradient series	
	Compound amount factor	Present worth factor	Compound amount factor	Sinking fund factor	Present worth factor	Capital recovery factor	Uniform series factor	Present worth factor
n	To find F given P F|P r,n	To find P given F P|F r,n	To find F given A F|A r,n	To find A given F A|F r,n	To find P given A P|A r,n	To find A given P A|P r,n	To find A given G A|G r,n	To find P given G P|G r,n
1	1.2214	0.8187	1.0000	1.0000	0.8187	1.2214	0.0000	0.0000
2	1.4918	0.6703	2.2214	0.4502	1.4891	0.6715	0.4502	0.6703
3	1.8221	0.5488	3.7132	0.2693	2.0379	0.4907	0.8675	1.7680
4	2.2255	0.4493	5.5354	0.1807	2.4872	0.4021	1.2528	3.1160
5	2.7183	0.3679	7.7609	0.1289	2.8551	0.3503	1.6008	4.5875
6	3.3201	0.3012	10.4792	0.0954	3.1563	0.3168	1.9306	6.0935
7	4.0552	0.2466	13.7993	0.0725	3.4029	0.2939	2.2255	7.5731
8	4.9530	0.2019	17.8545	0.0560	3.6048	0.2774	2.4929	8.9863
9	6.0496	0.1653	22.8076	0.0438	3.7701	0.2652	2.7344	10.3087
10	7.3891	0.1353	28.8572	0.0347	3.9054	0.2561	2.9515	11.5268
11	9.0250	0.1108	36.2463	0.0276	4.0162	0.2490	3.1459	12.6348
12	11.0232	0.0907	45.2713	0.0221	4.1069	0.2435	3.3194	13.6327
13	13.4637	0.0743	56.2945	0.0178	4.1812	0.2392	3.4736	14.5240
14	16.4446	0.0608	69.7583	0.0143	4.2420	0.2357	3.6102	15.3145
15	20.0855	0.0498	86.2030	0.0116	4.2918	0.2330	3.7307	16.0115
16	24.5325	0.0408	106.2885	0.0094	4.3326	0.2308	3.8367	16.6230
17	29.9641	0.0334	130.8211	0.0076	4.3659	0.2290	3.9297	17.1570
18	36.5982	0.0273	160.7853	0.0062	4.3933	0.2276	4.0110	17.6214
19	44.7012	0.0224	197.3835	0.0051	4.4156	0.2265	4.0819	18.0241
20	54.5981	0.0183	242.0849	0.0041	4.4339	0.2255	4.1435	18.3721
21	66.6863	0.0150	296.6829	0.0034	4.4489	0.2248	4.1970	18.6720
22	81.4508	0.0123	363.3694	0.0028	4.4612	0.2242	4.2432	18.9299
23	99.4843	0.0101	444.8203	0.0022	4.4713	0.2237	4.2831	19.1510
24	121.5103	0.0082	544.3049	0.0018	4.4795	0.2232	4.3175	19.3403
25	148.4130	0.0067	665.8154	0.0015	4.4862	0.2229	4.3471	19.5020
26	181.2721	0.0055	814.2290	0.0012	4.4918	0.2226	4.3724	19.6399
27	221.4062	0.0045	995.5017	0.0010	4.4963	0.2224	4.3942	19.7573
28	270.4260	0.0037	1216.9085	0.0008	4.5000	0.2222	4.4127	19.8572
29	330.2991	0.0030	1487.3352	0.0007	4.5030	0.2221	4.4286	19.9420
30	403.4282	0.0025	1817.6350	0.0006	4.5055	0.2220	4.4421	20.0138
31	492.7483	0.0020	2221.0640	0.0005	4.5075	0.2219	4.4536	20.0747
32	601.8442	0.0017	2713.8142	0.0004	4.5092	0.2218	4.4634	20.1262
33	735.0945	0.0014	3315.6607	0.0003	4.5105	0.2217	4.4717	20.1698
34	897.8464	0.0011	4050.7566	0.0002	4.5115	0.2216	4.4767	20.2065
35	1096.6321	0.0009	4948.6016	0.0002	4.5125	0.2216	4.4847	20.2375
40	2980.9551	0.0003	13459.4649	0.0001	4.5152	0.2215	4.5032	20.3328
45	8103.0703	0.0001	36594.3516	0.0000	4.5161	0.2214	4.5111	20.3727
50	22026.4414	0.0000	99481.5000	0.0000	4.5165	0.2214	4.5144	20.3891

TABLE B.17 Continuous Compounding: r = 25%

	Single payment		Uniform series				Gradient series	
	Compound amount factor	Present worth factor	Compound amount factor	Sinking fund factor	Present worth factor	Capital recovery factor	Uniform series factor	Present worth factor
n	To find F given P $F\|P\ r,n$	To find P given F $P\|F\ r,n$	To find F given A $F\|A\ r,n$	To find A given F $A\|F\ r,n$	To find P given A $P\|A\ r,n$	To find A given P $A\|P\ r,n$	To find A given G $A\|G\ r,n$	To find P given G $P\|G\ r,n$
1	1.2840	0.7788	1.0000	1.0000	0.7788	1.2840	0.0000	0.0000
2	1.6487	0.6065	2.2840	0.4378	1.3853	0.7218	0.4378	0.6065
3	2.1170	0.4724	3.9327	0.2543	1.8577	0.5383	0.8350	1.5513
4	2.7183	0.3679	6.0498	0.1653	2.2256	0.4493	1.1929	2.6549
5	3.4903	0.2865	8.7680	0.1141	2.5121	0.3981	1.5131	3.8009
6	4.4817	0.2231	12.2584	0.0816	2.7352	0.3656	1.7975	4.9166
7	5.7546	0.1738	16.7401	0.0597	2.9090	0.3438	2.0486	5.9592
8	7.3891	0.1353	22.4947	0.0445	3.0443	0.3285	2.2687	6.9066
9	9.4877	0.1054	29.8837	0.0335	3.1497	0.3175	2.4605	7.7498
10	12.1825	0.0821	39.3715	0.0254	3.2318	0.3094	2.6265	8.4885
11	15.6426	0.0639	51.5540	0.0194	3.2957	0.3034	2.7696	9.1278
12	20.0855	0.0498	67.1966	0.0149	3.3455	0.2989	2.8921	9.6755
13	25.7903	0.0388	87.2821	0.0115	3.3843	0.2955	2.9964	10.1408
14	33.1154	0.0302	113.0725	0.0088	3.4145	0.2929	3.0849	10.5333
15	42.5211	0.0235	146.1879	0.0068	3.4380	0.2909	3.1595	10.8626
16	54.5981	0.0183	188.7090	0.0053	3.4563	0.2893	3.2223	11.1373
17	70.1054	0.0143	243.3072	0.0041	3.4706	0.2881	3.2748	11.3655
18	90.0171	0.0111	313.4124	0.0032	3.4817	0.2872	3.3186	11.5544
19	115.5842	0.0087	403.4295	0.0025	3.4904	0.2865	3.3550	11.7101
20	148.4131	0.0067	519.0139	0.0019	3.4971	0.2860	3.3851	11.8381
21	190.5662	0.0052	667.4273	0.0015	3.5023	0.2855	3.4100	11.9431
22	244.6918	0.0041	857.9934	0.0012	3.5064	0.2852	3.4305	12.0289
23	314.1902	0.0032	1102.6851	0.0009	3.5096	0.2849	3.4474	12.0989
24	403.4285	0.0025	1416.8765	0.0007	3.5121	0.2847	3.4612	12.1560
25	518.0125	0.0019	1920.3054	0.0005	3.5140	0.2846	3.4725	12.2023
26	665.1411	0.0015	2438.3167	0.0004	3.5155	0.2845	3.4817	12.2399
27	854.0581	0.0012	3003.4590	0.0003	3.5167	0.2844	3.4892	12.2703
28	1096.6326	0.0009	3857.5183	0.0003	3.5176	0.2843	3.4953	12.2949
29	1408.1043	0.0007	4954.1484	0.0002	3.5183	0.2842	3.5002	12.3148
30	1808.0415	0.0006	6362.2539	0.0002	3.5189	0.2842	3.5042	12.3309
31	2321.5711	0.0004	8170.2969	0.0001	3.5193	0.2841	3.5075	12.3438
32	2980.9568	0.0003	10491.8711	0.0001	3.5196	0.2841	3.5101	12.3542
33	3827.6238	0.0003	13472.8281	0.0001	3.5199	0.2841	3.5122	12.3626
34	4914.7617	0.0002	17300.4492	0.0001	3.5201	0.2841	3.5139	12.3692
35	6310.6797	0.0002	22215.2188	0.0000	3.5203	0.2841	3.5153	12.3746

	Single payment		Uniform series				Gradient series	
	Compound amount factor	Present worth factor	Compound amount factor	Sinking fund factor	Present worth factor	Capital recovery factor	Uniform series factor	Present worth factor
n	To find F given P $F\|P\ r,n$	To find P given F $P\|F\ r,n$	To find F given A $F\|A\ r,n$	To find A given F $A\|F\ r,n$	To find P given A $P\|A\ r,n$	To find A given P $A\|P\ r,n$	To find A given G $A\|G\ r,n$	To find P given G $P\|G\ r,n$
1	1.3499	0.7408	1.0000	1.0000	0.7408	1.3499	0.0000	0.0000
2	1.3221	0.5488	2.3499	0.4256	1.2896	0.7754	0.4256	0.5488
3	2.4596	0.4066	4.1720	0.2397	1.6962	0.5896	0.8029	1.3620
4	3.3201	0.3012	6.6316	0.1508	1.9974	0.5007	1.1342	2.2655
5	4.4817	0.2231	9.9517	0.1005	2.2205	0.4503	1.4222	3.1581
6	6.0496	0.1653	14.4334	0.0693	2.3858	0.4191	1.6701	3.9846
7	8.1662	0.1225	20.4830	0.0488	2.5083	0.3987	1.8815	4.7193
8	11.0232	0.0907	28.6492	0.0349	2.5990	0.3848	2.0601	5.3543
9	14.8797	0.0672	39.6724	0.0252	2.6662	0.3751	2.2099	5.8920
10	20.0855	0.0498	54.5522	0.0183	2.7160	0.3682	2.3343	6.3401
11	27.1126	0.0369	74.6377	0.0134	2.7529	0.3633	2.4370	6.7089
12	36.5982	0.0273	101.7504	0.0098	2.7802	0.3597	2.5212	7.0095
13	49.4024	0.0202	138.3486	0.0072	2.8004	0.3571	2.5897	7.2524
14	66.6863	0.0150	187.7510	0.0053	2.8154	0.3552	2.6452	7.4473
15	90.0170	0.0111	254.4374	0.0039	2.8265	0.3538	2.6898	7.6028
16	121.5102	0.0082	344.4544	0.0029	2.8343	0.3528	2.7255	7.7263
17	164.0217	0.0061	465.9646	0.0021	2.8409	0.3520	2.7540	7.8238
18	221.4061	0.0045	629.9865	0.0016	2.8454	0.3514	2.7766	7.9006
19	298.8667	0.0033	851.3928	0.0012	2.8487	0.3510	2.7945	7.9608
20	403.4280	0.0025	1150.2603	0.0009	2.8512	0.3507	2.8086	8.0079
21	544.5708	0.0018	1553.6890	0.0006	2.8531	0.3505	2.8197	8.0447
22	735.0940	0.0014	2098.2608	0.0005	2.8544	0.3503	2.8283	8.0732
23	992.2730	0.0010	2833.3550	0.0004	2.8554	0.3502	2.8351	8.0954
24	1339.4285	0.0007	3825.6304	0.0003	2.8562	0.3501	2.8404	8.1126
25	1808.0383	0.0005	5165.0547	0.0002	2.8567	0.3501	2.8445	8.1258
26	2440.5967	0.0004	6973.0977	0.0001	2.8571	0.3500	2.8476	8.1361
27	3294.4622	0.0003	9413.6992	0.0001	2.8574	0.3500	2.8501	8.1440
28	4447.0547	0.0002	12708.1641	0.0001	2.8577	0.3499	2.8520	8.1500
29	6002.8945	0.0002	17155.2227	0.0001	2.8578	0.3499	2.8535	8.1547
30	8103.0625	0.0001	23158.1328	0.0000	2.8579	0.3499	2.8546	8.1583
31	10937.9922	0.0001	31261.2070	0.0000	2.8580	0.3499	2.8555	8.1610
32	14764.7500	0.0001	42199.2188	0.0000	2.8581	0.3499	2.8561	8.1631
33	19930.3281	0.0001	56963.9922	0.0000	2.8582	0.3499	2.8566	8.1647
34	26903.1133	0.0000	76894.2500	0.0000	2.8582	0.3499	2.8570	8.1660
35	36315.4141	0.0000	103797.4375	0.0000	2.8582	0.3499	2.8573	8.1669

SECTION II—CONTINUOUS COMPOUNDING, CONTINUOUS FLOW INTEREST FACTORS

TABLE B.19 Continuous Compounding: r = 4%

	Continuous flow, uniform series			
	Present worth factor	Capital recovery factor	Compound amount factor	Sinking fund factor
n	To find P given $\overline{A}$ $P\backslash\overline{A}\ r,n$	To find $\overline{A}$ given P $\overline{A}\backslash P\ r,n$	To find F given $\overline{A}$ $F\backslash\overline{A}\ r,n$	To find $\overline{A}$ given F $\overline{A}\backslash F\ r,n$
1	0.9803	1.0201	1.0203	0.9801
2	1.9221	0.5203	2.0822	0.4803
3	2.8270	0.3537	3.1874	0.3137
4	3.6964	0.2705	4.3378	0.2305
5	4.5317	0.2207	5.5351	0.1807
6	5.3345	0.1875	6.7812	0.1475
7	6.1054	0.1638	8.0782	0.1238
8	6.8463	0.1461	9.4282	0.1061
9	7.5581	0.1323	10.8332	0.0923
10	8.2420	0.1213	12.2956	0.0813
11	8.8991	0.1124	13.8177	0.0724
12	9.5304	0.1049	15.4018	0.0649
13	10.1370	0.0986	17.0507	0.0586
14	10.7198	0.0933	18.7668	0.0533
15	11.2797	0.0887	20.5529	0.0487
16	11.8177	0.0846	22.4120	0.0446
17	12.3346	0.0811	24.3469	0.0411
18	12.8312	0.0779	26.3608	0.0379
19	13.3083	0.0751	28.4569	0.0351
20	13.7668	0.0726	30.6385	0.0326
21	14.2072	0.0704	32.9092	0.0304
22	14.6304	0.0684	35.2725	0.0284
23	15.0370	0.0665	37.7322	0.0265
24	15.4277	0.0648	40.2924	0.0248
25	15.8030	0.0633	42.9570	0.0233
26	16.1636	0.0619	45.7304	0.0219
27	16.5101	0.0606	48.6170	0.0206
28	16.8430	0.0594	51.6213	0.0194
29	17.1628	0.0583	54.7483	0.0183
30	17.4701	0.0572	58.0029	0.0172
31	17.7654	0.0563	61.3903	0.0163
32	18.0491	0.0554	64.9150	0.0154
33	18.3216	0.0546	68.5855	0.0146
34	18.5835	0.0538	72.4048	0.0138
35	18.8351	0.0531	76.3800	0.0131
40	19.9526	0.0501	98.8258	0.0101
45	20.8675	0.0479	126.2411	0.0079
50	21.6166	0.0463	159.7263	0.0063
55	22.2299	0.0450	200.6253	0.0050
60	22.7321	0.0440	250.5793	0.0040
65	23.1432	0.0432	311.5933	0.0032
70	23.4797	0.0426	386.1160	0.0026
75	23.7553	0.0421	477.1379	0.0021
80	23.9809	0.0417	588.3128	0.0017
85	24.1657	0.0414	724.1018	0.0014
90	24.3159	0.0411	889.9553	0.0011
95	24.4407	0.0409	1092.5293	0.0009
100	24.5421	0.0407	1339.9531	0.0007

TABLE B.20 Continuous Compounding: r = 5%

	Continuous flow, uniform series			
	Present worth factor	Capital recovery factor	Compound amount factor	Sinking fund factor
n	To find P given $\bar{A}$ $P\|\bar{A}\ r,n$	To find $\bar{A}$ given P $\bar{A}\|P\ r,n$	To find F given $\bar{A}$ $F\|\bar{A}\ r,n$	To find $\bar{A}$ given F $\bar{A}\|F\ r,n$
1	0.9754	1.0252	1.0254	0.9752
2	1.9032	0.5254	2.1034	0.4754
3	2.7858	0.3590	3.2367	0.3090
4	3.6254	0.2758	4.4280	0.2258
5	4.4240	0.2260	5.6805	0.1760
6	5.1836	0.1929	6.9972	0.1429
7	5.9062	0.1693	8.3813	0.1193
8	6.5936	0.1517	9.8365	0.1017
9	7.2474	0.1380	11.3662	0.0880
10	7.8694	0.1271	12.9744	0.0771
11	8.4610	0.1182	14.6651	0.0682
12	9.0238	0.1108	16.4424	0.0608
13	9.5591	0.1046	18.3108	0.0546
14	10.0683	0.0993	20.2750	0.0493
15	10.5527	0.0948	22.3400	0.0448
16	11.0134	0.0908	24.5108	0.0408
17	11.4517	0.0873	26.7929	0.0373
18	11.8686	0.0843	29.1920	0.0343
19	12.2652	0.0815	31.7142	0.0315
20	12.6424	0.0791	34.3656	0.0291
21	13.0012	0.0769	37.1530	0.0269
22	13.3426	0.0749	40.0833	0.0249
23	13.6673	0.0732	43.1638	0.0232
24	13.9761	0.0716	46.4023	0.0216
25	14.2699	0.0701	49.8068	0.0201
26	14.5494	0.0687	53.3859	0.0187
27	14.8152	0.0675	57.1485	0.0175
28	15.0681	0.0664	61.1040	0.0164
29	15.3086	0.0653	65.2623	0.0153
30	15.5374	0.0644	69.6338	0.0144
31	15.7550	0.0635	74.2294	0.0135
32	15.9621	0.0626	79.0606	0.0126
33	16.1590	0.0619	84.1396	0.0119
34	16.3463	0.0612	89.4789	0.0112
35	16.5245	0.0605	95.0920	0.0105
40	17.2933	0.0578	127.7810	0.0078
45	17.8920	0.0559	169.7546	0.0059
50	18.3583	0.0545	223.6498	0.0045
55	18.7214	0.0534	292.8523	0.0034
60	19.0043	0.0526	381.7105	0.0026
65	19.2245	0.0520	495.8064	0.0020
70	19.3961	0.0515	642.3086	0.0016
75	19.5296	0.0512	830.4209	0.0012
80	19.6337	0.0509	1071.9624	0.0009
85	19.7147	0.0507	1382.1074	0.0007
90	19.7773	0.0506	1780.3406	0.0006
95	19.8270	0.0504	2291.5831	0.0004
100	19.8652	0.0503	2948.2608	0.0003

TABLE B.21 Continuous Compounding: r = 6%

	Continuous flow, uniform series			
	Present worth factor	Capital recovery factor	Compound amount factor	Sinking fund factor
n	To find P given $\bar{A}$ $P\|\bar{A}\ r,n$	To find $\bar{A}$ given P $\bar{A}\|P\ r,n$	To find F given $\bar{A}$ $F\|\bar{A}\ r,n$	To find $\bar{A}$ given F $\bar{A}\|F\ r,n$
1	0.9706	1.0303	1.0306	0.9703
2	1.8847	0.5306	2.1249	0.4706
3	2.7455	0.3642	3.2869	0.3042
4	3.5562	0.2812	4.5208	0.2212
5	4.3197	0.2315	5.8310	0.1715
6	5.0387	0.1985	7.2221	0.1385
7	5.7159	0.1750	8.6994	0.1150
8	6.3550	0.1574	10.2679	0.0974
9	6.9542	0.1438	11.9334	0.0838
10	7.5198	0.1330	13.7020	0.0730
11	8.0525	0.1242	15.5799	0.0642
12	8.5541	0.1169	17.5739	0.0569
13	9.0265	0.1108	19.6912	0.0508
14	9.4715	0.1056	21.9394	0.0456
15	9.8905	0.1011	24.3267	0.0411
16	10.2851	0.0972	26.8616	0.0372
17	10.6568	0.0938	29.5532	0.0338
18	11.0067	0.0909	32.4113	0.0309
19	11.3363	0.0882	35.4461	0.0282
20	11.6463	0.0859	38.6686	0.0259
21	11.9391	0.0838	42.0904	0.0238
22	12.2144	0.0819	45.7237	0.0219
23	12.4737	0.0802	49.5817	0.0202
24	12.7179	0.0786	53.6783	0.0186
25	12.9478	0.0772	58.0281	0.0172
26	13.1644	0.0760	62.6470	0.0160
27	13.3684	0.0743	67.5515	0.0148
28	13.5604	0.0737	72.7592	0.0137
29	13.7413	0.0728	78.2890	0.0128
30	13.9117	0.0719	84.1608	0.0119
31	14.0721	0.0711	90.3956	0.0111
32	14.2232	0.0703	97.0160	0.0103
33	14.3655	0.0696	104.0457	0.0096
34	14.4995	0.0690	111.5101	0.0090
35	14.6257	0.0684	119.4361	0.0084
40	15.1547	0.0660	167.0529	0.0060
45	15.5455	0.0643	231.3288	0.0043
50	15.8369	0.0631	318.0918	0.0031
55	16.0519	0.0623	435.2102	0.0023
60	16.2113	0.0617	593.3035	0.0017
65	16.3293	0.0612	806.7070	0.0012
70	16.4167	0.0609	1094.7717	0.0009
75	16.4815	0.0607	1483.6182	0.0007
80	16.5295	0.0605	2008.5054	0.0005
85	16.5651	0.0604	2717.0303	0.0004
90	16.5914	0.0603	3673.4382	0.0003
95	16.6109	0.0602	4964.4492	0.0002
100	16.6254	0.0601	6707.1406	0.0001

TABLE B.22 Continuous Compounding: r = 9%

n	Continuous flow, uniform series			
	Present worth factor	Capital recovery factor	Compound amount factor	Sinking fund factor
	To find P given $\bar{A}$ $P\|\bar{A}\ r,n$	To find $\bar{A}$ given P $\bar{A}\|P\ r,n$	To find F given $\bar{A}$ $F\|\bar{A}\ r,n$	To find $\bar{A}$ given F $\bar{A}\|F\ r,n$
1	0.9610	1.0405	1.0411	0.9605
2	1.8402	0.5411	2.1689	0.4611
3	2.6671	0.3749	3.3906	0.2949
4	3.4231	0.2921	4.7141	0.2121
5	4.1210	0.2427	6.1478	0.1627
6	4.7652	0.2099	7.7009	0.1299
7	5.3599	0.1866	9.3834	0.1066
8	5.9088	0.1692	11.2060	0.0892
9	6.4156	0.1559	13.1804	0.0759
10	6.8834	0.1453	15.3193	0.0653
11	7.3152	0.1367	17.6362	0.0567
12	7.7133	0.1296	20.1462	0.0496
13	8.0813	0.1237	22.8652	0.0437
14	8.4215	0.1187	25.8107	0.0387
15	8.7361	0.1145	29.0014	0.0345
16	9.0245	0.1108	32.4580	0.0308
17	9.2917	0.1076	36.2024	0.0276
18	9.5384	0.1048	40.2587	0.0248
19	9.7661	0.1024	44.6528	0.0224
20	9.9763	0.1002	49.4129	0.0202
21	10.1703	0.0983	54.5694	0.0183
22	10.3494	0.0966	60.1554	0.0166
23	10.5148	0.0951	66.2067	0.0151
24	10.6674	0.0937	72.7620	0.0137
25	10.8083	0.0925	79.8632	0.0125
26	10.9384	0.0914	87.5558	0.0114
27	11.0584	0.0904	95.8892	0.0104
28	11.1693	0.0895	104.9165	0.0095
29	11.2716	0.0887	114.6958	0.0087
30	11.3660	0.0880	125.2896	0.0080
31	11.4532	0.0873	136.7657	0.0073
32	11.5337	0.0867	149.1976	0.0067
33	11.6080	0.0861	162.6649	0.0061
34	11.6765	0.0856	177.2539	0.0056
35	11.7399	0.0852	193.0578	0.0052
40	11.9905	0.0834	294.1563	0.0034
45	12.1585	0.0822	444.9773	0.0022
50	12.2711	0.0815	669.9761	0.0015
55	12.3465	0.0810	1005.6348	0.0010
60	12.3971	0.0807	1506.3784	0.0007
65	12.4310	0.0804	2253.3999	0.0004
70	12.4538	0.0803	3367.8254	0.0003
75	12.4690	0.0802	5030.3477	0.0002
80	12.4792	0.0801	7510.5469	0.0001
85	12.4861	0.0801	11210.5703	0.0001
90	12.4907	0.0801	16730.3516	0.0001
95	12.4937	0.0800	24964.9024	0.0000
100	12.4958	0.0800	37249.4063	0.0000

TABLE B.23 Continuous Compounding: r = 10%

	Continuous flow, uniform series			
	Present worth factor	Capital recovery factor	Compound amount factor	Sinking fund factor
n	To find P given $\overline{A}$ $P\|\overline{A}\ r,n$	To find $\overline{A}$ given P $\overline{A}\|P\ r,n$	To find F given $\overline{A}$ $F\|\overline{A}\ r,n$	To find $\overline{A}$ given F $\overline{A}\|F\ r,n$
1	0.9516	1.0508	1.0517	0.9508
2	1.8127	0.5517	2.2140	0.4517
3	2.5918	0.3858	3.4986	0.2858
4	3.2968	0.3033	4.9182	0.2033
5	3.9347	0.2541	6.4872	0.1541
6	4.5119	0.2216	8.2212	0.1216
7	5.0341	0.1986	10.1375	0.0986
8	5.5067	0.1816	12.2554	0.0816
9	5.9343	0.1685	14.5960	0.0685
10	6.3212	0.1582	17.1828	0.0582
11	6.6713	0.1499	20.0417	0.0499
12	6.9881	0.1431	23.2012	0.0431
13	7.2747	0.1375	26.6930	0.0375
14	7.5340	0.1327	30.5520	0.0327
15	7.7687	0.1287	34.8169	0.0287
16	7.9810	0.1253	39.5303	0.0253
17	8.1732	0.1224	44.7395	0.0224
18	8.3470	0.1198	50.4964	0.0198
19	8.5043	0.1176	56.8589	0.0176
20	8.6466	0.1157	63.8905	0.0157
21	8.7754	0.1140	71.6617	0.0140
22	8.8920	0.1125	80.2501	0.0125
23	8.9974	0.1111	89.7418	0.0111
24	9.0928	0.1100	100.2317	0.0100
25	9.1792	0.1089	111.8248	0.0089
26	9.2573	0.1080	124.6372	0.0080
27	9.3279	0.1072	138.7972	0.0072
28	9.3919	0.1065	154.4463	0.0065
29	9.4498	0.1058	171.7412	0.0058
30	9.5021	0.1052	190.8551	0.0052
31	9.5495	0.1047	211.9792	0.0047
32	9.5924	0.1042	235.3249	0.0042
33	9.6312	0.1038	261.1260	0.0038
34	9.6663	0.1035	289.6404	0.0035
35	9.6980	0.1031	321.1541	0.0031
40	9.8168	0.1019	535.9805	0.0019
45	9.8889	0.1011	890.1697	0.0011
50	9.9326	0.1007	1474.1289	0.0007
55	9.9591	0.1004	2436.9136	0.0004
60	9.9752	0.1002	4024.2783	0.0002
65	9.9850	0.1002	6641.3984	0.0002
70	9.9909	0.1001	10956.3008	0.0001
75	9.9945	0.1001	18070.3750	0.0001
80	9.9965	0.1000	29799.4727	0.0000
85	9.9980	0.1000	49137.5078	0.0000
90	9.9988	0.1000	81020.5625	0.0000
95	9.9993	0.1000	133586.7500	0.0000
100	9.9995	0.1000	-209242.8750	0.0000

TABLE B.24 Continuous Compounding: r = 15%

	Continuous flow, uniform series			
	Present worth factor	Capital recovery factor	Compound amount factor	Sinking fund factor
n	To find P given $\overline{A}$ $P|\overline{A}\ r,n$	To find $\overline{A}$ given P $\overline{A}|P\ r,n$	To find F given $\overline{A}$ $F|\overline{A}\ r,n$	To find $\overline{A}$ given F $\overline{A}|F\ r,n$
1	0.9286	1.0769	1.0789	0.9269
2	1.7275	0.5787	2.3324	0.4287
3	2.4158	0.4139	3.7887	0.2639
4	3.0079	0.3325	5.4803	0.1825
5	3.5175	0.2843	7.4467	0.1343
6	3.9562	0.2528	9.7307	0.1028
7	4.3337	0.2307	12.3843	0.0807
8	4.6587	0.2147	15.4674	0.0647
9	4.9384	0.2025	19.0495	0.0525
10	5.1791	0.1931	23.2113	0.0431
11	5.3863	0.1857	28.0465	0.0357
12	5.5647	0.1797	33.6643	0.0297
13	5.7182	0.1749	40.1912	0.0249
14	5.8503	0.1709	47.7744	0.0209
15	5.9640	0.1677	56.5349	0.0177
16	6.0619	0.1650	66.8211	0.0150
17	6.1461	0.1627	78.7140	0.0127
18	6.2186	0.1608	92.5315	0.0108
19	6.2810	0.1592	108.5851	0.0092
20	6.3348	0.1579	127.2368	0.0079
21	6.3810	0.1567	148.9069	0.0067
22	6.4208	0.1557	174.0840	0.0057
23	6.4550	0.1549	203.3358	0.0049
24	6.4845	0.1542	237.3214	0.0042
25	6.5099	0.1536	276.8069	0.0036
26	6.5317	0.1531	322.6826	0.0031
27	6.5505	0.1527	375.9827	0.0027
28	6.5667	0.1523	437.9082	0.0023
29	6.5806	0.1520	509.8560	0.0020
30	6.5926	0.1517	593.4468	0.0017

TABLES B.25 Continuous Compounding: r = 20%

	Continuous flow, uniform series			
	Present worth factor	Capital recovery factor	Compound amount factor	Sinking fund factor
n	To find P given $\overline{A}$ $P\|\overline{A}\ r,n$	To find $\overline{A}$ given P $\overline{A}\|P\ r,n$	To find F given $\overline{A}$ $F\|\overline{A}\ r,n$	To find $\overline{A}$ given F $\overline{A}\|F\ r,n$
1	0.9063	1.1033	1.1070	0.9033
2	1.0484	0.6066	2.4591	0.4066
3	2.2559	0.4433	4.1106	0.2433
4	2.7534	0.3632	6.1277	0.1632
5	3.1605	0.3164	8.5914	0.1164
6	3.4940	0.2862	11.6006	0.0862
7	3.7670	0.2655	15.2760	0.0655
8	3.9905	0.2506	19.7652	0.0506
9	4.1735	0.2396	25.2482	0.0396
10	4.3233	0.2313	31.9453	0.0313
11	4.4460	0.2249	40.1251	0.0249
12	4.5464	0.2200	50.1159	0.0200
13	4.6286	0.2160	62.3187	0.0160
14	4.6959	0.2129	77.2232	0.0129
15	4.7511	0.2105	95.4276	0.0105
16	4.7962	0.2085	117.6625	0.0085
17	4.8331	0.2069	144.8204	0.0069
18	4.8634	0.2056	177.9911	0.0056
19	4.8881	0.2046	218.5057	0.0046
20	4.9084	0.2037	267.9905	0.0037
21	4.9250	0.2030	328.4314	0.0030
22	4.9386	0.2025	402.2542	0.0025
23	4.9497	0.2020	492.4212	0.0020
24	4.9589	0.2017	602.5518	0.0017
25	4.9663	0.2014	737.0652	0.0014
26	4.9724	0.2011	901.3636	0.0011
27	4.9774	0.2009	1102.0315	0.0009
28	4.9815	0.2007	1347.1314	0.0007
29	4.9849	0.2006	1646.4966	0.0006
30	4.9876	0.2005	2012.1423	0.0005

TABLE B.26 Continuous Compounding: r = 25%

	Continuous flow, uniform series			
	Present worth factor	Capital recovery factor	Compound amount factor	Sinking fund factor
n	To find P given $\overline{A}$ $P\|\overline{A}\ r,n$	To find $\overline{A}$ given P $\overline{A}\|P\ r,n$	To find F given $\overline{A}$ $F\|\overline{A}\ r,n$	To find $\overline{A}$ given F $\overline{A}\|F\ r,n$
1	0.8843	1.1302	1.1361	0.8802
2	1.5739	0.6354	2.5949	0.3854
3	2.1105	0.4738	4.4680	0.2238
4	2.5285	0.3955	6.8731	0.1455
5	2.8540	0.3504	9.9614	0.1004
6	3.1075	0.3218	13.9268	0.0718
7	3.3049	0.3026	19.0184	0.0526
8	3.4587	0.2891	25.5562	0.0391
9	3.5784	0.2795	33.9509	0.0295
10	3.6717	0.2724	44.7300	0.0224
11	3.7443	0.2671	58.5705	0.0171
12	3.8009	0.2631	76.3421	0.0131
13	3.8449	0.2601	99.1613	0.0101
14	3.8792	0.2578	128.4617	0.0078
15	3.9059	0.2560	166.0842	0.0060
16	3.9267	0.2547	214.3925	0.0047
17	3.9429	0.2536	276.4214	0.0036
18	3.9555	0.2528	356.0684	0.0028
19	3.9654	0.2522	458.3369	0.0022
20	3.9730	0.2517	589.6524	0.0017
21	3.9790	0.2513	758.2647	0.0013
22	3.9837	0.2510	974.7673	0.0010
23	3.9873	0.2508	1252.7615	0.0008
24	3.9901	0.2506	1609.7146	0.0006
25	3.9923	0.2505	2068.0496	0.0005
26	3.9940	0.2504	2656.5647	0.0004
27	3.9953	0.2503	3412.2319	0.0003
28	3.9964	0.2502	4382.5274	0.0002
29	3.9972	0.2502	5628.4141	0.0002
30	3.9978	0.2501	7228.1641	0.0001

SECTION III—GEOMETRIC SERIES FACTORS

TABLE B.27 Continuous Compounding: r = 5%

n	Geometric series present worth factor $(P|A_1\ r,c,n)\infty$				
	$c = 4\%$	$c = 6\%$	$c = 8\%$	$c = 10\%$	$c = 15\%$
1	0.9512	0.9512	0.9512	0.9512	0.9512
2	1.8930	1.9120	1.9314	1.9512	2.0025
3	2.8254	2.8825	2.9415	3.0025	3.1643
4	3.7485	3.8627	3.9823	4.1077	4.4484
5	4.6624	4.8527	5.0548	5.2695	5.8674
6	5.5673	5.8527	6.1600	6.4909	7.4357
7	6.4631	6.8628	7.2988	7.7749	9.1690
8	7.3500	7.8830	8.4723	9.1248	11.0845
9	8.2281	8.9134	9.6816	10.5439	13.2015
10	9.0975	9.9542	10.9276	12.0357	15.5412
11	9.9582	11.0055	12.2117	13.6040	18.1269
12	10.8103	12.0673	13.5348	15.2527	20.9845
13	11.6540	13.1398	14.8982	16.9860	24.1427
14	12.4893	14.2231	16.3032	18.8081	27.6331
15	13.3162	15.3173	17.7509	20.7236	31.4905
16	14.1350	16.4225	19.2427	22.7374	35.7536
17	14.9455	17.5388	20.7800	24.8544	40.4651
18	15.7481	18.6663	22.3640	27.0799	45.6721
19	16.5426	19.8051	23.9964	29.4196	51.4267
20	17.3292	20.9554	25.6784	31.8792	57.7865
21	18.1080	22.1172	27.4116	34.4649	64.8152
22	18.8791	23.2907	29.1977	37.1832	72.5831
23	19.6425	24.4760	31.0381	40.0408	81.1679
24	20.3982	25.6732	32.9346	43.0450	90.6556
25	21.1465	26.8825	34.8883	46.2032	101.1412
26	21.8873	28.1039	36.9026	49.5233	112.7295
27	22.6208	29.3376	38.9777	53.0136	125.5366
28	23.3469	30.5836	41.1159	56.6829	139.6906
29	24.0658	31.8422	43.3193	60.5403	155.3332
30	24.7776	33.1135	45.5893	64.5955	172.6210
31	25.4823	34.3975	47.9295	68.8586	191.7269
32	26.1800	35.6944	50.3404	73.3403	212.8423
33	26.8707	37.0044	52.8247	78.0518	236.1783
34	27.5546	38.3275	55.3847	83.0048	261.9685
35	28.2316	39.6640	58.0226	88.2118	290.4712
36	28.9019	41.0138	60.7409	93.6857	321.9717
37	29.5656	42.3772	63.5420	99.4403	356.7849
38	30.2226	43.7544	66.4283	105.4899	395.2595
39	30.8732	45.1453	69.4026	111.3497	437.7805
40	31.5172	46.5503	72.4675	118.5356	484.7737
41	32.1548	47.9694	75.6257	125.5643	536.7090
42	32.7861	49.4027	78.8800	132.9533	594.1062
43	33.4111	50.8504	82.2335	140.7212	657.5403
44	34.0299	52.3127	85.6891	148.8874	727.6455
45	34.6425	53.7897	89.2500	157.4722	805.1240
46	35.2490	55.2815	92.9193	166.4972	890.7507
47	35.8495	56.7883	96.7003	175.9849	985.3831
48	36.4441	58.3103	100.5965	185.9591	1089.9680
49	37.0327	59.8475	104.6113	196.4447	1205.5523
50	37.6154	61.4003	108.7484	207.4678	1333.2925

TABLE B.28 Continuous Compounding: r = 5%

	Geometric series future worth factor $(F|A_1\ r, c, n)^\infty$				
n	$c = 4\%$	$c = 6\%$	$c = 8\%$	$c = 10\%$	$c = 15\%$
1	1.0000	1.0000	1.0000	1.0000	1.0000
2	2.0921	2.1131	2.1346	2.1564	2.2131
3	3.2826	3.3489	3.4175	3.4884	3.6764
4	4.5784	4.7179	4.8640	5.0171	5.4332
5	5.9867	6.2310	6.4905	6.7662	7.5339
6	7.5150	7.9003	8.3151	8.7618	10.0372
7	9.1716	9.7387	10.3575	11.0332	13.0114
8	10.9650	11.7600	12.6392	13.6126	16.5362
9	12.9043	13.9790	15.1837	16.5361	20.7041
10	14.9992	16.4117	18.0166	19.8435	25.6231
11	17.2601	19.0753	21.1659	23.5792	31.4185
12	19.6977	21.9881	24.6620	27.7923	38.2363
13	22.3237	25.1699	28.5381	32.5373	46.2464
14	25.1503	28.6419	32.8305	37.8748	55.6462
15	28.1905	32.4267	37.5786	43.8719	66.6654
16	31.4579	36.5489	42.8255	50.6030	79.5711
17	34.9673	41.0345	48.6178	58.1505	94.6740
18	38.7340	45.9116	55.0067	66.6058	112.3351
19	42.7743	51.2102	62.0476	76.0704	132.9743
20	47.1057	56.9626	69.8011	86.6565	157.0799
21	51.7464	63.2032	78.3329	98.4886	185.2190
22	56.7159	69.9691	87.7147	111.7043	218.0515
23	62.0347	77.2999	98.0243	126.4565	256.3438
24	67.7245	85.2381	109.3466	142.9142	300.9871
25	73.8085	93.8290	121.7739	161.2648	353.0173
26	80.3111	103.1214	135.4065	181.7155	413.6379
27	87.2579	113.1674	150.3534	204.4960	484.2481
28	94.6764	124.0227	166.7333	229.8604	566.4734
29	102.5954	135.7470	184.6752	258.0901	662.2034
30	111.0455	148.4042	204.3194	289.4968	773.6338
31	120.0590	162.0627	225.8182	324.4253	903.3157
32	129.6702	176.7956	249.3374	363.2566	1054.2146
33	139.9152	192.6811	275.0569	406.4138	1229.7756
34	150.8322	209.8028	303.1726	454.3638	1434.0025
35	162.4617	228.2502	333.8970	507.6235	1671.5471
36	174.8465	248.1190	367.4607	566.7654	1947.8142
37	188.0318	269.5115	404.1152	632.4221	2269.0879
38	202.0653	292.5369	444.1326	705.2944	2642.6638
39	216.9977	317.3123	487.8091	786.1567	3077.0222
40	232.8822	343.9624	535.4658	875.8662	3582.0191
41	249.7754	372.6211	587.4524	975.3711	4169.0977
42	267.7366	403.4304	644.1475	1085.7195	4851.5703
43	286.8294	436.5435	705.9627	1208.0718	5644.8867
44	307.1199	472.1226	773.3450	1343.7107	6567.0078
45	328.6787	510.3421	846.7798	1494.0547	7638.8008
46	351.5801	551.3877	926.7932	1660.6736	8884.5078
47	375.9026	595.4578	1013.9573	1845.3023	10332.3008
48	401.7290	642.7642	1108.8926	2049.8591	12014.9063
49	429.1470	693.5337	1212.2720	2276.4688	13970.3516
50	458.2493	748.0078	1324.8269	2527.4751	16242.8203

TABLE B.29 Continuous Compounding: r = 8%

n	Geometric series present worth factor $(P|A_1\ r,c,n)\infty$				
	c = 4%	c = 6%	c = 8%	c = 10%	c = 15%
1	0.9231	0.9231	0.9259	0.9231	0.9231
2	1.8100	1.8280	1.8519	1.8649	1.9132
3	2.6622	2.7149	2.7778	2.8257	2.9750
4	3.4809	3.5842	3.7037	3.8059	4.1138
5	4.2675	4.4364	4.6296	4.8059	5.3352
6	5.0233	5.2716	5.5556	5.8261	6.6452
7	5.7495	6.0904	6.4815	6.8669	8.0501
8	6.4471	6.8929	7.4074	7.9287	9.5570
9	7.1175	7.6795	8.3333	9.0120	11.1730
10	7.7615	8.4506	9.2593	10.1172	12.9063
11	8.3803	9.2064	10.1852	11.2447	14.7652
12	8.9748	9.9472	11.1111	12.3949	16.7589
13	9.5460	10.6733	12.0370	13.5685	18.8972
14	10.0948	11.3851	12.9630	14.7657	21.1905
15	10.6221	12.0828	13.8889	15.9871	23.6501
16	11.1287	12.7666	14.8148	17.2332	26.2881
17	11.6155	13.4370	15.7407	18.5044	29.1173
18	12.0832	14.0940	16.6667	19.8013	32.1517
19	12.5325	14.7380	17.5926	21.1245	35.4060
20	12.9642	15.3693	18.5185	22.4743	38.8964
21	13.3790	15.9881	19.4444	23.8514	42.6398
22	13.7775	16.5946	20.3704	25.2564	46.6546
23	14.1604	17.1892	21.2963	26.6897	50.9606
24	14.5283	17.7719	22.2222	28.1520	55.5788
25	14.8817	18.3431	23.1481	29.6438	60.5318
26	15.2213	18.9030	24.0741	31.1658	65.8440
27	15.5476	19.4518	25.0000	32.7185	71.5413
28	15.8611	19.9898	25.9259	34.3026	77.6518
29	16.1623	20.5171	26.8519	35.9186	84.2053
30	16.4517	21.0339	27.7778	37.5674	91.2339
31	16.7297	21.5405	28.7037	39.2494	98.7723
32	16.9968	22.0371	29.6296	40.9654	106.8571
33	17.2535	22.5239	30.5556	42.7161	115.5283
34	17.5001	23.0010	31.4815	44.5021	124.8281
35	17.7370	23.4687	32.4074	46.3242	134.8023
36	17.9647	23.9271	33.3333	48.1831	145.4997
37	18.1834	24.3764	34.2593	50.0796	156.9727
38	18.3935	24.8168	35.1852	52.0144	169.2777
39	18.5954	25.2485	36.1111	53.9883	182.4748
40	18.7894	25.6717	37.0370	56.0020	196.6288
41	18.9758	26.0865	37.9630	58.0565	211.8091
42	19.1548	26.4930	38.8889	60.1524	228.0902
43	19.3269	26.8916	39.8148	62.2907	245.5517
44	19.4922	27.2822	40.7407	64.4722	264.2793
45	19.6510	27.6651	41.6667	66.6977	284.3648
46	19.8036	28.0404	42.5926	68.9682	305.9065
47	19.9502	28.4083	43.5185	71.2846	329.0105
48	20.0910	28.7689	44.4444	73.6477	353.7896
49	20.2264	29.1223	45.3704	76.0586	380.3652
50	20.3564	29.4688	46.2963	78.5182	408.8679

TABLE B.30 Continuous Compounding: r = 8%

	Geometric series future worth factor $(F|A_1 \ r, c, n)^\infty$				
n	$c = 4\%$	$c = 6\%$	$c = 8\%$	$c = 10\%$	$c = 15\%$
1	1.0000	1.0000	1.0000	1.0000	1.0000
2	2.1241	2.1451	2.1666	2.1835	2.2451
3	3.3843	3.4513	3.5205	3.5921	3.7820
4	4.7937	4.9359	5.0850	5.2412	5.6653
5	6.3664	6.6183	6.8856	7.1695	7.9592
6	8.1181	8.5194	8.9509	9.4154	10.7391
7	10.0654	10.6623	11.3125	12.0217	14.0932
8	12.2269	13.0722	14.0054	15.0367	18.1246
9	14.6224	15.7771	17.0683	18.5146	22.9543
10	17.2735	18.8071	20.5443	22.5162	28.7235
11	20.2040	22.1956	24.4809	27.1098	35.5975
12	23.4395	25.9790	28.9308	32.3718	43.7692
13	27.0078	30.1972	33.9520	38.3881	53.4643
14	30.9392	34.8937	39.6090	45.2546	64.9459
15	35.2667	40.1162	45.9728	53.0789	78.5212
16	40.0261	45.9170	53.1219	61.9814	94.5487
17	45.2562	52.3530	61.1429	72.0967	113.4465
18	50.9993	59.4865	70.1315	83.5754	135.7022
19	57.3013	67.3856	80.1932	96.5858	161.8842
20	64.2121	76.1247	91.4445	111.3159	192.6548
21	71.7856	85.7851	104.0136	127.9762	228.7860
22	80.0308	96.4553	118.0421	146.8011	271.1768
23	89.1614	108.2321	133.6860	168.0527	320.8750
24	99.0967	121.2213	151.1168	192.0235	379.1001
25	109.9618	135.5382	170.5238	219.0398	447.2725
26	121.8385	151.3085	192.1153	249.4654	527.0457
27	134.8153	168.6693	216.1205	283.7063	620.3440
28	148.9883	187.7704	242.7917	322.2151	729.4080
29	164.4620	208.7748	272.4063	365.4961	856.8447
30	181.3495	231.8604	305.2698	414.1113	1005.6873
31	199.7737	257.2210	341.7180	468.6868	1179.4649
32	219.8679	285.0679	382.1201	529.9204	1382.2839
33	241.7767	315.6311	426.8816	598.5684	1618.9207
34	265.6567	349.1619	476.4485	675.5557	1894.9295
35	291.6787	385.9334	531.3108	761.7847	2216.7742
36	320.0271	426.2427	592.0066	858.3469	2591.9695
37	353.9019	470.4143	659.1275	966.4343	3029.2527
38	384.5205	518.8010	733.3223	1087.3733	3538.7886
39	421.1182	571.7871	815.3037	1222.6384	4132.3867
40	460.9507	629.7910	905.8545	1373.8706	4823.7969
41	504.2952	693.2676	1005.8328	1542.8943	5628.9844
42	551.4514	762.7127	1116.1814	1731.7376	6566.5234
43	602.7459	838.6653	1237.9341	1942.6543	7658.0039
44	658.5313	921.7124	1372.2249	2178.1519	8928.5156
45	719.1909	1012.4922	1520.2979	2441.0142	10407.2422
46	785.1399	1111.6995	1683.5166	2734.3374	12128.0859
47	856.8284	1220.0896	1863.3775	3061.5567	14130.4727
48	934.7446	1338.4841	2061.5215	3426.4910	16460.2149
49	1019.4177	1467.7769	2279.7454	3833.3838	19170.5664
50	1111.4214	1608.9395	2520.0191	4286.9414	22323.4219

TABLE B.31 Continuous Compounding: r = 10%

	Geometric series present worth factor $(P|A_1 \ r,c,n)\infty$				
n	c = 4%	c = 6%	c = 8%	c = 10%	c = 15%
1	0.9048	0.9048	0.9048	0.9091	0.9048
2	1.7570	1.7742	1.7918	1.8182	1.8561
3	2.5595	2.6095	2.6611	2.7273	2.8561
4	3.3153	3.4120	3.5133	3.6364	3.9073
5	4.0271	4.1830	4.3485	4.5455	5.0125
6	4.6974	4.9239	5.1673	5.4545	6.1743
7	5.3287	5.6356	5.9698	6.3636	7.3957
8	5.9232	6.3195	6.7564	7.2727	8.6798
9	6.4831	6.9765	7.5275	8.1818	10.0295
10	7.0104	7.6078	8.2832	9.0909	11.4487
11	7.5070	8.2143	9.0241	10.0000	12.9405
12	7.9746	8.7971	9.7502	10.9091	14.5088
13	8.4151	9.3570	10.4620	11.8182	16.1576
14	8.8298	9.8949	11.1597	12.7273	17.8908
15	9.2205	10.4118	11.8435	13.6364	19.7129
16	9.5883	10.9084	12.5138	14.5455	21.6285
17	9.9348	11.3855	13.1709	15.4545	23.6422
18	10.2611	11.8439	13.8149	16.3636	25.7592
19	10.5684	12.2843	14.4462	17.2727	27.9848
20	10.8577	12.7075	15.0650	18.1818	30.3244
21	11.1303	13.1141	15.6715	19.0909	32.7840
22	11.3869	13.5047	16.2660	20.0000	35.3697
23	11.6287	13.8800	16.8488	20.9091	38.0880
24	11.8563	14.2406	17.4200	21.8182	40.9457
25	12.0707	14.5871	17.9799	22.7273	43.9498
26	12.2726	14.9199	18.5287	23.6364	47.1080
27	12.4627	15.2397	19.0667	24.5455	50.4281
28	12.6418	15.5470	19.5939	25.4545	53.9185
29	12.8104	15.8423	20.1108	26.3636	57.5878
30	12.9692	16.1259	20.6174	27.2727	61.4452
31	13.1188	16.3984	21.1140	28.1818	65.5004
32	13.2597	16.6603	21.6008	29.0909	69.7635
33	13.3923	16.9119	22.0779	30.0000	74.2452
34	13.5172	17.1536	22.5455	30.9091	78.9567
35	13.6349	17.3858	23.0039	31.8182	83.9097
36	13.7457	17.6089	23.4533	32.7273	89.1167
37	13.8501	17.8233	23.8937	33.6364	94.5907
38	13.9483	18.0293	24.3254	34.5455	100.3453
39	14.0409	18.2272	24.7486	35.4545	106.3949
40	14.1280	18.4173	25.1634	36.3636	112.7547
41	14.2101	18.6000	25.5699	37.2727	119.4436
42	14.2874	18.7755	25.9685	38.1818	126.4693
43	14.3602	18.9442	26.3591	39.0909	133.8583
44	14.4288	19.1062	26.7420	40.0000	141.6262
45	14.4934	19.2619	27.1173	40.9091	149.7924
46	14.5542	19.4115	27.4852	41.8182	158.3773
47	14.6114	19.5552	27.8458	42.7273	167.4023
48	14.6654	19.6932	28.1992	43.6364	176.8900
49	14.7162	19.8259	28.5457	44.5455	186.8642
50	14.7640	19.9533	28.8853	45.4545	197.3498

TABLE B.32 Continuous Compounding: r = 10%

	Geometric series future worth factor $(F|A_1 \ r, c, n)\infty$				
n	$c = 4\%$	$c = 6\%$	$c = 8\%$	$c = 10\%$	$c = 15\%$
1	1.0000	1.0000	1.0000	1.0000	1.0000
2	2.1460	2.1670	2.1885	2.2103	2.2670
3	3.4550	3.5224	3.5921	3.6642	3.8553
4	4.9458	5.0901	5.2412	5.3994	5.8291
5	6.6395	6.8967	7.1695	7.4591	8.2642
6	8.5592	8.9718	9.4154	9.8923	11.2504
7	10.7306	11.3488	12.0217	12.7548	14.8932
8	13.1823	14.0643	15.0367	16.1100	19.3172
9	15.9458	17.1595	18.5146	20.0299	24.6689
10	19.0562	20.6802	22.5162	24.5960	31.1208
11	22.5521	24.6773	27.1098	29.9011	38.8755
12	26.4767	29.2074	32.3718	36.0500	48.1710
13	30.8773	34.3336	38.3881	43.1615	59.2868
14	35.8067	40.1259	45.2546	51.3701	72.5508
15	41.3232	46.6624	53.0789	60.8280	88.3472
16	47.4914	54.0295	61.9814	71.7070	107.1264
17	54.3825	62.3235	72.0967	84.2015	129.4162
18	62.0759	71.6514	83.5754	98.5310	155.8341
19	70.6589	82.1317	96.5858	114.9432	187.1030
20	80.2284	93.8963	111.3159	133.7177	224.0686
21	90.8917	107.0915	127.9762	155.1700	267.7195
22	102.7672	121.8799	146.8011	179.6555	319.2117
23	115.9861	138.4415	168.0527	207.5751	379.8963
24	130.6938	156.9764	192.0235	239.3801	451.3506
25	147.0507	177.7065	219.0398	275.5789	535.4180
26	165.2344	200.8777	249.4654	316.7444	634.2493
27	185.4415	226.7630	283.7063	363.5203	750.3562
28	207.8892	255.6650	322.2151	416.6318	886.6692
29	232.8180	287.9190	365.4961	476.8941	1046.6074
30	260.4934	323.8970	414.1113	545.2236	1234.1585
31	291.2100	364.0112	468.6868	622.6509	1453.9729
32	325.2925	408.7183	529.9204	710.3335	1711.4734
33	363.1004	458.5247	598.5884	809.5723	2012.9807
34	405.0313	513.9907	675.5557	921.8284	2365.8623
35	451.5249	575.7383	761.7847	1048.7420	2778.7039
36	503.0674	644.4553	858.3469	1192.1546	3261.5088
37	560.1963	720.9043	966.4343	1354.1326	3825.9312
38	623.5056	805.9299	1087.3733	1536.9951	4485.5430
39	693.6526	900.4668	1222.6384	1743.3423	5256.1563
40	771.3633	1005.5510	1373.8706	1976.0942	6156.1875
41	857.4412	1122.3289	1542.8943	2238.5198	7207.0703
42	952.7744	1252.0701	1731.7376	2534.2879	8433.7578
43	1058.3440	1396.1799	1942.6543	2867.5071	9865.3164
44	1175.2356	1556.2146	2178.1519	3242.7839	11535.5625
45	1304.6487	1733.8960	2441.0142	3665.2815	13483.8633
46	1447.9092	1931.1311	2734.3374	4140.7774	15756.0274
47	1606.4837	2150.0286	3061.5567	4675.7500	18405.3789
48	1781.9920	2392.9263	3426.4910	5277.4531	21493.9453
49	1976.2271	2662.4080	3833.3838	5953.9961	25093.9102
50	2191.1680	2961.3311	4286.9414	6714.4727	29289.2383

TABLE B.33 Continuous Compounding: r = 15%

	Geometric series present worth factor $(P|A_1 \; r,c,n) \infty$				
n	c = 4%	c = 6%	c = 8%	c = 10%	c = 15%
1	0.8607	0.8607	0.8607	0.8607	0.8696
2	1.6318	1.6473	1.6632	1.6794	1.7391
3	2.3225	2.3663	2.4115	2.4582	2.6087
4	2.9413	3.0233	3.1092	3.1991	3.4783
5	3.4956	3.6238	3.7597	3.9037	4.3478
6	3.9922	4.1726	4.3662	4.5741	5.2174
7	4.4370	4.6742	4.9317	5.2117	6.0870
8	4.8356	5.1326	5.4590	5.8182	6.9565
9	5.1926	5.5515	5.9507	6.3952	7.8261
10	5.5124	5.9344	6.4091	6.9440	8.6957
11	5.7989	6.2844	6.8365	7.4660	9.5652
12	6.0556	6.6042	7.2350	7.9626	10.4348
13	6.2855	6.8965	7.6066	8.4350	11.3043
14	6.4915	7.1636	7.9530	8.8843	12.1739
15	6.6760	7.4078	8.2761	9.3117	13.0435
16	6.8413	7.6309	8.5773	9.7183	13.9130
17	6.9894	7.8348	8.8581	10.1050	14.7826
18	7.1220	8.0212	9.1199	10.4729	15.6522
19	7.2409	8.1915	9.3641	10.8229	16.5217
20	7.3473	8.3472	9.5917	11.1557	17.3913
21	7.4427	8.4895	9.8040	11.4724	18.2609
22	7.5281	8.6195	10.0019	11.7736	19.1304
23	7.6047	8.7383	10.1864	12.0601	20.0000
24	7.6732	8.8470	10.3584	12.3326	20.8696
25	7.7346	8.9462	10.5189	12.5918	21.7391
26	7.7897	9.0369	10.6684	12.8384	22.6087
27	7.8389	9.1198	10.8079	13.0730	23.4783
28	7.8831	9.1956	10.9379	13.2961	24.3478
29	7.9227	9.2649	11.0591	13.5084	25.2174
30	7.9581	9.3282	11.1722	13.7103	26.0870
31	7.9898	9.3860	11.2776	13.9023	26.9565
32	8.0183	9.4389	11.3759	14.0850	27.8261
33	8.0438	9.4872	11.4675	14.2588	28.6957
34	8.0666	9.5313	11.5529	14.4241	29.5652
35	8.0870	9.5717	11.6326	14.5813	30.4348
36	8.1053	9.6086	11.7069	14.7309	31.3043
37	8.1218	9.6423	11.7761	14.8732	32.1739
38	8.1365	9.6731	11.8407	15.0085	33.0435
39	8.1496	9.7013	11.9009	15.1372	33.9130
40	8.1614	9.7270	11.9570	15.2597	34.7826
41	8.1720	9.7505	12.0094	15.3762	35.6522
42	8.1814	9.7720	12.0582	15.4870	36.5217
43	8.1899	9.7916	12.1037	15.5924	37.3913
44	8.1975	9.8096	12.1461	15.6926	38.2609
45	8.2043	9.8260	12.1856	15.7880	39.1304
46	8.2104	9.8410	12.2225	15.8787	40.0000
47	8.2159	9.8547	12.2569	15.9650	40.8696
48	8.2208	9.8672	12.2890	16.0471	41.7391
49	8.2252	9.8787	12.3189	16.1252	42.6087
50	8.2291	9.8891	12.3468	16.1995	43.4783

TABLE B.34 Continuous Compounding: r = 15%

	Geometric series future worth factor $(F/A_1\ r, c, n)_\infty$				
n	$c = 4\%$	$c = 6\%$	$c = 8\%$	$c = 10\%$	$c = 15\%$
1	1.0000	1.0000	1.0000	1.0000	1.0000
2	2.2026	2.2237	2.2451	2.2670	2.3237
3	3.6424	3.7110	3.7820	3.8553	4.0496
4	5.3594	5.5088	5.6653	5.8291	6.2732
5	7.4002	7.6716	7.9592	8.2642	9.1106
6	9.8192	10.2630	10.7391	11.2504	12.7020
7	12.6795	13.3572	14.0932	14.8932	17.2172
8	16.0546	17.0408	18.1246	19.3172	22.8612
9	20.0300	21.4147	22.9543	24.6689	29.8810
10	24.7048	26.5963	23.7235	31.1208	38.5742
11	30.1947	32.7226	35.5975	38.8755	49.2986
12	36.6340	39.9531	43.7692	48.1710	62.4837
13	44.1787	48.4733	53.4643	59.2868	78.6454
14	53.0103	58.4994	64.9459	72.5508	98.4016
15	63.3399	70.2829	78.5212	88.3472	122.4924
16	75.4126	84.1167	94.5487	107.1264	151.8037
17	89.5134	100.3414	113.4465	129.4162	187.3939
18	105.9736	119.3532	135.7022	155.8341	230.5277
19	125.1782	141.6133	161.3842	187.1030	282.7146
20	147.5745	167.6580	192.6548	224.0686	345.7554
21	173.6827	198.1109	228.7860	267.7195	421.7959
22	204.1069	233.6974	271.1768	319.2117	513.3931
23	239.5493	275.2610	320.8750	379.8963	623.5901
24	280.8257	323.7825	379.1001	451.3506	756.0088
25	328.8848	380.4024	447.2725	535.4180	914.9551
26	384.8279	446.4463	527.0457	634.2493	1105.5471
27	449.9353	523.4553	620.3440	750.3562	1333.8650
28	525.6948	613.2214	729.4080	886.6692	1607.1275
29	613.8352	717.8272	856.8447	1046.6074	1933.9006
30	716.3648	839.6934	1005.6873	1234.1585	2324.3504
31	835.6172	981.6343	1179.4649	1453.9729	2790.5278
32	974.3042	1146.9202	1382.2839	1711.4734	3346.7151
33	1135.5767	1339.3521	1618.9207	2012.9807	4009.8399
34	1323.0952	1563.3477	1894.9295	2365.8623	4799.9414
35	1541.1135	1824.0415	2216.7742	2778.7039	5740.7578
36	1794.5728	2127.3999	2591.9695	3261.5088	6860.3750
37	2089.2175	2480.3567	3029.2527	3825.9312	8192.0274
38	2431.7168	2890.9712	3538.7886	4485.5430	9775.0156
39	2829.8240	3368.6047	4132.3867	5256.1563	11655.8125
40	3292.5454	3924.1423	4823.7969	6156.1875	13889.3555
41	3830.3455	4570.2227	5628.9844	7207.0703	16540.5586
42	4455.3789	5321.5469	6566.5234	8433.7578	19686.1055
43	5181.7774	6195.1836	7658.0039	9865.3164	23416.5625
44	6025.9531	7210.9766	8928.5156	11535.5625	27838.8477
45	7006.9727	8391.9727	10407.2422	13483.8633	33079.2188
46	8146.9883	9764.9609	12128.0859	15756.0274	39286.6289
47	9471.7461	11361.0664	14130.4727	18405.3789	46636.8242
48	11011.1524	13216.4531	16460.2149	21493.9453	55337.1367
49	12799.9570	15373.1406	19170.5664	25093.9102	65632.0000
50	14878.5274	17879.9570	22323.4219	29289.2383	77809.6875

ANSWERS TO EVEN-NUMBERED PROBLEMS

CHAPTER 2

2. Choose turret lathe when $x \geq 43$ units.

4. $x^* = 1826$ units/year

6. Continue to operate since $TR = \$625,000 > VC = \$568,750$.

8. (a) $x^* = 430$ hours/years   (b) Savings/year $= \$114$

10. Net loss before taxes $= \$3025$

12. (a) Total cost $= \$15,186.46$ (b) Unit selling price $= \$3.95$

14. Burden rates are \$2.44/hr., \$2.73/hr., \$4.91/hr., and \$2.36/hr. for cost centers A, B, C, and D, respectively.

CHAPTER 3

2. (a) \$308, \$333, \$359 (b) \$80, \$55, \$29

4. 20%

6. \$4467.20

8. (a) \$3833.80   (b) \$4063.83

10. (a) \$11,559   (b) \$12,228.83

12. \$1296.10

14. (a) \$1759.98   (b) \$2394.45

16. \$10,000

18. \$2358.20

20. $8618.46

22. $14,108.60

24. − $979.50

26. $669.10

28. $2570.45

30. $1021.84

32. $1360.19

34. $53.20

36. $5730.40

38. $3090.60

40. 14 years

42. 42.576%

44. $107,654.07

46. $1325.45

48. $6025.70

50. $3480.91

52. −$809.88

54. $1000.22

56. $6377.60

58. $1788.70

60. 6.68%

62. $1543.79

64. −$1049.18

66. $9558.69

68. $28,583.80

70. $5847.09

72. (a) $6665.07   (b) 7.45%   (c) 8.325%

74. $E_1 = 887.05$ $I_1 = 299.95$
    $E_2 = 940.22$ $I_2 = 246.78$
    $E_3 = 996.61$ $I_3 = 190.39$
    $E_4 = 1056.43$ $I_4 = 130.57$
    $E_5 = 1119.82$ $I_5 = 67.18$

76. $5539.79

78. $1408.10

80. $2848.41

82. $1858.36

84. (a) $4303.45   (b) 5.6%

86. $P = \$13,889.75$ $A = \$3321.84$

88. $6303.00

90. $5373.84

92. $2956.83

94. 14.36%

96. $2115.75

98. (a) $4290.60   (b) 12.61%

100. $37,921.78

102. $23,880.00

CHAPTER 4

2. A and D
   $PW = \$4252.50$

4. Do nothing.

6. $AW(X) = \$18,031$
   $AW(W$ and $Z) = \$6364$
   $AW(Y$ and $Z) = \$8818$
   $AW(Z) = \$0$

8. 0%

10. $6840.40

12. $6068.20

14. (a) Do nothing; A; B;
D; C and D.
(c) PW(do nothing) $= 0$
$PW(A) = \$269,470$
$PW(B) = \$303,365$
$PW(D) = \$240,945$
$PW(C, D) = \$431,890$

16. $-\$12,330$

18. (a) $EUAC_A(0\%) = \$2108$ (b) $EUAC_A(8\%) = \$2565$ (c) $EUAC_A(10\%) = \$2693$
       $EUAC_B(0\%) = \$2032$ $EUAC_B(8\%) = \$2764$ $EUAC_B(10\%) = \$2962$
       $EUAC_C(0\%) = \$2525$ $EUAC_C(8\%) = \$3460$ $EUAC_C(10\%) = \$3713$

20. $EUAC_A(10\%) = \$2924.29$
$EUAC_B(10\%) = \$4066.36$

22. $EUAC_A(10\%) = \$7383.64$
$EUAC_B(10\%) = \$8458.73$

24. Holding pond yields an equivalent annual savings of \$8905 when compared to the storage tank alternative.

26. Alternative 2

28. Two-inch insulation

30. Compressor B is preferred.

32. (a) (i–iv) contract
(b) (i, ii) purchase equipment.
(b) (iii, iv) contract
(c) (i–iv) purchase equipment.
(d) (i–iv) purchase equipment.

34. Compressor Y has the largest PW.
$PW_X(10\%) = -\$70,356$
$PW_Y(10\%) = -\$64,608$

36. $i^*_{B-A} = 10\% < MARR$
Choose A.

38. Choose B.

40. \$64,570 end of year lease payment or \$58,700 beginning of year lease payment.

42. Select proposal C; $AW = \$12,300$.

44. Leasing is preferred by \$4500/year.

46. \$9095 end of year lease payment or \$8268 beginning of year lease payment.

48. (a) Keep.
(b) Replace.

50. Replace.

52. Yes.

54. Purchase X; annual savings = $12,695.

56. Replace old mixer; annual savings = $352.80.

58. $EUAC$ (old lathe) = $8182.80.
$EUAC$ (new lathe) = $7082.80.
$EUAC$ (subcontract) = $6681.00.
Subcontract!

60. PW (overhaul) = $-$24,963.50.
PW (replace) = $-$20,540.60.

62. Subcontract; annual savings = $1894.40.

64. (a) Replace every 10 years with infinite planning horizon.
(b) Replace after 9 years service; keep the replacement for 6 years.
(c) Replace every 7 years with infinite planning horizon; replace after 7 or 8 years and keep the replacement for the balance of the planning horizon.
(d) The greater the discount rate, the greater the economic life of an asset.

66. (a) Replace every 10 or 11 years.
(b) The maximum life of 12 years is the economic life

CHAPTER 5

2. (a) $20,000, $0 (no switch to straight line)
(b) $24,116.89, $4116.89
(c) $20,000, $0

4. $25,000, $12,500, $6250, $6250

6. $1306.30

8. $2250, $1200, $600, $450

10. (a) $2816
(b) $5500
(c) $17,500
(d) $473,500

12. $2.2237/unit

14. $PW(12) = $930.34. Yes, invest.

16. $12,418.40

18. $PW(10) = $288.01. Yes, invest.

20. (a) A: $11,448.20 B: $17,310.25
(b) A: $12,475.60 B: $15,286.00
(c) A: $11,448.20 B: $15,286.00

22. $191.76

24. Figures for year 2 only are shown.
(a) $9482, $3138, $30,000, $-8138, $4069, $16,449
(b) $10,000, $3,000, $30,000, $-8000, $-4069, $16,000

26. $5559.76/year

28. $1375, $1350

30. $ − 100,000, $20,000, $20,000, $20,000, $20,000, $63,000

32. $33,261.58

34. $PW(10) = \$4156.80$. Yes, purchase incinerator.

36. 10 years, $2336.85/year

38. $42,765.09

40. Keep: $AW(25) = \$11,827.10$/year.
Subcontract: $AW(25) = \$15,690.94$/year.
Prefer to subcontract.

42. (a) $70,000   (b) $68,250

CHAPTER 6

2. $3.19/hour

4. 0.7412

6. Route B; Route C

8. $B − C = \$30,351.40$/year. Yes, build the new addition.

10. Projects A and B. Opportunity cost slightly less than $i = 20\%$.

12. $B − C = \$1,453,751.30$ for preferred project C at $i = 5\%$.
$B − C = \$51,562.40$ for preferred project A at $i = 15\%$.
Higher discounting emphasized higher yearly benefits of project A during early years.

14. $B/C = 1.6667$; $B − C = \$58,400$/year. Yes, build.

16. $B − C = \$6351.40$/year. Yes, build.

18. The fallacy is that the 9000 persons receiving benefits of $1.75 will not patronize the facility. Also, the entrance fee was not deducted from the benefits. True $B/C = 0.50$. True $B − C = \$9000$.
Do not implement the $2.00 fee.

20. (a) 10; (b) 1.1286; (c) $45,000; (d) Prefer $B − C$ as it would remain invariant.

22. Design 1 best. Present worth cost $= \$31,530,320$.

CHAPTER 7

2. Either 30 or 31 years

4. (a) 37.3%
   (b) 60.7%
   (c) 88.0%
   (d) 117.8%

6. $7344

8. If annual usage is less than 902.43 hours, use Z.
If annual usage is greater than 1,017.81 hours, use X.
Otherwise, use Y.

10. $132.10 per year.

12. If production volume is less than 7.5, use B.
Otherwise, use A.

14. Pessimistic value $= 156.1\%$.
Optimistic value $= 53.2\%$.

16. Sensitivity analysis yields the following breakeven equation:
$Y = -0.0711 + 0.3664X$.
Combinations of X and Y above the breakeven line result in $AW > 0$.

18. 0.25

20. 0.29

22.

PW	$p(PW)$
−2000	0.0016
−1000	0.0160
0	0.0696
1000	0.1720
2000	0.2641
3000	0.2580
4000	0.1566
5000	0.0540
6000	0.0081

24. (a) $E[PW] = 14,000$
$V[PW] = 896 \times 10^4$
$Pr(PW \geq 0) = 1.000$
$Pr(AW \geq 0) = Pr(PW \geq 0)$
(b) $E[PW] = -\$9762$.
$V[PW] = 664.804 \times 10^4$
$Pr(PW \geq 0) = 0.0001$

26. Probability A is best $= 0.2269$.

28. Cumulative average $PW = 662.50$.

30.

Trial	ERR	Trial	ERR
1	16.17%	6	15.83%
2	6.83%	7	23.71%
3	14.02%	8	27.07%
4	17.88%	9	24.74%
5	21.87%	10	21.07%

32. $E[PW] = -\$18,097.73$

CHAPTER 8

2. If $0 \le \alpha < 1/3$, choose A_1.
$\alpha = 1/3$, choose either A_1 or A_3.
$1/3 \le \alpha < 1/2$, choose A_3.
$\alpha = 1/2$, choose either A_3 or A_2.
$1/2 \le \alpha \le 1.0$, choose A_2.

4. Laplace principle: Choose A_1 with $E(A_1) = \$13,000$.
Maximin principle: Choose either A_1 or A_5 with value = $8000.
Maximax principle: Choose A_2 with value = $20,000.
Savage principle: Choose A_1 with minimax regret value = $4000.
Hurwicz principle: Choose A_1 or A_5 if $0 \le \alpha < 1/3$.
$\qquad\qquad\qquad$ Choose either A_1, A_2, or A_5 if $\alpha = 1/3$
$\qquad\qquad\qquad$ Choose A_2 if $1/3 < \alpha \le 1.0$.

6. Minimax principle: Choose a_2 with minimax value = 60.
Minimin principle: Choose a_1 with minimin value = 10.
Minimax regret: Choose either a_2 or a_3 with minimax regret value = 50.
Expected cost: Choose a_2 with $E(a_2) = 60$ and zero variance.
Hurwicz principle: Choose a_2 with Hurwicz value = 60.

8. Minimax principle: Choose a_2 with minimax value = 200.
Minimin principle: Choose a_1 with minimin value = 100.
Minimax regret: Choose a_3 with minimax regret value = 75.

10. Keep present equipment with $E(\text{Keep}) = \$84,000$. The most probable future principle would select the purchase alternative, ignoring the $100,000 loss possibility.

12. (a) $p(S_1) = 0.216$; $p(S_2) = 0.432$; $p(S_3) = 0.288$; $p(S_4) = 0.064$
(b)

	0.216	0.432	0.288	0.064	
	S_1	S_2	S_3	S_4	$E(a_i)$
Guess S_1	$3	−$1	−$1	−$1	−0.136
Guess S_2	−$1	$2	−$1	−$1	0.296
Guess S_3	−$1	−$1	$3	−$1	0.152
Guess S_4	−$1	−$1	−$1	$4	−0.680

Both the expectation principle and most probable future principle selects the "Guess S_2" alternative.

14. A solution can be obtained using the conditional probability theorem directly. However, using Bayes' theorem, Supplier C is the most likely producer of Model R with $P(C|R) = 0.507$ and, similarly, $P(A|R) = 0.338$ and $P(B|R) = 0.155$.

16. (a) $P(S_1|O_1) = 0.8235$; $P(S_2|O_1) = 0.1765$
$P(S_1|O_2) = 0.2258$; $P(S_2|O_2) = 0.7742$
$P(S_1|O_3) = 1.0000$; $P(S_2|O_3) = 0.0000$
(c) Choose A_1 with $E(A_1) = 1.2$ versus $E(\text{send recon. flight}) = 2.232$ and $E(A_2) = 8.4$.

18. (b) Keep present truck, with expected value = $-\$21,700$

20. Choose A_1 with $E(A_1) = 0.9078 > E(A_2) = 0.8799$.

INDEX